Smart Nanomaterials Technology

Series Editors

Azamal Husen, Wolaita Sodo University, Wolaita, Ethiopia

Mohammad Jawaid, Laboratory of Biocomposite Technology, Universiti Putra Malaysia, INTROP, Serdang, Selangor, Malaysia

Nanotechnology is a rapidly growing scientific field and has attracted a great interest over the last few years because of its abundant applications in different fields like biology, physics and chemistry. This science deals with the production of minute particles called nanomaterials having dimensions between 1 and 100 nm which may serve as building blocks for various physical and biological systems. On the other hand, there is the class of smart materials where the material that can stimuli by external factors and results a new kind of functional properties. The combination of these two classes forms a new class of smart nanomaterials, which produces unique functional material properties and a great opportunity to larger span of application. Smart nanomaterials have been employed by researchers to use it effectively in agricultural production, soil improvement, disease management, energy and environment, medical science, pharmaceuticals, engineering, food, animal husbandry and forestry sectors.

This book series in Smart Nanomaterials Technology aims to comprehensively cover topics in the fabrication, synthesis and application of these materials for applications in the following fields:

- Energy Systems—Renewable energy, energy storage (supercapacitors and electrochemical cells), hydrogen storage, photocatalytic water splitting for hydrogen production
- Biomedical—controlled release of drugs, treatment of various diseases, biosensors,
- Agricultural—agricultural production, soil improvement, disease management, animal feed, egg, milk and meat production/processing,
- Forestry—wood preservation, protection, disease management
- Environment—wastewater treatment, separation of hazardous contaminants from wastewater, indoor air filters.

Archana (Joshi) Bachheti ·
Rakesh Kumar Bachheti · Azamal Husen
Editors

Carbon-Based Nanomaterials

Synthesis, Agricultural, Biomedical, and Environmental Interventions

Springer

Editors
Archana (Joshi) Bachheti
Department of Environmental Science
Graphic Era (Deemed to be University)
Dehradun, Uttarakhand, India

Azamal Husen
Department of Biotechnology
Smt. S. S. Patel Nootan Science
and Commerce College
Sankalchand Patel University
Visnagar, Gujarat, India

Rakesh Kumar Bachheti
Department of Industrial Chemistry
Addis Ababa Science and Technology
University
Addis Ababa, Ethiopia

Department of Allied Sciences
Graphic Era Hill University
Dehradun, Uttarakhand, India

ISSN 3004-8273 ISSN 3004-8281 (electronic)
Smart Nanomaterials Technology
ISBN 978-981-97-0239-8 ISBN 978-981-97-0240-4 (eBook)
https://doi.org/10.1007/978-981-97-0240-4

This Springer imprint is published by the registered company Springer Nature Singapore Pte Ltd.
The registered company address is: 152 Beach Road, #21-01/04 Gateway East, Singapore 189721, Singapore

Preface

Scientific research on carbon nanomaterials is increasing due to their numerous applications. Several new nanomaterials with inherent and intrinsic properties for their viable application in the development of sensing and biosensing devices have been reported. Carbon nanomaterials are essential in environmental pollution control because they absorb heavy metals, antibiotics, and toxic gases. Carbon nanotubes, fullerenes, carbon nanoparticles, and carbon nanohorns are other examples of carbon nanomaterials that have great application potential in agriculture by increasing crop production and plant protection. With its deep dive into science and technology, carbon-based nanomaterials offers a comprehensive and singular perspective on how these materials have transformed a multitude of research and application domains. These nanomaterials have become lights of hope and innovation in a world of ever-growing difficulties, providing answers to some of the most important problems confronting humanity right now.

The book consists of 17 chapters. Staring chapters "Carbon-Based Smart Nanomaterials: An Overview" to "Functionalization of Carbon-Based Nanoparticles for Various Applications", gives a general overview of carbon nanoparticles and discuss the synthesis, functionalization, characterization, and application of carbon nanoparticles. After that, chapters "Smart Carbon Nanomaterials and Their Effect on the Antioxidant System of Plants" to "Role of Carbon Nanotubes for Herbicide Detection and Remediation" discuss the impact of carbon nanomaterials on plants. These chapters explain the role of carbon nanomaterials in the prevention of plant disease, the effect of carbon nanomaterials on growth and development, and their effect on the antioxidant system of plants. Biomedical applications of carbon nanomaterials are discussed in chapters "Drug Delivery Using Carbon Nanomaterials" and "Functionalized Carbon Nanotubes Biomedical Applications and Toxicological Implications". Drug delivery using carbon nanomaterials and biomedical applications of 1D and 2D carbon-based nanomaterials along with the role of graphene-based nanomaterial drug delivery and tissue engineering have been discussed here. Also, functionalized of carbon nanotubes for their biomedical applications and toxicological effect also explained in this section. The function of carbon nanoparticles in cleaning up oil spills and air pollution is covered in chapters "Role of Carbon Nanomaterials in Air Pollution Remediation" and "Carbon Nanomaterial for Oil Spill Clean-Up". The remaining chapters "Environmental Toxicity of Engineered Carbon Nanoparticles" and "Role of Carbon Nanomaterials in Energy Generation, Storage, and Conversion" explore the toxicity of manufactured carbon nanoparticles to the environment, offering insights into the potential and difficulties of using these materials for environmental preservation. Besides this, there is a focus on the role of carbon nanomaterials in energy generation, storage, and conversion.

Overall, this book provides the recent development of carbon nanomaterial in agricultural, biomedical, and environmental applications.

Moreover, this book also discusses the toxicity of engineered carbon nanoparticles in the environment as well as in plant and animal systems. We hope this book inspires further research, innovation, and collaboration in the realm of nanomaterials, and ultimately contributes to a better and more sustainable future for all. We are highly thankful to all eminent authors who have contributed chapters and provided their valuable time and knowledge for this edited book. We also wish to pay thanks to Dr. Ramesh Nath Premnath, Editor Applied Sciences (*Engineering, Materials Science, Energy, Water, and Climate*), Springer Singapore; and all his associates, for their cooperation at every stage of the book production. We shall be happy to receive comments and criticism, if any, from subject experts and general readers of this book.

Dehradun, India — Archana (Joshi) Bachheti
Addis Ababa, Ethiopia — Rakesh Kumar Bachheti
Visnagar, India — Azamal Husen

Contents

Editors and Contributors

About the Editors

Archana (Joshi) Bachheti completed her B.Sc. in 1997 and M.Sc. in 1999 from HNB Garhwal University. She earned her Ph.D. from Forest Research Institute, Dehradun, India, in 2006. Throughout her career, she has been actively involved in research projects and consultancy work in various areas, including ecorestoration and development of wasteland, physico-chemical properties of Jatropha curcus seed oil in relation to altitudinal variation, and serving as a consultant ecologist for a project funded by a government agency. Currently, Archana is a Professor at Graphic Era University in Dehradun, India. With an extensive experience of more than 17 years, she has served in different capacities within academia in India and has also provided expertise internationally. Her teaching portfolio includes subjects like Ecology and Environment, Environmental Science, Freshwater Ecology, Disaster Management, and Bryophytes and Pteridophytes. Archana's research interests cover a broad and interdisciplinary field of plant ecology. Her focus areas include ecorestoration, green chemistry, especially the synthesis of nanomaterials, and exploring the medicinal properties of plants. Her research has encompassed ecological amelioration of degraded lands, studying physical and chemical properties of plant oils, and delving into plant-based nanomaterials. Throughout her career, she has been actively involved in mentoring students and researchers. She guided one Ph.D. student and supervised three scholars. Additionally, she has provided guidance to graduate and undergraduate students in their research projects. Driven by her fascination with forest biodiversity, Archana has maintained a passion for exploring the values of biodiversity and how it can contribute to social upliftment. As an editor, she has contributed to six books and has published over 80 research articles in both international and national journals. She has also authored 16 book chapters. Furthermore, she has organized several National seminars/conferences at Graphic Era University, India.

Rakesh Kumar Bachheti graduated from Hemwati Nandan Bahuguna University, Garhwal, India, in 1996. He completed his M.Sc. in Organic Chemistry from the same university in 1998. Later, in 2001, he underwent a one-year Post Graduate Diploma in Pulp and Paper Technology from Forest Research Institute, Dehradun, India. Subsequently, he pursued his Ph.D. in Organic Chemistry from Kumaun University, Nainital, India, which he obtained in 2007. Bachheti has a total research and teaching experience of 20 Years. Before joining Addis Ababa Science and Technology University (AASTU) in Ethiopia, he worked as Dean of Project (Assistant) at Graphic Era University in Dehradun, India. It's worth noting that Graphic Era University is accredited with an 'A+' grade by the National Assessment and Accreditation Council (NAAC). Bachheti is also a Research Fellow at INTI International University Persiaran Perdana BBN, Putra Nilai, Nilai Negeri Sembilan, Malaysia, and Adjunct Faculty at the Department of Allied Sciences, Graphic Era Hill University, Dehradun, India. Bachheti has an impressive research background and has presented papers at various international (Malaysia, Thailand, and India) and national conferences. He has also been an active member of important committees like the Internal Quality Assurance Cell (IQAC) and the Anti-ragging Committee. His research interests primarily focus on natural products for industrial applications, biofuels and bioenergy, green synthesis of nanoparticles and their applications, and pulp and paper technology. His passion for natural products is evident in all of his research endeavors. As an academic mentor, Bachheti has successfully advised 40 M.Sc. and six Ph.D. students to completion. Additionally, numerous undergraduates have conducted research in his laboratory under his guidance. He actively contributes to curriculum development for B.Sc./M.Sc./Ph.D. programs. Bachheti is also recognized for his editorial work, having edited six books and authored 100 publications on various aspects of natural product chemistry and nanotechnology. He has contributed twenty book chapters published by prestigious publishers like Springer, Elsevier, Tylor and Francis and Nova Publisher.

Azamal Husen is presently working as a Professor at Sankalchand Patel University, Visnagar, India; and Adjunct Professor at Graphic Era (Deemed to be University), Dehradun, Uttarakhand, India. He is also working as a Visiting Professor at University Putra Malaysia, Selangor, Malaysia. Previously, he served as Professor and Head of the Department of Biology, University of Gondar, Ethiopia; and worked as a Foreign Delegate at Wolaita Sodo University, Wolaita, Ethiopia. He also worked as a Visiting Faculty of the Forest Research Institute and the Doon College of Agriculture and Forest at Dehra Dun, India. His research and teaching experience of 25 years encompasses biogenic nanomaterial fabrication and application; plant responses to nanomaterials; plant adaptation to harsh environments at the physiological, biochemical, and molecular levels; herbal medicine; and clonal propagation for improvement of tree species. He has conducted research sponsored by the World Bank, the National Agricultural Technology Project, the Indian Council of Agriculture Research, the Indian Council of Forest Research Education, and the Japan Bank for International Cooperation. Husen has published extensively (over 300) and served on the Editorial Board and as reviewer of reputed journals published by Elsevier, Frontiers Media, Taylor & Francis, Springer Nature, RSC, Oxford University Press, Sciendo, the Royal Society, CSIRO, PLOS, MDPI, John Wiley & Sons, and UPM Journals. He is on the advisory board of Cambridge Scholars Publishing, UK. He is a fellow of the Plantae group of the American Society of Plant Biologists, and a member of the International Society of Root Research, Asian Council of Science

Editors, and International Natural Product Sciences. He is Editor-in-Chief of the *American Journal of Plant Physiology*, and a Series Editor of *Exploring Medicinal Plants* (Taylor & Francis Group, USA); *Plant Biology, Sustainability, and Climate Change* (Elsevier, USA); and *Smart Nanomaterials Technology* (Springer Nature, Singapore). He has been achieved the distinguished honor of being recognized as one of the "World's Top 2% Scientists" for the year 2022, and again for the 2023 by Stanford University, USA. This recognition has also been prominently featured in the Elsevier Data Repository.

Contributors

Kalash Aggrawal School of Biotechnology, Gautam Buddha University, Greater Noida, Uttar Pradesh, India

Archana (Joshi) Bachheti Department of Environment Science, Graphic Era (Deemed to be University), Dehradun, Uttarakhand, India

Rakesh Kumar Bachheti Department of Industrial Chemistry, College of Natural and Applied Sciences, Addis Ababa Science and Technology University, Addis Ababa, Ethiopia;
Department of Allied Sciences, Graphic Era Hill University, Dehradun, Uttarakhand, India

Saswati Barman Department of Biotechnology, School of Basic and Applied Sciences, Sanskriti University, Mathura, Uttar Pradesh, India

Gaurav Bhardwaj School of Basic and Applied Sciences, Sanskriti University, Mathura, Uttar Pradesh, India

Noureddine Chaachouay Agri-Food and Health Laboratory (AFHL), Faculty of Sciences and Techniques of Settat, Hassan First University, Settat, Morocco

Kundan Kumar Chaubey Department of Biotechnology, School of Basic and Applied Sciences, Sanskriti University, Mathura, Uttar Pradesh, India

Sonam Chawla Department of Biotechnology, Jaypee Institute of Information Technology, Noida, India

Versha Dixit Department of Life Sciences, Christ University, Bangalore, Karnataka, India

Megha Gangwar Department of Biotechnology, School of Basic and Applied Sciences, Sanskriti University, Mathura, Uttar Pradesh, India

Yilma Hunde Gonfa Department of Chemistry, Ambo University, Ambo, Ethiopia

Urvashi Goyal Department of Microbiology, Sanjay Gandhi Postgraduate Institute of Medical Sciences, Lucknow, India

Kaan Hürkan Department of Agricultural Biotechnology, Faculty of Agriculture, Iğdır University, Iğdır, Türkiye

Azamal Husen Department of Biotechnology, Smt. S. S. Patel Nootan Science and Commerce College, Sankalchand Patel University, Gujarat, India;
Department of Biotechnology, Graphic Era (Deemed to Be University), Dehradun, Uttarakhand, India;
Wolaita Sodo University, Wolaita, Ethiopia

N. C. Joshi Division of Research and Innovation, School of Applied and Life Sciences, Uttaranchal University, Dehradun, Uttarakhand, India

Divya Kapoor Department of Microbiology, CCS Haryana Agricultural University, Hisar, Haryana, India

Gurpreet Kaur Department of Biotechnology, Chandigarh College of Technology (CCT), Chandigarh Group of Colleges (CGC), Landran, Mohali, Punjab, India

Sevgi Kemeç Aslan Department of Chemical Engineering, Faculty of Engineering, and Natural Sciences, Bursa Technical University, Bursa, Türkiye

Noureen Amir Khan Institute of Chemical Sciences, University of Peshawar, Peshawar, Pakistan

Tahmeena Khan Department of Chemistry, Integral University, U.P., Lucknow, India

Teeya Khandelwal Department of Biotechnology, School of Basic and Applied Sciences, Sanskriti University, Mathura, Uttar Pradesh, India

Piyush Kumar Department of Biotechnology, Jaypee Institute of Information Technology, Noida, India

Rahul Kumar Department of Zoology, CCS Haryana Agricultural University, Hisar, Haryana, India

Anupriya Kumari School of Basic and Applied Sciences, Sanskriti University, Mathura, Uttar Pradesh, India

Priyambada Kumari College of Biotechnology, DUVASU, Mathura, Uttar Pradesh, India

Atul Loyal Department of Genetics and Plant Breeding, CCS Haryana Agricultural University, Hisar, Haryana, India

Faizan ul Haque Nagrami College of Biotechnology, DUVASU, Mathura, Uttar Pradesh, India

Anis Kumar Pal Division of Research and Innovation, School of Applied and Life Sciences, Uttaranchal University, Dehradun, Uttarakhand, India

Anish Kumar Pal Division of Research and Innovation, School of Applied and Life Sciences, Uttaranchal University, Premnagar, Dehradun, Uttarakhand, India

Brijesh Pandey Department of Biotechnology, School of Life Sciences, Mahatma Gandhi Central University, Motihari, Bihar, India

Saumya Pandey Department of Biotechnology, IILM University, Greater Noida, India

Shreya Pandey Department of Biotechnology, School of Basic and Applied Sciences, Sanskriti University, Mathura, Uttar Pradesh, India

Anand Prakash Department of Biotechnology, School of Life Sciences, Mahatma Gandhi Central University, Motihari, Bihar, India

Gul Rahman Institute of Chemical Sciences, University of Peshawar, Peshawar, Pakistan

Qazi Inamur Rahman Department of Chemistry, Integral University, U.P., Lucknow, India

Sanjana Rajawat Department of Biotechnology, School of Basic and Applied Sciences, Sanskriti University, Mathura, Uttar Pradesh, India

Saman Raza Department of Chemistry, Isabella Thoburn College, U.P., Lucknow, India

Prakash Kumar Sarangi College of Agriculture, Central Agricultural University, Imphal, India

Mayank Saxena Department of Biotechnology, IILM University, Greater Noida, India

Vanshika Saxena School of Basic and Applied Sciences, Sanskriti University, Mathura, Uttar Pradesh, India

Lalit Kumar Sharma School of Basic and Applied Sciences, Sanskriti University, Mathura, Uttar Pradesh, India

Pankaj Sharma Department of Microbiology, CCS Haryana Agricultural University, Hisar, Haryana, India;
Department of Ophthalmology and Visual Sciences, University of Illinois Chicago, Chicago, IL, USA

Mayur Mukut Murlidhar Sharma Department of Agriculture and Industry, Kangwon National University, Chuncheon, Republic of Korea

Shalini Sharma Department of Veterinary Physiology and Biochemistry, Lala Lajpat Rai University of Veterinary and Animal Sciences, Hisar, Haryana, India

Shweta Sharma College of Biotechnology, DUVASU, Mathura, Uttar Pradesh, India

Soni Sharma School of Basic and Applied Sciences, Sanskriti University, Mathura, Uttar Pradesh, India

Shivani Shivu School of Basic and Applied Sciences, Sanskriti University, Mathura, Uttar Pradesh, India

Akhilesh Kumar Singh Department of Biotechnology, School of Life Sciences, Mahatma Gandhi Central University, Motihari, Bihar, India

Krishan Raj Singh School of Basic and Applied Sciences, Sanskriti University, Mathura, Uttar Pradesh, India

Maharudra Pratap Singh Department of Biotechnology, School of Life Sciences, Mahatma Gandhi Central University, Motihari, Bihar, India

Parul Singh College of Biotechnology, DUVASU, Mathura, Uttar Pradesh, India

Rishi Kumar Singh School of Biosciences, IMS Institute University Courses Campus, Uttar Pradesh, Ghaziabad, India

Ritu Singh Department of Biotechnology, Jaypee Institute of Information Technology, Noida, India

Vineeta Singh Department of Biotechnology, Institute of Engineering and Technology, Dr. A.P.J. Abdul Kalam Technical University, Lucknow, India

Vijay Laxmi Tripathi College of Biotechnology, DUVASU, Mathura, Uttar Pradesh, India

Ümran Ünlü Department of Chemical Engineering, Faculty of Engineering, Gebze Technical University, Kocaeli, Türkiye

Addisu Tamir Wassie Department of Industrial Chemistry, College of Natural and Applied Sciences, Addis Ababa Science and Technology University, Addis Ababa, Ethiopia;
Department of Chemistry, College of Natural Sciences, Wollo University, Dessie, Ethiopia

Sonali Yadav School of Basic and Applied Sciences, Sanskriti University, Mathura, Uttar Pradesh, India

Lahcen Zidane Animal Productions and Agro-Industry Laboratory, Department of Biology, Faculty of Sciences, Ibn Tofail University, Kenitra, Morocco

Carbon-Based Smart Nanomaterials: An Overview

Yilma Hunde Gonfa, Archana (Joshi) Bachheti, Azamal Husen, and Rakesh Kumar Bachheti

Abstract Carbon-based nanomaterials (CNMs) have different structures, such as tube-like, spherical, horn-like, or ellipsoidal. They are mainly classified into three types of dimensionality: zero-dimensional, one-dimensional, and two-dimensional. CNMs show unique chemical, biological, optical, thermal, and mechanical properties. CNMs can be synthesized and tailored toward smart materials with characteristic applications. Currently, CNMs have attracted the interest of many researchers because of their diversified shapes and applications. Physicochemical parameters responsible for toxicity, such as size, shape and morphology, surface charge, composition, coatings, surface roughness, aggregation, and biological and chemical reactivity, are required to monitor the properties of CNMs. This chapter summarizes the types, synthesis, characterization, challenges of fabrication, current research trends, future aspects, toxicity, and environmental impacts of CNMs and their regulatory bodies, risk assessment, and mitigation, emphasizing their applications.

Y. H. Gonfa
Department of Chemistry, Ambo University, P. O. Box: 19, Ambo, Ethiopia
e-mail: yilma.hunde@ambou.edu.et

A. Bachheti
Department of Environment Science, Graphic Era (Deemed to be University), Dehradun, Uttarakhand 248002, India
e-mail: rkbachheti@gmail.com

A. Husen
Department of Biotechnology, Smt. S. S. Patel Nootan Science and Commerce College, Sankalchand Patel University, Visnagar, Gujarat 384315, India

Department of Biotechnology, Graphic Era (Deemed to be University), Dehradun, Uttarakhand 248002, India

Wolaita Sodo University, Wolaita, Ethiopia

R. K. Bachheti (✉)
Department of Industrial Chemistry, College of Applied Sciences, Addis Ababa Science and Technology University, P.O. Box: 16417, Addis Ababa, Ethiopia
e-mail: rkbachheti@gmail.com

Department of Allied Sciences, Graphic Era Hill University, Society Area, Clement Town, Dehradun, Uttarakhand 248002, India

A. Bachheti et al. (eds.), *Carbon-Based Nanomaterials*, Smart Nanomaterials Technology, https://doi.org/10.1007/978-981-97-0240-4_1

Keywords Smart materials · Nanomaterials · Carbon nanotubes · Carbon nanofibers · Carbon dots · Graphene · Synthesis · Applications · Environmental impacts

1 Introduction

The advancement of nanoscience and nanotechnology over a few decades has brought invaluable opportunities to develop smart nanomaterials [1]. Nanomaterials with nano-scaled dimensionality (having at least one dimension less than 100 nm) have extraordinary properties due to their unique optical, magnetic, electrical, thermal, and higher surface area-to-volume ratio natures [2–4]. The development of various dimensions of nanomaterials, including carbon-based nanomaterials (CNMs), metal nanomaterials (MNMs), ceramic nanomaterials (CNMs), semiconductor nanomaterials (SNMs), and polymeric and lipid-based nanomaterials has played a decisive role in the improvement of the lifestyle of human being and its environment [5, 6]. Nanomaterials can be synthesized through 'top-down' and 'bottom-up' approaches, which involve three methods such as physical, chemical, and biological (biogenic or green) synthesis [7–13]. The size, shape, and stability of nanomaterials during their synthesis depend upon the types of reductants, capping, and stabilizing agents used [14]. Nanomaterials are highly applicable in food processing, agriculture, marine, environmental, biomedicine, pharmaceutics, textile, catalysis, sensors, mechanics, and electronics fields [15–26]. For example, the development of smart materials used for imaging and disease treatment purposes in biomedical science fields named 'nanomedicine' has been transformed due to the revolutionization of nanotechnology [27]. Smart devices, especially carbon-based nanomaterials, such as graphene quantum dots (GQDs), are increasing daily with impressive properties because of their quantum nature and absorption quality [28].

Among nanomaterials, carbon-based nanomaterials (CNMs) are highly important smart materials used as sensors [29], chemical signaling in plants [30], anti-corrosion [31], smart construction materials (cement) filler [32], detection of mycotoxins [33], eco-friendly supercapacitors [34], and tissue engineering agents [35]. Carbon-based nanoparticles have lower toxicity than other nanoparticles, such as metal nanoparticles, silica nanoparticles, and lipid and polymer-based nanoparticles [36]. The main aim of this book chapter is to concisely review the current research literature on the types of carbon-based nanomaterials and their synthesis methods, potential applications, and environmental impacts.

2 Types of Carbon-Based Nanomaterials

CNMs are mainly classified as carbon nanotubes (CNTs), carbon nanofibers (CNFs), graphenes, graphene oxide (GO), fullerenes, carbon quantum dots (CQDs), nanocellulose, and graphite-like carbon nitride (g-C_3N_4) [37]. As per their structural dimensions, CNMs include (1) zero-dimensional (0D), for example, carbon quantum dots, carbon nanodots, and fullerenes, (2) one-dimensional (1D), for example; carbon nanotubes (CNTs) and carbon nanofibers (CNFs), and (3) two-dimensional (2D), example; single-layered graphene classes [38–41]. CNMs have highly enhanced mechanical, physical, chemical, and electronic properties compared to their bulk carbon form [42]. High thermal conductivity and large charge carrier mobility and density of CNMs result from *sp*2 hybridized carbon atoms in their structures [43].

The advantages of CNMs are raised from their small size (average size < 10 nm), high ability to create polar derivatives and high biocompatibility properties [44]. Energy storage, catalyst, energy conversion, drug delivery, biosensors, DNA sequencing, artificial tissue engineering, electromagnetic shielding, and bioimaging are the potential application areas for CNMs [38]. CNMs have significant applications in various areas, such as the enhancement of oil recovery [45], the reinforcement of engineering materials [46, 47], strain sensors [48], external pressure sensors [49], nanocarriers [50], drug delivery [51], wastewater treatment [52], phase change materials (PCM) [53], and additives of drilling fluids [42]. For instance, carbon nanotubes (CNTs) and graphene showed indispensable drug delivery systems for the effective release of drugs at the cancer site because of their interesting characteristics such as enhanced surface area, high drug loading capacity, and easily modifiable surface [51]. Oxidized CNTs demonstrated high efficiency in removing heavy metals from the environment due to their adsorption capacity for metal ions with fast kinetics [52]. CNTs, CNFs, and graphenes are highly important nanomaterials used in cement composites due to their high surface area, mechanical strength, and electrical conductivity [32]. The structures of some carbon-based nanomaterials are given in Fig. 1a–l.

Fullerenes

Fullerenes are one of the focus areas of researchers due to their abundant reserves and environmentally benign materials [54]. Fullerenes consist of different shapes, such as spheres, ellipsoids, and tubes containing hexagonal and pentagonal rings. The fullerene family contains different numbers of carbon atoms (C_n; examples are C_{20}, C_{60}, C_{70}, C_{76}, C_{78}, and C_{82}) [27]. The most abundant and stable form of fullerene is Buckminsterfullerene (C_{60}). Fullerenes are widely used in biomedical sciences and medicinal chemistry upon functionalizing their structures [55, 56]. They are also reported to be used as efficient gas adsorbents, inorganic and organic catalyst supporters, drug carriers, imaging agents, and environmental remediation agents [57]. Fullerenes, the symmetric cage with *sp*2 hybridized carbons, are applied to reinforce polymers with poor mechanical properties [35].

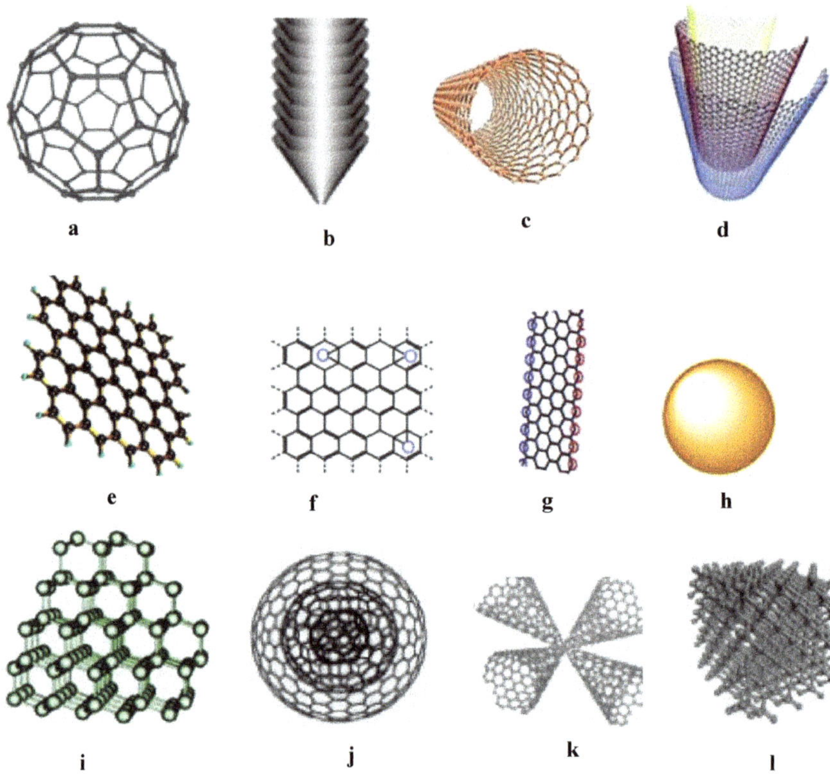

a. Buckminster fullerene (0D), b. Carbon nanofiber (1D), c. SWCNT (1D), d. MWCNT (1D), e. Graphene (2D), f. Graphene oxide (2D), g. Carbon nanoribbons (2D), h. Carbon dots (0D), i. Carbon nanodiamond (0D), j. Carbon nano-onion (0D), k. Carbon nanohorn (1D), l. Carbon nanodot (0D)

Fig. 1. Some classes of carbon-based nanomaterials with their dimensions [53, 70, 79–82]

Carbon Dots

Carbon dots (C dots) are zero-dimensional dominated carbon-based nanomaterials with less than 20 nm nanosize, mainly consisting of sp^2 and sp^3 carbon hybridization [58]. For instance, carbon quantum dots (CQDs) are known nanomaterials with the quantum confinement effect, which results in superior optoelectronic properties [59]. CQDs have emerged as a promising area for visible lighted-activated antimicrobial agents [60]. CQDs are also reported for their potential application for the human coronavirus (HCoV-229E) diagnosis [56]. Carbon dots are mixed with various polymers and used for energy storage applications [61]. Carbon dots are also reported to be used as biosensing agents [38].

Carbon Nanotubes (CNTs)

Carbon nanotubes (CNTs) are carbon-based materials with low dimensionality and high crystalline purity, which are currently the focus area of the researchers [62]. CNTs possess various unique structural, chemical, mechanical, and electromechanical properties. CNTs are tubular shapes and are the most widely used CNMs due to their excellent electrical conductivity and low cost [63]. Multi-walled carbon-based nanotubes (MWCNTs), which range from 2 to 30 nm in diameter, and single-walled carbon nanotubes (SWCNTs), which range from 1 to 2 nm in diameter, were discovered by Iijima in 1991 and 1993, respectively [35]. CNTs have demonstrated a positive effect on the germination and growth of tomato plants because of their ability to penetrate the seed coat and enhance the water uptake capacity [64]. CNTs are novel support materials that might interact with novel metal nanoparticles (MNPs) and develop novel catalysts [65]. CNTs are reportedly used as chemical detectors in the area of agriculture, including H_2O_2, Ca^{2+}, and NO in plants under biotic and abiotic stress [30]. MWCNTs decorated with Au are used as a detector of the quality of ethanol with possible interference from benzene [66]. MWCNTs are also reported as versatile devices integrated into the layer-by-layer assembly for electronic (for example, on ITO-coated PET substrates) and biomaterial surfaces [67].

Carbon Nanofibers

Carbon nanofibers (CNFs) are complex carbon-based nanomaterials that form fibrous microcrystallites with graphite-like hexagonal networks which resemble hollow cylindrical bars with 50–200 nm diameter [68]. The ability to highly graphitize at low temperature and their low cost of mass production make CNFs promising anodic materials for Li-ion batteries [69]. Excellent physicochemical activities such as chemical and mechanical stability, electrical conductivity, reinforcement capability, and antifouling properties of CNFs make them highly important carbon-based nanomaterials. Inherent chemical inertness, thermal stability, and hydrophobicity of CNFs were also reported as required properties for energy-efficient techniques for separating the oil–water mixture [70].

Graphene and Graphene Oxide

Graphene is the popular carbon nanofiller with a planar sheet and carbon sp2 hybridization in the honeycomb-like lattice [71]. Graphene is a carbon-based innovative nanomaterial with low dimensionality and high crystalline purity nature and highly attracting the interest of researchers [62]. Graphene is a semiconductor material with zero band gaps, an ambipolar electric field, and significant charge carrier mobility at room temperature [72]. Graphene and its derivatives have been exploited to introduce unique functional nanomaterials for different environment applications [73]. Graphene and graphene oxide have attracted the scientific community's attention due to their applicability in making polymer composites with superior anticorrosion properties [31]. Graphene and its oxide are necessary photothermal materials synthesized because of their strong absorption in visible and near-infrared light and rapidly transforming solar energy into heat energy [74]. The disruption of the *sp*2 bonding network in their structure makes graphene oxide an electrically insulating material [35]. The quantum confinement and edge effects are the source of remarkable electrical and magnetic properties of pentagonal graphene nanoribbons [62].

Other Carbon Nanomaterials

Nanogels, the cross-linked polymer networks, have porous structures with diverse applications such as bioimaging, biosensing, and drug delivery [75]. Carbon nanohorns and carbon nanodiamonds have been reported for their analytical sample extraction in the last decades [76]. Carbon nano-onions are employed as the lubricant additives for the oil and water-based coolant fluids to improve the drilling process [42]. Carbon nano-onions are some of the carbon-based nanomaterials utilized for constructing high-performance volatile organic compound sensors [77]. Carbon nanodiamonds are highly stable in corrosion media, chemically inert, optically transparent, and biocompatible materials [36]. Carbon nanohorns can be a gas-detecting material with improved sensitivity [78].

3 Synthesis and Characterization of Carbon-Based Nanomaterials

Synthesis Methods of Some Carbon-Based Nanomaterials

Various fabrication techniques for producing carbon-based nanomaterials (Fig. 2) [55, 83]. The 'top-down' method mainly cuts graphite-based materials with perfect *sp*2 carbon structures with oxidation and exfoliation processes in synthesizing carbon-based nanomaterials. The 'bottom-up' method also involves the cross-linking reactions of compounds rich in active functional groups through polymerization and carbonization [54]. The *sp*/*sp*2/*sp*3 hybridization ratio in carbon-based nanomaterials determines the synthesis of CNTs, graphene, fullerenes, CQDs, and CNFs [33]. Electrospinning and vapor growth are also essential synthesis methods for nanofibers [75].

The oxidation process of graphene in an acidic medium produces graphene oxides and increases the hydrophobicity of their surface [35]. Carbon nanohorns can be synthesized from pure graphite using high-rate laser ablations [84]. According to Adorinni and co-workers [80], carbon nano-onions (CNOs) are formed by concentric fullerenes, while carbon nanohorns (CNHs) are clusters of nanocones [80]. Carbon nano-onions (CNOs) are also reported to be formed from carbon nanodiamonds (CNDs) under a vacuum [55]. Carbon nanoribbons were reported to be synthesized when the reaction proceeds at room temperature [85]. There are few research focuses on the production of carbon nanomaterials by inserting fullerene or graphene nanoribbons into CNTs, encapsulating CNTs with carbon nanoribbons, and incorporating CNTs among graphene nanoplatelets [78]. Some specific examples of the synthesis of carbon-based nanomaterials are provided in Fig. 3a–d.

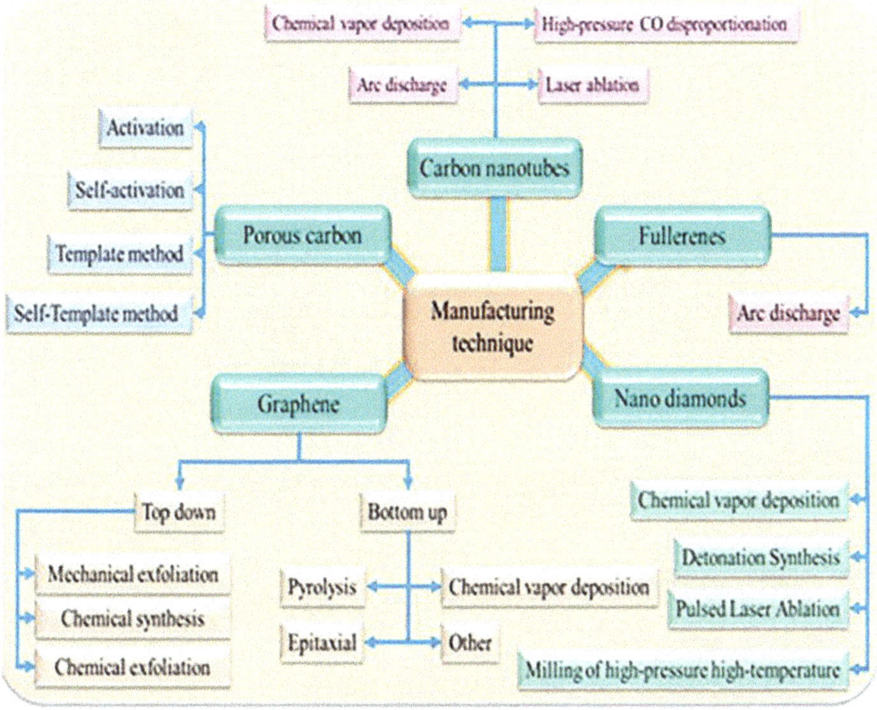

Fig. 2. Adaptable techniques for the synthesis of carbon-based nanomaterials [55]

Characterization Techniques for Carbon-Based Nanomaterials

Various analytical methods are used for the characterization of synthesized carbon-based nanomaterials. Scanning electron microscope (SEM) and Raman analyses are wisely employed for the detailed microstructural characterizations of carbon-based nanomaterials [75]. Surface morphologies of carbon-based nanomaterials are analyzed using the SEM technique [36]. The morphologies of the carbon-based nanomaterials are also characterized using field-emission transmission electron microscopy (TEM) [29]. Dynamic light scattering (DLS) technique is highly used to determine the size of carbon-based nanomaterials [28]. The surface area of nanomaterials is responsible for the adsorption–desorption activities and is mainly characterized using Brunauer–Emmett–Teller (BET) technique [88]. The surface roughness of nanomaterials can be evaluated by the atomic force microscopy (AFM) technique [89]. The inductively coupled plasma mass spectrometry (ICP-MS) technique is applied to quantify synthesized carbon-based nanotubes [90]. Powerful characterizations of carbon-based nanomaterials are combined with theoretical calculations [91]. Energy dispersive X-ray spectroscopy (EDX) or Energy dispersive spectroscopy (EDS) techniques are used for the determination of elemental composition in carbon-based nanomaterials [46, 92]. The chemical functional groups covalently bonded to

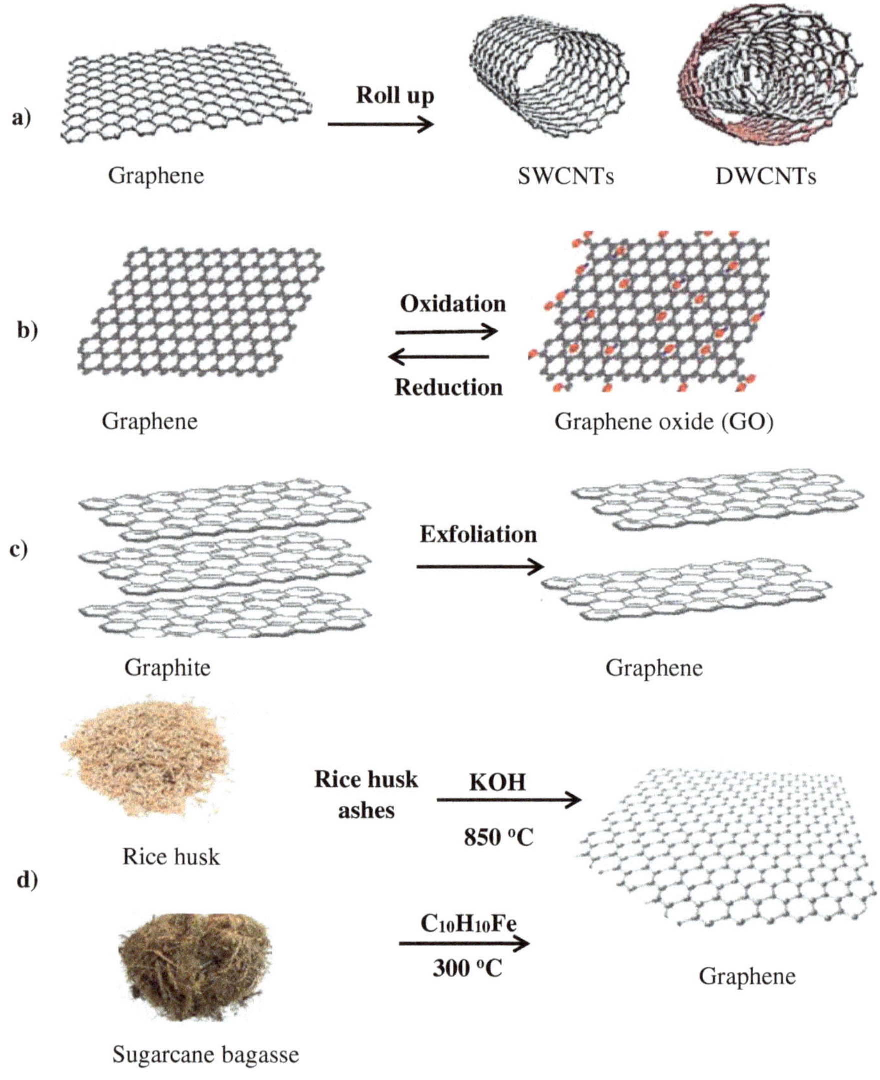

Fig. 3 **a–d** Synthesis methods of some specific carbon-based nanomaterials [35, 86, 87]

the surface of carbon-based nanomaterials are characterized by the Fourier Transform Infrared spectroscopy (FT-IR) method [93].

Challenges in Synthesis and Characterization

One of the main challenges of nanomaterials synthesis was reported as the innovation of novel systems, including simple, eco-friendly, low cost, and readily available techniques in industrial-scale fabrication [94]. For instance, producing CNTs on a large scale for water purification membranes is expensive and may need to be more cost-efficient [70]. In the synthesis of nanomaterials, controlling the physicochemical properties such as size, shape and morphology, surface charge, composition, coatings, surface roughness, aggregation, and biological and chemical reactivity is a tough and challenging task [52]. Reducing toxicity, enhancing biocompatibility, improving dispersibility, and enhancing water solubility are challenging tasks in carbon-based nanomaterials [41]. According to [95], CNMs production might suffer from problems derived from the difficulty of their fabrication [95]. The structural instability resulting from materials' low-dimensional shape and size is a possible challenge in synthesizing some carbon-based nanomaterials [78]. Another big challenge of carbon-based nanomaterial synthesis is the optimizing process for selecting and tuning materials to the required net electrochemical performance [58].Carbon-based nanomaterials have some drawbacks due to their improper dispersion effect as a result of the hydrophobicity nature [46]. Fabrication of large stretchability and high-sensitivity strain sensors using carbon-based nanomaterials is reported as a great challenge that needs improvement [75]. Carbon-based nanomaterials may suffer from the challenges derived from the high commercial production costs [95]. Some additional challenges of nanomaterials, including some carbon-based nanomaterials, are their separation from the aqueous solution and the limitation of their practical application due to their high price and low abundance [91]. Characterization of nanomaterials, including carbon-based nanomaterials, requires advanced instruments and skilled human power. As a result, the characterization of carbon-based nanomaterials needs expensive and advanced technology [96].

4 Potential Applications, Current Research Trends, and Future Prospects of Carbon-Based Nanomaterials

Different carbon-based nanomaterials have wide-spectrum applications in areas such as catalysts, electronic devices, energy production, environmental treatment, plant protection and production, and biomedical uses [84, 97]. Novel applications of smart nanomaterials are changing the current innovation patterns regarding carbon-based nanomaterials' trace trajectories to forecast the possible effects on future research trends and industrial processes across the globe [98]. The hot research trend in the area is the production of diverse shapes, sizes, and various compositions of carbon-based nanomaterials using their surface functionalization (Fig. 4). The small size and tailored functionalization of the surface of carbon-based nanomaterials allow for favoring drug delivery and penetrating through the cell membranes [99]. Surface functionalization, capability for drug loading, biocompatibility, and carbon-based nanomaterials' antiviral properties attract current researchers' attention [41]. One of the goals of carbon-based nanomaterials technology is to ensure the production of efficient and eco-friendly smart nanomaterials [42]. Carbon quantum dots are promising candidates for the future generation of supercapacitors [59].

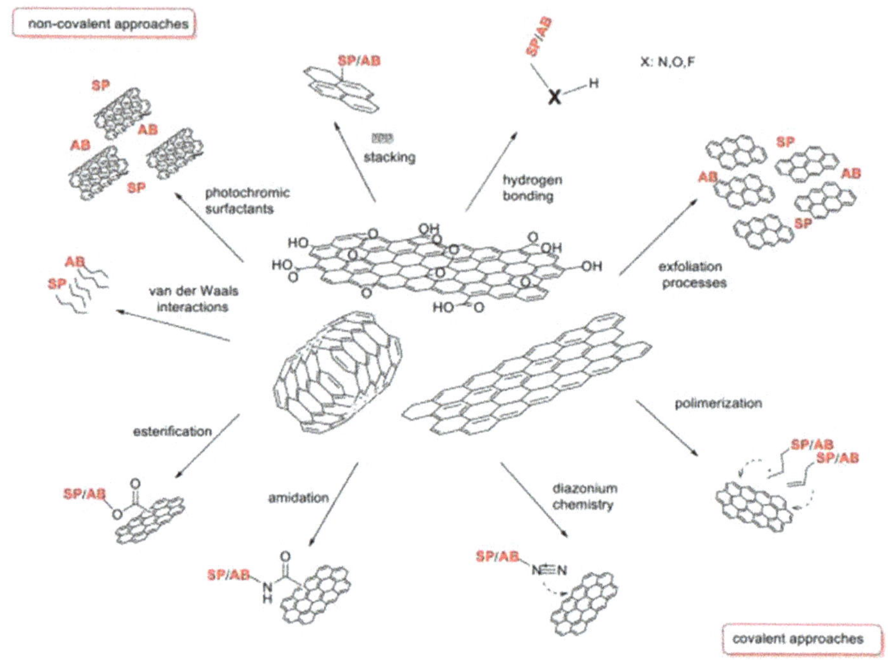

Fig. 4. Non-covalent and covalent approaches for carbon-based nanomaterials functionalization [101]

Similarly, preclinical investigation demonstrated that in vivo studies of carbon-based quantum dots have no side effects on living cells. Therefore, carbon quantum dots are reported as potential candidates for photodynamic therapy in biomedical applications [58]. Carbon-based nanomaterials production is mainly aimed at replacing the major drawbacks of heavy metals' toxicity and disposal problems [100].

5 Toxicity and Environmental Impact of Carbon-Based Nanomaterials

The toxicity of residues of carbon-based nanomaterials can cause serious health problems when released into the environment [70]. It is important to monitor the toxicity nature of nanomaterials, including carbon-based nanomaterials, and engineer their structures with smart approaches during their synthesis. This effect of nanomaterials has resulted in the emergence of the dedicated field known as nanotoxicology [27]. Carbon-based nanoparticles are reported with a lower toxicity nature to the living system when compared with metal nanoparticles, silica-based Cornell dots, and lipid and polymer-based nanoparticles [36]. Earlier reports also showed that, generally, carbon quantum dots have low cytotoxicity [99, 102]. For instance, some studies reported that the functionalized carbon-based nanomaterials showed enhanced biocompatibility with reduced in vivo and in vitro toxicity [50, 103]. The regulatory framework for the safe use and disposal of nanomaterials consists of well-designed guidelines and methods. International organizations have established important regulatory policies and standards for the safe use and storage of nanomaterials [104]. Regulatory bodies encourage research and development of nanotechnology-based products to ensure their safety for human health and the environment [105]. Even though legally binding regulatory rules have yet to be established for carbon-based nanomaterials in many countries, the maximum allowable exposure concentrations of CNTs and CNFs (inhalable dust materials) are specified by many institutions and organizations [106]. The risk must be assessed by carefully justifying the acquired benefits versus associated risks [27]. Designing well-defined risk assessments and risk mitigation strategies is highly important to monitor the risks related to the use of carbon-based nanomaterials. Risk assessment models and techniques for various types of nanosized materials share some common steps, such as lung dose estimation, interspecies scaling of dose, and temporal extrapolation of dose and response [107]. Considering the existing knowledge, reliable assessment, and scientific data is pivotal to monitoring the harmful environmental impacts of the wide use of carbon-based nanomaterials [108].

6 Conclusion and Future Directions

Carbon-based nanomaterials are one of the most promising products of nanotechnology, with diversified properties that enable them to be candidates for a broad spectrum of applications from electronics to drug delivery. CNMs contain three main types of nanostructures, such as 0-, 1-, and 2-dimensionality, with various properties and applications. The synthesis of CNMs occurs by two main techniques: bottom-up and top-down approaches. Advanced physical and chemical characterization methods are widely employed to produce CNMs. CNMs have wide applications as electronics, water treatment, medical, pharmaceutical, and agricultural smart delivery agents. It is important to widely investigate the challenges regarding the synthesis, characterization, toxicity, and environmental impacts of CNMs. In the future, the research on the CNMs should focus on obtaining their wide field applications and large-scale production. Future studies should also concentrate on determining the concentration of CNMs in the living system and its environment. Compared to other nanomaterials, the regulatory rules for risk management and risk mitigation strategies for using and disposing of CNMs require strong attention.

References

1. Serag E, El-Maghraby A, El Nemr A (2022) Recent developments in the application of carbon-based nanomaterials in implantable and wearable enzyme-biofuel cells. Carbon Lett 32:395–412
2. Godeto YG, Ayele A, Ahmed IN, Husen A, Bachheti RK (2023) Medicinal plant-based metabolites in nanoparticles synthesis and their cutting-edge applications: an overview. In: Bachheti RK, Bachheti A (eds) Secondary metabolites from medicinal plants. CRC Press, Boca Raton, USA, pp 1–34
3. Gonfa YH, Tessema FB, Tadesse MG, Bachheti A, Bachheti RK (2023) Medicinally important plant roots and their role in nanoparticles synthesis and applications. In: Bachheti RK, Bachheti A (eds) Secondary metabolites from medicinal plants. CRC Press, Boca Raton, USA, pp 227–242
4. Mengstu A, Esubalew S, Abate L, Husen RKBA, Bachheti A (2023) Medicinal plant-based flavonoid-mediated nanoparticles synthesis, characterization, and applications. In: Bachheti RK, Bachheti A (eds) Secondary metabolites from medicinal plants. CRC Press, Boca Raton, USA, pp 35–52
5. Nazeruddin GM, Prasad NR, Prasad SR, Shaikh YI, Waghmare SR, Adhyapak P (2014) Coriandrum sativum seed extract assisted in situ green synthesis of silver nanoparticle and its anti-microbial activity. Ind Crops Prod 60:212–216
6. Patil S, Chandrasekaran R (2020) Biogenic nanoparticles: a comprehensive perspective in synthesis, characterization, application and its challenges. J Genetic Eng Biotechnol 18:1–23
7. Bachheti RK, Konwarh R, Gupta V, Husen A, Joshi A (2019) Green synthesis of iron oxide nanoparticles: cutting edge technology and multifaceted applications. In: Nanomaterials and plant potential. Springer, Cham, pp 239–259. https://doi.org/10.1007/978-3-030-05569-1_9
8. Bachheti RK, Fikadu A, Bachheti A, Husen A (2020) Biogenic fabrication of nanomaterials from flower-based chemical compounds, characterization and their various applications: a review. Saudi J Biol Sci 27(10):2551–2562
9. Bachheti RK, Godebo Y, Bachheti A, Yassin MO, Husen A (2020) Root-based fabrication of metal/metal-oxide nanomaterials and their various applications. In: Husen A, Jawaid M (eds) Nanomaterials for agriculture and forestry applications. Elsevier, pp 135–166. https://doi.org/10.1016/B978-0-12-817852-2.00006-8
10. Bachheti A, Bachheti RK, Abate L, Husen A (2022) Current status of Aloe-based nanoparticle fabrication, characterization and their application in some cutting-edge areas. S Afr J Bot 147:1058–1069
11. Bachheti RK, Abate L, Bachheti A, Madhusudhan A, Husen A (2021) Algae-, fungi-, and yeast-mediated biological synthesis of nanoparticles and their various biomedical applications. In: Handbook of greener synthesis of nanomaterials and compounds. Elsevier, Amsterdam, pp 701–734. https://doi.org/10.1016/B978-0-12-821938-6.00022-0
12. Husen A, Rahman QI, Iqbal M, Yassin MO, Bachheti RK (2019) Plant-mediated fabrication of gold nanoparticles and their applications. In: Nanomaterials and plant potential. Springer, Cham, pp 71–110. https://doi.org/10.1007/978-3-030-05569-1_3
13. Husen A, Siddiqi KS (2014) Phytosynthesis of nanoparticles: concept, controversy and application. Nanoscale Res Lett 9(229):1–24
14. Ayinde WB, Gitari WM, Samie A (2019) Optimization of microwave-assisted synthesis of silver nanoparticle by Citrus paradisi peel and its application against pathogenic water strain. Green Chem Lett Rev 12:225–234
15. Bachheti RK, Bachheti A, Husen A (2023) Metal and metal-oxide based nanomaterials (Synthesis, Agricultural, Biomedical and Environmental Interventions). Springer Nature Singapore Pte Ltd., 152 Beach Road, #21–01/04 Gateway East, Singapore 189721, Singapore. https://link.springer.com/book/9789819976720
16. Chawla S, Singh S, Husen A (2023) Smart nanomaterials targeting pathological hypoxia. Springer Nature Singapore Pte Ltd., 152 Beach Road, #21–01/04 Gateway East, Singapore 189721, Singapore. https://doi.org/10.1007/978-981-99-1718-1

17. Dahlan AS (2019) Smart and functional materials based nanomaterials in construction styles in nano-architecture. SILICON 11:1949–1953
18. Husen A, Siddiqi KS (2014) Carbon and fullerene nanomaterials in plant system. J Nanobiotechnol 12(1):1–16
19. Husen A, Siddiqi KS (2023) Advances in smart nanomaterials and their applications. Elsevier Inc., 50 Hampshire St., 5th Floor, Cambridge, MA 02139, USA
20. Husen A (2022) Engineered nanomaterials for sustainable agricultural production, soil improvement and stress management. Elsevier Inc., 50 Hampshire St., 5th Floor, Cambridge, MA 02139, USA
21. Husen A (2023) Nanomaterials and nanocomposites exposures to plants (Response, Interaction, Phytotoxicity and Defense Mechanisms). Springer Nature Singapore Pte Ltd., 152 Beach Road, #21-01/04 Gateway East, Singapore 189721, Singapore. https://doi.org/10.1007/978-981-99-2419-6
22. Husen A (2023) Nanomaterials from agricultural and horticultural products. Springer Nature Singapore Pte Ltd., 152 Beach Road, #21-01/04 Gateway East, Singapore 189721, Singapore. https://doi.org/10.1007/978-981-99-3435-5
23. Husen A, Bachheti RK, Bachheti A (2023) Current trends in green nano-emulsions (Food, Agriculture and Biomedical Sectors). Springer Nature Singapore Pte Ltd., 152 Beach Road, #21-01/04 Gateway East, Singapore 189721, Singapore
24. Huq MA, Ashrafudoulla M, Rahman MM, Balusamy SR, Akter S (2022) Green synthesis and potential antibacterial applications of bioactive silver nanoparticles: a review. Polymers 14:1–22
25. Madhusudhan A, Purohit SD, Prasad R, Husen A (2023) Functional smart nanomaterials and their theranostics approaches. Springer Nature Singapore Pte Ltd., 152 Beach Road, #21–01/04 Gateway East, Singapore 189721, Singapore. https://link.springer.com/book/9789819965960
26. Singh R, Sharma A, Saji J, Umapathi A, Kumar S, Daima HK (2022) Smart nanomaterials for cancer diagnosis and treatment. In Nano Convergence 9:1–39
27. Umapathi A, Kumawat M, Daima HK (2022) Engineered nanomaterials for biomedical applications and their toxicity-a review. Environ Chem Lett 20:445–468
28. de Menezes FD, Alencar LMR, dos Santos CC, da Silva MIB, Santos-Oliveira R (2020) Using graphene quantum dots for treating radioactive liquid waste. Environ Sci Pollut Res 27:3508–3512
29. Tang J, Wu Y, Ma S, Yan T, Pan Z (2022) Sensing mechanism of a flexible strain sensor developed directly using electrospun composite nanofiber yarn with ternary carbon nanomaterials. IScience 25:1–20
30. Zhu L, Chen L, Gu J, Ma H, Wu H (2022) Carbon-based nanomaterials for sustainable agriculture: their application as light converters, nanosensors, and delivery tools. Plants 11:1–12
31. Ramezanzadeh B, Karimi B, Ramezanzadeh M, Rostami M (2019) Synthesis and characterization of polyaniline tailored graphene oxide quantum dot as an advance and highly crystalline carbon-based luminescent nanomaterial for fabrication of an effective anti-corrosion epoxy system on mild steel. J Taiwan Inst Chem Eng 95:369–382
32. Guo R, Suo Y, Xia H, Yang Y, Ma Q, Yan F (2021) Study of piezoresistive behavior of smart cement filled with graphene oxide. Nanomaterials 11:1–12
33. Ma X, Li X, Zhang W, Meng F, Wang X, Qin Y, Zhang M (2021) Carbon-based nanocomposite smart sensors for the rapid detection of mycotoxins. Nanomaterials 11:1–24
34. Landi G, La Notte L, Palma AL, Sorrentino A, Maglione MG, Puglisi G (2022) A comparative evaluation of sustainable binders for environmentally friendly carbon-based supercapacitors. Nanomaterials 12:1–16
35. De Armentia SL, Del Real JC, Paz E, Dunne N (2020) Advances in biodegradable 3D printed scaffolds with carbon-based nanomaterials for bone regeneration. Materials 13:1–49
36. Mitura K, Kornacka J, Kopczyńska E, Kalisz J, Czerwińska E, Affeltowicz M, Kaczorowski W, Kolesińska B, Frączyk J, Bakalova T, Svobodová L, Louda P (2021) Active carbon-based nanomaterials in food packaging. Coatings 11:1–31

37. Lu D, Zhong J (2022) Carbon-based nanomaterials engineered cement composites: a review. J Infrastructure Preservation and Resilience 3:1–20
38. Castro KPR, Colombo RNP, Iost RM, da Silva BGR, Crespilho FN (2023) Low-dimensionality carbon-based biosensors: the new era of emerging technologies in bioanalytical chemistry. Anal Bioanal Chem 415:3879–3895
39. Kumar R, Aadil KR, Ranjan S, Kumar VB (2020) Advances in nanotechnology and nanomaterials based strategies for neural tissue engineering. J Drug Deliv Sci Technol 57:1–21
40. Liang X, Li N, Zhang R, Yin P, Zhang C, Yang N, Liang K, Kong B (2021) Carbon-based SERS biosensor: from substrate design to sensing and bioapplication. NPG Asia Mater 13:1–36
41. Riley PR, Narayan RJ (2021) Recent advances in carbon nanomaterials for biomedical applications: a review. Curr Opin Biomed Eng 17:1–9
42. Rana A, Khan I, Saleh TA (2021) Advances in carbon nanostructures and nanocellulose as additives for efficient drilling fluids: trends and future perspective-a review. Energy Fuels 35:7319–7339
43. Deshmukh MA, Park SJ, Hedau BS, Ha TJ (2021) Recent progress in solar cells based on carbon nanomaterials. Sol Energy 220:953–990
44. Zadeh Mehrizi T, Eshghi P (2021) Investigation of the effect of nanoparticles on platelet storage duration 2010–2020. Int Nano Lett 1–31
45. Shayan Nasr M, Esmaeilnezhad E, Choi HJ (2021) Effect of carbon-based and metal-based nanoparticles on enhanced oil recovery: a review. J Mol Liq 338:1–17
46. Sheikh TM, Anwar MP, Muthoosamy K, Jaganathan J, Chan A, Mohamed AA (2021) The mechanics of carbon-based nanomaterials as cement reinforcement-a critical review. Constr Build Mater 303:1–21
47. Srivastava VK, Jain PK, Kumar P, Pegoretti A, Bowen CR (2020) Smart manufacturing process of carbon-based low-dimensional structures and fiber-reinforced polymer composites for engineering applications. J Mater Eng Perform 29:4162–4186
48. Her SC, Liang YM (2022) Carbon-based nanomaterials thin film deposited on a flexible substrate for strain sensing application. Sensors 22:1–12
49. Kang K, Park J, Kim K, Yu KJ (2021) Recent developments of emerging inorganic, metal and carbon-based nanomaterials for pressure sensors and their healthcare monitoring applications. Nano Res 14:3096–3111
50. Jha R, Singh A, Sharma PK, Porwal O, Fuloria NK (2021) Graphene-based nanomaterial system: a boon in the era of smart nanocarriers. J Pharm Invest 51:245–280
51. Navya PN, Kaphle A, Srinivas SP, Bhargava SK, Rotello VM, Daima HK (2019) Current trends and challenges in cancer management and therapy using designer nanomaterials. Nano Convergence 6:1–30
52. Villaseñor MJ, Ríos Á (2018) Nanomaterials for water cleaning and desalination, energy production, disinfection, agriculture and green chemistry. Environ Chem Lett 16:11–34
53. Olabi AG, Wilberforce T, Elsaid K, Sayed ET, Ramadan M, Atiqure Rahman SM, Abdelkareem MA (2021) Recent progress on carbon-based nanomaterial for phase change materials: prospects and challenges. Thermal Sci Eng Progr 23:1–16
54. Tao S, Feng T, Zheng C, Zhu S, Yang B (2019) Carbonized polymer dots: a brand new perspective to recognize luminescent carbon-based nanomaterials. [Review-article]. J Phys Chem Lett 10:5182–5188
55. Debnath SK, Srivastava R (2021) Drug delivery with carbon-based nanomaterials as versatile nanocarriers: progress and prospects. Front Nanotechnol 3:1–22
56. Thapa S, Singh KRB, Verma R, Singh J, Singh RP (2022) State-of-the-art smart and intelligent nanobiosensors for SARS-CoV-2 diagnosis. Biosensors 12:1–27
57. Marinescu L, Ficai D, Oprea O, Marin A, Ficai A, Andronescu E, Holban AM (2020) Optimized synthesis approaches of metal nanoparticles with antimicrobial applications. J Nanomater 2020:1–14
58. Jovanović S, Marković Z, Budimir M, Prekodravac J, Zmejkoski D, Kepić D, Bonasera A, Marković BT (2023) Lights and dots toward therapy-carbon-based quantum dots as new agents for photodynamic therapy. Pharmaceutics 15:1–40

59. Permatasari FA, Irham MA, Bisri SZ, Iskandar F (2021) Carbon-based quantum dots for supercapacitors: recent advances and future challenges. Nanomaterials 11:1–34
60. Heidari G, Hassanpour M, Nejaddehbashi F, Sarfjoo MR, Yousefiasl S, Sharifi E, Bigham A, Agarwal T, Borzacchiello A, Lagreca E, Natale CD, Nikfarjam N, Vasseghian Y (2022) Biosynthesized nanomaterials with antioxidant and antimicrobial properties. Mater Chem Horizons 2022:35–48
61. Diez-Pascual AM (2020) Carbon-based polymer nanocomposites for high-performance applications. Polymers 12:1–6
62. Pawar S, Duadi H, Fixler D (2023) Recent advances in the spintronic application of carbon-based nanomaterials. Nanomaterials 13:1–19
63. Lee HK, Nam IW, Tafesse M, Kim HK (2019) Fluctuation of electrical properties of carbon-based nanomaterials/cement composites: case studies and parametric modeling. Cement Concr Compos 102:55–70
64. Athanassiou CG, Kavallieratos NG, Benelli G, Losic D, Usha Rani P, Desneux N (2017) Nanoparticles for pest control: current status and future perspectives. J Pest Sci 10:1–15
65. Alshammari A, Kalevaru VN, Martin A (2012) Metal nanoparticles as emerging green catalysts. Intech, pp 1–35
66. Tomić M, Šetka M, Vojkůvka L, Vallejos S (2021) Vocs sensing by metal oxides, conductive polymers, and carbon-based materials. Nanomaterials 11:1–34
67. Zan X, Bai H (2021) Review-novel carbon nanomaterials based flexible electrochemical biosensors. J Electrochem Soc 168:1–15
68. Dobrzański LA, Dobrzańska-Danikiewicz AD (2019) Why are carbon-based materials important in civilization progress and especially in the industry 4.0 stage of the industrial revolution. Mater Perform Characterization 8:1–25
69. Roselin LS, Juang RS, Hsieh CT, Sagadevan S, Umar A, Selvin R, Hegazy HH (2019) Recent advances and perspectives of carbon-based nanostructures as anode materials for Li-ion batteries. Materials 12:1–46
70. Noamani S, Niroomand S, Rastgar M, Sadrzadeh M (2019) Carbon-based polymer nanocomposite membranes for oily wastewater treatment. Npj Clean Water 2:1–14
71. Sanivada UK, Esteves D, Arruda LM, Silva CA, Moreira IP, Fangueiro R (2022) Joule-heating effect of thin films with carbon-based nanomaterials. Materials 15:1–16
72. Bagyalakshmi S, Sivakami A, Pal K, Sarankumar R, Mahendran C (2022) Manufacturing of electrochemical sensors via carbon nanomaterials novel applications: a systematic review. J Nanopart Res 24:1–28
73. Dutta V, Verma R, Gopalkrishnan C, Yuan MH, Batoo KM, Jayavel R, Chauhan A, Lin KYA, Balasubramani R, Ghotekar S (2022) Bio-inspired synthesis of carbon-based nanomaterials and their potential environmental applications: a state-of-the-art review. Inorganics 10:1–33
74. Zhao Y, Ji H, Lu M, Tao J, Ou Y, Wang Y, Chen Y, Huang Y, Wang J, Mao Y (2022) Thermochromic smart windows assisted by photothermal nanomaterials. Nanomaterials 12:1–16
75. Yan T, Wang Z, Pan ZJ (2018) Flexible strain sensors fabricated using carbon-based nanomaterials: a review. Curr Opin Solid State Mater Sci 22:213–228
76. Armenta S, Esteve-Turrillas FA, Garrigues S, de la Guardia M (2021) Smart materials for sample preparation in bioanalysis: a green overview. Sustain Chem Pharm 21:1–20
77. Yin F, Yue W, Li Y, Gao S, Zhang C, Kan H, Niu H, Wang W, Guo Y (2021) Carbon-based nanomaterials for the detection of volatile organic compounds: a review. Carbon 180:274–297
78. Verma G, Gupta A (2022) Recent development in carbon nanotubes based gas sensors. J Mater NanoSci 2022:3–12
79. Abd-Elsalam KA (2019) Carbon nanomaterials: 30 years of research in agroecosystems. In: Carbon nanomaterials for agri-food and environmental applications, pp 1–18
80. Adorinni S, Rozhin P, Marchesan S (2021) Smart hydrogels meet carbon nanomaterials for new frontiers in medicine. Biomedicines 9:1–22
81. Rahmati M, Mozafari M (2019) Biological response to carbon-family nanomaterials: interactions at the nano-bio interface. Front Bioeng Biotechnol 7:1–22

82. Sharma A, Sharma N, Kumari A, Lee HJ, Kim TY, Tripathi KM (2020) Nano-carbon based sensors for bacterial detection and discrimination in clinical diagnosis: a junction between material science and biology. Appl Mater Today 18:1–28
83. Belay T, Worku LA, Bachheti RK, Bachheti A, Husen A (2023) Nanomaterials: introduction, synthesis, characterization, and applications. In: Husen A, Siddiqi KS (eds) Advances in smart nanomaterials and their applications. Elsevier, Amsterdam, Netherlands, pp 1–21
84. Hosu O, Tertiş M, Cernat A, Feier B, Săndulescu R (2020) Recent approaches to the synthesis of smart nanomaterials for nanodevices in disease diagnosis. In: Nanomaterials in diagnostic tools and devices, pp 1–55
85. Zare EN, Padil VVT, Mokhtari B, Venkateshaiah A, Wacławek S, Černík M, Tay FR, Varma RS, Makvandi P (2020) Advances in biogenically synthesized shaped metal and carbon-based nanoarchitectures and their medicinal applications. Adv Coll Interface Sci 283:1–17
86. Onyancha RB, Ukhurebor KE, Aigbe UO, Osibote OA, Kusuma HS, Darmokoesoemo H (2022) A Methodical review on carbon-based nanomaterials in energy-related applications. Adsorpt Sci Technol 2022:1–21
87. Thangadurai D, Naik J, Sangeetha J, Shettar AK, Adetunji JB (2020) Nanomaterials from agrowastes: past, present, and the future. In: Handbook of nanomaterials and nanocomposites for energy and environmental applications, pp 1–18
88. Khan I, Saeed K, Khan I (2019) Nanoparticles: properties, applications and toxicities. Arab J Chem 12:908–931
89. Restrepo CV, Villa CC (2021) Synthesis of silver nanoparticles, influence of capping agents, and dependence on size and shape: a review. Environ Nanotechnol, Monit Manage 15:1–11
90. Patra JK, Baek KH (2014) Green nanobiotechnology: factors affecting synthesis and characterization techniques. J Nanomater 2014:1–13
91. Fu Y, Zeng G, Lai C, Huang D, Qin L, Yi H, Liu X, Zhang M, Li B, Liu S, Li L, Li M, Wang W, Zhang Y, Pi Z (2020) Hybrid architectures based on noble metals and carbon-based dots nanomaterials: a review of recent progress in synthesis and applications. Chem Eng J 399:1–22
92. Piaskowski K, Zarzycki PK (2020) Carbon-based nanomaterials as promising material for wastewater treatment processes. Int J Environ Res Public Health 17:1–14
93. de Oliveira TC, Ferreira FV, de Menezes BRC, da Silva DM, dos Santos AS, Kawachi EY, Simonetti EAN, Cividanes LS (2021) Engineering the surface of carbon-based nanomaterials for dispersion control in organic solvents or polymer matrices. Surf Interfaces 24:1–10
94. Eisa WH, Zayed MF, Anis B, Abbas LM, Ali SSM, Mostafa AM (2019) Clean production of powdery silver nanoparticles using *Zingiber officinale*: The structural and catalytic properties. J Clean Prod 241:1–12
95. Campuzano S, Yáñez-Sedeño P, Pingarrón JM (2019) Carbon dots and graphene quantum dots in electrochemical biosensing. Nanomaterials 9:1–18
96. Zaytseva O, Neumann G (2016) Carbon nanomaterials: production, impact on plant development, agricultural and environmental applications. Chem Biol Technol Agric 3:1–26
97. Husen A (2023) Secondary metabolites based green synthesis of nanomaterials and their applications. Springer Nature Singapore Pte Ltd., 152 Beach Road, #21–01/04 Gateway East, Singapore 189721, Singapore
98. Sohail MI, Waris AA, Ayub MA, Usman M, Zia ur Rehman M, Sabir M, Faiz T (2019) Environmental application of nanomaterials: a promise to sustainable future. Comprehensive Anal Chem 87:1–54
99. Innocenzi P, Stagi L (2020) Carbon-based antiviral nanomaterials: graphene, C-dots, and fullerenes. A perspective. Chem Sci 11:6606–6622
100. Lochab A, Sharma R, Kumar S, Saxena R (2020) Recent advances in carbon based nanomaterials as electrochemical sensor for toxic metal ions in environmental applications. Mater Today: Proc 45:3741–3753
101. Cardano F, Frasconi M, Giordani S (2018) Photo-responsive graphene and carbon nanotubes to control and tackle biological systems. Front Chem 6:1–17

102. Alavi M, Jabari E, Jabbari E (2021) Functionalized carbon-based nanomaterials and quantum dots with antibacterial activity: a review. Expert Rev Anti Infect Ther 19:35–44
103. Mahor A, Singh PP, Bharadwaj P, Sharma N, Yadav S, Rosenholm JM, Bansal KK (2021) Carbon-based nanomaterials for delivery of biologicals and therapeutics: a cutting-edge technology. J Carbon Res 7:19
104. Demirbas AK, Cevik S (2020) Regulatory policies for safety of nanomaterials. Open J Nano 5:1–16
105. Kumari R, Suman K, Karmakar S, Mishra V, Lakra SG, Saurav GK, Mahto BK (2023) Regulation and safety measures for nanotechnology-based agri-products. Front Genome Editing 5:1–12
106. Tsunemi K, Yoshida M, Kawamoto A (2022) Screening risk assessment at the production and use stage of carbon nanomaterials generated in hydrogen manufacture by methane decomposition. Sustainability 14:1–12
107. Kuempel ED, Geraci CL, Schulte PA (2012) Risk assessment and risk management of nanomaterials in the workplace: translating research to practice. Ann Occup Hyg 56:491–505
108. Mohan VB (2019) Handling and risk mitigation of nanoscale graphene and related materials: some considerations and recommendations. C-J Carbon Res 5:1–9
109. Bachheti RK, Sharma A, Bachheti A, Husen A, Shanka GM, Pandey DP (2020) Nanomaterials from various forest tree species and their biomedical applications. In: Husen A, Jawaid M (eds) Nanomaterials for agriculture and forestry applications. Elsevier, pp 81–106. https://doi.org/10.1016/B978-0-12-817852-2.00004-4
110. Husen A (2023b) Smart nanomaterials from agricultural and horticultural products. Springer Nature Singapore Pte Ltd., 152 Beach Road, #21–01/04 Gateway East, Singapore 189721, Singapore
111. Husen A (2020) Interactions of metal and metal-oxide nanomaterials with agricultural crops: an overview. In: Husen A, Jawaid M (eds) Nanomaterials for agriculture and forestry applications. Elsevier Inc. 50 Hampshire St., 5th Floor, Cambridge, MA 02139, USA, pp 167–197 https://doi.org/10.1016/B978-0-12-817852-2.00007-X
112. Husen A (2020) Carbon-based nanomaterials and their interactions with agricultural crops. In: Husen A, Jawaid M (eds) Elsevier Inc. 50 Hampshire St., 5th Floor, Cambridge, MA 02139, USA, pp 199–218 https://doi.org/10.1016/B978-0-12-817852-2.00008-1
113. Jeerapan, & Ma. (2019) Challenges and opportunities of carbon nanomaterials for biofuel cells and supercapacitors: personalized energy for futuristic self-sustainable devices. C -J Carbon Res 5:1–27
114. Liu X, Huang D, Lai C, Qin L, Zeng G, Xu P, Li B, Yi H, Zhang M (2019) Peroxidase-like activity of smart nanomaterials and their advanced application in colorimetric glucose biosensors, pp 1–27
115. Mandal AK, Ghorai S, Husen A (2023) Functionalized smart nanomaterials for point-of-care testing. Springer Nature Singapore Pte Ltd., 152 Beach Road, #21–01/04 Gateway East, Singapore 189721, Singapore. https://link.springer.com/book/9789819957866
116. Wang G, Wang Y, Luo Y, Luo S (2018) Carbon nanomaterials based smart fabrics with selectable characteristics for in-line monitoring of high-performance composites. Materials 11:1–12
117. Wang H, Chen Q, Zhou S (2018) Carbon-based hybrid nanogels: a synergistic nanoplatform for combined biosensing, bioimaging, and responsive drug delivery. Chem Soc Rev 47:4198–4232

Graphene-Based Nanomaterial Synthesis, Characterization, and Applications

Ümran Ünlü and Kaan Hürkan

Abstract Graphene, a nanomaterial, is recognized as a two-dimensional (2D) carbon form, one of the universe's most prevalent elements. Despite its short history since its discovery, it exhibits extraordinary thermal, electrical, optical, and mechanical properties thanks to its high surface area, high electron mobility, biocompatible structure, and high adsorption capacity. These characteristics enable graphene and graphene-based nanomaterials to be remarkable in a variety of applications. According to the application area, the desired properties should be considered in the choice of synthesis method. Graphene synthesis methods are basically divided into the top-down method, which is based on the principle of exfoliation of layers, and the bottom-up method, which is based on the principle of layer growth. While synthesis methods provide a range of dimensional, physical, chemical, and morphological properties, these attributes are assessed using diverse characterization techniques, including X-ray photoelectron spectroscopy, Raman spectroscopy, X-ray diffraction, transmission electron microscopy, Brunauer–Emmett–Teller analysis, and thermogravimetric analysis. Graphene-based nanomaterials exhibit significant potential in catalyst application, aiding in air and water purification as well as water splitting for H_2 production due to their remarkable adsorption capacity. Moreover, they can also be used in energy storage because of their high electron mobility and conductivity and in various biomedical application areas because of their biocompatible structure. Looking ahead, these graphene-based nanomaterials hold significant promise, offering numerous technological opportunities in the future. This section initially highlights the structure and brief history of graphene, followed by examinations of the production processes, characterization procedures, and application fields for graphene-based nanomaterials.

Ü. Ünlü (✉)
Department of Chemical Engineering, Faculty of Engineering, Gebze Technical University, 41400 Kocaeli, Türkiye
e-mail: umranunlu@gtu.edu.tr

K. Hürkan
Department of Agricultural Biotechnology, Faculty of Agriculture, Iğdır University, 76000 Iğdır, Türkiye

A. Bachheti et al. (eds.), *Carbon-Based Nanomaterials*, Smart Nanomaterials Technology, https://doi.org/10.1007/978-981-97-0240-4_2

Keywords Graphene · Reduced graphene oxide · Synthesis · Graphene oxide · Characterization · Application

1 Introduction

In recent years, worldwide interest in nanostructured materials has been increasing due to their remarkable chemical, mechanical, and physical characteristics [1–12]. Graphene is a nanomaterial known as one of the two-dimensional (2D) allotropes of carbon, one of the universe's most abundant elements [13]. The graphene structure has a honeycomb lattice arrangement of carbon atoms, consisting of sp^2 hybridization because of three σ bonds and one π bond. Furthermore, the term "graphene" also refers to the monolayer of the three-dimensional (3D) graphite structure, which has been known for almost 500 years. Graphene also serves as the basis for wrapped 0D buckyball fullerenes and rolled 1D carbon nanotubes (CNT). The graphene structure is around 0.35 nm in size [14–17]. 0D, 1D, and 2D structures are given in Fig. 1. In 1994, Boehm, Setton, and Stumpp suggested the term "graphene" to describe single layers of graphite with the -ene suffix [18].

The earliest investigations into the presence of graphene date back to theoretical investigations into energy bands in graphite by Wallace conducted in the early 1900s [19]. However, the isolation of a single layer of graphene from graphite was first achieved in 2004. In the investigation conducted by Geim and Novoselov, the scotch tape micromechanical exfoliation method was implemented to obtain a 2D graphene structure successfully. Research on graphene and graphene-based nanomaterials groundbreakingly increased after Geim and Novoselov were awarded the Nobel Prize in Physics "*for groundbreaking experiments regarding the two-dimensional material*

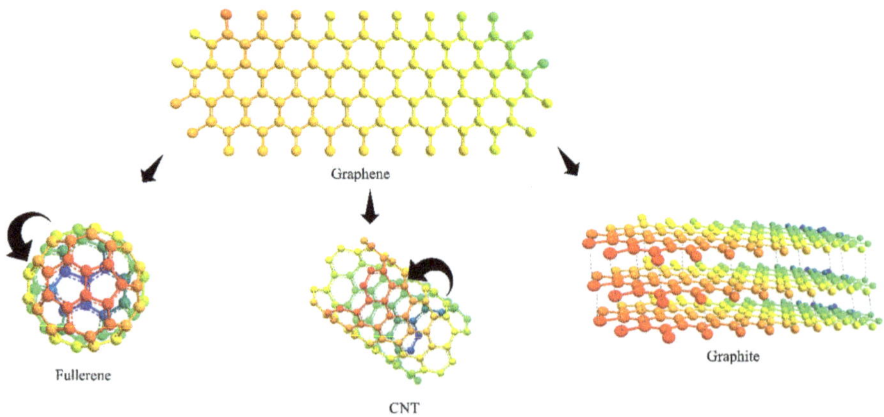

Fig. 1. Illustration of 0D buckyballs, 1D CNT, 2D graphene, 3D graphite structure

Table 1. Properties of graphene

Properties		References
Thermal conductivity	2000–5000 W/m K	Balandin et al. [25]
Mechanical strength	1000 GPa Young's Module	Lee et al. [26]
Optical transmittance	~97.7%	Nair et al. [27]
Surface area	2630 m^2g^{-1}	Stoller et al. [28]
Electron mobility	(>200.000 cm^2/V.s)	Morozov et al. [29]

graphene" in 2010 and described as one of the thinnest and strongest substances in the universe [20].

Graphene-based nanomaterials encompass graphene oxide (GO), reduced graphene oxide (rGO), and graphene quantum dots (GQD). Additionally, graphene-based nanomaterials might include graphene nanosheets, graphene nanoribbons, monolayer graphene, and few-layer graphene with various morphologies. Studies on graphene oxide and reduced graphene oxide date back further, in contrast to graphene [17, 21, 22]. The discovery of graphene oxide and intercalation using various oxidants by Brodie in 1859 [23] followed Schafhaetl's 1840 study on the intercalation and exfoliation of graphite by treating the graphite structure with acid [24].

Graphene-based nanomaterials are distinguished from other nanomaterials and become remarkable in a broad range of application areas as a result of their superior characteristics. For instance, having non-cytotoxicity or very low cytotoxicity, biocompatibility, and high surface area enable high drug loading capacity in drug delivery systems. In addition, graphene is significant in numerous application fields, including catalysis, conductor material, energy storage systems, optoelectronics, sensor technology, and biological imaging, due to its excellent thermal, mechanical, optical, and electrical characteristic features as well as excellent electron mobility at ambient temperature [17]. Further details of graphene properties are given in Table 1.

The structure and characteristics of graphene vary depending on the fabrication process [30]. This necessitates research on different methods of production for various application areas. A brief overview of the production processes, application areas, and graphene-based nanomaterial characterization is provided in the following section.

2 Graphene-Based Nanomaterial Synthesis

A variety of techniques have been enhanced for graphene synthesis, which is a single sheet of graphite. These methods can be categorized as bottom-top [31, 32] and top-down [33–35] or chemical and physical [24, 36] (Fig. 2). The growth of graphene sheets using the assembly of fundamental structural components is known as the bottom-up approach. Conversely, the top-down approach is based on breaking weak

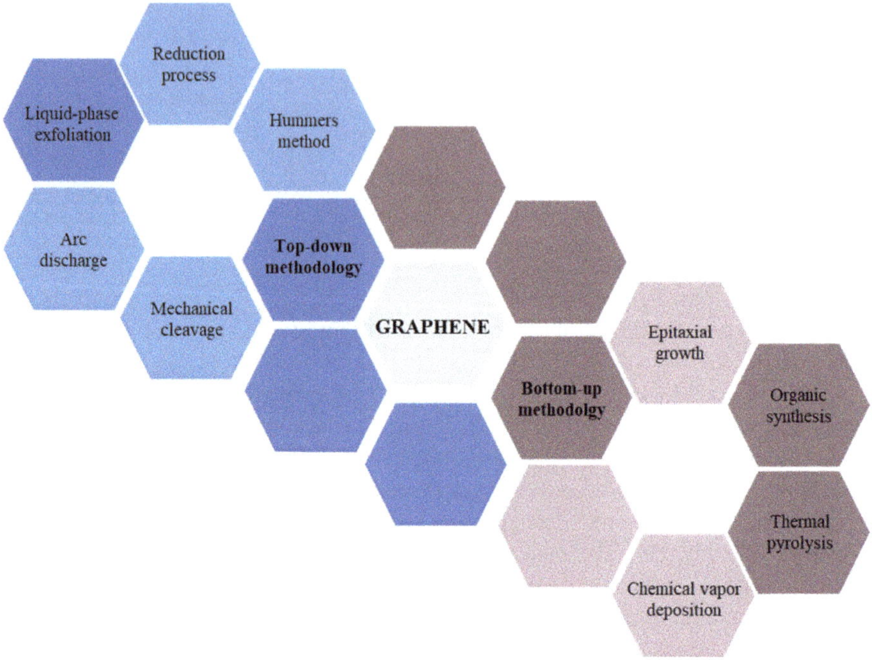

Fig. 2. Schematic illustration of graphene synthesis methods

intermolecular interactions between graphite layers by mechanical and chemical exfoliation [13]. Different production techniques have various positive and negative aspects. While the bottom-up approach has advantages such as affordable cost, material savings, controllable morphology, and thickness, the method nevertheless has drawbacks such as challenging synthesis steps and material dimensions that are often smaller than needed. Additionally, this method provides material with a more homogenous structure, making this technique preferable for biomedical applications. The top-down approach necessitates long periods of time, high energy consumption, and high costs but provides industrial-scale production [17].

2.1 Hummers Method (Top-Down Method)

The Hummers method is employed to produce graphite oxide. Hummers utilized a combination of H_2SO_4, $NaNO_3$, and $KMnO_4$ as an alternative to the oxidants used in the discovery of graphene oxide in 1957, and this process emerged as a more reliable method according to Brodie's method [37]. The temperature value is one of the crucial factors in this procedure. Therefore, Hummers and Offman [37] reported that as long as the temperature value was kept below 45 °C, graphitic oxide could be obtained reliably by treating graphite with H_2SO_4, $NaNO_3$, and $KMnO_4$ for 2 h in a water-free atmosphere. Although this approach is still widely used today, modified Hummers methods are being researched, such as adding new agents or eliminating oxidizing agents to enhance production conditions, increase efficiency, and save costs [38].

The hydrophilic feature of graphene oxide, in contrast to the hydrophobic property of graphene, enables it to disperse in some organic solvents and water; thanks to the functional groups in the structure [30]. Additionally, due to disrupted sp^2 bondings, graphene oxide and graphite oxide are both electrically insulating.

2.2 Reduction of Graphene Oxide (Top-Down Method)

This process involves generating reduced graphene oxide (rGO) using several methods such as chemical [33], thermal [39], and electrochemical [40] reduction of the oxygen content in graphene oxide. This method is also known as graphene production and is remarkable due to its high efficiency. Furthermore, this approach improves electrical conductivity and makes it possible to fix the honeycomb structure, which is harmed during graphene oxidation [30, 41].

In chemical reduction, hydrazine monohydrate might be utilized as a reducing agent in the colloidally dispersed system. Hydrazine monohydrate holds prominence as the most extensively employed and one of the earliest recognized reducing agents in the reduction process of graphene oxide [41, 42]. However, excessive use of reducing agents may cause contamination of the final products.

In electrochemical reduction, two alternative ways are presented to obtain graphene. The electrochemical reduction might be accepted as a more environmentally friendly method and also offers a more controllable and sensitive route. One of these approaches uses an aqueous buffer electrolyte solution and an external power source (potentiostat) in an electrochemical cell, while the other, known as a two-stage technique, involves the coating stage of graphene oxide on the electrode surface [43–45].

In thermal reduction, high heat is used to decrease GO; however, the annealing temperature used impacts the reduction process's effectiveness. Compared to conventional chemical reduction, it can achieve higher reduction degrees. While an imperfect lattice structure may form depending on the rate of increase in temperature, if the temperature rises more slowly, the method requires a long time, and the application of high heat raises energy costs. Therefore, the system requires precisely accurate control of application conditions [39, 46].

2.3 Liquid Phase Exfoliation (LPE) (Top-Down Method)

In the liquid phase exfoliation method, the exfoliation of these layers is accomplished via sonication because of the weak Van der Waals interactions between graphite layers. Graphite must first be dispersed in an appropriate solvent for this purpose. Therefore, the selection of solvent in the LPE method is crucial. The obtained graphene layers must be purified following the sonication-assisted exfoliation procedure. Sonication parameters are essential in this synthesis method because they can affect the quality of the graphene sheets to be obtained. This technique enables the manufacture of graphene at high production volumes more affordably after determining the appropriate solvent choice and sonication parameters [35, 47].

2.4 Mechanical Cleavage (MC) (Top-Down Method)

Mechanical cleavage was first reported for graphene production by Geim and Novoselov. Small forces are sufficient to break the weak Van der Waals interactions that exist between graphene layers. Different techniques, including normal force [48] and shear force [49], are applied in various directions to break these interactions. These methods have a low-cost requirement and are easy to apply. Nevertheless, while being the first method developed for producing graphene, drawbacks such as difficulties managing the number of layers and low yield limit the use of this method in commercial production [17, 50–52].

2.5 Arc Discharge (Top-Down Method)

The arc discharge approach was first employed to produce certain other carbon allotropes, but it was subsequently suggested that it could also be used to produce graphene. This method uses a reaction chamber with a liquid or gaseous atmosphere for the anode (carbon precursor) and cathode (graphite rod). When the power source creates an arc, the arc current generates plasma, which sublimates the carbon precursor in the anode. The impact of the temperature gradient causes sublimated carbon precursors to accumulate at the cathode from the vapor phase [53]. The procedure used to create the graphene material has superior electrical conductivity and enables a less defective graphene structure compared to chemically treated approaches [17].

2.6 Chemical Vapor Deposition (CVD) (Bottom-Up Method)

The CVD process is regarded as a highly significant and promising approach, particularly in considering high-capacity graphene production. The procedure is complicated by the requirement to accurately manage process variables such as pressure value, precursor type, deposition time, and high annealing temperatures $\geq$1000 °C [3, 54, 55]. This technique includes depositing graphene on a metal surface with a large surface area under appropriate conditions. A variety of metal substrates, such as Nickel (Ni) [56], Copper (Cu) [57], and alloys [58], can be employed. However, the necessity of separating these metal substrates from the deposited layer causes high costs [59].

2.7 Epitaxial Growth (Bottom-Up Method)

This method can be expressed as the high-vacuum and high-temperature deposition of graphene on an insulating substrate such as silicon carbide (SiC). The same parameters that are crucial for CVD were also applied to this method, including temperature and vacuum. Additionally, unlike the CVD technique, the created graphene layer in this technique does not require transfer to a separate substrate. In some applications, this allows for direct production on the sheet and enables an easier process than CVD [60, 61].

In addition to these techniques, there is another bottom-up strategy known as organic synthesis, in which smaller chemical compounds are used to create polycyclic aromatic carbons [17].

3 Characterization Techniques

Different characterization methods are used to examine the structure, thickness, and morphology of graphene, GO, and graphene layers. The most common among these methods are Raman Spectroscopy, X-Ray Diffraction (XRD), Transmission Electron Microscopy (TEM), Raman Spectroscopy, Fourier-Transform Infrared Spectroscopy (FTIR), and Thermogravimetric Analysis (TGA).

In Raman Spectroscopy, the fundamental working principle is to measure the frequency changes resulting from the vibration, rotation, or movement of molecules caused by the light scattered from this material after the material is exposed to light (Li et al. 2023b). It is a technique for characterizing graphene in which its chemical composition and structural characteristics are examined. Graphene has different peak regions due to sp^2 hybridization in its molecular structure. These peaks are called G-band (Graphene band), D-band (Defect band), and 2D-band (G'). G-band is located at a wavelength of $\sim$1580 cm^{-1} and indicates the greater crystallinity of graphene, while D-band is located between wavelengths of 1300 $\sim$ 1400 cm^{-1} and indicates defects in the structure. Since the intensity of these wavelengths varies depending on the production methods of graphene, it allows the effectiveness of production methods to be examined. It can also disclose details regarding the number of layers. Application parameters in different production methods affect these values [62, 63]. For example, according to Sayahi et al., the number of graphene layers increased when varied solution concentrations were utilized to produce graphene using the electrochemical exfoliation method [64].

In X-Ray Diffraction (XRD), depending on the structure of the material, X-ray beams that are sent to the material scatter in various directions [65]. Especially when providing information about the distance and structural arrangement between graphene layers, it also clearly shows the distinction between graphene, GO, and graphite [66]. Stobinski et al. stated that the diffraction peak reached a lower value at 2θ in the GO structure and this value increased in the rGO structure, and they related this with the distance between the graphene layers [67].

In Fourier Transformed Infrared Spectroscopy (FTIR), functional groups in the structure are determined by measuring the IR light absorbed or scattered after the beams are transmitted to the Material. O–H bonds offer stretching absorption in the wavelength range of 3590–3650 cm^{-1}, –COOH, –COH, and C–O–C are detected at wavelengths of 1727 cm^{-1}, 1225 cm^{-1}, and 1052 cm^{-1}, respectively [50, 62]. Evaluation of the effectiveness of the applied procedure is possible with the reduction in functional groups in the structure when rGO is obtained from graphene oxide. Sharma et al. reported the observation of the presence of many functional groups in graphite and the intensities of these peaks decreased in rGO [68].

In Thermogravimetric Analysis (TGA), the functional groups' existence can be investigated using this technique. With the increase in temperature over time, the separation of oxygen groups and moisture from the structure allows the presence of these structures to be detected [24]. Li et al. investigated the formation of the catalytic active site and stated that high temperatures damage the structure and, therefore, contribute to the formation of the active site, taking into account the TGA results [69]. Consequently, the structural analysis of this characterization approach presents an opportunity to comment on the material's applicability in the field of application.

Transmission Electron Microscopy (TEM) enables a more sensitive morphological observation and analysis of structural features such as agglomerations and pore structures for graphene and its derivatives [62]. Various techniques, such as Atomic Force Microscopy (AFM) [43], Scanning Electron Microscope (SEM) [57], and High-Resolution Transmission Electron Microscopy (HR-TEM), can be utilized in addition to this microscopic observation characterization.

Many characterization processes can be applied except the method mentioned above such as Brunauer–Emmett–Teller (BET) to measure pore and surface areas, X-ray Photoelectron Spectroscopy (XPS) to determine the quantity of oxygen and the existence of chemical species [63], and Scanning Tunneling Microscopy (STM) [70] to examine the surface at the atomic level [71].

4 Graphene-Based Nanomaterials Application

Graphene and graphene-based nanomaterials can be used in many different application areas with their distinctive characteristics. Among these application areas, biomedical and energy storage applications, both of which are crucial presently, are mentioned in detail in this section (Fig. 3).

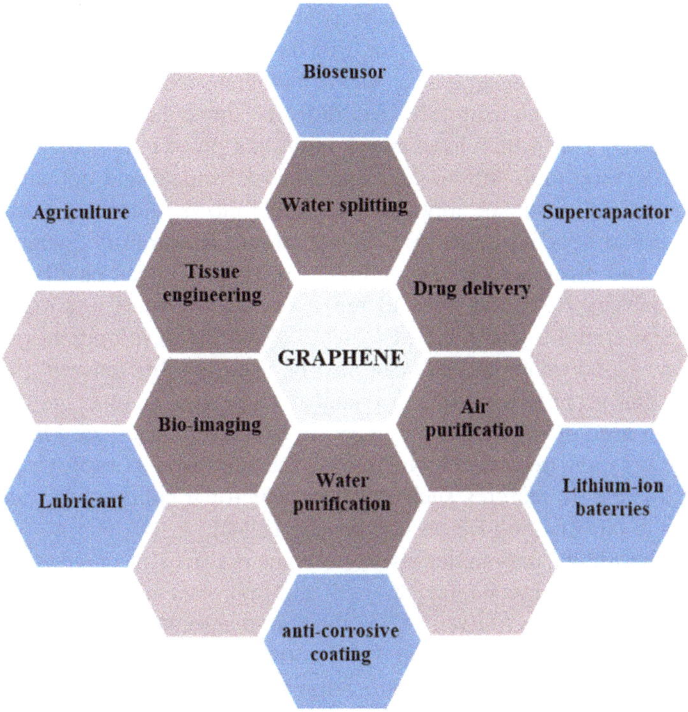

Fig. 3. Schematic illustration of graphene-based nanomaterials application area

In catalytic applications, Graphene-based nanomaterial is a promising material in many catalyst applications [71]. Some of these application areas include water splitting [72] to ensure H_2 production, oxygen reduction reaction (ORR) [73], water [74], and air [75] purification due to adsorption capacity. In the water-splitting process, Graphene-based nanomaterials serve as electrocatalysts owing to their electrical conductivity, mechanical strength, and high surface area properties. Water splitting offers a more environmentally friendly and sustainable energy source production [76]. For this purpose, graphene-based nanomaterial and rGO, which is especially active against ORR, can be utilized and also serve as a support material [77, 78].

Energy storage application: Due to depleting energy resources, interest in renewable energy and energy storage continues to increase day by day. Fuel cells [79], supercapacitors [80], lithium-ion batteries [81], and solar cells are among the application areas developed for this renewable energy [82]. The performance of these applications depends on the structural properties of the materials. The development of electrical devices can be made possible by the conductivity, high electron mobility, surface area, transparency, and especially flexibility and lightness of graphene-based nanomaterials [13].

In biomedical applications, graphene-based nanomaterials are biocompatible, offer high drug loading capacity with a high surface area, and have the required mechanical strength and chemical stability within the body. These characteristic properties make these nanomaterials for drug delivery applications [17]. Some drug delivery applications using graphene-based nanomaterials are DOX [83] hydroxyurea [84], and Ibuprofen drug [85]. Additionally, these nanomaterials are used in biosensor applications, which have a very important place in the early diagnosis of diseases due to their high surface area and electrochemical and optical properties. Some of the recent biosensor applications are as follows: detection of cancer cells [86], hemoglobin concentration [87], Severe Acute Respiratory Syndrome Coronavirus [88], and glucose [89]. Tissue engineering has fundamentally emerged as an interdisciplinary field that aims to repair damaged tissues and organs. Nanomaterials to be added to the scaffold material allow the development of this field, and graphene-based nanomaterials can also be used for this purpose [90]. Sharifi et al. stated that graphene oxide's mechanical properties increase its compressive strength when used in bone tissue engineering [91]. Graphene-based nanomaterials enable more sensitive detection by increasing the contrast and improving the image with its characteristic features such as optics, electrical conductivity, and thermal conductivity in bio-imaging for disease diagnosis [92].

In addition, adding nanomaterials to polymer matrices in *composite materials* increases the product's mechanical strength and improves many properties, such as electrical and heat conductivity. These composite materials can be used in different areas, such as sports footwear [93] and anti-corrosive coatings [94].

Along with the aforementioned application areas, graphene-based nanomaterials can be utilized in many different application areas such as transistors[95], lubricants [96], agriculture [97], and radio frequency absorbers [98].

5 Conclusions

In this chapter, beginning with the discovery of graphene, the brief history, superior properties, synthesis methods, characterization techniques, and graphene-based nanomaterials application areas are presented. Graphene-based nanomaterials have remained a fascinating topic in several application fields since their discovery due to their high performance, biocompatible structure, and optical, thermal, electrical, and mechanical characteristics. According to the application area, the synthesis process must be chosen by taking into account the required properties. Synthesis methods offer different dimensional, physical, chemical, and morphological properties. The bottom-up method offers a more uniform structure based on the principle of growing layers, is low-cost, and enables a controllable structure. These features make bottom-up methods preferable in more sensitive applications such as drug delivery systems and optoelectronics. However, these production methods also have disadvantages, such as complex production stages and low-scale production. Top-down includes methods based on the principle of exfoliation of layers and offers a more applicable process in high-scale production. These bottom-up and top-down methods are still being developed for graphene and graphene-based nanomaterials, which is a material that has attracted attention, especially in the last 10 years. Raman spectroscopy, AFM, TEM, XRD, FTIR, and TGA are important characterization techniques used to analyze and confirm the produced graphene and nanomaterials. Graphene-based nanomaterials provide an excellent study topic for energy storage systems, biological applications, and sub-branches of these application areas, which have lately become significant issues. As application areas need to be developed, interest in graphene-based nanomaterials will also increase.

References

1. Bachheti RK, Bachheti A, Husen A (eds) (2023) Nanomaterials for environmental and agricultural sectors. Springer, Singapore. https://doi.org/10.1007/978-981-99-2874-3
2. Bachheti RK, Bachheti A, Husen A (2023) Metal and metal-oxide based nanomaterials (Synthesis, Agricultural, Biomedical and Environmental Interventions). Springer Nature Singapore Pte Ltd., 152 Beach Road, #21-01/04 Gateway East, Singapore 189721, Singapore
3. Bahri M, Gebre SH, Elaguech MA, Dajan FT, Sendeku MG, Tlili C, Wang D (2023) Recent advances in chemical vapour deposition techniques for graphene-based nanoarchitectures: from synthesis to contemporary applications. Coord Chem Rev 475
4. Chawla S, Singh S, Husen A (2023) Smart nanomaterials targeting pathological hypoxia. Springer Nature Singapore Pte Ltd., 152 Beach Road, #21-01/04 Gateway East, Singapore 189721, Singapore. https://doi.org/10.1007/978-981-99-1718-1
5. Husen A (2023) Smart nanomaterials from agricultural and horticultural products. Springer Nature Singapore Pte Ltd., 152 Beach Road, #21-01/04 Gateway East, Singapore 189721, Singapore
6. Husen A, Siddiqi KS (2023) Advances in smart nanomaterials and their applications. Elsevier Inc., 50 Hampshire St., 5th Floor, Cambridge, MA 02139, USA
7. Husen A (2023) Secondary metabolites based green synthesis of nanomaterials and their applications. Springer Nature Singapore Pte Ltd., 152 Beach Road, #21-01/04 Gateway East, Singapore 189721, Singapore
8. Husen A (2023) Nanomaterials and nanocomposites exposures to plants (Response, interaction, phytotoxicity and defense mechanisms). Springer Nature Singapore Pte Ltd., 152 Beach Road, #21-01/04 Gateway East, Singapore 189721, Singapore. https://doi.org/10.1007/978-981-99-2419-6
9. Husen A (2023) Nanomaterials from agricultural and horticultural products. Springer Nature Singapore Pte Ltd., 152 Beach Road, #21-01/04 Gateway East, Singapore 189721, Singapore. https://doi.org/10.1007/978-981-99-3435-5
10. Husen A, Bachheti RK, Bachheti A (2023) Current trends in green nano-emulsions (Food, agriculture and biomedical sectors). Springer Nature Singapore Pte Ltd., 152 Beach Road, #21-01/04 Gateway East, Singapore 189721, Singapore
11. Jun LY, Khan FSA, Mubarak NM, Yon LS, Bing CH, Khalid M, Abdullah EC (2021) Importance of nanomaterials in engineering application. Eng Mater 1–20
12. Madhusudhan A, Purohit SD, Prasad R, Husen A (2023) Functional smart nanomaterials and their theranostics approaches. Springer Nature Singapore Pte Ltd., 152 Beach Road, #21-01/04 Gateway East, Singapore 189721, Singapore
13. Siqueira JR, Oliveira ON (2017) 9 - Carbon-based nanomaterials
14. Allen MJ, Tung VC, Kaner RB (2010) Honeycomb carbon: a review of graphene. Chem Rev 110:132–145

15. Dasari Shareena TP, McShan D, Dasmahapatra AK, Tchounwou PB (2018) A review on graphene-based nanomaterials in biomedical applications and risks in environment and health. Nano Micro Lett 10
16. Geim AK, Novoselov KS (2007) The rise of graphene. Nat Mater 6(3):183–191
17. Li J, Zeng H, Zeng Z, Zeng Y, Xie T (2021) Promising graphene-based nanomaterials and their biomedical applications and potential risks: a comprehensive review. ACS Biomater Sci Eng 7:5363–5396
18. Boehm H, Setton R, Stumpp E (1994) 93; Germany); Chatillon C (1989–93; France); Clark JB (1985–91; South Africa); Coutures J-P (1985–87; France); J Drowart (1985–87; Belgium). Pure & Appl Chem 66
19. Wallace PR, Ace PRW (1947) The band theory of graphite. 71
20. Geim AK (2009) Graphene: status and prospects. Science 324(5934):1530–1534
21. Gill FS, Uniyal D, Prasad B, Saluja S, Mishra A, Bachheti RK, Juyal S (2022) Investigation of increased electrical conductivity by rGO in rGO/PVDF/PMMA/PTFE nanocomposites. J Mol Struct 1267:133541
22. Singh S, Hasan MR, Sharma P, Narang J (2022) Graphene nanomaterials: The wondering material from synthesis to applications. Sens Int 3
23. Brodie BC (1859) XIII. On the atomic weight of graphite. Philos Trans R Soc 149:249–259
24. Spyrou K, Rudolf P (2014) An introduction to graphene. Functionalization of Graphene 9783527335510:1–20
25. Balandin AA, Ghosh S, Bao W, Calizo I, Teweldebrhan D, Miao F, Lau CN (2008) Superior thermal conductivity of single-layer graphene. Nano Lett 8:902–907
26. Lee C, Wei X, Kysar JW, Hone J (2008) Measurement of the elastic properties and intrinsic strength of monolayer graphene. Science (1979) 321:385–388
27. Nair RR, Blake P, Grigorenko AN, Novoselov KS, Booth TJ, Stauber T, Peres NMR, Geim AK (1979) Fine structure constant defines visual transparency of graphene. Science 320:1308
28. Stoller MD, Park S, Yanwu Z, An J, Ruoff RS (2008) Graphene-based ultracapacitors. Nano Lett 8:3498–3502
29. Morozov SV, Novoselov KS, Katsnelson MI, Schedin F, Elias DC, Jaszczak JA, Geim AK (2008) Giant intrinsic carrier mobilities in graphene and its bilayer. Phys Rev Lett 100
30. Obodo RM, Ahmad I, Ezema FI, Obodo RM, Ahmad I, Ezema FI (2019) Introductory chapter: graphene and its applications. Graphene and its derivatives—synthesis and applications
31. Ramírez C, Shamshirgar AS, Pérez-Coll D, Osendi MI, Miranzo P, Tewari GC, Karppinen M, Hussainova I, Belmonte M (2023) CVD nanocrystalline multilayer graphene coated 3D-printed alumina lattices. Carbon N Y 202:36–46
32. Zhou T, Hou J, Tai M, Shi J, Mi X, Hu B, Liu C, Yan L, Liu L (2023) Polyethyleneimine-induced in-situ chemical epitaxial growth ultrathin 2D/2D graphene carbon nitride intralayer heterojunction with elevating photocatalytic activity: Performances and mechanism insight. Int J Hydrogen Energy
33. Fan YJ, Wu SF (2016) A graphene-supported copper-based catalyst for the hydrogenation of carbon dioxide to form methanol. J CO_2 Utilization 16:150–156
34. Hu B, Xu C, Aslam MK, Cen Y, Hu J, Li Y, Liu Y, Guo C, Yu D, Chen C (2020) La-doped V2O5·nH2O@OAB and flexible Fe2O3@rGO as binder-free thin film electrodes for asymmetric supercapacitors. Chem Eng J 389
35. Tung TT, Yoo J, Alotaibi FK, Nine MJ, Karunagaran R, Krebsz M, Nguyen GT, Tran DNH, Feller JF, Losic D (2016) Graphene oxide-assisted liquid phase exfoliation of graphite into graphene for highly conductive film and electromechanical sensors. ACS Appl Mater Interfaces 8:16521–16532
36. Basak S, Packirisamy G (2021) Graphene-based nanomaterials for biomedical, catalytic, and energy applications. ChemistrySelect 6:9669–9683
37. Hummers WS, Offeman RE (1958) Preparation of graphitic oxide
38. Ikram R, Jan BM, Ahmad W (2020) An overview of industrial scalable production of graphene oxide and analytical approaches for synthesis and characterization. J Market Res 9:11587–11610

39. Kumar Singh P (2023) Impact of high-temperature thermal reduction on microstructure, crystallinity and surface topography of reduced graphene oxide. Mater Today Proc
40. Kauppila J, Kunnas P, Damlin P, Viinikanoja A, Kvarnström C (2013) Electrochemical reduction of graphene oxide films in aqueous and organic solutions. Electrochim Acta 89:84–89
41. Dreyer DR, Park S, Bielawski CW, Ruoff RS (2010) The chemistry of graphene oxide. Chem Soc Rev 39:228–240
42. Stankovich S, Dikin DA, Piner RD, Kohlhaas KA, Kleinhammes A, Jia Y, Wu Y, Nguyen SBT, Ruoff RS (2007) Synthesis of graphene-based nanosheets via chemical reduction of exfoliated graphite oxide. Carbon N Y 45:1558–1565
43. Karačić D, Gutić SJ, Vasić B, Mirsky VM, Skorodumova NV, Mentus SV, Pašti IA (2022) Electrochemical reduction of thin graphene-oxide films in aqueous solutions—restoration of conductivity. Electrochim Acta 410
44. Toh SY, Loh KS, Kamarudin SK, Daud WRW (2014) Graphene production via electrochemical reduction of graphene oxide: synthesis and characterisation. Chem Eng J 251:422–434
45. Zhou A, Bai J, Hong W, Bai H (2022) Electrochemically reduced graphene oxide: preparation, composites, and applications. Carbon N Y 191:301–332
46. De Silva KKH, Huang HH, Joshi R, Yoshimura M (2020) Restoration of the graphitic structure by defect repair during the thermal reduction of graphene oxide. Carbon N Y 166:74–90
47. Lee XJ, Hiew BYZ, Lai KC, Lee LY, Gan S, Thangalazhy-Gopakumar S, Rigby S (2019) Review on graphene and its derivatives: synthesis methods and potential industrial implementation. J Taiwan Inst Chem Eng 98:163–180
48. Zhou Q, Xia G, Du M, Lu Y, Xu H (2019) Scotch-tape-like exfoliation effect of graphene quantum dots for efficient preparation of graphene nanosheets in water. Appl Surf Sci 483:52–59
49. Zhao W, Fang M, Wu F, Wu H, Wang L, Chen G (2010) Preparation of graphene by exfoliation of graphite using wet ball milling. J Mater Chem 20:5817–5819
50. Liu J, Chen S, Liu Y, Zhao B (2022) Progress in preparation, characterization, surface functional modification of graphene oxide: a review. J Saudi Chem Soc 26:101560
51. Liu N, Tang Q, Huang B, Wang Y (2022) Graphene synthesis: method, exfoliation mechanism and large-scale production. Crystals (Basel) 12
52. Yi M, Shen Z (2015) A review on mechanical exfoliation for the scalable production of graphene. J Mater Chem A Mater 3:11700–11715
53. Arora N, Sharma NN (2014) Arc discharge synthesis of carbon nanotubes: comprehensive review. Diam Relat Mater 50:135–150
54. Coroş M, Pogăcean F, Măgeruşan L, Socaci C, Pruneanu S (2019) A brief overview on synthesis and applications of graphene and graphene-based nanomaterials. Front Mater Sci 13:23–32
55. Fauzi FB, Ismail E, Ani MH, Syed Abu Bakar SN, Mohamed MA, Majlis BY, Md Din MF, Azam Mohd Abid MA (2018) A critical review of the effects of fluid dynamics on graphene growth in atmospheric pressure chemical vapor deposition. J Mater Res 33:1088–1108
56. Banciu CA, Nastase F, Istrate AI, Veca LM (2022) 3D Graphene foam by chemical vapor deposition: synthesis, properties, and energy-related applications. Molecules 27
57. Li N, Li D, Zhen Z, Zhang R, Mu R, Xu Z, He L (2023) Nucleation and growth of graphene at different temperatures by plasma enhanced chemical vapor deposition. Mater Today Commun 36:106568
58. Al-Hilfi SH, Derby B, Martin PA, Whitehead JC (2020) Chemical vapour deposition of graphene on copper-nickel alloys: the simulation of a thermodynamic and kinetic approach. Nanoscale 12:15283–15294
59. Novoselov KS, Fal'Ko VI, Colombo L, Gellert PR, Schwab MG, Kim K (2012) A roadmap for graphene. Nature 490:192–200
60. Chang H, Wu H (2013) Graphene-based nanomaterials: synthesis, properties, and optical and optoelectronic applications. Adv Funct Mater 23:1984–1997

61. Yazdi GR, Iakimov T, Yakimova R (2016) Epitaxial graphene on SiC: a review of growth and characterization. Crystals (Basel) 6
62. Fraga TJM, Da Motta Sobrinho MA, Carvalho MN, Ghislandi MG (2020) State of the art: synthesis and characterization of functionalized graphene nanomaterials. Nano Express 1
63. Zhang Y, Lin H, Zhang L, Peng S, Weng Z, Wang J, Wu L, Zheng L (2023) Mechanical exfoliation assisted with carbon nanospheres to prepare a few-layer graphene for flexible strain sensor. Appl Surf Sci 611
64. Sayahi H, Mehrvar Z, Mohsenzadeh F, Darabi HR, Aghapoor K, Taherzadeh H (2023) Room-temperature defect-controlled fabrication of graphene via sustainable electrochemical exfoliation: an ultra-performance supercapacitor. J Energy Storage 68:107646
65. Khan H, Yerramilli AS, D'Oliveira A, Alford TL, Boffito DC, Patience GS (2020) Experimental methods in chemical engineering: x-ray diffraction spectroscopy—XRD. Can J Chem Eng 98:1255–1266
66. Sayfo P, Pirityi DZ, Pölöskei K (2023) Characterization of graphene-rubber nanocomposites: a review. Mater Today Chem 29:101397
67. Stobinski L, Lesiak B, Malolepszy A, Mazurkiewicz M, Mierzwa B, Zemek J, Jiricek P, Bieloshapka I (2014) Graphene oxide and reduced graphene oxide studied by the XRD, TEM and electron spectroscopy methods. J Electron Spectros Relat Phenomena 195:145–154
68. Sharma A, Jain P, Bajpai R, Patel P, Tiwari A (2023) Synthesis of ultra-thin graphene oxide/reduced graphene oxide nanocomposites and examine their properties. Mater Today Proc
69. Li B, Yuan Z, Xu Y, Liu J (2016) N-doped graphene as an efficient electrocatalyst for lithium-thionyl chloride batteries. Appl Catal A Gen 523:241–246
70. Deng D, Pan X, Yu L, Cui Y, Jiang Y, Qi J, Li WX, Fu Q, Ma X, Xue Q, Sun G, Bao X (2011) Toward N-doped graphene via solvothermal synthesis. Chem Mater 23:1188–1193
71. Hu M, Yao Z, Wang X (2017) Graphene-based nanomaterials for catalysis. Ind Eng Chem Res 56:3477–3502
72. Saha S (2023) 2D stacking of graphene and boron nitride for efficient metal-free overall water splitting. Int J Hydrogen Energy
73. Zhang R, Shehzad N, Zhang L, Ali A, Amin B, Shahid I (2023) Efficient electrocatalysts of single metal atom supported on defective graphene for oxygen reduction reaction (ORR): a first principles study. Chem Phys 570
74. AlAnazi M, Ghrib T, Ercan F, Kotb E, Bubshait R, Almilad F, Elibol E, Ercan İ (2023) Comparison between NiO and MnO_2 nanoparticles deposited on graphene-coated nickel foam for water purification applications. J Water Process Eng 55
75. Chen C, Li W, Liu X, Yu J, Xing S, Yang J, Han Q (2023) Silver nanoparticles/graphene oxide arranged on polytetrafluoroethylene substrate hydrophilic modified with TiO_2 to construct efficient air purification material. J Environ Chem Eng 110848
76. Jin H, Guo C, Liu X, Liu J, Vasileff A, Jiao Y, Zheng Y, Qiao SZ (2018) Emerging Two-dimensional nanomaterials for electrocatalysis. Chem Rev 118:6337–6408
77. Lee I, Joo JB, Shokouhimehr M (2015) Graphene derivatives supported nanocatalysts for oxygen reduction reaction. Cuihua Xuebao/Chinese J Catal 36:1799–1810
78. Rivera LM, García G, Pastor E (2018) Novel graphene materials for the oxygen reduction reaction. Curr Opin Electrochem 9:233–239
79. Lee FC, Ismail MS, Zhang K, Ingham DB, Aldakheel F, Hughes KJ, Ma L, El-Kharouf A, Pourkashanian M (2023) Optimisation and characterisation of graphene-based microporous layers for polymer electrolyte membrane fuel cells. Int J Hydrogen Energy
80. Dong X, Wang L, Wang D, Li C, Jin J (2012) Layer-by-layer engineered Co-Al hydroxide nanosheets/graphene multilayer films as flexible electrode for supercapacitor. Langmuir 28:293–298
81. Min W, Chen X, Huang S, Liao Y, Liang Z, Lei Y, Xu J (2023) High performance lithium ion battery cathode based reduced holey graphene oxides from spent lithium ion batteries. Carbon N Y 210
82. Boateng E, Thiruppathi AR, Hung CK, Chow D, Sridhar D, Chen A (2023) Functionalization of graphene-based nanomaterials for energy and hydrogen storage. Electrochim Acta 452

83. Le QH, Neila F, Smida K, Li Z, Abdelmalek Z, Tlili I (2023) PH-responsive anticancer drug delivery systems: insights into the enhanced adsorption and release of DOX drugs using graphene oxide as a nanocarrier. Eng Anal Bound Elem 157:157–165
84. Gber TE, Etinwa B, Benjamin I, Ekereke E, Offiong OE, Adeyinka AS, Louis H (2023) Functionalized boron doped graphene (BGP) as smart nanocarrier for delivery of hydroxyurea (HU) drug. Chem Phys Impact 7
85. Esfahani S, Akbari J, Soleimani-Amiri S, Mirzaei M, Ghasemi Gol A (2023) Assessing the drug delivery of ibuprofen by the assistance of metal-doped graphenes: insights from density functional theory. Diam Relat Mater 135
86. Ibrahimi KM, Kumar R, Pakhira W (2023) A graphene/Au/TiO_2 coated dual-core PCF SPR biosensor with improved design and performance for early cancer cell detection of with high sensitivity. Optik (Stuttg) 288
87. Ansari G, Pal A, Srivastava AK, Verma G (2023) Detection of hemoglobin concentration in human blood samples using a zinc oxide nanowire and graphene layer heterostructure based refractive index biosensor. Opt Laser Technol 164
88. Kadadou D, Tizani L, Wadi VS, Banat F, Alsafar H, Yousef AF, Hasan SW (2023) Detection of SARS-CoV-2 in clinical and environmental samples using highly sensitive reduced graphene oxide (rGO)-based biosensor. Chem Eng J 453
89. Zou L, Wang SS, Qiu J (2020) Preparation and properties of a glucose biosensor electrode based on an ionic liquid-functionalized graphene/carbon nanotube composite. Xinxing Tan Cailiao/New Carbon Materials 35:12–19
90. Ghosal K, Mondal P, Bera S, Ghosh S (2021) Graphene family nanomaterials-opportunities and challenges in tissue engineering applications. FlatChem 30
91. Sharifi S, Ebrahimian-Hosseinabadi M, Dini G, Toghyani S (2023) Magnesium-zinc-graphene oxide nanocomposite scaffolds for bone tissue engineering. Arab J Chem 16
92. Johnson AP, Gangadharappa HV, Pramod K (2020) Graphene nanoribbons: a promising nanomaterial for biomedical applications. J Control Release 325:141–162
93. Lunchev AV, Kashcheev A, Tok A ling Y, Lipik V (2022) Mechanical characteristics of poly (ethylene vinyl acetate) foams with graphene for the applications in sport footwear. Polym Test 113
94. Guan J, Du X (2024) Improving the anti-corrosive performance of epoxy/zinc composite coatings with N-doped 3D reduced graphene oxide. Prog Org Coat 186:107959
95. Abrantes M, Rodrigues D, Domingues T, Nemala SS, Monteiro P, Borme J, Alpuim P, Jacinto L (2022) Ultrasensitive dopamine detection with graphene aptasensor multitransistor arrays. J Nanobiotechnol 20
96. Sun J, Ge C, Wang C, Li S (2022) Tribological behavior of graphene oxide-Fe3O4 nanocomposites for additives in water-based lubricants. Fullerenes, Nanotubes, Carbon Nanostruct 30:863–872
97. Liang Y, Yang D, Cui J (2017) A graphene oxide/silver nanoparticle composite as a novel agricultural antibacterial agent against Xanthomonas oryzae pv. oryzae for crop disease management. New J Chem 41:13692–13699
98. Wu B, Tuncer HM, Naeem M, Yang B, Cole MT, Milne WI, Hao Y (2014) Experimental demonstration of a transparent graphene millimetre wave absorber with 28% fractional bandwidth at 140 GHz. Sci Rep 4
99. Husen A, Siddiqi KS (2014) Phytosynthesis of nanoparticles: concept, controversy and application. Nanoscale Res Lett 9(229):1–24
100. Husen A, Siddiqi KS (2014) Carbon and fullerene nanomaterials in plant system. J Nanobiotechnol 12(1):1–16
101. Husen A (2020) Interactions of metal and metal-oxide nanomaterials with agricultural crops: an overview. In: Husen A, Jawaid M (eds) Nanomaterials for agriculture and forestry applications. Elsevier Inc. 50 Hampshire St., 5th Floor, Cambridge, MA 02139, USA, pp 167–197. https://doi.org/10.1016/B978-0-12-817852-2.00007-X
102. Husen A (2020) Carbon-based nanomaterials and their interactions with agricultural crops. In: Husen A, Jawaid M (eds) Elsevier Inc. 50 Hampshire St., 5th Floor, Cambridge, MA 02139, USA, pp 199–218. https://doi.org/10.1016/B978-0-12-817852-2.00008-1

103. Husen A (2022) Engineered nanomaterials for sustainable agricultural production, soil improvement and stress management. Elsevier Inc., 50 Hampshire St., 5th Floor, Cambridge, MA 02139, USA
104. Li Z, Deng L, Kinloch IA, Young RJ (2023) Raman spectroscopy of carbon materials and their composites: graphene, nanotubes and fibres. Prog Mater Sci 135:101089
105. Mandal AK, Ghorai S, Husen A (2023) Functionalized smart nanomaterials for point-of-care testing. Springer Nature Singapore Pte Ltd., 152 Beach Road, #21–01/04 Gateway East, Singapore 189721, Singapore

Synthesis, Characterization, and Applications of Carbon Nanotube

Parul Singh, Shweta Sharma, Versha Dixit, Kundan Kumar Chaubey, Shalini Sharma, Kalash Aggrawal, Faizan ul Haque Nagrami, Priyambada Kumari, and Vijay Laxmi Tripathi

Abstract Carbon nanotubes, which resemble graphite and diamond in their interesting chemical and physical characteristics, have an unusual atomic structure. Recent discoveries of different sorts of these structures have sparked research into the use of carbon nanostructures in several domains. They could be useful in the domains of genetics, medicine, and drug delivery. Currently, a variety of methods, including functionalization, filling, doping, and chemical modification, can be used to create carbon nanotubes (CNTs). Individual CNTs may now be identified, distinguished, and controlled. The reactive capacity of carbon nanotubes is greatly influenced by factors including structure, dimension, charge on the surface, distribution of sizes, surface chemistry, aggregation state, and sample purity. In addition, the strength and flexibility of carbon nanotubes make them potentially helpful for modifying various minuscule structures, signifying that they will play a significant role in the advancement of nanotechnology. This chapter's major objective is to examine all of the potential uses of CNTs for improving human health, with an emphasis on nano carriers and biological applications in particular. The first focus of this chapter is on the fundamental characteristics of CNTs, and it is then broadened to detail their synthesis methods and the many difficulties associated with their functioning,

P. Singh · S. Sharma · F. H. Nagrami · P. Kumari · V. L. Tripathi
College of Biotechnology, DUVASU, Mathura, Uttar Pradesh, India

V. Dixit
Department of Life Sciences, Christ University, Bangalore, Karnataka 560029, India

K. K. Chaubey (✉)
Division of Research and Innovation, School of Applied and Life Sciences, Uttaranchal University, Dehradun, India
e-mail: kundan2006chaubey@gmail.com

S. Sharma
Department of Veterinary Physiology and Biochemistry, Lala Lajpat Rai University of Veterinary and Animal Sciences, Hisar, Haryana 125004, India

K. Aggrawal
School of Biotechnology, Gautam Buddha University, Greater Noida, Uttar Pradesh 201312, India

K. K. Chaubey
School of Basic and Applied Sciences, Sanskriti University, Mathura, Uttar Pradesh 281401, India

A. Bachheti et al. (eds.), *Carbon-Based Nanomaterials*, Smart Nanomaterials Technology, https://doi.org/10.1007/978-981-97-0240-4_3

dispersion, and toxicity. Our discussion also places a strong emphasis on the future possibilities for these young disciplines of study.

Keywords SWNTs · MWNTs · Carbon nanotubes · Synthesis · TEM · SEM · Drug delivery · Toxicity

1 Introduction

The element that is most prevalent on Earth is carbon, having an atomic number of 6, and is one of the six elements in the order of elements. Carbon can hybridize to generate the sp, sp2, and sp3 forms [5]. Small-sized, recently found sp2 carbon-associated materials including the substance graphene (Ouyang et al. 2015) and nanotubes made of carbon [23] have sparked interest in this topic. The two major allotropic types of the carbon atom are carbon nanotubes and diamond, both of which have a three-dimensional structure (Fig. 1). A large sheet of the material graphene is rolled to form hollow, hexagonal cylinders having an outside diameter that is approximately a nanometer (nm), a length of a micrometer, and linking carbon atoms together via sp2 bonds. These cylinders are then cut into desired shapes. CNTs have unique properties like high mechanical strength, they can be metallic and semiconducting having a wide range of electrical properties, good thermal stability, a lot of surface area, small weight, and chemical stability [80]. However, it's crucial to keep in mind that most of those outstanding characteristics were intended for a fundamentally perfect ideal CNT, which is extremely dissimilar from the actual nanotubes of carbon we are now generating. We have made great strides in nanotube research over the years, but we are still unable to create enough CNTs by using a technique that is both effective and economical [62]. CNTs presently exceed all carbon and metal-based nanomaterials due to their unique and remarkable qualities, such as their high durability and lightweight design [71]. CNTs were discovered by Sumio Iijima (1991) of Japan. Due to their hollow shape, which has numerous new techniques, these carbon carrier molecules are useful in a variety of uses which are pertaining to apparatus, electrical in nature, physiological in nature, pertaining to structure, thermal form and visual, and a variety for medical, engineering and science [80].

The nanotubes of carbon (CNTs) come in two varieties: single-walled nanotubes made of carbon (SWCNTs) and also multi-walled nanotubes made of carbon (MWCNTs). Because they are lighter and more elastic than steel, CNTs perform better with regard to specific strength. The various properties of carbon nanotubes make them intriguing prospects for a range of applications. For instance, their enormous surface area and high perspective ratio make them ideal for therapeutic uses including enzyme delivery and medication administration. The advantages of interconnect applications include stable temperature, high electrical conductivity, and big current capacity [81]. CNTs are thought to be the foundation of nanotechnology for instance. They are usually chosen as a result of their great cell penetration and high

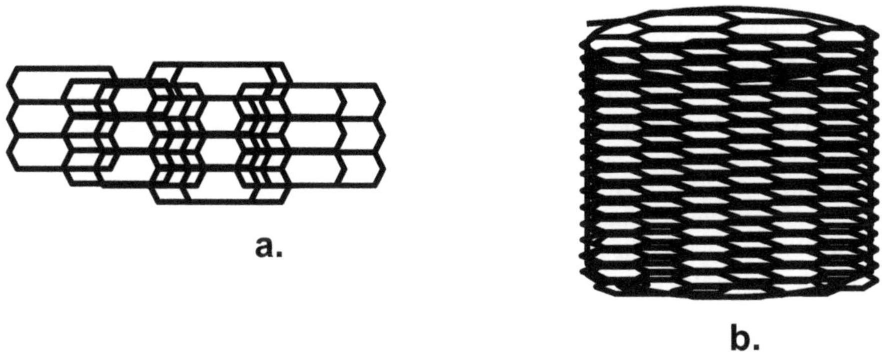

Fig. 1. Graphene's structure and the carbon nanotube's structure

cargo molecule loading, which enable them to carry drug substances into cells that are cancerous enabling target killing and minimizing the dispersion of medications to healthy cells nearby, helping to prevent cell toxicity, especially in cancerous cells [81] (Fig. 2).

Types of CNT

The form of the nanotube that is often referenced is the nanotube made of carbon (CNT), which has attracted a lot of interest from scientists in a range of fields in recent years. The three basic types of CNTs that are classed are single-walled CNTs (SWCNTs), double-walled CNTs (DWCNTs), and multi-walled CNTs (MWCNTs). SWCNTs are made from a single layer of graphite which gets bent into a tube-like shape, as opposed to MWCNTs, which comprise a pattern of SWCNTs that are primarily arranged one inside the other to replicate the ring-like structures that make up a tree trunk [80]. Thus, a sheet of rolled-up graphite can be used to represent

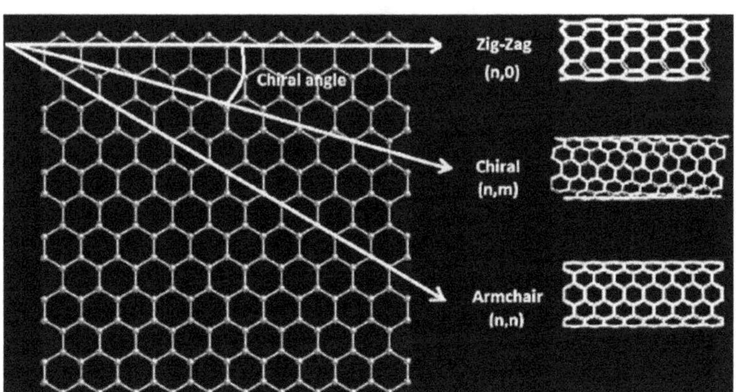

Fig. 2. Chirality and carbon nanotube types based on the roll-up vector

carbon nanotubes (Chau). Contrary to diamonds (sp2 hybridization), Graphite (sp3 hybridization) develops as a two-dimensional model of carbon molecules grouped in the shape of a hexagonal array, in contrast to diamond, which acquires its shape as a three-dimensional diamond cube-shaped crystalline framework with each of the four closest neighboring carbon atoms organized in a tetrahedron. In this case, each carbon atom has three close neighbors. Carbon nanotubes are made by "rolling" graphite sheets into cylindrical forms. The atomic arrangement (How the graphite sheet is "rolled"), the tube's dimensions, and the form of its tube or nanostructure all affect the properties of nanotubes made from carbon. By using different synthesis methods and precise process settings, it is possible to produce carbon nanotubes with a variety of morphologies and properties, opening the door to numerous scientific fields for their potential multifunctional uses (Kumar et al. 2017).

2 Synthesis of CNTs

The creation of the nanomaterial is the first step in the process of developing modern nanotechnology systems. Utilizing carbon nanotubes can be done in three ways (Fig. 3, Table 1):

Arc discharge, laser ablation method, chemical vapor deposition (CVD), and other techniques are used to make carbon nanotubes in a variety of ways. For previous ablation using lasers and discharge of arc processes, a high temperature (41,700 °C) is frequently required during synthesis. Nevertheless, CVD has replaced these methods

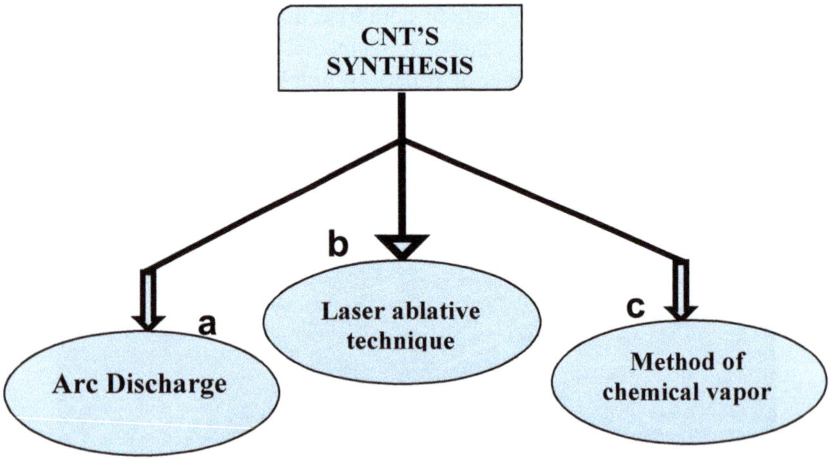

Fig. 3. **a** and **b** are physical production procedures for carbon nanotubes, whereas **c** is a chemical approach

Table 1. Ways of utilization for carbon nanotubes

Discharge of arc	Ablation by laser	Deposition of chemical vapor
1. The original and most fundamental approach	Extensive carbon nanotube synthesis	Provides the manufactured carbon nanotubes with an intrinsically high level of alignment
2. Offers a huge supply of carbon nanotubes	Compared to arc procedures, the method's elements may be controlled and repeated more precisely	It is only possible to make a very modest amount of carbon nanotubes
3. Requires precise environmental control using inert gases	Cost-cutting	Very high purity
4. The provision of an electrode cooling system is required	Requires complex facilities (lasers, ovens)	There could be a lot of crystallographic flaws in the carbon nanotubes that are created
5. It is very fast	It moves slowly	It moves slowly

because it can be done at 800 °C or lower temperatures. Along with these well-established processes, certain unconventional approaches have also been used, such as pyrolysis and hydrothermal treatment [48].

By making the physical and chemical processes that take place during the series of events more significant, the parameters are employed to maximize the synthesis capability (Kumar et al. 2017).

- Warmth condition,
- Low or high Pressure,
- Noblet or not-noblet gas,
- synthesis period,
- Level of Voltage,
- Catalysts (yttrium, cobalt, nickel, molybdenum, etc.), as well as basic materials like graphite are the main parameters used during the synthesis process.

It is helpful to list the properties of the synthesis methods before describing them in detail [48].

2.1 Arc Discharge

This technique has been widely applied for the direct current, also referred to as arc discharge from CNT production. Using graphite, CNTs are created when both the anode and cathode electrodes are together in a vacuum space while a neutral gas (which could be argon or helium) is present; this process also produces MWNTs

(Szabó et al. 2017). In most cases, the arc discharge method employs two high-purity electrodes made of graphite (Fig. 4). Either metals or pure graphite makes up the anode. The latter method involves drilling a hole by using the negative electrode center and adding the metals after mixing them with the graphite powder. An arc is struck when the electrodes briefly come into contact. Low pressure, reactive or noble gas is used during the synthesis, which is carefully supervised. The highly elevated temperatures in the inter-electrode area causes the anode, which is the positive electrode that is being utilized to release carbon [13]. By altering the anode's position, the space between the anode and cathode is kept constant. Between the electrodes, a plasma forms [10]. By altering the distance between the electrodes, it is possible to keep the plasma for a long reaction time. This is accomplished by using the voltage range (25–40 V). The range of the reaction time is between 30 and 60 s and 2 and 10 min. The unit's several components each produce a range of distinct products: (1) The electrolytic storage is encased in a porous collaret, (2) a firm, gray deposition at the end of the cathode electrode (3) a dense coating of leathery smoldering on the reactor walls, (4) the cathode and the chamber walls were connected by web-like structures [12]. The elements iron, nickel, cobalt, Molybdenum, and Yttrium were most frequently utilized, either individually or in combination [51, 53]. The chamber's temperature (Su and Zhang 2015), the concentration and specific make-up of the catalyst that is used, the proportion of carbon vapor that exists within the noble gas, and the amount of promoters [63], are several process variables that affect the amount and caliber of carbon nanotubes that are created during the arc-discharge process; including the presence of hydrogen [52]. Numerous recent research have used silicon substrates, the element nickel grid catalysts that act as stainless steel electrodes, and non-vacuum local electrical discharges to study the formation of CNTs utilizing ethanol, which is (as the carbon precursor). It should be emphasized that using stainless electrodes reduces the likelihood of electrode degradation, it raises the standard regarding the CNTS (Aslam et al. 2021) [105].

The following parameters were used to create the arc between the two electrodes [64, 68, 85]:

- 20–30 V.
- Current: 60–120A (AC and DC both refer to current flow are two different arc discharge topologies that may be used).
- Electrodes made of pure or doped graphite.
- 1–3 mm is separated by the electrode outer layer that is accessible to the arc.
- Noble gas, such as argon or helium (To prevent electrode oxidation, the manufacturing chamber's pressure is controlled, and after the method is complete, a vacuum is obtained.).
- 10–60 s for discharge.

This explanation is the only macroscopic one available for the arc discharge phenomenon, other hypotheses use atomic models to describe the formation of carbon nanotubes during arc discharge. It is important to distinguish between theoretical models, which are essential for learning about the physical and chemical processes underlying carbon nanotube production, and models (mostly semi-empiric) helpful

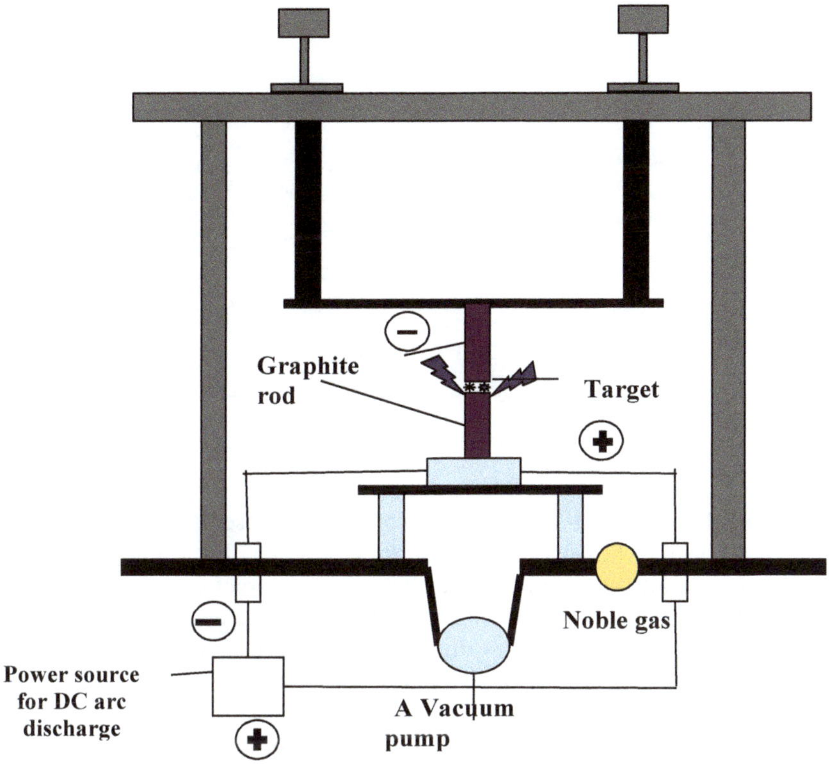

Fig. 4. Depicts an example of a typical experimental setup for arc method synthesis

for the creation of carbon nanotubes through engineering [68, 85]. From a technological perspective, the results of the synthesis strongly depend on a variety of criteria. To understand a carbon nanotube development strategy at the microscopic level, in-depth parametric studies and academic research are necessary [104].

2.2 *Laser Ablation*

Manufacturing of MWCNTs and SWCNTs using the ablation by laser process is an intriguing technique. Guo et al. (1995) are the authors of the first report on this technique for CNT synthesis. Figure 5 displays a schematic for laser ablation (Maser et al. 1998) and [104] had both employed this technique as a source for the ultrafine particles and clusters that In: Bernstein (1990) had found.

This technology improves the arc discharge approach. There are two common laser methods used:

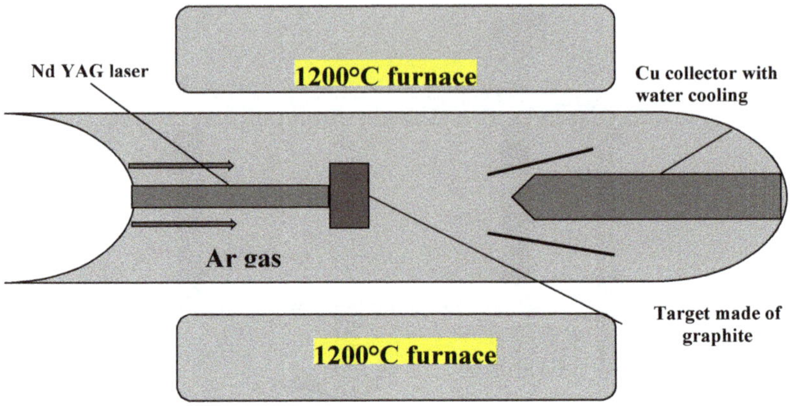

Fig. 5. Illustrative diagram of laser ablation

1. Nd Yag: Using the Nd-Yag process, a carbon rod is laser-ablated at a temperature of 1200 degrees Celsius in a passive flow of argon. The carbon rod can either be only graphite or modified. The above carbon target was exposed to a green Nd-Yag laser pulse at a wavelength of 532 nm with an energy of 50 mJ per pulse and a pulse duration of 6–7 ns, followed by an IR laser pulse at the wavelength of 1064 nm with a latency of 50 ns, this caused the growth of carbon nanotubes (Terrones 2003) [72].
2. CO2: A carbon-based rod that is either natural or modified, is laser-ablated at ambient warmth (no heater is needed) in a neutral flow of argon. The single continual CO2-based laser with a wavelength of 1064 nm and a power range of 400–900 Watts was applied to illuminate the carbon target mentioned above. The main goal initiated is to produce carbon nanotubes [8].

It is constructed from a tube made of quartz that lies within a heating chamber (CO_2 removal is not necessary). The tube contains a safe gas reservoir, which is completely sealed and is connected to a vacuum system. Utilizing a solitary orifice that has been vacuum clamped, the laser beam penetrates the quartz tube. The carbon-derived target is positioned in the middle of the quartz tube, with the target's axis perpendicular to the laser's beam. With the help of a short quartz tube which is conformal to the metal collector that is cooling with water and placed at the other end of the outer tube, it is possible to adjust the distance between the laser beam and the carbon-based target [18, 58].

Following is a succinct summary of the laser ablation procedure:

- The beam of laser light is possibly two times pulsed or continuous, and the target material is carbon.
- A water-cooled metallic collector (usually made of aluminum or copper) is the recipient of the smallest carbon-based surge generated by a harmless gas flow supplied to it.

The surface of the collector is coated with nanostructures that include fullerenes, carbon nanotubes, and other carbon-based components. Ablation with a laser has shown to be the most efficient way to create high-purity nanotubes of carbon. As a result of the catalyst dose, oven temperatures, gas supply, tension, and energy provided, among other factors, several publications that examine the effects of these parameters concentrate on improving the production process [102]). For instance, experiments demonstrate that utilizing a bimetallic catalytic combination (similar to Ni/Co) is superior to using one type of metal. The production rate from carbon nanostructures, and in particular, the arrangement of their diameters are directly influenced by the boiler's temperature as well as gaseous conditions (flow and pressure) [30]. The development of large nanotube diameters is encouraged by an increase in laser intensity. The temperature plays a significant effect in the synthesis as well. Particularly, temperatures below 800–900 °C result in a decrease in the yield of the generated nanostructures, which enhances the quality of the component of noncrystalline carbon that accumulated on the collector [4, 88]. It is crucial to keep in mind that the value of the temperature and the amount of energy transferred to reach the carbon target are related. It is frequently found that the propensity to prefer the production of exceptionally high nanotube carbon is a sign of both heat and a laser level of intensity. Every single one optimized their laser intensity, thus causing an "optimal warming" on a carbon target, enhancing the effectiveness of the syntheses. This is also compatible with the fact that the continuous laser spot's double pulse configuration maximizes the energy absorbed by the target surface, increasing the yield of carbon nanotubes [79, 87].

2.3 *Chemical Vapor Deposition (CVD)*

Deposition by chemical vapor is the most effective technique for producing carbon nanotubes. The breakdown of the precursor carbon in this process and the formation of the carbon nanotube structure are catalyzed by the surfaces of the catalyst particles. The CVD using heat and plasma enhancement (PECVD) are the two key CVD methods for the production of CNT. By changing the diameter of the catalytic particles, which is possible at comparatively low temperatures, the diameter of the CNTs that are used can be controlled. CNTs are produced in vast quantities via CVD. It is performed by employing a gaseous supply of carbon and a form of energy, such as a resistively hot string, an airborne carbon molecule with heat transmission [100]. It essentially takes two stages to create CNTs using CVD. The actual synthesis of nanotubes comes after a step of catalyst preparation. The most common methods used to create the catalyst are sputtering, dip coating, and physical vapor deposition (PVD). The substrate is then heated in a gaseous atmosphere that is rich in carbon. Temperatures between 500 and 1000 degrees Celsius are typical for the formation of nanostructures using this approach [25, 44, 56].

A simplified diagram of a fundamental CVD arrangement for the creation of CNTs is is shown in Fig. 6. A vaporized hydrocarbon is passed through a furnace constructed of tubes that contains an activating substance that has been warmed to and its temperature (commonly within 600 and 1200 °C) high enough to break down the hydrocarbon molecules (often ranging among 15 and 60 min). As soon as the system has reached room temperature, CNTs expand over the catalyst and are gathered. In order to convey a vapor form into the process of reaction heater when utilizing a fluid hydrocarbon (for example, benzene, alcohol, etc.), the fluid must first begin heating up in a vessel before an inert gas is driven through it. The liquid or gaseous state of the carbon intermediates depends on how the CVD reactor is built (Dillon et al. 2003) [78]. In liquid carbon precursors, the reactants are frequently vaporized in a bubbler before being transported into the CVD reactor by a carrier gas (usually a reactive gas like hydrogen or a noble gas like Helium or Argon) [39, 60, 100].

In CVD, the hydrocarbons, the chemical catalyst that triggers them, and the process temperature are the three key factors that influence CNT growth. According to the findings, reactions at elevated temperatures (900–1200 °C) encourage the creation of SWNTs, while lower temperatures of CVD (600–900 °C) often yield MWNTs [2]. This suggests that SWNTs have a greater energy of creation, likely because of their tiny diameters, which produce high curvature and high strain energy [39]. It has been found that the diameter of a nanotube depends on the size of a catalyst particle [6, 74, 76]. Compared to the first two techniques, at lower temperatures and air pressure, CVD is a speedy and inexpensive method for producing CNTs, but it sacrifices crystallinity. According to [18, 93, 100], it is a flexible process because it can grow CNTs in a variety of shapes, including dry chemicals, fine or firm films, organized or connected flat or coiled structures, or just the liked framework of tiny tubes at predetermined locations on a specific substrate. It additionally operates with a variety of hydrocarbons in any state (solid, fluid, as well as gas). Additionally, it provides improved control over factors that affect growth.

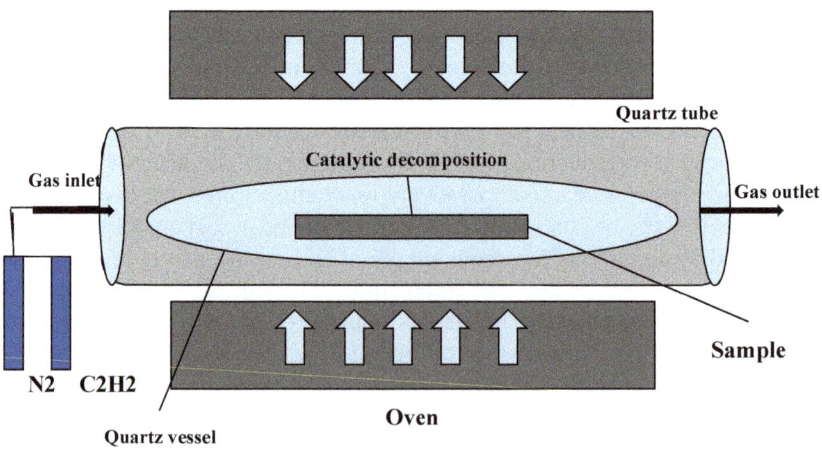

Fig. 6. Schematic representation of CVD

3 Characterization of CNTs

CNTs are intriguing and sought-after candidates for a variety of remedies because of their inherent features. Characterization is greatly influenced by the particular properties of CNTs. There are many methods as well as approaches used for analyzing CNTs have been the subject of numerous comments and discussions written throughout the years [11, 65]. Four categories of techniques are employed to characterize CNTs: thermal, spectroscopic, separation, microscopy, and diffraction techniques (Herrero-Latorre et al. 2015). The characterization of the nanomaterials and nanostructures used (which have been synthesized, purified, etc.) is an integral part of every stage in the development of a sophisticated nanotechnology mechanism or equipment. It is necessary to examine the "nanomaterials" that are part of the convoluted framework in question for their core characteristics and attributes. Although it is challenging to examine and analyze nanostructures with small-scale and nano-metric dimensions, it is essential to create a precise standard methodology for each stage, including preparing sample processes, analytical standards, assessment and analysis methods, and archived sample methods.

To increase the capacity to compare various samples used as raw materials, standard procedures are absolutely necessary for describing nanostructures. A standard approach with high levels of dependability that can be used to perform a quick "nano-elements" characterization must be used to describe the needed protocol [82]. Magnetic measurement techniques are occasionally used in research as an additional group. It should be noted that many techniques are commonly used to characterize CNTs, techniques used alone are neither fully characterized nor completely quantified. Although electron microscopy can be used to do a qualitative study of CNTs, examining the nanostructure of the tubes is done using a scanning electron microscope (SEM) and a more complete inspection is performed using transmission electron microscopy (TEM), which typically identifies CNT flaws [44, 91]. The available techniques are illustrated in this section, along with examples of how they are typically used.

3.1 Raman Spectroscopy

One of the key techniques for characterizing structural flaws in CNTs is Raman spectroscopy [3]. At 1350 cm21 (D band) and 1580 cm21 (G band), respectively, the Raman spectra of CNTs exhibit two significant peaks [34]. The G band is connected to planar charcoal hexagonal lattice structures with sp2 hybridization, The D band, on the other hand, is coupled to noncrystalline carbon (like sp3 diamond) [33]. There is occasionally a shoulder-like structure on the G band near 1620 cm21 (D 0 spectra), and it has something to do with sp3 hybridized carbon that isn't symmetric in the plane of graphite [3, 31]. Since the amount of faulty carbon is associated with the amplitude of the D band, the ratio of the total D to G band frequency ranges (ID/IG) is frequently used as a gauge of structural disorder. Increased ID/IG denotes faults brought on by the switch from CC sp2 to CC sp3 hybridization, which results in a loss of aromaticity in CNT rings [92]. For CNTs, many investigations have revealed a range of 0.27 to 0.90 for the ID/IG ratio [17]. To enhance the crystal quality of the nanotubes, the post-process of graphite conditioning is frequently used. By heating deficient the carbon (CC sp3) to arranged charcoal (CC sp2) at elevated temperatures (2000 °C), this technique lowers the ID/IG ratio. To create well-graphitized CNTs, MWCNTs have been heat-treated at approximately 2800 °C for half an hour at room air. The faulty sites have changed into organized graphite as evidenced by the ID and IG ratio dropping from 0.54 to 0.18 following annealing [97]. As a result of the nanostructures' enhanced purity and crystallinity after heat treatment, other researchers have noticed a decrease in their ID and IG ratio between 0.09 and 0.01 in this case [92]. In disordered graphite, the G band's location can also change (defect). These imperfections in graphitic formations give rise to two pentagonal rings, a polygonal ring, and a Stone Wales emptiness [61, 70]. An increased full width at the half-width highest, or FWHM, of the G band, is another sign of disordered graphite [30]. After 2.5 h of the graphite formation procedure at 2100 C, MWCNT defects were assessed using Raman spectroscopy by Boncel and Konziol [13]. The ID/IG ratio dropped from 0.44 to 0.32 following annealing, as seen by the researchers. Additionally, the FWHM of the G band was reduced by 28%, which could indicate the presence of larger graphitic structures [15]. Also, chirality and CNTwall tally can be determined by Raman spectroscopy. As previously mentioned [50], the numeral index (n, m) is used to identify SWCNT varieties. Electronic and vibrational characteristics, as well as the absolute concentration of SWCNTs, are revealed by the characterization of (n, m) species. Gontijo et al. (2017) used the Raman excitation spectra of the circular breathing patterns in addition to analyzing the spectra of optical absorption to evaluate the corresponding n, m proportions and SWCNT abundance. It is now possible to count SWCNTs that are present in a CNT layer using an innovative technique disclosed according to [3]. It combines the G frequency wavelength with accurate mass measurements and ocular absorbance assessments. The number of sp2 carbon molecules and the Raman spectra strength were identified to be correlated by the authors.

3.2 Fourier Transform IR and Infrared Spectroscopies

Utilizing infrared technology, impurities from synthesis have been evaluated as well as organic molecules that have been cap-coated on nanotube surfaces [11, 89]. It is common practice to functionalize CNTs by attaching various functional groups to their sidewalls via chemical or physical interactions in order to improve their solubility and selectivity [69], and numerous applications of IR and FTIR have been made to analyze these surface modifications [38]. utilizing substances like aminobenzothiazole to create MWCNTs and using FTIR to prove this [67], TREN stands for tris(2-aminoethyl) amine [26], ethylenediamine [103], and chitosan [84] are a few examples of this application. The required functionalization is demonstrated and the degree of functionalization is determined by its presence in the vibrational spectrums associated with the linked functional group. Because of their simplicity, velocity, cost-effectiveness, accessible availability, and relatively easy sample preparation, Infrared and FTIR are advantageous for characterizing functionalized nanotubes. Various methods for spectroscopy, such as in order to evaluate the success of the restructuring and functionality of carbon nanotubes and NMR (nuclear magnetic resonance) research, have also been used [73, 90].

3.3 Ultraviolet–visible Spectroscopy

As previously stated, most applications require dispersing CNTs in water or other solvents. To assess how CNTs are distributed in the end product, UV–vis spectroscopy was performed. Three techniques have been employed to accomplish this using data from optical spectroscopy. Herrero-Latorre et al. (2015) state that the value obtained at a specific wavelength was connected to the tiny tubes distribution in some cases, that absorbance/resonance proportions were used to evaluate a wavelength range in other cases, and that the full spectrum in order were taken into consideration in other cases. These are some instances of the first strategy. Here are some illustrations of the first strategy. The amount of scattered nanotubes in solution was directly associated with the absorbance at 253 nm by Meng et al. (2009). While using a different wavelength (273 nm), Attal et al. (2006) used a comparable approach to get the same result. Wenseleers et al. (2004) computed When evaluating a wide range of detergents, especially salts obtained from bile, for SWCNT solubilization, CNT dispersions utilize a spectrum of absorption of bands at 972 and 1710 nm. In a parallel inquiry, Haggenmueller et al. (2008) proposed $l = 891$ nm as the determination's wavelength. Tan and Resasco (2005) evaluated the CoMoCAT method's bundle splitting and dispersibility of SWCNTs, which results in bulk samples with limited length and chirality distributions indicated by their optical spectra [59]. Both parameters—"resonance proportion" and "normalized width"—were calculated using the wavelength region between 539 and (S11, a transition peak) at 628 nm have been used to quantify dispersions. The wavelength range below 800 nm is perfect for measuring the level of concentration of uniformly dispersed nanotubes in suspension, which is the best technique to do so. Grossiord et al. (2005) demonstrated how Ultraviolet–visible readings could be used to trace the removal of clusters and clumps of nanotubes composed of carbon in a water-based solution during ultrasonic treatment using the third and final technique, this utilizes the full spectrum or a variety of wavelengths. The greater percentage of disintegration is connected with the biggest increase in frequency within the range of wavelengths (which is expressed from 200 to 1100 nm), and the rise in abrasion is negatively correlated with an enhancement with the UV–vis wavelength of the SWCNT solution. To ascertain when the exfoliation is finished, UV–vis spectroscopy can also be used. For MWCNTs, similar outcomes were attained [101]. The immaculate state of SWCNTs can be ascertained via UV–vis spectroscopy, according to a recent method proposed by Goak et al. (2014). After p-plasmon removal, it entails fusing metallic carbon nanotubes with absorbance bands caused by interband shifts in the semiconducting nature's electrical structure. How helpful this technique is was demonstrated by looking at high-quality SWCNTs with different kinds and amounts of carbonaceous pollutants.

3.4 X-ray Photoelectron Spectroscopy

By using X-ray photoelectron spectroscopy, it is possible to see the functional groups linked to nanotube walls as small as 10 nm when examining components on CNT surfaces. Chong et al. [24] when exposed to X-rays, CNTs produce photoelectrons. By calculating the binding affinity of these electrons, we may determine the surface composition (Cano et al. 2017). This is comparable to the stiff photoelectron showing a straight route, which can be used to determine the fundamental composition of a surface possibly with the exclusion of hydrogen. The fact that neighboring atoms can easily change the binding energy makes X-ray photoelectron spectroscopy (XPS) a useful technique for learning about additional elements that are present on surfaces. The sample's required huge bulk presents the only challenge for XPS.

3.5 Fluorescence Spectroscopy

Semiconducting SWCNTs make up about 66% of all SWCNTs, and when they are semiconducting SWCNTs individually, they produce illumination that is unique to the diameter and handedness of the CNTs [75]. When semiconducting nanotubes are activated using visible wavelengths (between 500 and 900 nm), their photoluminescent emission occurs in the near-infrared range [45]. The band gaps of semiconducting nanotubes have directly been observed to undergo fluorescence. FS is a valuable method for describing bulk materials because of the connection across the band disparity and the degree of handedness (n, m) as well as the width of the SWCNTs. A sizeable p-system made up of p electrons is present in CNTs. The wide range of electronic states in this system has led to the observation of van Hove's singularities. Please refer to Belin and Epron's work [11], and also the fascinating analysis by [95] for a more in-depth description of the light-emitting method for SWCNTs. In the bands connected to valence and conduction, there are shifts among Van Hove anomalies that are connected to photoluminescence characteristics. It should be noted, nonetheless, that a sizable sample frequently contains both conductive and semi-conductive nanotubes and the presence of photo-luminescence is absent when both of these kinds of nanotubes become bundled in bundles (since Van der Waals effects are present) [21]. So, sonication and the usage of surfactants are essential for fluorescence characterization. The main benefit of FS is that it makes it possible to measure fluorimetric signals in order to obtain a quantitative (nm) abundance distribution [7, 86].

3.6 Scanning and Transmission Electron Microscopies

When it comes to CNTs, microscopy techniques are crucial for the quick identification of impurities that alter the structure and other characteristics. This is so that leftover metals can be incorporated between CNT layers and amorphous carbon can be adsorbed on the CNT wall side using the majority of production techniques. In addition to measuring the inner as well as employing low magnification, measuring the CNTs' dimensions and quantity, and determining the outermost diameter, TEM, or transmission electron microscopy, has a number of advantages. These advantages include being able to differentiate between these contaminants based on variations in the emission features (Hyung et al. 2017). The intershell positioning, which is more than the interplanar range of graphite, was found by [57] to vary between 0.34 and 0.39 nm in diameter based on the CNT diameter using high-magnification TEM (Belin and Epron 2004). This difference is due to the bending of the graphene layer, which increases the repulsive force as well as the size effect. It was suggested that the target area only be exposed briefly.

The well-known methods SEM and TEM are frequently used for studying the ultrastructure of all organic, inorganic, and biological materials. Using a laser-focused electron microscope that cooperates with CNT atoms to scan the sample, SEM creates pictures. This technique is frequently used to examine the shape and length of nanotubes [22, 83], and in some cases, to establish the presence of a process of functional chemical responses and to evaluate the quality produced CNTs in their natural state [51]. Transmission electron microscopy (TEM) is employed when the necessary measurements demand more resolution than SEM can provide (1–20 nm) [37, 91]. The material is imaged by leveraging the ability of CNTs to transfer electrons at high energies (in the range of 300 keV) under TEM. High-resolution images from TEM can be used to identify MWCNTs' minor characteristics, such as size, the number of layers, as well as the difference between layers [57]. Furthermore, TEM has been used to assess the functionalization process when it is obvious that the surface of nanotubes has changed as a result of CNTs' interaction with various inorganic as well as organic functional groups [90, 96]. SEM and TEM have both been used to analyze the fundamental structural integrity of CNTs and to examine the incomplete surface functionalization produced through acidic and basic oxidation reactions to form groups containing oxygen, such as hydroxyl molecules, which are carbonyl, and groups made of carboxylic acid (Datsyuk et al. 2008).

3.7 Scanning Probe Microscopies

Nanotube characterization has recently been carried out using microscopy techniques such as scanning probe microscopies (SPMs), that produce multifaceted surface images at an exceptionally high level of resolution (in a wavelength range of 0.01–0.1 nm, according to the technology). Due to their ability to provide details regarding the elasticity and electrical characteristics of small structures, the two types of SPMs that are most often used in this industry are AFM (atomic force microscopy) as well as scanning tunneling microscopy (STM). The morphology of the test sample (0.01 nm in-depth resolution), STM also covers in great detail the local electronic concentration of states (LDOS), which causes regional differences in DOS, and the electrical supply of values (DOS), which characterizes the overall amount of energy levels which are easily accessible for electron occupancy rates in a certain system. These concepts can frequently be caused by distortions in the initial setup [14, 49]. The outcome, STM makes it possible to separate conductive and semiconducting CNTs, as well as to look at the relationships between harmonic and electronic properties [66, 98]. Despite the fact that they are not commonly used, STM and AFM simultaneously involve complicated techniques. Additionally, creating samples is more challenging in SEM (Caplovicova et al. 2007).

4 Applications of CNTs

The qualities of nanotubes are undoubtedly remarkable; in recent years, numerous research have indicated possible CNT applications and have demonstrated a plethora of intriguing applications when these recently discovered materials are mixed with conventional goods [9, 47]. CNTs are used as reactive templates in the production of nanorods [35, 43]. Nanotubes are used in a wide range of sectors and industries, including technological devices, production, construction, manufacturing, and nanotechnology. High-strength composites [28, 46] actuators [1], energy storage and energy conversion devices [29], nano-probes and sensors [27], Examples of uses to be aware of include electrical components [99], catalyst [16], and hydrogen medium for storage [40]. However, the parts that follow exclusively cover current CNT uses in the biomedical sector.

4.1 CNT-Based Drug and Delivery of Genes

A method for constantly delivering an active medicinal ingredient to a specific location of the body is called drug delivery. The absence of efficiency, systemic adverse effects, poor cell distribution throughout the body, limited solubility, drugs' cellular barriers cannot be crossed, and there are no available clinical techniques for diagnosing resistant to multiple drugs (MDR) tumors are just a few of the difficulties related to standard administration of chemotherapy medications [41] (Jabr-Milane et al. 2008). Researchers have created a wide range of alternative ways to deliver drugs, such as polymer compounds, nanoparticles of silica, quantum-dot formulations, dendrimers, and liposomes as molecular conjugates and micelles in an attempt to circumvent these problems [36]. The delivery of medicines, peptides, as well as nucleic acids to the cells is a top priority for CNTs because of their extraordinarily large surface area. By attaching the appropriate drug or gene to the outside walls and ends of the CNTs and finding cancer-specific sites on the cell surface, CNTs can penetrate the outermost layer of an organelle in mammals through endocytosis or by other means [106] and more effectively deliver therapeutic drugs or genes to cells that weren't previously accessible [19]. The SWNT-based tumor-targeted medicine delivery system (DDS), which combines chemotherapy medication, engineered SWNTs, and tumor-targeting proteins, has recently been created by scientists to be more effective and safe. By locating cancerous-specific sites on the outermost covering of cancer cells, if this process interacts with cancer cells, it can trigger endocytosis through receptors for transporting and releasing chemotherapeutic drugs efficiently and selectively.

4.2 Artificial Implants

Due to their appealing chemical and physical features, nano-materials have potential and promise in regenerative medicine [66]. Implant rejection is typically accompanied by post-administration discomfort, therefore, attaching nanotubes to proteins and amino acids has shown promise in preventing this rejection. Without triggering host rejection, single as well as multilayer WNT carbon nanotube implants can be used to create prosthetic joints and other implants. CNTs can also act as implants and bone replacements if they have been loaded with calcium and organized in the framework of a bone due to their special properties, such as their high tensile strength. As a result of research into how SWCNT and MWCNT composites can improve cellular adhesion and proliferation, they have been added to both natural and manmade materials to create nano-composites.

4.3 Tissue Engineering

The goal of tissue engineering is to replace diseased or injured tissue with biological substitutes that can restore and retain its original structure and normal function. The fields of material studies and engineering have made significant strides that have helped the hoped-for advancements in tissue repair medicine and engineering. Four applications of carbon nanotubes in tissue engineering include improving tissue matrices, boosting cellular behavior, and monitoring and labeling of cells [94]. Tracking transplanted cells and the growth of newly formed tissues in vivo without the use of an invasive technique is possible using cell tracking and labeling. The labeling of implanted cells aids in understanding the distribution of transplanted cells throughout the body, their movements, relocation, and movement paths, along with assisting in the assessment of the survival of the targeted tissue. The usage of noninvasive procedures is becoming more and more common as people realize how time-consuming and difficult to utilize conventional techniques, like flow cytometry. It has been established that carbon nanotubes are effective contrast agents in imaging using magnetically, visually, and radiotracer modalities.

4.4 Field Emission Source in Electronic Devices

When a field of electricity is applied between the anode and the nanotube surface, electrons discharge from the CNT tips. Carbon nanotubes (CNTs) are therefore applicable to a variety of electronic devices, particularly X-ray systems (Rinzler et al. 1995). The benefits of using carbon nanotubes as electromagnetic transmitters include their long lifetime, stable operation of the generated field, which does not require a high vacuum to work, and their high current capacity (up to 4 A cm2).

5 Conclusion

The production, characterization, and research applications of CNTs are summarized in this chapter. We make the following deductions from the discussion above. Carbon nanotubes' electrical and physical properties have garnered a lot of interest. Several synthesis methods, such as electrode discharge and laser ablation, are used to create them, depending on the necessary physical parameters, which include diameter and length. They are effective drug delivery vehicles because of their enormous surface area, but to prevent toxicity, a purifying procedure is needed after synthesis. We cover some effective methods for characterizing nanotubes. It is possible to determine the purpose of the nanotubes utilizing X-ray photoelectron spectroscopy (XPS). Both the FWHM and binding energy undergo significant modifications. The additional substances and carbon atoms bind with one another to form new peaks. In order to gather three-dimensional images and information on the electrical properties of the nanotubes, scanning tunneling microscopy (STM) is a helpful technique. Transmission electron microscopy, also known as TEM, is a technique that can be used to measure the intershell gap, chiral indices, and helicity. Unwanted byproducts of synthesis or molecules stuck onto the nanotube surface are frequently found using infrared spectroscopy. However, Raman spectroscopy continues to be the most effective method for characterizing nanotubes since it allows for quick, non-destructive examination without the need for sample preparation. Additionally, simulations of

Raman spectra are carried out on several nanotube geometries. All of the techniques discussed here must be utilized in concert rather than independently in order to accurately characterize nanotubes. Carbon nanotubes that have been processed or refined are very likely to find new applications in a variety of medical disciplines. The incorporation of additional materials would further boost the likelihood of employing nanotubes made of carbon for biomedical applications. The physical structures and substances of the twenty-first century, according to some academics, will be "nano-elements." Though it will take significant economic, scientific, and technical endeavors, the application of Nano scale research to human existence will ultimately turn into a reality. Universities, businesses, academics, and scientists will work closely together to make this possible.

References

1. Ajayan PM, Zhou OZ (2001) Applications of carbon nanotubes. Carbon nanotubes: synthesis, structure, properties, and applications, 391–425
2. Allouche H, Monthioux M, Jacobsen RL (2003) Chemical vapor deposition of pyrolytic carbon on carbon nanotubes: Part 1. Synthesis and Morphology. Carbon 41(15):2897–2912
3. Anoshkin IV, Nefedova II, Lioubtchenko DV, Nefedov IS, Räisänen AV (2017) Single walled carbon nanotube quantification method employing the Raman signal intensity. Carbon 116:547–552
4. Arepalli S, Nikolaev P, Holmes W, Scott CD (2000) Diagnostics of laser-produced plume under carbon nanotube growth conditions. Appl Phys A 70:125–133
5. Armano A, Agnello S (2019) Two-dimensional carbon: a review of synthesis methods, and electronic, optical, and vibrational properties of single-layer graphene. C 5(4):67
6. Avouris P, Chen Z, Perebeinos V (2007) Carbon-based electronics. Nat nanotechnol 2(10):605–615
7. Bachilo SM, Strano MS, Kittrell C, Hauge RH, Smalley RE, Weisman RB (2002) Structure-assigned optical spectra of single-walled carbon nanotubes. Science 298(5602):2361–2366
8. Bandaru PR (2007) Electrical properties and applications of carbon nanotube structures. J Nanosci Nanotechnol 7(4–5):1239–1267
9. Banerjee S, Naha S, Puri IK (2008) Molecular simulation of the carbon nanotube growth mode during catalytic synthesis. Appl Phys Lett 92(23)
10. Belgacem AB, Hinkov I, Yahia SB, Brinza O, Farhat S (2016) Arc discharge boron nitrogen doping of carbon nanotubes. Mater Today Commun 8:183–195
11. Belin T, Epron F (2005) Characterization methods of carbon nanotubes: a review. Mater Sci Eng: B 119(2):105–118
12. Berkmans J, Jagannatham M, Reddy R, Haridoss P (2015) Synthesis of thin bundled single walled carbon nanotubes and nanohorn hybrids by arc discharge technique in open air atmosphere. Diam Relat Mater 55:12–15
13. Biró LP, Bernardo CA, Tibbetts GG, Lambin P (eds) (2012) Carbon filaments and nanotubes: common origins, differing applications
14. Biró LP, Gyulai J, Lambin P, Nagy JB, Lazarescu S, Márk GI, Lucas AA (1998) Scanning tunneling microscopy (STM) imaging of carbon nanotubes. Carbon 36(5–6):689–696
15. Boncel S, Koziol KK (2014) Enhanced graphitization of c-CVD grown multi-wall carbon nanotube arrays assisted by removal of encapsulated iron-based phases under thermal treatment in argon. Appl Surf Sci 301:488–491
16. Bower C, Rosen R, Jin L, Han J, Zhou O (1999) Deformation of carbon nanotubes in nanotubes–polymer composites. Appl Phys Lett 74(22):3317–3319

17. Bulmer JS, Gspann TS, Barnard JS, Elliott JA (2017) Chirality-independent characteristic crystal length in carbon nanotube textiles measured by Raman spectroscopy. Carbon 115:672–680
18. Burakov VS, Bokhonov AF, Nedelko MI, Savastenko NA, Tarasenko NV (2003) Optical emission characteristics of carbon plasma produced by single and double pulse laser ablation. In: Proceedings of international conference on phenom in ionized gases. Greifswald, vol 4, pp 207–208
19. Chen J, Chen S, Zhao X, Kuznetsova LV, Wong SS, Ojima I (2008) Functionalized single-walled carbon nanotubes as rationally designed vehicles for tumor-targeted drug delivery. J Am Chem Soc 130(49):16778–16785
20. Chen M, Chen CM, Chen CF (2002) Preparation of high yield multi-walled carbon nanotubes by microwave plasma chemical vapor deposition at low temperature. J Mater Sci 37:3561–3567
21. Chernov AI, Obraztsova ED (2010) Photoluminescence of single-wall carbon nanotube films pss (b) 247(11–12):2805–2809
22. Chiang IW, Brinson BE, Huang AY, Willis PA, Bronikowski MJ, Margrave JL, Hauge RH (2001) Purification and characterization of single-wall carbon nanotubes (SWNTs) obtained from the gas-phase decomposition of CO (HiPco process). J Phys Chem B 105(35):8297–8301
23. Chico L, Crespi VH, Benedict LX, Louie SG, Cohen ML (1996) Pure carbon nanoscale devices: nanotube heterojunctions. Phys Rev Lett 76(6):971
24. Chong CT, Tan WH, Lee SL, Chong WWF, Lam SS, Valera-Medina A (2017) Morphology and growth of carbon nanotubes catalytically synthesized by premixed hydrocarbon-rich flames. Mater Chem Phys 197:246–255
25. Collins PG, Arnold MS, Avouris P (2001) Engineering carbon nanotubes and nanotube circuits using electrical breakdown. J Sci 292(5517):706–709
26. Cui Y, Liu S, Hu ZJ, Liu XH, Gao HW (2011) Solid-phase extraction of lead (II) ions using multiwalled carbon nanotubes grafted with tris (2-aminoethyl) amine. Mikrochim Acta 174:107–113
27. Dai H, Hafner JH, Rinzler AG, Colbert DT, Smalley RE (1996) Nanotubes as nanoprobes in scanning probe microscopy. Nature 384(6605):147–150
28. Dai H, Wong E, Lu W, Yuan Z, Fan S, Lieber CM (1995) Synthesis and characterization of carbide nanorods. Nature 375(6534):769–772
29. De Heer WA (2004) Nanotubes and the pursuit of applications. MRS Bull 29(4):281–285
30. Dervishi E, Li Z, Xu Y, Saini V, Biris AR, Lupu D, Biris AS (2009) Carbon nanotubes: synthesis, properties, and applications. Part Sci Technol 27(2):107–125
31. Dresselhaus MS, Jorio A, Souza Filho AG, Saito R (2010) Defect characterization in graphene and carbon nanotubes using Raman spectroscopy. Philos Trans Royal Soc A PHILOS T R SOC A 368(1932):5355–5377
32. Faraji S, Yildiz O, Rost C, Stano K, Farahbakhsh N, Zhu Y, Bradford PD (2017) Radial growth of multi-walled carbon nanotubes in aligned sheets through cyclic carbon deposition and graphitization. Carbon 111:411–418
33. Ferreira FV, Brito FS, Franceschi W, Simonetti EAN, Cividanes LS, Chipara M, Lozano K (2018) Functionalized graphene oxide as reinforcement in epoxy based nanocomposites. Surf Interfaces 10:100–109
34. Ferreira FV, Cividanes LDS, Brito FS, Menezes BRCD, Franceschi W, Nunes Simonetti EA, Thim GP (2016) Functionalization of graphene and applications. Functionalizing graphene and carbon nanotubes: a review 1–29
35. Fonseca A, Hernadi K, Piedigrosso P, Colomer JF, Mukhopadhyay K, Doome R, Nagy JB (1998) Synthesis of single-and multi-wall carbon nanotubes over supported catalysts. Appl Phys A 67:11–22
36. Goldstein D, Nassar T, Lambert G, Kadouche J, Benita S (2005) The design and evaluation of a novel targeted drug delivery system using cationic emulsion–antibody conjugates. JCR 108(2–3):418–432

37. Gommes C, Blacher S, Masenelli Varlot K, Bossuot C, McRae E, Fonseca A, Pirard JP (2003) Image analysis characterization of multi-walled carbon nanotubes. Carbon 41(13):2561–2572
38. Grazhulene SS, Red'kin AN, Telegin GF, Bazhenov AV, Fursova TN (2010) Adsorption properties of carbon nanotubes depending on the temperature of their synthesis and subsequent treatment. J Anal Chem 65:682–689
39. Han JH, Choi SH, Lee TY, Yoo JB, Park CY, Jung T, Kim JM (2003) Growth characteristics of carbon nanotubes using platinum catalyst by plasma enhanced chemical vapor deposition. Diam Relat Mater 12(3–7):878–883
40. Han W, Fan S, Li Q, Hu Y (1997) Synthesis of gallium nitride nanorods through a carbon nanotube-confined reaction. J Sci 277(5330):1287–1289
41. Heister E, Neves V, Tîlmaciu C, Lipert K, Beltrán VS, Coley HM, McFadden J (2009) Triple functionalization of single-walled carbon nanotubes with doxorubicin, a monoclonal antibody, and a fluorescent marker for targeted cancer therapy. Carbon 47(9):2152–2160
42. Herrero Latorre C, Álvarez Méndez J, Barciela García J, García Martín S, Peña Crecente RM (2015) Characterization of carbon nanotubes and analytical methods for their determination in environmental and biological samples: a review. Anal Chim Acta 853:77–94
43. Hersam MC (2008) Progress towards monodisperse single-walled carbon nanotubes. Nat Nanotechnol 3(7):387–394
44. Hou PX, Liu C, Cheng HM (2008) Purification of carbon nanotubes. Carbon 46(15):2003–2025
45. Huang H, Zou M, Xu X, Liu F, Li N, Wang X (2011) Near-infrared fluorescence spectroscopy of single-walled carbon nanotubes and its applications. TrAC Trends Anal Chem 30(7):1109–1119
46. Huang X, McLean RS, Zheng M (2005) High-resolution length sorting and purification of DNA-wrapped carbon nanotubes by size-exclusion chromatography. Anal Chem 77(19):6225–6228
47. Husen A, Siddiqi KS (2014) Carbon and fullerene nanomaterials in plant system. J Nanobiotechnol 12(1):1–16
48. Hutchison JL, Kiselev NA, Krinichnaya EP, Krestinin AV, Loutfy RO, Morawsky AP, Zakharov DN (2001) Double-walled carbon nanotubes fabricated by a hydrogen arc discharge method. Carbon 39(5):761–770
49. Ichimura K, Osawa M, Nomura K, Kataura H, Maniwa Y, Suziki S, Achiba Y (2002) Tunneling spectroscopy on carbon nanotubes using STM. Phys B Condens Matter 323(1–4):230–232
50. Iijima S, Ichihashi T (1993) Single-shell carbon nanotubes of 1-nm diameter. Nature 363(6430):603–605
51. Imasaka K, Kanatake Y, Ohshiro Y, Suehiro J, Hara M (2006) Production of carbon nano onions and nanotubes using an intermittent arc discharge in water. Thin Solid Films 506:250–254
52. van Jabr Milane LS, Vlerken LE, Yadav S, Amiji MM (2008) Multi-functional nanocarriers to overcome tumor drug resistance. Cancer Treat Rev 34(7):592–602
53. Jimeno A, Goyanes S, Eceiza A, Kortaberria G, Mondragon I, Corcuera MA (2009) Effects of amine molecular structure on carbon nanotubes functionalization. J Nanosci Nanotechnol 9(10):6222–6227
54. Catherine J, Patrick B (1998) Production of carbon nanotubes. Appl Phys A 67:1–9. https://doi.org/10.1007/s003390050731
55. Journey C, Maser WK, Bernier P, Loiseau A, de La Chapelle ML, Lefrant DS, Fischer JE (1997) Large-scale production of single-walled carbon nanotubes by the electric-arc technique. Nature 388(6644):756–758
56. Jung KH, Boo JH, Hong B (2004) Synthesis of carbon nanotubes grown by hot filament plasma-enhanced chemical vapor deposition method. Diam Relat Mater 13(2):299–304
57. Kiang CH, Endo M, Ajayan PM, Dresselhaus G, Dresselhaus MS (1998) Size effects in carbon nanotubes. Phys Rev Lett 81(9):1869
58. Kingston CT, Simard B (2003) Fabrication of carbon nanotubes. Anal Lett 36(15):3119–3145

59. Kitiyanan B, Alvarez WE, Harwell JH, Resasco DE (2000) Controlled production of single-wall carbon nanotubes by catalytic decomposition of CO on bimetallic Co–Mo catalysts. Chem Phys Lett 317(3–5):497–503
60. Kong J, Cassell AM, Dai H (1998) Chemical vapor deposition of methane for single-walled carbon nanotubes. Chem Phys Lett 292(4–6):567–574
61. Kudin KN, Ozbas B, Schniepp HC, Prud'Homme RK, Aksay IA, Car R (2008) Raman spectra of graphite oxide and functionalized graphene sheets. Nano Lett 8(1):36–41
62. Kumar M (2011) Carbon nanotube synthesis and growth mechanism. INTECH. https://doi.org/10.5772/19331
63. Lange H, Sioda M, Huczko A, Zhu YQ, Kroto HW, Walton DRM (2003) Nanocarbon production by arc discharge in water'. Carbon 41:1617–1623
64. Lee SJ, Baik HK, Yoo JE, Han JH (2002) Large scale synthesis of carbon nanotubes by plasma rotating arc discharge technique. Diam Relat Mater 11(3–6):914–917
65. Lehman JH, Terrones M, Mansfield E, Hurst KE, Meunier V (2011) Evaluating the characteristics of multiwall carbon nanotubes. Carbon 49(8):2581–2602
66. Li J, He Y, Han Y, Liu K, Wang J, Li Q, Jiang K (2012) Direct identification of metallic and semiconducting single-walled carbon nanotubes in scanning electron microscopy. Nano Lett 12(8):4095–4101
67. Li R, Chang X, Li Z, Zang Z, Hu Z, Li D, Tu Z (2011) Multiwalled carbon nanotubes modified with 2-aminobenzothiazole modified for uniquely selective solid-phase extraction and determination of Pb (II) ion in water samples. MCA 172:269–276
68. Li Y, Xie S, Zhou W, Tang D, Liu Z, Zou X (2001) Small diameter carbon nanotubes synthesized in an arc-discharge. Carbon (New York NY) 39(9):1429–1431
69. Ma PC, Siddiqui NA, Marom G, Kim JK (2010) Dispersion and functionalization of carbon nanotubes for polymer-based nanocomposites: a review. Compos Part A Appl Sci Manuf 41(10):1345–1367
70. Maio A, Fucarino R, Khatibi R, Rosselli S, Bruno M, Scaffaro R (2015) A novel approach to prevent graphene oxide re-aggregation during the melt compounding with polymers. Compos Sci Technol 119:131–137
71. Mostafa AM, Mwafy EA, Awwad NS, Ibrahium HA (2021) Synthesis of multi-walled carbon nanotubes decorated with silver metallic nanoparticles as a catalytic degradable material via pulsed laser ablation in liquid media. Colloids Surf A Physicochem Eng Asp 626:126992
72. Nagy JB, Bister G, Fonseca A, Mehn D, Konya Z, Kiricsi I, Biro LP (2004) On the growth mechanism of single-walled carbon nanotubes by catalytic carbon vapor deposition on supported metal catalysts. J Nanosci Nanotechnol 4(4):326–345
73. Nelson DJ, Kumar R (2013) Characterizing covalently sidewall-functionalized single-walled carbon nanotubes by using 1H NMR spectroscopy. J Phys Chem C 117(28):14812–14823
74. Obraztsov AN, Zolotukhin AA, Ustinov AO, Volkov AP, Svirko YP (2003) Chemical vapor deposition of carbon films: in-situ plasma diagnostics. Carbon 4(41):836–839
75. O'connell MJ, Bachilo SM, Huffman CB, Moore VC, Strano MS, Haroz EH, Smalley RE (2002) Band gap fluorescence from individual single-walled carbon nanotubes. J Sci 297(5581):593–596
76. Okazaki T, Shinohara H (2003) Synthesis and characterization of single-wall carbon nanotubes by hot-filament assisted chemical vapor deposition. Chem Phys Lett 376(5–6):606–611
77. Ouyang M, Huang JL, Cheung CL, Lieber CM (2001) Atomically resolved single-walled carbon nanotube intramolecular junctions. J Sci 291(5501):97–100
78. Park D, Kim YH, Lee JK (2003) Synthesis of carbon nanotubes on metallic substrates by a sequential combination of PECVD and thermal CVD. Carbon 41(5):1025–1029
79. Purohit R, Purohit K, Rana S, Rana RS, Patel V (2014) Carbon nanotubes and their growth methods. Procedia Mater Sci 6:716–728
80. Rahamathulla M, Bhosale RR, Osmani RA, Mahima KC, Johnson AP, Hani U, Gangadharappa HV (2021) Carbon nanotubes: current perspectives on diverse applications in targeted drug delivery and therapies. J Mater 14(21):6707

81. Regi M (2007) Synthesis, characterization and application of carbon nanotubes: the case of aerospace engineering. In: Nanofib and nanotech in text. Woodhead Publishing, pp 113–193
82. Regi M, Mancia F (2004) Synthesis and characterization of carbon-nanotubes and their application in the aerospace engineering. In: Proceedings polymer fibers UMIST Conference Centre Manchester UK
83. Rinzler AG, Liu J, Dai H, Nikolaev P, Huffman CB, Rodriguez Macias FJ, Smalley RE (1998) Large-scale purification of single-wall carbon nanotubes: process, product, and characterization. Appl Phys A 67(1)
84. Salam MA, Makki MS, Abdelaal MY (2011) Preparation and characterization of multi-walled carbon nanotubes/chitosan nanocomposite and its application for the removal of heavy metals from aqueous solution. J Alloys Compd 509(5):2582–2587
85. Sano N, Wang H, Chhowalla M, Alexandrou I, Amaratunga GA (2001) Synthesis of carbon'onions' in water. Nature 414(6863):506–507
86. Schierz A, Parks AN, Washburn KM, Chandler GT, Ferguson PL (2012) Characterization and quantitative analysis of single-walled carbon nanotubes in the aquatic environment using near-infrared fluorescence spectroscopy. J Environ Sci Technol 46(22):12262–12271
87. Scott CD, Arepalli S, Nikolaev P, Smalley RE (2001) Growth mechanisms for single-wall carbon nanotubes in a laser-ablation process. Appl Phys A 72:573–580
88. Scott CD, Arepalli S, Nikolaev P, Smalley RE (2002) Growth mechanisms for single-wall carbon nanotubes in a laser ablation process. Appl Phys A 74(1):11–12
89. Stobinski L, Lesiak B, Kövér L, Tóth J, Biniak S, Trykowski G, Judek JJJOA (2010) Multiwall carbon nanotubes purification and oxidation by nitric acid studied by the FTIR and electron spectroscopy methods. J Alloys Compd 501(1):77–84
90. Sun YP, Fu K, Lin Y, Huang W (2002) Functionalized carbon nanotubes: properties and applications. Acc Chem Res 35(12):1096–1104
91. Täschner C, Pacal F, Leonhardt A, Spatenka P, Bartsch K, Graff A, Kaltofen R (2003) Synthesis of aligned carbon nanotubes by DC plasma-enhanced hot filament CVD. Surf Coat Technol 174:81–87
92. Varga M, Izak T, Vretenar V, Kozak H, Holovsky J, Artemenko A, Kromka A (2017) Diamond/carbon nanotube composites: Raman, FTIR and XPS spectroscopic studies. Carbon 111:54–61
93. Wang X, Hu Z, Chen X, Chen Y (2001) Preparation of carbon nanotubes and nano-particles by microwave plasma-enhanced chemical vapor deposition. Scr Mater 44(8–9):1567–1570
94. Wang Y, Da S, Kim MJ, Kelly KF, Guo W, Kittrell C, Smalley RE (2004) Ultrathin "bed-of-nails" membranes of single-wall carbon nanotubes. J Am Chem Soc 126(31):9502–9503
95. Weisman RB (2010) Fluorimetric characterization of single-walled carbon nanotubes. Anal Bioanal Chem 396:1015–1023
96. Wepasnick KA, Smith BA, Bitter JL, Howard Fairbrother D (2010) Chemical and structural characterization of carbon nanotube surfaces. Anal Bioanal Chem 396:1003–1014
97. Xue Y, Zheng S, Sun Z, Zhang Y, Jin W (2017) Alkaline electrochemical advanced oxidation process for chromium oxidation at graphitized multi-walled carbon nanotubes. Chemosphere 183:156–163
98. Yasuda S, Ikeda K, Yu L, Murakoshi K (2012) Characterization of isolated individual single-walled carbon nanotube by electrochemical scanning tunneling microscopy. Jpn J Appl Phys 51(8S3):08KB06
99. Ye XR, Lin Y, Wang C, Wai CM (2003) Supercritical fluid fabrication of metal nanowires and nanorods templated by multiwalled carbon nanotubes. J Adv Mater 15(4):316–319
100. Yokomichi H, Sakai F, Ichihara M, Kishimoto N (2001) Attempt to synthesize carbon nanotubes by hot-wire chemical vapor deposition. Thin Solid Films 395(1–2):253–256
101. Yu J, Grossiord N, Koning CE, Loos J (2007) Controlling the dispersion of multi-wall carbon nanotubes in aqueous surfactant solution. Carbon 45(3):618–623
102. Yudasaka M, Komatsu T, Ichihashi T, Iijima S (1997) Single-wall carbon nanotube formation by laser ablation using double-targets of carbon and metal. Chem Phys Lett 278(1–3):102–106
103. Zang Z, Hu Z, Li Z, He Q, Chang X (2009) Synthesis, characterization and application of ethylenediamine-modified multiwalled carbon nanotubes for selective solid-phase extraction and preconcentration of metal ions. J Hazard Mater 172(2–3):958–963

104. Zhang H, Ding Y, Wu C, Chen Y, Zhu Y, He Y, Zhong S (2003) The effect of laser power on the formation of carbon nanotubes prepared in CO_2 continuous wave laser ablation at room temperature. Phys B Condens Matter 325:224–229
105. Zhang H, Xue X, Wang D, He Y, Peng S (1999) The effect of different kinds of inert gases and their pressures on the preparation of carbon nanotubes by carbon arc method. Mater Chem Phys 58(1):1–5
106. Zhang X, Meng L, Lu Q, Fei Z, Dyson PJ (2009) Targeted delivery and controlled release of doxorubicin to cancer cells using modified single wall carbon nanotubes. Biomater Res 30(30):6041–6047

Functionalization of Carbon-Based Nanoparticles for Various Applications

Shweta Sharma, Vijay Laxmi Tripathi, Versha Dixit, Faizan ul Haque Nagrami, Priyambada Kumari, Parul Singh, Kundan Kumar Chaubey, Anis Kumar Pal, and N. C. Joshi

Abstract Functionalized carbon nanomaterials are emerging as novel and innovative biomaterials with several applications in different research and biological sciences. The utilization of carbon-based nanoparticles is due to their intensive strength, stiffness, and property to get functionalized via different forms of reaction such as amidation and esterification of the nano-based particles. Furthermore, based on their size and structure, these carbon-based nanomaterials are characterized in different forms like one-dimensional, two-dimensional (Carbon nanotubes, Graphenes), and three-dimensional (Dendrimers). CNTs are nano-carriers in various biological, cancer diagnosis, and therapy applications. The potential efficacy and enhanced functional properties are due to their capability to move across membranes of biological systems and cause relatively low cytotoxicity. Recent advancement in nanotechnology has brought about the formulation of more effective nanocomposites, which have proven activity in bioremediation biomedical applications such as bio-imaging, biosensors, disease diagnostics, and pesticide management. This chapter focuses on functionalizing carbon nanoparticles and their role in different fields.

Keywords Carbon nanotubes · Graphenes · Fullerenes · Dendrimers · Nanocomposites

---

S. Sharma · V. L. Tripathi · F. H. Nagrami · P. Kumari · P. Singh
College of Biotechnology, DUVASU, Mathura, Uttar Pradesh, India

V. Dixit
Department of Life Sciences, Christ University, Bangalore, Karnataka 560029, India

K. K. Chaubey (✉) · A. K. Pal · N. C. Joshi
Division of Research and Innovation, School of Applied and Life Sciences, Uttaranchal University, Dehradun, Uttarakhand 248007, India
e-mail: kundan2006chaubey@gmail.com

K. K. Chaubey
School of Basic and Applied Sciences, Sanskriti University, Mathura, Uttar Pradesh 281401, India

A. Bachheti et al. (eds.), *Carbon-Based Nanomaterials*, Smart Nanomaterials Technology, https://doi.org/10.1007/978-981-97-0240-4_4

1 Introduction

A new and precise technology, nanotechnology operates at the nanoscale. The Latin word "Nano" (meaning "dwarf") is the source of the word "Nano." Since the term "nano" refers to sizes down to 103 millionths of a unit, a nanometer is equal to 103 millionths of a meter (1n = 109 m) [1]. There is a significant amount of potential in research and development that includes modeling, imaging, and measuring. It pertains to extremely minute details and can be applied in various scientific and research fields or considered in cross-disciplinary fields like biotechnology, engineering, chemistry, electronics, physics, materials science, medicine, informatics and energy. In the last ten years, many studies have focused on using nano-based particles as carriers for both small and large molecules, leading to major advancements in nanotechnology. The current nanotechnology revolution will create nanomaterials with traits and abilities to improve people's lives and advance the medical field [2, 3]. Based on their dimensions nanomaterials can be classified as nanoparticles, nanotubes/wires, nanoplates/ribbons, and inorganic or organic nanomaterials, depending on their chemical makeup [4, 5]. Interest in nanoparticles produced using materials science-based methods is rising due to the successful development of nanotechnology and associated synthetic techniques [6, 7]. Nanomaterials provide a vast range of potent commercial uses based on this adjustable functionality. The composition of molecules, structure, size, and chemical properties of the surface of nanomaterials may be precisely regulated to customize their capabilities for utilization in target-specified devices. For instance, nanomaterials have become more potent to be used as building blocks for the development of cross-functional textiles that display higher range of hydrophobicity, thermochromicity, photochromicity, and a greater activity against microbial agents and still offering users a level of comfort that is acceptable [8, 9]. Multicomponent nanomaterials with hierarchical nanostructure have also made significant advancements in creating wearable electrochemical energy storage devices and high-performance electrodes.

In the last few years, severe health issues have been caused by the growing resistance of microbes to antibiotics. At least one antibiotic often used to treat infections is resistant to the majority of infection-causing microorganisms [10]. This issue motivates researchers to look into novel compounds that can successfully stop microbial development. In superconductors, optical devices, fuel cells, drug, biosensors, catalysts and gene delivery, and other technologies, nanomaterials have been studied for usage [11–14]. Nanomaterials have been used as innovative vehicles for drug delivery to enhance pharmaceuticals' physicochemical and therapeutic potency. Similarly, nanotechnology in microbiology and pharmaceuticals demonstrated interesting applications to address the issue of antibiotic resistance [15]. Researchers have recently investigated a variety of nano-sized antibacterial agents, including metal and metal oxide nanoparticles. Various types of metal NPs and metal oxide-based nanoparticles exhibit antimicrobial action, including silver (Ag), dioxide of titanium (TiO_2), oxide of silver (Ag_2O), silica (Si), oxide of zinc (ZnO), and oxides of copper, magnesium [16–27]. Carbon has distinct solid-state allotropes with diverse structures and properties, like sp^3 bonded diamond and sp^2 bonded graphite. New carbon nanomaterials have been created in recent years that are entirely made of sp^2 hybridized carbon atoms and have distinguished dimensions, varying from fullerenes of 0-dimension to carbon nanotubes (CNTs) of 1-dimension to carbon nanotubes of 2-dimensional (2D), i.e. graphene (Fig. 3) [28, 29]. In general, the atomic structures of carbon nanomaterials and interfacial interactions with materials in other phases significantly affect the characteristics of carbon nanomaterials. Therefore, designing and functionalizing nanomaterials of carbon based on nanoscale has become a preeminent approach to obtaining desired properties in specific applications. Nanoscale functionalized carbon nanomaterials have various chemical and physical properties, such as chemical stability, good mechanical properties, and thermal conductivity, superelectric conductivity, and enhanced optical properties. Due to these characteristics, it has gained much attention and shown exceptional potential in environmental science, energy storage, biology, and medicine [29]. Early reports showed that graphene oxide nanoparticles (GO), fullerenes, and SWCNTs can kill microbial agents.

2 Types of Nanomaterials

According to Hett (2004), the nanomaterials were classified into three dimensions, which are discussed below:

2.1 1-D Nanoparticles

Thin films or artificial surfaces, utilized in electronics, chemistry, and technology for decades, are examples of one-dimensional systems. These monolayers or thin films (sizes 1–100 nm) are widely used in information storage systems, biological and chemical sensors, solar cells, and other technical applications. They are also frequently used in the field of catalysis.

2.2 2-D Nanoparticles-Carbon Nanotubes

This is a layer of graphite composed of hexagonal networks of C atoms that have a length of 100 nm and 1 nm diameter. CNTs come in two varieties: SWCNTs, or single-walled carbon nanotubes, and MWCNTs, or multi-walled carbon nanotubes (Figs 1 and 2). Carbon nanotubes are a special class of materials due to their short size and exceptional mechanical, physical, and electrical characteristics (Kohler et al. 2004). Based on wrapping of carbon sheet around itself, they have metallic or semiconductive qualities. Carbon nanotubes are sixty times more mechanically robust

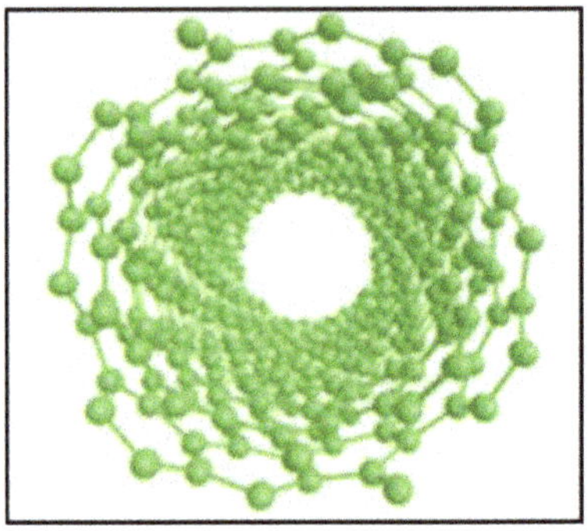

Fig. 1 Typical structure of multi-walled carbon nanotubes

when compared with the best steels. Carbon nanotubes also have a three-dimensional form and excellent molecule absorption capacity.

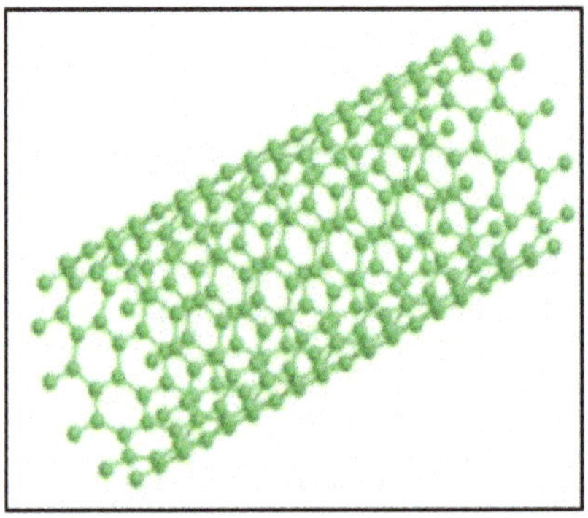

Fig. 2 Typical structure single-walled carbon nanotubes

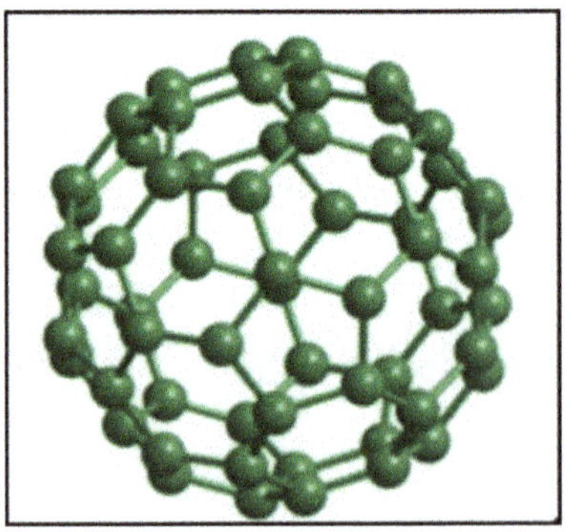

Fig. 3 Typical structure fullerenes

2.3 0-D Nanoparticles-Fullerenes

These are spherical cages made up of 28 to 100 or more C60 carbon atoms. This is a hollow, soccer-ball-like spherical made of pentagons and hexagons joined by carbon. Nanomaterials called fullerenes have peculiar physical characteristics. Extreme pressure can be applied to them; they can revert to their original shape upon releasing the pressure. They can be used to fill a variety of biological substances due to their hollowness and resemblance in size to several biologically active compounds, which makes them useful in medical applications [30].

2.4 Dendrimers

Dendrimers are a novel family of nanometric-sized polymers with controllable structures. Dendrimers are good carriers for targeted delivery of drugs [31] because of their numerous functional groups on their surface and average diameter range of 10–100 nm. Dendrimers are employed in drug delivery and imaging. Dendrimer structure and function have received extensive research. Today's dendrimers can be extremely specialized and contain functional molecules in their core, such as medicinal or compounds used for diagnostic purposes [32]. Large-scale production of inorganic and organic nanostructures having dimensions ranging from 1 to 100 nm are considered as the fundamental building blocks [30]. They can also be made into metallic nanotubes and nanostructures and are highly suitable for organic structures like DNA. They also have the ability to encapsulate substances (Table 1).

3 Functionalization of Carbon-Based Nanoparticles-

The carbon-based nanoparticles are functionalized through several methods, in which incorporation of such nano-based components is done, and it has a role in the different aspects of biological sciences. The Functionalization is done via exohedral and endohedral strategies [40].

3.1 Exohedral Functionalization

This method retains the core structure of CNTs. Furthermore, carbon-based nanomaterials formed through this method exhibit both the properties of functional groups and CNTs. The exohedral approach is classified into two major methods, namely covalent and non-covalent methods of Functionalization [40].

Table 1 Applications of various carbon-based nanomaterials

Types of carbon-based nanomaterial	Applications in respective fields	References
Carbon quantum dots (CQDs) and graphene quantum dots (GrQDs)	As photosensitizers in photodynamic therapy due to their excellent biocompatibility and bioluminescence property	[33]
Graphene nanosheets	Cardiac tissue regeneration	[34]
Carbon nanotubes, graphene	Bioremediation of soil polluted with heavy metals	[35]
SWCNTs, MWCNTs, graphene, carbon nanoparticles	Enzyme-based biofuel cells (EFC) for generation of high energy in implantable biomedical devices	[36]
Carbon-based quantum dots	Photocatalytic technology in which these materials are used to convert CO_2 and thus reducing global warming	[37]
Carbon-based nanozymes made up of graphene, fullerene, and other carbon nanomaterials	Carbon nanomaterial-based enzymes have excellent stability, abundant active sites, and good biological safety. These properties of carbon-based nanomaterials show great potential in biomedical sciences	[38]
Carbon-based nanoparticles	They show enhanced capability of seed germination, pest control, and drug delivery in target specific manner, leading to higher yields of plant based food material	[39]

3.1.1 Covalent Functionalization

Functional groups are directly bound to the graphitic surface through reactions like adding fluorine, hydrogen, and 1,3-dipolar, or by using COOH or OH functional groups to nanotubes for later derivation (produced by amide or ester group addition).

Direct interaction of functional groups results in covalent Functionalization.

(A) Fluorination

Margrave and colleagues invented fluorination of CNTs [41]. Molecular fluorine is typically used for fluorination when temperatures are between 150 and 600 °C. Different levels of fluorination are possible based on the reaction temperature. With enhanced solubility in organic solvents like alcohols, this approach results in CNTs that have been highly functional [42]. The Carbon-Fluorine valence vibrations at 1220–1250 cm^{-1} have been used by IR spectroscopy to demonstrate the covalent bond synthesis with fluorine. When fluorination is done at temperatures above 250 °C, CNTs undergo a significant shift in their electrical characteristics, becoming insulators. However, hydrazine processing of such processed nanotubes leads to defunctionalization, allowing for the recovery of intact CNTs.

(B) Hydrogenation

Pekker and colleagues showed that CNTs were hydrogenated in ammonia through birch reduction. Khare et al. [43] later employed a gas phase method to functionalize CNTs using atomic hydrogen. Compared to wet chemical procedures, which modify the chemical and physical characteristics of CNTs, this method is clean and low temperature, which makes it helpful. This method can attach a broad range of entities (that are chemically active in nature) on CNTs and be created using a light discharge to produce the necessary precursors. Atomic hydrogen was used to carry out the Functionalization.

(C) Dipolar 1,3 cycloadditions

Prato and colleagues described 1,3-dipolar cycloadditions of azomethine ylides for functionalizing single-walled (SW) and MWCNTs [44]. The end products of this method of Functionalization are highly soluble in both water and organic solvents like acetone, chloroform, and methanol. The fundamental advantage of this type of reaction is the simple attachment of the pyrrolidine rings replaced with chemical characteristics to the outer surfaces of CNTs, creating new substances with different applications. Thus, changed amino acids could also result in various functionalized materials. In contrast, substitution in 2 pyrrolidine moieties that are present on the side walls of CNTs can be achieved by functionalized aldehyde groups.

(D) Aryl diazonium salts are functionalized via the reduction.

It is a known fact that aryl diazonium salts can cause the arylation of aromatic substances when they interact with olefins. Thus, a number of modified aryl diazonium salts have been reduced via electrochemistry or at room temperature to functionalize SWCNTs with aryl moieties. It was discovered that 4-tert-butylbenzene-modified CNTs have significantly better solubility in organic liquids. But heating in an argon environment can eliminate the functionalizing moieties [45].

(E) Functionalization of reactive species

According to Holzinger et al., this sort of Functionalization is accomplished by direct reaction between two different entities, one that is highly reactive groups, namely carbenes, nitrenes, and radicals, and another entity that are CNTs [46]. For instance, the heat reaction of alkyl azidoformate produces derivatized CNTs from alkoxycarbonyl aziridino that are soluble in organic solvents like DMSO and N2 extrusion.

(F) Functionalization through electrophilic group addition

A report on an electrophilic addition to SWCNTs has shown the addition of electrophiles by utilizing a gas phase catalytic technique. It was found that a chloroform molecule was attached to the CNTs side walls following a mechanochemical process that involves the utilization of a Lewis acid, like aluminum trichloride (AlC13). It was also conceivable to replace the Cl with OH groups that were then esterified in a subsequent step to produce the appropriate esters [47].

The advancements in this area of research have led to the development of several nanomaterials that are mainly based on carbon and serve various attributes of different fields, such as chemistry biological sciences. Further processes are still in series to be explored and are continuously progressing.

3.1.2 Non-covalent Functionalization

Non-covalent method for the Functionalization of CNTs is an effective strategy because the adsorption of different groups to the surface of CNT can be done without creating any hindrance in the π system of the sheets made up of graphene [48]. These methods include the interaction of adsorbed molecules with the sidewalls of nanotubes via their hydrophobic entities via weaker bonds like van der Waal forces, CH–π, π–π, and other bonds. The non-covalent Functionalization can be done by surfactant functionalization, polymer wrapping, polymer absorption, and encapsulation [40].

3.2 Endohedral Functionalization

The interior of carbon nanotubes is thought of as a nanometer-sized test tube that offers a remarkably uncharted scope for discovery of innovative reactions that do not occur outside the tubes, making endohedral Functionalization of special interest [49] Similar to what would be expected for a molecule near the concave internal surface of the tube, the interior surface of CNTs likewise demonstrates a higher energy pattern for molecules that binds at this side in comparison to the planar carbon surface [50]. Additionally, the inside chamber of CNTs provides room for storing and defending visitor molecules.

3.2.1 Functionalization via Encapsulation of Metal

The protective carbon coatings on CNT-encapsulated metallic materials currently have everyone's attention because they stop naked metal nanoparticles from oxidizing quickly. Encapsulation can often be accomplished in two ways: either during the in situ CNT manufacturing process or by the chemically utilized way to fill and open the nanotubes. Both procedures have received much research. Several methods have been devised for the in situ synthesis of metal-filled CNTs, including modified arc deposition, catalytic pyrolysis, and the common carbon arc method [40].

3.2.2 Functionalization via Encapsulation of Fullerene

A particularly spectacular illustration of the endohedral chemistry of SWCNTs by inserting fullerenes like C60 and C70 or metal-based fullerenes. This integration is carried out at specific, end- or sidewall-located fault spots. Non-covalent forces such as Van der Waals often cause the enclosed fullerenes to form chains. These arrays are occasionally referred to as ™bucky peapods, which are thought to be one of the most intriguing and unusual materials in the case of novel kinds of composites formed of nanotubes [40].

4 Applications of Carbon-Based Nanoparticles in Different Areas

The research community actively seeks numerous nanotechnology applications, including those in the well-known fields of bio-nanotechnology, nanostructured catalysts, and carbon-based nanomaterials. It is one of the most eco-friendly and adaptable elements, and carbon is abundant in our planet's many geological phases. Carbon-based materials, including carbon elements in various allotropic forms and carbon composites, greatly aid the development of nanotechnology. Surface functionalization, distinctive shape, and chemical adaptability are the main benefits of materials made of carbon [51]. Due to these properties, C-based nanomaterials exhibit a preeminent role in different areas of science, which are discussed below in detail.

4.1 Applications of Functionalized Carbon Nanotubes

As a result of the extensive and improper use of antibiotics, microorganisms have evolved resistivity to the common antimicrobial medications. It is thus imperative to develop effective antibacterial medications. Recent studies have discovered that nanomaterials based on carbon, like SWCNTs, GO nanocomposites, fullerenes, and carbon nanotubes (CNTs), all possess potent antibacterial properties [52]. Lijima (1991) is credited with creating CNTs, which are nanoscale hollow cylinders of carbon origin [53]. After that, CNTs are being used in a broader manner in research and technologies. The first study to demonstrate SWCNTs' potent antibacterial action against Escherichia coli was Kang et al.'s from 2007. SWCNTs have been shown to degrade membranes and cause cell death significantly. In a distinct investigation, they presented the first evidence that the size of carbon nanotubes significantly affected their antibacterial activity. The effectiveness of the SWCNTs and MWCNTs in preventing the growth of *E. coli* was examined [52]. They discovered that SWCNTs significantly outperformed MWCNTs in case of bacteria toxicity. Furthermore, the authors noted that *E. coli*'s morphological, metabolical, and integrity of membrane were all affected by direct cell interaction with CNTs. Since SWCNTs have smaller nanotube diameters than MWCNTs, the researchers claim they can invade the cell wall more successfully. Furthermore, the increase in the surface area of SWCNTs led to better cell surface interaction.

In recent studies, it has been reported that the solution casting made with three components poly(butylene adipate-co-terephthalate (PBAT) with tannic acid (TA) and CNPs, shows greater antimicrobial efficacy against food-borne pathogens like *S. aureus* and *E. coli* [54]. Recent research in this field has given evidence that nanoparticles of titanium dioxide show antimicrobial activity properties. These nanoparticles are highly effective in the treatment of infections caused by bacteria. The cell count was done after testing the minimum inhibitory concentration, and it showed a huge difference in the cell number [55]. In order to attach to SWNTs while retaining their functional characteristics, a variety of molecules (biological in nature) are designed both in a covalent and non-covalent manner. 1-pyrenebutanoic acid succinimidyl ester is one such example of a biological molecule and is non-covalently and irreversibly adsorbed on the hydrophobic surface of SWNTs [56]. Then, using a nucleophilic substitution process, the protein can be joined to this moiety by replacing the succinimidyl group with an amine group from the protein.

Additionally, this increases the specificity of SWNT biological identification. Covalent CNT attachment to proteins and DNA is achieved through the use of carbodiimide chemistry [57–59]. In order to covalently bind enzymes, proteins, and antibodies, carbon nanotubes are functionalized with carboxylic acid, and the functionalized moiety subsequently undergoes amidation, which is triggered by diimide [60]. Functionalized SWNT fiber with protein bovine serum albumin (BSA) can illustrate this context. To verify this functionalization, indirect immunochemistry is employed. Firstly, a primary antibody is applied against BSA, and later, a secondary antibody, conjugated with a fluorescent molecule, is applied against the primary antibody. Fluorescent microscopy is used to see the complex. Similar to this, amidation processes can be utilized to link CNTs' carboxylic acid functional groups with amine-terminated DNA. Furthermore, developing such modified CNTs uses the great specificity of the biotin-avidin/streptavidin interaction. There are numerous examples of how biotin and streptavidin can immobilize SWNTs [61].

The interaction of nanoparticles with live organisms needs to be studied because nanotechnology is developing quickly. Numerous research in this area has demonstrated that nanoparticles have an impact on physiological processes, including DNA damage (Ghosh et al.2011), [62], changes in the expression of genes [63, 64], and an increase in ROS generation [65, 66]. These reactions fluctuate depending on the plant species [65, 67], variety [68, 69], and stage of growth [70]. Hydroponics or SWCNT-supplied culture media were mostly used to study the interactions of SWCNTs with immature seedlings. According to reports, applying SWCNTs improves seedling growth in a variety of plant species, including maize [64], fig plants [71], and tomato seedlings [72] Tall fescue (*F. arundinacea*), salvia (*S. macrosiphon*), and pepper *(C. annuum*) have all shown a dose-dependent response to SWCNTs [73] By immediately germinating the seeds on media containing SWNTs, SWNTs were applied [64]. In other instances, seedlings were moved to hydroponic media containing MWNTs following germination [65]. The exposure to SWNTs or MWNTs was typically administered for 15 days at various doses up to 2000 mg/L. According to these investigations, seedling biomass increased from 10 to 30 mg L1 of SWCNTs, however, seedling development suffered at 40 mg L1 of SWCNTs. MWCNTs have a favorable impact on shoot and root elongation during seedling development and early growth. Numerous plant species have reported such effects, including maize, soybean, wheat, mustard, tomato, and black lentil. Beyond germination and early growth, MWCNTs also impact plant development as they affect the reproductive phase. Applying MWCNTs to the soil has been shown to increase tomato flower setting and yield twofold [74], these effects were not observed when control soils were treated with charcoal.

Carbon nanotubes are an ideal candidate for biosensing due to their distinctive structural characteristics and small size. Because of their enormous surface-to-volume ratio, SWNTs can interact with various biological substances. Due to SWNTs' distinctive electronic structure and high conductivity, we can also choose electrical detection. When different tiny biomolecules are present, SWNT conductance significantly changes [75]. In order to identify adsorbed cytochrome c that is present in mitochondria, SWNT is employed as a redox catalyst [76]. Utilizing biotin functionalized SWNTs, the binding of biotin and streptavidin has also been electronically monitored [75]. In SWNTs, this binding lowers conductance. When SWNTs interact with analytes, charge transfer processes likely play a role in the changes that occur in SWNTs that are detected electronically. Alkaline phosphatase (ALP) enzyme-functionalized CNTs have been used to detect DNA at extremely low concentrations (1 pg/L) electronically [77]. Using optically transparent conductive SWNT sheets, it is also feasible to detect molecules using light [78]. Utilizing CNT nanoelectrode assemblies, glucose may also be selectively and sensitively detected [79]. Hydrogen peroxide [80], dopamine [81], and reduced nicotinamide adenine dinucleotide (NADH) are only a few examples of the electrochemical properties of several compounds that have been studied in relation to SWNTs in biology. Because of their structural characteristics, CNTs [82, 83] are the best vectors for delivering molecular probes like DNA and proteins into mammalian cells. Binding CNTs to various biological molecules, including proteins and DNA, is beneficial. A number of proteins, including streptavidin, interact with MWNTs [84], and DNA molecules can also bind to their surface [84–86]. The cavity of CNTs can also be used to introduce small protein molecules like b-Lactase I, cytochrome c, and others [87]. Nano-pills are created by introducing a tiny molecule into the interior cavity of CNTs with one end shut and the other left open, then sealing the two ends together. The potential of SWNTs for intracellular drug delivery has been successfully determined [82, 83]. In order to enable its tracking, FITC was tagged to SWNT [82]. Specialized cell lines of fibroblast cells (Human and murine origin) were exposed to such SWNTs, and SWNT-FITC began to build up inside the cells. The potential for employing CNTs for intracellular drug delivery was demonstrated in a way that was based on the work tested on SWNT [82, 83].

Fe nanoparticles with a carbon coating were used to treat in vitro-grown plants [88]. Precision farming is another benefit of nanotechnology that helps to improve agricultural practices [88]. Numerous studies have found that carbon nanomaterials benefit plant growth. Such fertilizers are more likely to decrease nutrient losses, increase the availability of nutrients to plants, and promote plant development. The effectiveness of carbon nanomaterials in enhancing plant health and grain quality was evident [89]. A faster industrial boom and agricultural expansion have caused numerous chemical discharges of hazardous compounds and components into the environment. The biggest contributors to environmental pollution include the careless usage of pesticides and medicines, the release of industrial waste into waterways, and the unrestricted use of hazardous materials in various consumer items. When these unidentified systemic pollutants or chemicals penetrate the food chain and accumulate in more developed species at higher levels in the food chain through a process known as biomagnification, they pose a major threat to all living things. These substances are naturally non-biodegradable [90]. Petroleum hydrocarbon pollution has become a global issue and a topic of scientific interest due to its hazardous nature, high persistence in environmental matrices, and limited degradability [91]. One option to cope with this is to combine remediation techniques that work around the constraints of traditional physio-chemical and bionano-remediation strategies. This direction's improved idea of bioremediation, called nano-bioremediation, provides a productive, affordable, and environmentally beneficial way to reduce petroleum toxins. Using NPs, a number of environmental contaminants in water may be quickly eliminated [92]. One of the most effective ways to get rid of pesticides and other toxic substances is through the development of functional nanomaterials. Hazardous pollutants have been removed from environmental samples using various metal–organic framework-based nanocomposites [93]. Covalent organic frameworks with NPs have been employed to detect and adsorb a variety of pollutants from water bodies. TiO2 and SiO2, two nanostructured carbon compounds with photocatalytic characteristics, are employed to eliminate the contaminant from the environment [94]. Biologically produced NPs provide a variety of useful characteristics for the detection and destruction of hazardous contaminants. Because of the discharge of contaminants from textile or other industries, the water bodies have become poisonous, carcinogenic, allergenic, and resistant to degradation. Thus, textile dyes pose a severe threat to the environment. Textile dyes can be categorized as acidic, basic, dispersive, and direct [95]. Therefore, wastewater resources containing dye should be properly treated using eco-friendly technology to avoid harmful human and environmental health effects. Using nanocarbon to eliminate these toxins or dyes from water streams has resulted in successful elimination [96]. To increase the adsorptive properties of the persistent reactive yellow azo dye, calcium carbonate NPs were doped with EDTA, which increased the decolorization effectiveness [97]. Methylene blue, heavy metals, and auramine are removed from wastewater using zinc ferrite NPs as adsorbents [98]. Up to 96% of chlorinated pesticides were removed from wastewater by titanium oxide NPs [99]. Acetamiprid was photocatalytically destroyed by graphene oxide NPs in 60 min at pH-3 [100]. Chlorpyrifos and imidacloprid were broken down by phytocatalysis in titanium and chitosan NPs composites, with up to 50% efficacy in eliminating them from wastewater [94]. The effectiveness and potential of the decontamination process are considerably enhanced by nano-bioremediation. Researchers have recently become interested in this relatively new approach, which has produced encouraging results in using NPs to regulate harmful contaminants from soil and water. Microbes and various nanomaterials can work together to remove pollutants more quickly. According to recent studies, immobilization of microbial enzymes employing various nano-supports has drawn significant attention in treating waste biomass [101]. In the bioconversion and remediation procedure, microbial enzymes are also used in conjunction with NPs. Using various nanoparticles in combination with bacteria, nano-bioremediation offers environmentally benign methods for removing dangerous environmental pollutants [94].

The adsorption of proteins through non-covalent interactions is one of the most popular methods for immobilization. This immobilization technique depends on surface residues that bind with SWCNTs via van der Waals, hydrophobic, and stacking interactions. Protein immobilization is essential for numerous biological and medicinal applications, including medication administration, electrical signaling, and optical detection [102]. Despite being very simple, this non-covalent method produces heterogeneous conjugates with different structures and protein loading. Furthermore, these conjugates remain rather unstable over time, especially when exposed to in vivo denaturing conditions. The creation and use of covalent-based immobilization techniques are motivated by these drawbacks. Adding functional groups to SWCNTs that can be utilized to conjugate proteins or protein derivatives is one method of covalently immobilizing proteins. As mentioned above, the covalent fusion of carboxylated SWCNTs with amino groups on proteins has received the majority of attention in recent investigations. By activating EDC or EDC/NHS, carboxylated SWCNTs create a semi-stable intermediate interacting with the free amino groups on a protein surface [103, 104]. With an enzyme activity of up to 64.7% and a comparatively good yield of 52.1%, hydroxyquinol dioxygenase has been immobilized using this technique [104]. Despite the relative simplicity of the SWCNT carboxylation, this approach frequently causes the entire or partial denaturation of enzyme structure and activity [102].

As a result, by putting more space between the proteins and the SWCNTs, lengthy polymer chains have been used to reduce inactivation [105]. Linking the SWCNTs with an active group that the proteins can more easily access, such as maleimide derivatives [106], is another way to boost activity. The most common polymers used in these ligations are those based on polyethylene glycol (PEG), but other polymers have since been modified to work with these techniques [107]. Other functional handles can be utilized in addition to carboxylated SWCNTs to target particular protein groups. The science of nanomedicine is rapidly developing, opening up new possibilities for better illness diagnostics and drug delivery. Carbon nanotubes (CNTs), which have an unusual one-dimensional structure and intriguing intrinsic mechanical, physical, and chemical properties, have been thoroughly investigated for a variety of applications in biology and medicine among the different nanomaterials used in nanomedicine [108]. One-dimensional (1-D) flexible nanotubes may bend to enable numerous functionalized nanotube binding sites to one cell, resulting in a multi-valence effect and enhanced binding affinity of nanotubes conjugated with targeted ligands. SWNTs' inherent optical and electrical features can be used for multimodality imaging and therapy. SWNTs operate as quasi 1-D quantum wires with strong electronic states (electronic DOS) densities at the van Hove singularities due to quantum confined effects, giving SWNTs unique optical capabilities. Having substantial optical absorption in the near-infrared (NIR) region (800–1600 nm), SWNTs are highly absorbent materials. These wavelengths range from 800 to 1400 nm in the electromagnetic spectrum's tissue-transparent zone, allowing radiation to pass through without significantly heating up or harming tissues. SWNTs can, therefore, be used for photothermal therapy and photoacoustic imaging. Therapeutic medications with externally controlled release capabilities can also be delivered via SWNTs [109]. Chemotherapy is still the most effective treatment plan for cancer therapy because it uses carbon nanotube micro-focus to destroy unsuspected tumors and release cancer cells. CNTs' broad superficial extend and capacity for alterations with their shallows and sensual dimensions were reconnaissance to be employed as a photothermal killing of malignant cells in addition to their capability to function as all-purpose Nano-carriers for pharmaceutical medications. The delivery of peptides, antigens, and nucleic acids to cancer cells is made possible by CNTs [110]. CNTs are efficient at delivering peptides, antigens, and nucleic acids to cancer cells, to name just a few. Researchers are putting much effort into developing cancer theranostics using CNT-based drug delivery devices, focusing on the extracellular components of the tumor microenvironment and the intracellular components of lump cells [110].

4.2 Applications of Functionalized Quantum Dots

GQDs are carbon family members and have become cutting-edge carbon-based fluorescent materials because of their high biocompatibility, fluorescence properties, and non-toxicity. Compared to other semiconductor nanomaterials, these characteristics of GQDs make them the perfect nanomaterial for biological applications. Because of their special properties and the necessary synergism for creating biosensors, nanomaterials like GQDs exhibit remarkable qualities both on their own and when combined with multiple other nanomaterials to create hybrid nanocomposites. These biosensing systems have provided many options to diagnose various illnesses over the past ten years, including cancer, cardiovascular, infectious, and neurodegenerative diseases. When determining target analytes through a label-free method, GQDs can function as electro- or nano-catalysts for hydrogen peroxide (H2O2) catalysis [111]. They have catalytic properties similar to peroxidase (POD), which causes a redox process that occurs within H2O2 along with an electron-donating precursor. Horseradish peroxidase (HRP) is a common enzyme used in the biosensor industry to label secondary bioreceptors for the purpose of determining the related analyte, which makes the assays more time-consuming and expensive [112]. GQDs are used in place of HRP-conjugated secondary bioreceptors to complete these procedures more quickly and affordably. GQDs work as attractive fluorophores or fluorescent labels, quenchers, energy sources, and charge donors [113]. They are luminous nanoscale graphene segments that result in quantum-size phenomena and exciton confinement in particles with a size between 3 and 20 nm [114]. Pentaethylenehexamine and graphene quantum dots with histidine functionalization (PEHA-GQD-His). The molecular beacon double-cycle amplification method was used to create a microRNA (miRNA) fluorescence biosensing nanoplatform using the PEHA-GQD-His as the fluorescence probe [112]. These functionalized dots have shown better activity as fluorophores.

4.3 Applications of Functionalized Dendrimers

A particular class of hyperbranched polymer is called a dendrimer. Dendrimers have a unique molecular structure with a central core radiating repetitive branching units, the number of which determines a dendrimer's "generation" (designated as G0, G1, G2, G3) and whose termini form the dendrimer's peripheral or surface. The core, branching units, and terminal groups all contribute to the dendrimer's physicochemical characteristics, which include size, hydrophilicity, charge on the surface, functionality, and conformational flexibility. By appropriately manipulating these components chemically, these qualities can be adjusted. As a result, numerous dendrimer kinds have been reported as of late according to the kind of desired use. The simplicity of dendrimer preparation is a crucial factor to take into account. Dendrimers have the benefit of being monodisperse substances with a known molecular weight. As a result, their pharmacodynamic characteristics exhibit the repeatability that is frequently preferred for biological applications. Dendrimers were very early examined for their potential application as imaging agents, instruments in biological tests, medicines, and as a route of drug delivery because of their unique properties [115]. Dendrimers' unique structural characteristics and the variety of alterations that can be made allow them to function to be targeted DDS with lower drug dosages and, as a result, fewer side effects. By acting as dissolving agents or drug stabilizers and by providing safe storage spaces for moving into the bloodstream, dendrimers can improve drug bioavailability by encapsulating hydrophobic, hydrophilic, or amphiphilic compounds in their internal cavities through non-covalent interactions. A wide variety of novel dendrimers have been developed by specifically functionalizing the dendrimer periphery (e.g.s with -OH, -SH, -NH2, -COOH, azido, and allyl groups) and conjugating carrier species such polysaccharides, peptides, particular ligands, or functional groups. The carrier-mediated transport method opened the BBB through Functionalization with nutrition like glucose, maltose, or amino acids like histidine. It has been used to functionalize with a nutrient like glucose, maltose, or an amino acid-like histidine to cause the carrier-mediated mechanism for transport to cross the BBB. Alternatives include functionalizing with a positively charged group for activating non-selective absorptive-mediated transcytosis or anchoring a ligand like transferrin (Tf), sialic acid (SA), or lactoferrin (Lf) on the dendrimer to activate selective receptor-mediated transport [116].

PAMAM (polyamidoamine) dendrimers, along with related poly(propylene imine) (PPI) and poly-L-lysine (PLL) systems, have received widespread acceptance from biologists and medicinal chemists. They are primarily used as solubility enhancers for hydrophobic molecules because the internal part of the dendrimer contains hydrophobic cavities that can trap incoming molecules at a physiological pH, where both host and guest molecules are ionized. The number of voids increases with dendrimer size or production, resulting in a higher capacity for drug loading. PAMAM dendrimers may be made in practical numbers very cheaply, and the branching amide bonds in which they are embedded contain a peptide skeleton that enables them to be worked with in accordance with techniques common to protein chemistry. Higher-generation dendrimers exhibit increased toxicity. Because of this, it has frequently been discovered that using older generations helps to improve in vivo biocompatibility. In order to create polyanionic half-generation PAMAM dendrimers having carboxylate groups on the surface, peripheral amine groups can be functionalized by carboxylation, acetylation, or PEGylation. This process results in charge neutralization or less toxic PAMAM dendrimers [116].

4.4 Applications of Functionalized Fullerenes

If fullerenes are functionalized with hydrophilic groups to increase their solubility, photodynamic therapy (PDT) may be able to use them as photosensitizers [117]. Superoxide is indeed formed when water-soluble fullerenes are combined with biological reducing agents, and this process is susceptible to cells of microbial origin compared to cells of mammalians [118, 119]. After being exposed to photoirradiation, fulleropyrrolidinium salts were examined for their antibacterial activity and their findings demonstrated that 99.9% or, even more fungal cells and bacterial were destroyed [117]. After being exposed to white light, cationic-substituted fullerene derivatives were found to be highly effective at killing a variety of cells of microbial system, according to [120]. A new derivatives group of fullerene that was synthesized was tested for its ability to combat gram-negative, gram-positive, and fungal, bacteria. They claimed that the main contributing component was an increase in quaternary cationic groups, which were evenly distributed all over the cage of fullerene in order to prevent aggregation. In contrast to *E. coli* and *S. aureus*, the most susceptible microorganisms, *C. albicans* was found to be the most resilient species [52]. *E. coli* growth was found to be constrained by both graphene and GO.

Numerous biotic and abiotic elements continually stress plants due to the constantly shifting environmental circumstances. Numerous abiotic factors, particularly drought stress, can affect crop growth, development, and production [121]. Plants' physiological and molecular characteristics are altered by water content restrictions, including decreased photosynthesis, increased biomass, and an excess of reactive oxygen species (ROS). The excessive creation of ROS results in damage to DNA, proteins, and cellular membranes [121]. Controlling ROS generation and maintaining ROS levels' equilibrium is crucial and essential. Numerous carbon nanoparticles, including fullerene and its derivatives, are recognized to be essential for a plant's ability to withstand abiotic stress [122, 123]. According to several studies, fullerol helps plants fight against the negative effects of osmotic stress at optimal concentrations [124, 125].

5 Conclusion

Functionalized carbon nanomaterials have unleashed many directions to revolutionize the health, medical, and environmental sectors. The Functionalized carbon-based nanomaterial shows greater strength and flexibility. Thus imparting a mysterious role in each sector. The Functionalized SWNTs and other Nanocomposites formed with different conjugates have reoriented their functions as therapeutic agents. Further research in this area can allow distinct applications to be accomplished, which were so far not possible. The rising rate of pollution and exposure of xenobiotic compounds to the soil have indeed given birth to several hazardous issues. The Nano remediation through using carbon-based nanomaterials has come up with some solutions to circumvent such effects. Innovative techniques utilized in the Functionalization of carbon-based nanomaterials such as CNT, Fullerenes, quantum dots, and dendrimers have allowed us to study various insights, and continuous progress in this area of research can lead to the intervention in the field of carbon-based nanotechnology and may allow scientists or researchers to rectify many perplexed environmental and health conditions.

References

1. Lochab A, Sharma R, Kumar S, Saxena R (2021) Recent advances in carbon based nanomaterials as electrochemical sensor for toxic metal ions in environmental applications. Mater Today Proc 45:3741–3753
2. Godeto YG, Ayele A, Ahmed IN, Husen A, Bachheti RK (2023) Medicinal plant-based metabolites in nanoparticles synthesis and their cutting-edge applications: an overview. In: Bachheti RK, Bachheti A (eds) Secondary metabolites from medicinal plants. CRC Press, Boca Raton, USA, pp 1–34
3. Kumar R, Kumar M, Luthra G (2023) Fundamental approaches and applications of nanotechnology: a mini review. Mater Today Proc
4. Lv P, Feng YY, Zhang P, Chen HM, Zhao N, Feng W (2011) Increasing the interfacial strength in carbon fiber/epoxy composites by controlling the orientation and length of carbon nanotubes grown on the fibers. Carbon 49(14):4665–4673
5. Wang X, Li Y, Zhao Y, Liu J, Tang S, Feng W (2010) Synthesis of PANI nanostructures with various morphologies from fibers to micromats to disks doped with salicylic acid. Synthet Metals 160(17–18):2008–2014
6. Feng W, Bai XD, Lian YQ, Liang J, Wang XG, Yoshino K (2003) Well-aligned polyaniline/carbon-nanotube composite films grown by in-situ aniline polymerization. Carbon 41(8):1551–1557
7. Wang X, Tang S, Zhou C, Liu J, Feng W (2009) Uniform TiO2–PANI composite capsules and hollow spheres. Synth Met 159(17–18):1865–1869
8. Wang W, Li Y, Feng Y, Han J, Zhang F, Long P, Feng W (2018) Asymmetric self-supporting hybrid fluorinated carbon nanotubes/carbon nanotubes sponge electrode for high-performance lithium-polysulfide battery. Chem Eng J 349:756–765
9. Zhang F, Feng Y, Qin M, Ji T, Lv F, Li Z, Feng W (2019) Stress-sensitive thermally conductive elastic nanocomposite based on interconnected graphite-welded carbon nanotube sponges. Carbon 145:378–388

10. Allahverdiyev AM, Abamor ES, Bagirova M, Rafailovich M (2011) Antimicrobial effects of TiO2 and Ag_2Onanoparticles against drug-resistant bacteria and leishmania parasites. Future microbiol 6(8):933–940
11. Adibkia K, Omidi Y, Siahi MR, Javadzadeh AR, Barzegar-Jalali M, Barar J, Nokhodchi A (2007) Inhibition of endotoxin-induced uveitis by methylprednisolone acetate nanosuspension in rabbits. J Ocul Pharmacol Ther 23(5):421–432
12. Bahrami K, Nazari P, Nabavi M, Golkar M, Almasirad A, Shahverdi AR (2014) Hydroxyl capped silver-gold alloy nanoparticles: characterization and their combination effect with different antibiotics against *Staphylococcus aureus*. J Nanomed 1(3):155–161
13. Tiwari PM, Vig K, Dennis VA, Singh SR (2011) Functionalized gold nanoparticles and their biomedical applications. Nanomaterials (Basel) 1(1):31–63
14. Zinjarde S (2012) Bio-inspired nanomaterials and their applications as antimicrobial agents. Chronicles Young Sci 3(1):74–74
15. Marambio-Jones C, Hoek EM (2010) A review of the antibacterial effects of silver nanomaterials and potential implications for human health and the environment. J Nanopart Res 12:1531–1551
16. Belay T, Worku LA, Bachheti RK, Bachheti A, Husen A (2023) Nanomaterials: introduction, synthesis, characterization, and applications. In: Husen A, Siddiqi KS (eds) Advances in smart nanomaterials and their applications. Elsevier, Amsterdam, Netherlands, pp 1–21
17. Husen A (2019a) Natural product-based fabrication of zinc oxide nanoparticles and their application. In: Husen A, Iqbal M (eds) N*anomaterials* and plant potential. Springer International Publishing AG, Gewerbestrasse 11, 6330 Cham, pp 193–291. https://doi.org/10.1007/978-3-030-05569-1_7
18. Husen A (2019b) Medicinal plant-product based fabrication nanoparticles (Au and Ag) and their anticancer effect. In: Kintzios SE, Barberaki M, Flampouri (eds) Plants that fight cancer, 2nd edn. Taylor & Francis/CRC Press, pp 133–147
19. Husen A, Rahman QI, Iqbal M, Yassin MO (2019) Plant-mediated fabrication of gold nanoparticles and their application in some cutting-edge areas. In: Husen A, Iqbal M (eds) N*anomaterials* and plant potential. Springer International Publishing AG, Gewerbestrasse 11, 6330 Cham, pp 71–110. https://doi.org/10.1007/978-3-030-05569-1_3
20. Husen A, Siddiqi KS (2014) Phytosynthesis of nanoparticles: concept, controversy and application. Nanoscale Res Lett 9:229. https://doi.org/10.1186/1556-276X-9-229
21. Joshi A, Sharma A, Bachheti RK, Husen A, Mishra VK (2019) Plant-mediated synthesis of copper oxide nanoparticles and their biological applications. In: Husen A, Iqbal M (eds) N*anomaterials* and plant potential. Springer International Publishing AG, Gewerbestrasse 11, 6330 Cham, pp 221–237. https://doi.org/10.1007/978-3-030-05569-1_8
22. Siddiqi KS, Husen A (2016a) Fabrication of metal and metal oxide nanoparticles by algae and their toxic effects. Nanoscale Res Lett 11:363. https://doi.org/10.1186/s11671-016-1580-9
23. Siddiqi KS, Husen A (2016b) Fabrication of metal nanoparticles from fungi and metal salts: scope and application. Nanoscale Res Lett 11:98. https://doi.org/10.1186/s11671-016-1311-2
24. Siddiqi KS, Husen A (2017) Recent advances in plant-mediated engineered gold nanoparticles and their application in biological system. J Trace Elem Med Biol 40:10–23. https://doi.org/10.1016/j.jtemb.2016.11.012
25. Siddiqi KS, Husen A (2020) Current status of plant metabolite-based fabrication of copper/copper oxide nanoparticles and their applications: A review. Biomat Res 24:11. https://doi.org/10.1186/s40824-020-00188-1
26. Siddiqi KS, Husen A, Rao RAK (2018) A review on biosynthesis of silver nanoparticles and their biocidal properties. J Nanobiotechnol 16:14. https://doi.org/10.1186/s12951-018-0334-5
27. Worku LA, Nigussie Y, Bachheti A, Bachheti RK, Husen A (2023) Antimicrobial activities of nanomaterials. In: Husen A, Siddiqi KS (eds) Advances in smart nanomaterials and their applications. Elsevier, Amsterdam, Netherlands, pp 127–148
28. Feng W, Fujii A, Ozaki M, Yoshino K (2005) Perylene derivative sensitized multi-walled carbon nanotube thin film. Carbon 43(12):2501–2507

29. Li Z, Wang L, Li Y, Feng Y, Feng W (2019) Carbon-based functional nanomaterials: preparation, properties and applications. Composi Sci Technol 179:10–40
30. Tomalia DA (2004) Birth of a new macromolecular architecture: dendrimers as quantized building blocks for nanoscale synthetic organic chemistry. Aldrichimica Acta 2:39–57
31. Wiener EC, Brechbiel MW, Brothers H, Magin RL, Gansow OA, Tomalia DA, Lauterbur PC (1994) Dendrimer-based metal chelates: a new class of magnetic resonance imaging contrast agents. Magn Reson Med 31(1):1–8
32. Li Y, Cheng Y, Xu T (2007) Design, synthesis and potent pharmaceutical applications of glycodendrimers: a mini review. Curr Drug Discov Technol 4(4):246–254
33. Karagianni A, Tsierkezos NG, Prato M, Terrones M, Kordatos KV (2023) Application of carbon-based quantum dots in photodynamic therapy. Carbon 203:273–310
34. Jalilinejad N, Rabiee M, Baheiraei N et al (2022) Electrically conductive carbon-based (bio)-nanomaterials for cardiac tissue engineering. Bioeng Transl Med 8(1):e10347
35. Fu T, Zhang B, Gao X et al (2023) Recent progresses, challenges, and opportunities of carbon-based materials applied in heavy metal polluted soil remediation. Sci Total Environ 856(Pt 1):158810
36. ul Haque S, Nasar A, Duteanu N, Pandey S (2023) Carbon based-nanomaterials used in biofuel cells–a review. Fuel 331:125634
37. Wang J, Jiang J, Li F, Zou J, Xiang K, Wang H, Li X (2023) Emerging carbon-based quantum dots for sustainable photocatalysis. Green Chem 25(1):32–58
38. Sun Y, Xu B, Pan X, Wang H, Wu Q, Li S, Liu H (2023) Carbon-based nanozymes: Design, catalytic mechanism, and bioapplication. Coord Chem Reviews 475:214896
39. Upadhyaya S, Devi M, Sarma NS (2023) Carbon and silver nanoparticles for applications in agriculture. In: Microbiomes for the management of agricultural sustainability. Springer Nature Switz, Cham, pp 297–316
40. Dubey R, Dutta D, Sarkar A, Chattopadhyay P (2021) Functionalized carbon nanotubes: synthesis, properties and applications in water purification, drug delivery, and material and biomedical sciences. Nanoscale Adv 3(20):5722–5744
41. Mickelson ET, Huffman CB, Rinzler AG, Smalley RE, Hauge RH, Margrave JL (1998) Fluorination of single-wall carbon nanotubes. Chem Phys Lett 296:188–194
42. Mickelson ET, Chiang IW, Zimmerman JL, Boul PJ, Lozano J, Liu J, Smalley RE, Hauge RH, Margrave J (1999) Solvation of fluorinated single-wall carbon nanotubes in alcohol solvents. Phys Chem B 103:4318–4322
43. Khare BN, Meyyappan M, Cassell AM, Nguyen CV, Han J (2002) Functionalization of carbon nanotubes using atomic hydrogen from a glow discharge. Nano Lett 2(1):73–77
44. Georgakilas V, Kordatos K, Prato M, Guldi DM, Holzinger M, Hirsch A (2002) Organic functionalization of carbon nanotubes. J Am Chem Soc 124(5):760–761
45. Bahr JL, Yang J, Kosynkin DV, Bronikowski MJ, Smalley RE, Tour JM (2001) Functionalization of carbon nanotubes by electrochemical reduction of aryl diazonium salts: a bucky paper electrode. J Am Chem Soc 123(27):6536–6542
46. Holzinger M, Abraham J, Whelan P, Graupner R, Ley L, Hennrich F, Hirsch, A (2003) Functionalization of single-walled carbon nanotubes with (R-)oxycarbonyl nitrenes. J Am Chem Soc 125(28):8566–8580
47. Tagmatarchis N, Georgakilas V, Prato M, Shinohara H (2002) Sidewall functionalization of single-walled carbon nanotubes through electrophilic addition. Chem comm 11(18):2010–2011
48. Zhou Y, Fang Y, Ramasamy RP (2019) Non-covalent functionalization of carbon nanotubes for electrochemical biosensor development. Sensors (Basel) 19(2):392
49. Pichler T, Kuzmany H, Kataura H, Achiba Y (2001) Metallic polymers of C(60) inside single-walled carbon nanotubes. Phys Rev Lett 87(26):267401
50. Maddox MW, Gubbins KE (1997) A molecular simulation study of freezing/melting phenomena for Lennard-Jones methane in cylindrical nanoscale pores. J Chem Phys 107(22):9659–9667

51. Basha JS (2015) Applications of functionalized carbon-based nanomaterials. In: Chemical functionalization of carbon nanomaterials. CRC Pr, pp 598–613
52. Dizaj SM, Mennati A, Jafari S, Khezri K, Adibkia K (2015) Antimicrobial activity of carbon-based nanoparticles. Advan Pharma Bull 5(1):19
53. Meyyappan M (2004) Carbon nanotubes: science and applications. CRC Press
54. Venkatesan R, Sivaprakash P, Kim I, Eldesoky GE, Kim SC (2023) Tannic acid as a crosslinking agent in poly (butylene adipate-co-terephthalate) composite films enhanced with carbon nanoparticles: processing, characterization, and antimicrobial activities for food packaging. J Environ Chem Eng 11(4):110194
55. Younis AB, Haddad Y, Kosaristanova L, Smerkova K (2023) Titanium dioxide nanoparticles: recent progress in antimicrobial applications. Wiley Interdiscip Rev Nanomed Nanobiotechnol 15(3):e1860
56. Chen RJ, Zhang Y, Wang D, Dai H (2001) Noncovalent sidewall functionalization of single-walled carbon nanotubes for protein immobilization. J Am Chem Soc 123(16):3838–3839
57. Dwyer C, Guthold M, Falvo M, Washburn S, Superfine R, Erie D (2002) DNA-functionalized single-walled carbon nanotubes. Nanotechnology 13(5):601
58. Huang W, Taylor S, Fu K, Lin Y, Zhang D, Hanks TW, Sun YP (2002) Attaching proteins to carbon nanotubes via diimide-activated amidation. Nano Lett 2(4):311–314
59. Williams KA, Veenhuizen PT, de la Torre BG, Eritja R, Dekker C (2002) Nanotechnology: carbon nanotubes with DNA recognition. Nature 420(6917):761
60. Azamian BR, Davis JJ, Coleman KS, Bagshaw CB, Green ML (2002) Bioelectrochemical single-walled carbon nanotubes. J Am Chem Soc 124(43):12664–12665
61. Majeed N, Panigrahi KC, Sukla LB, John R, Panigrahy M (2020) Application of carbon nanomaterials in plant biotechnology. Mater Today Proc 30:340–345
62. Katti DR, Sharma A, Pradhan SM, Katti KS (2015) Carbon nanotube proximity influences rice DNA. Chem Phys 455:17–22
63. Ghosh M, Bhadra S, Adegoke A, Bandyopadhyay M, Mukherjee A (2015) MWCNT uptake in Allium cepa root cells induces cytotoxic and genotoxic responses and results in DNA hyper-methylation. Mutat Res 774:49–58
64. Yan S, Zhao L, Li H, Zhang Q, Tan J, Huang M, He S, Li L (2013) Single-walled carbon nanotubes selectively influence maize root tissue development accompanied by the change in the related gene expression. J Hazard Mater 246–247:110–118
65. Begum P, Ikhtiari R, Fugetsu B, Matsuoka M, Akasaka T, Watari F (2012) Phytotoxicity of multi-walled carbon nanotubes assessed by selected plant species in the seedling stage. Appl Surf Sci 262:120–124
66. Liu Q, Zhang X, Zhao Y, Lin J, Shu C, Wang C, Fang X (2013) Fullerene-induced increase of glycosyl residue on living plant cell wall. Environ Sci Technol 47(13):7490–7498
67. Cañas JE, Long M, Nations S, Vadan R, Dai L, Luo M, Ambikapathi R, Lee EH, Olszyk D (2008) Effects of functionalized and nonfunctionalized single-walled carbon nanotubes on root elongation of select crop species. Environ Toxicol Chem 9:1922–1931
68. Lahiani MH, Dervishi E, Chen J, Nima Z, Gaume A, Biris AS, Khodakovskaya MV (2013) Impact of carbon nanotube exposure to seeds of valuable crops. ACS Appl Mater Interfaces 5(16):7965–7973
69. Tiwari DK, Dasgupta-Schubert N, Villaseñor Cendejas LM, Villegas J, Carreto Montoya L, Borjas García SE (2014) Interfacing carbon nanotubes (CNT) with plants: enhancement of growth, water and ionic nutrient uptake in maize (Zea mays) and implications for nanoagriculture. Appl Nanosci 4:577–591
70. Stampoulis D, Sinha SK, White JC (2009) Assay-dependent phytotoxicity of nanoparticles to plants. Environ Sci Technol 43(24):9473–9479
71. Flores D, Chaves JS, Chacón R, Schmidt A (2013) A novel technique using SWCNTs to enhanced development and root growth of fig plants (Ficus carica). In: Technical proceedings of the NSTI nanotechnology conference and expo (NSTI-Nanotech'13). pp 167–170
72. Khodakovskaya MV, de Silva K, Nedosekin DA et al (2011) Complex genetic, photothermal, and photoacoustic analysis of nanoparticle-plant interactions. Proc Natl Acad Sci U S A 108(3):1028–1033

73. Pourkhaloee A, Haghighi M, Saharkhiz MJ, Jouzi H, Doroodmand MM (2011) Carbon nanotubes can promote seed germination via seed coat penetration. Seed Techn 155–169
74. Khodakovskaya MV, Kim BS, Kim JN, Alimohammadi M, Dervishi E, Mustafa T, Cernigla CE (2013) Carbon nanotubes as plant growth regulators: effects on tomato growth, reproductive system, and soil microbial community. Small 9(1):115–123
75. Star A, Gabriel JCP, Bradley K, Grüner G (2003) Electronic detection of specific protein binding using nanotube FET devices. Nano Lett 3(4):459–463
76. Broussard S, Tao NJ, Zhang R, Hopson T, Nagahara LA (2003) In situ detection of cytochrome c adsorption with single-walled carbon nanotube device. Chem Commun 13:1502–1503
77. Wang J, Liu G, Jan MR (2004) Ultrasensitive electrical biosensing of proteins and DNA: carbon-nanotube derived amplification of the recognition and transduction events. J Am Chem Soc 126(10):3010–3011
78. Wu Z, Chen Z, Du X, Logan JM, Sippel J, Nikolou M, Kamaras K, Reynolds JR, Tanner DB, Hebard AF, Rinzler AG (2004) Transparent, conductive carbon nanotube films. Science 305(5688):1273–1276
79. Lin Y, Lu F, Tu Y, Ren Z (2004) Glucose biosensors based on carbon nanotube nanoelectrode ensembles. Nano Lett 4(2):191–195
80. Wang J, Musameh M (2003) Carbon nanotube/teflon composite electrochemical sensors and biosensors. Anal Chem 75(9):2075–2079
81. Britto PJ, Santhanam KSV, Ajayan PM (1996) Carbon nanotube electrode for oxidation of dopamine. Bioelectrochem Bioenerg 41(1):121–125
82. Pantarotto D, Briand JP, Prato M, Bianco A (2004) Translocation of bioactive peptides across cell membranes by carbon nanotubes. Chem Commun (Camb) (1):16–17
83. Shi Kam NW, Jessop TC, Wender PA, Dai H (2004) Nanotube molecular transporters: internalization of carbon nanotube-protein conjugates into Mammalian cells. J Am Chem Soc 126(22):6850–6851
84. Balavoine F, Schultz P, Richard C, Mallouh V, Ebbesen TW, Mioskowski C (1999) Helical crystallization of proteins on carbon nanotubes: a first step towards the development of new biosensors. Angew Chem Int Ed Engl 38(13–14):1912–1915
85. Guo Z, Sadler PJ, Tsang SC (1998) Immobilization and visualization of DNA and proteins on carbon nanotubes. Adv Mater 10(9):701–703
86. Tsang SC, Guo Z, Chen YK, Green ML, Hill HAO, Hambley TW, Sadler PJ (1997) Immobilization of platinated and iodinated oligonucleotides on carbon nanotubes. Angewandte Chemie Internat Edition Eng 36(20):2198–2200
87. Davis JJ, Green ML, Hill HAO, Leung YC, Sadler PJ, Sloan J, Tsang SC (1998) The immobilisation of proteins in carbon nanotubes. Inorgan Chim Acta 272(1–2):261–266
88. González-Melendi P, Fernández-Pacheco R, Coronado MJ et al (2008) Nanoparticles as smart treatment-delivery systems in plants: assessment of different techniques of microscopy for their visualization in plant tissues. Ann Bot 101(1):187–195
89. Mukherjee A, Majumdar S, Servin AD, Pagano L, Dhankher OP, White JC (2016) Carbon nanomaterials in agriculture: a critical review. Front Plant Sci 7:172
90. Ripanda AS, Rwiza MJ, Nyanza EC, Njau KN, Vuai SA, Machunda RL (2021) A review on contaminants of emerging concern in the environment: a focus on active chemicals in Sub-Saharan Africa. Appl Sci 12(1):56
91. Chauhan P, Imam A, Kanaujia PK, Suman SK (2023) Nano-bioremediation: an eco-friendly and effective step towards petroleum hydrocarbon removal from environment. Environ Res 231(Pt 2):116224
92. Shahi MP, Kumari P, Mahobiya D, Shahi SK (2021) Nano-bioremediation of environmental contaminants: applications, challenges, and future prospects. Biorem Environ Sustain 83–98
93. Yadav GK, Ahmaruzzaman M (2021) Recent advances in the development of nanocomposites for effective removal of pesticides from aqueous stream. J Nanoparticle Res 23(9):213
94. Chaudhary P, Ahamad L, Chaudhary A, Kumar G, Chen WJ, Chen S (2023) Nanoparticle-mediated bioremediation as a powerful weapon in the removal of environmental pollutants. J Envir Chem Eng 109591

95. Correia VM, Stephenson T, Judd SJ (1994) Characterisation of textile wastewaters-a review. Envir Techn 15(10):917–929
96. Liang C, Shi Q, Feng J, Yao J, Huang H, Xie X (2022) Adsorption behaviors of cationic methylene blue and anionic reactive blue 19 dyes onto nano-carbon adsorbent carbonized from small precursors. Nanomaterials (Basel) 12(11):1814
97. Safari M, Khataee A, Darvishi Cheshmeh Soltani R, Rezaee R (2018) Ultrasonically facilitated adsorption of an azo dye onto nanostructures obtained from cellulosic wastes of broom and cooler straw. J Colloid Interface Sci 522:228–241
98. Zhao X, Baharinikoo L, Farahani MD, Mahdizadeh B, Farizhandi AAK (2022) Experimental modelling studies on the removal of dyes and heavy metal ions using ZnFe2O4 nanoparticles. Sci Rep 12(1):5987
99. Hadei M, Mesdaghinia A, Nabizadeh R, Mahvi AH, Rabbani S, Naddafi K (2021) A comprehensive systematic review of photocatalytic degradation of pesticides using nano TiO2. Environ Sci Pollut Res Int 13055–13071
100. Tabasum A, Alghuthaymi M, Qazi UY, Shahid I, Abbas Q, Javaid R, Nadeem N, Zahid M (2020) UV-Accelerated Photocatalytic Degradation of Pesticide over Magnetite and Cobalt Ferrite Decorated Graphene Oxide Composite. Plants (Basel) 10(1):6
101. Kadam AA, Saratale GD, Ghodake GS, Saratale RG, Shahzad A, Magotra VK, Sung JS (2022) Recent advances in the development of laccase-based biosensors via nano-immobilization techniques. Chemosensors 10(2):58
102. Wang H, Boghossian AA (2022) Covalent conjugation of proteins onto fluorescent single-walled carbon nanotubes for biological and medical applications. Mater Adv 4(3):823–834
103. Bartha-Vári JH, Toşa MI, Irimie FD (2015) Immobilization of phenylalanine ammonia-lyase on single-walled carbon nanotubes for stereoselective biotransformations in batch and continuous-flow modes. ChemCatChem 7(7):1122–1128
104. Suma Y, Kang CS, Kim HS (2016) Noncovalent and covalent immobilization of oxygenase on single-walled carbon nanotube for enzymatic decomposition of aromatic hydrocarbon intermediates. Environ Sci Pollut Res Int 23(2):1015–1024
105. Bencze LC, Bartha-Vári JH, Katona G, Toşa MI, Paizs C, Irimie FD (2016) Nanobioconjugates of Candida Antarctica lipase B and single-walled carbon nanotubes in biodiesel production. Bioresour Technol 200:853–860
106. Boros K, Moisă ME, Nagy CL, Paizs C, Toşa MI, Bencze LC (2021) Robust, site-specifically immobilized phenylalanine ammonia-lyases for the enantioselective ammonia addition of cinnamic acids. Catal Sci Technol 11(16):5553–5563
107. Pantarotto D, Partidos CD, Hoebeke J, Brown F, Kramer E, Briand JP, Muller S, Prato M, Bianco A (2003) Immunization with peptide-functionalized carbon nanotubes enhances virus-specific neutralizing antibody responses. Chem Biol 10(10):961–966
108. Wang X, Liu Z (2012) Carbon nanotubes in biology and medicine: an overview. Chinese Sci Bull 57:167–180
109. Liu Z (2009) Biomedical applications of carbon nanotubes. Stanford University
110. Cheung W, Pontoriero F, Taratula O, Chen AM, He H (2010) DNA and carbon nanotubes as medicine. Adv Drug Deliv Rev 62(6):633–649
111. Savas S, Altintas Z (2019) Graphene quantum dots as nanozymes for electrochemical sensing of Yersinia enterocolitica in milk and human serum. Materials (Basel) 12(13):2189
112. Mansuriya BD, Altintas Z (2020) Applications of graphene quantum dots in biomedical sensors. Sensors (Basel) 20(4):1072
113. Zheng P, Wu N (2017) Fluorescence and sensing applications of graphene oxide and graphene quantum dots: a review. Chem Asian J 12(18):2343–2353
114. Wang L, Zhu SJ, Wang HY, Qu SN, Zhang YL, Zhang JH, Chen QD, Xu HL, Han W, Yang B, Sun HB (2014) Common origin of green luminescence in carbon nanodots and graphene quantum dots. ACS Nano 8(3):2541–2547
115. Villalonga-Barber C, Micha-Screttas M, Steele BR, Georgopoulos A, Demetzos C (2008) Dendrimers as biopharmaceuticals: synthesis and properties. Curr Top Med Chem 8(14):1294–1309

116. Arbez-Gindre C, Steele BR, Micha-Screttas M (2023) Dendrimers in Alzheimer's disease: recent approaches in multi-targeting strategies. Pharmaceutics 15(3):898
117. Tegos GP, Demidova TN, Arcila-Lopez D, Lee H, Wharton T, Gali H, Hamblin MR (2005) Cationic fullerenes are effective and selective antimicrobial photosensitizers. Chem Biol 12(10):1127–1135
118. Lu Z, Dai T, Huang L, Kurup DB, Tegos GP, Jahnke A, Wharton T, Hamblin MR (2010) Photodynamic therapy with a cationic functionalized fullerene rescues mice from fatal wound infections. Nanomedicine (Lond) 5(10):1525–1533
119. Sharma SK, Chiang LY, Hamblin MR (2011) Photodynamic therapy with fullerenes in vivo: reality or a dream? Nanomedicine (Lond) 6(10):1813–1825
120. Mizuno K, Zhiyentayev T, Huang L, Khalil S, Nasim F, Tegos GP, Gali H, Jahnke A, Wharton T, Hamblin MR (2011) Antimicrobial photodynamic therapy with functionalized fullerenes: quantitative structure-activity relationships. J Nanomed Nanotechnol 2(2):1–9
121. Duhan JS, Kumar R, Kumar N, Kaur P, Nehra K, Duhan S (2017) Nanotechnology: the new perspective in precision agriculture. Biotechnol Rep (Amst) 15:11–23
122. Borišev M, Borišev I, Župunski M, Arsenov D, Pajević S, Ćurčić Ž, Vasin J, Djordjevic A (2016) Drought impact is alleviated in sugar beets (Beta vulgaris L.) by foliar application of fullerenol nanoparticles. PLoS One 11(11):e0166248
123. Khan MN, Mobin M, Abbas ZK, AlMutairi KA, Siddiqui ZH (2017) Role of nanomaterials in plants under challenging environments. Plant Physiol Biochem 110:194–209
124. Panova GG, Ktitorova IN, Skobeleva OV, Sinjavina NG, Charykov NA, Semenov KN (2016) Impact of polyhydroxy fullerene (fullerol or fullerenol) on growth and biophysical characteristics of barley seedlings in favourable and stressful conditions. Plant Growth Regulat 79:309–317
125. Trpkovic A, Todorovic-Markovic B, Trajkovic V (2012) Toxicity of pristine versus functionalized fullerenes: mechanisms of cell damage and the role of oxidative stress. Arch Toxicol 86(12):1809–1827

Smart Carbon Nanomaterials and Their Effect on the Antioxidant System of Plants

Anish Kumar Pal, Kalash Aggrawal, Kundan Kumar Chaubey, Sonali Yadav, Soni Sharma, Anupriya Kumari, Vanshika Saxena, Shivani Shivu, and Lalit Kumar Sharma

Abstract Carbon nanotubes (CNTs) are allotropes of carbon having a diameter in the nanometer range. The carbon nanotubes are differentiated by the design of graphite during its creation process. Smart carbon nanomaterials include carbon nanotubes (CNTs) with graphene, fullerene, carbon quantum dots (CDs), and mesoporous carbon nanomaterials (CNMs), which affect plant growth. CNMs have been shown to penetrate seed coatings, enter plant cells, and shift to various parts of the plant. Exposure reduces seed germination, root growth, and root architecture. CN inhibits seeds' growth and alters plants' genetic, biochemical, molecular, nutritional, and genetic levels. Using carbon-based nanomaterials is an ideal way to promote resistance to different stress levels in tomatoes and other crops. Antioxidants are part of the plant antioxidant defense system and are regarded as proteins. CNMs, such as carbon nanotubes, are known to induce the formation of antioxidant enzymes that produce protein accumulation. All these aspects have been thoroughly reviewed in this chapter, focusing on the recent updates on the role of CNMs in preventing or delaying plant growth. Concluding remarks have been added to propose future directions of research on the CNM plants interaction and also to sound a warning on the use of CNMs in agriculture.

Keywords Carbon-based nanomaterials · Growth responses · Plant technology · Translocation · Bioaccumulation · Fullerene · Carbon nanotubes · Antioxidants

---

A. K. Pal
Division of Research and Innovation, School of Applied and Life Sciences, Uttaranchal University, Arcadia Grant, P.O. Chandanwari, Premnagar, Dehradun, Uttarakhand 248007, India

K. Aggrawal
School of Biotechnology, Gautam Buddha University, Greater Noida, Uttar Pradesh 201312, India

K. K. Chaubey (✉) · S. Yadav · S. Sharma · A. Kumari · V. Saxena · S. Shivu · L. K. Sharma
School of Basic and Applied Sciences, Sanskriti University, Mathura, Uttar Pradesh 281401, India
e-mail: kundan2006chaubey@gmail.com

A. Bachheti et al. (eds.), *Carbon-Based Nanomaterials*, Smart Nanomaterials Technology, https://doi.org/10.1007/978-981-97-0240-4_5

1 Introduction

Plant growth, protection and stress tolerance is increased by the use of nanomaterials made from carbon (CNMs) in plant systems that contain carbon dots (CDs), graphene (GRA), graphene oxide (GO), nano horns (CNHs), and carbon nanofibers (CNFs) [1–4]. Carbon-based nanomaterials' potential impact on plants as a basic element of natural and cultivation ecosystem food chains, especially plant life, is expected to increase due to the intentional use of carbon nanomaterials in agriculture and the environment. CNTs have unique rigidity, resistance, and elasticity properties compared to other fibrous materials due to their structure [5]. Carbon nanotubes (CNTs) can be mono- and multi-walled and are suitable for electronics, nano-composting, and drug delivery because of their distinctive electronics and mechanical properties. Graphene is a single carbon-atom layer placed in a two-dimensional honeycomb grid. Smart graphite-based nanomaterials can be designed to react to external stimuli such as light, heat, or chemicals, making them a promising candidate for sensors, actuators, and flexible electronics [6]. The development, handling, and utilization of nanomaterials—a subject with exciting potential for a wide range of applications—formed the basis of the carbon nanomaterials process of nanotechnology.

Nanotechnology is a rapidly growing business with several uses in energy, materials, computer chipsets, manufacturing, medical care diagnosis, agriculture, forestry and environment [7–18]. Carbon semiconductors, metal oxide lipids, zero-valent metal quantum dots, nanopolymers, and various density materials like nanowires, as nanosheets in shape and nanofibers make up the major class of nanomaterials [19]. The production of nanomaterials usually involves direct and synthetic processes, resulting in particles in the nanoscale range, which are then subjected to grinding, high-pressure homogenization, and sound to reduce their size. At the same time, the bottom-up process for the synthesis of nanomaterials involved reactive precipitation and solvent displacement. Graphene is made up of a single layer of carbon atoms arranged in a two-dimensional honeycomb grid (Zann et al.). It is a reliable and small heat and electrical current conductor. Graphene oxide is an oxidized form of graphene containing an oxygen functional group. Graphite compounds are frequently used in many different applications, including water filtration, medicine delivery systems, and sensors. Antioxidants can be used in various ways to increase stress resistance and improve plant growth [20]. Various applications of antioxidants have several common direct effects on the chemical and enzyme systems of native antioxidants, from improvements in the ability or flexibility of photosynthetic enzymes to the modification of the signals of phytohormones to the metabolism of the antioxidants applied. [21]

2 Overview of Smart Carbon Nanomaterials

Despite being comprised of carbon atoms, carbon nanomaterials differ from one another in terms of their physical characteristics. Allotropes, or allotrope modifications of a specific chemical element, are what these are. Only two natural carbon allotropes, diamond and graphite, were recognized until recently [22]. On the other hand, different new allotropic forms, such as carbon nanomaterials, have been described. Materials that contain carbon are relatively rare chemical elements, including silicon that can be bonded to the atomic level. Since the outer layer of an electron has four electrons that have four valances, they can be connected to other elements to form single, double, and triple-covalent bonds, respectively [23]. Carbon can exist in various molecular forms composed of the same type of particle. All carbon-atom nanomaterials are called carbon-based nanomaterials. Smart carbon-based nanomaterials are found most frequently according to the shape of their geometric structure, including particles in tubes, horns, spheres, and ellipse-shaped shapes [24, 25]. The horn-shaped nanoparticles are known as nano horns, the spheres and ellipses are members of the fullerene carbon nanomaterials, and the tube-shaped nanoparticles are known as carbon nanotubes. CNT is among the best carbon nanotubes for technical applications, among other carbon nanomaterials. Carbon nanotubes are composed of round graphite sheets with a diameter of several nanometers. Circular graphene sheet symmetry, length, diameter, and layer count are all variables in carbon nanotubes. Single-walled carbon nanotubes (SWCNTs) and multi-walled carbon nanotubes (MWCNTs) are the two primary types of CNTs [26] Rese Bihar Cher developed another type of nanotube called double-walled carbon nanotubes. SWCNTs are generally 1–3 nm in diameter and several micrometers in length, while MWCNTs have a diameter of 5–40 nm and a length of about 10 m. Fullerene is a molecular form of carbon that has undergone an alteration transition to become carbon. The fullerene family consists of n clusters ($n > 20$) covered in carbon atoms. The tips of the pentagon and hexagon, which are on the surface of the sphere, are often where carbon atoms are found. Covalent bonds hold the carbon atoms of fullerene together, where they are typically sp2 hybrids. The most prevalent variety of fullerene is fullerene C60, and the molecules are very symmetric, consisting of 20 hexagonal and 12 pentagonal vertices along with 60 carbon atoms [27]. The diameter of the C60 is 0.7 nm. Smart carbon nanomaterials have been widely used in biomedical and medical chemistry for their unique mechanical and structural properties. Carbon nanotubes are highly sensitive to physical properties. They are commonly used in optics, nanomedicine, solar cells, and electronics for high-efficiency renewable energy production repair, biosensor pollutants, and contamination degradation. The interaction of the three components (plant, carbon nanomaterial, and growth media) as a complex system determines the effect of nanomaterials on developing plants (Fig. 1).

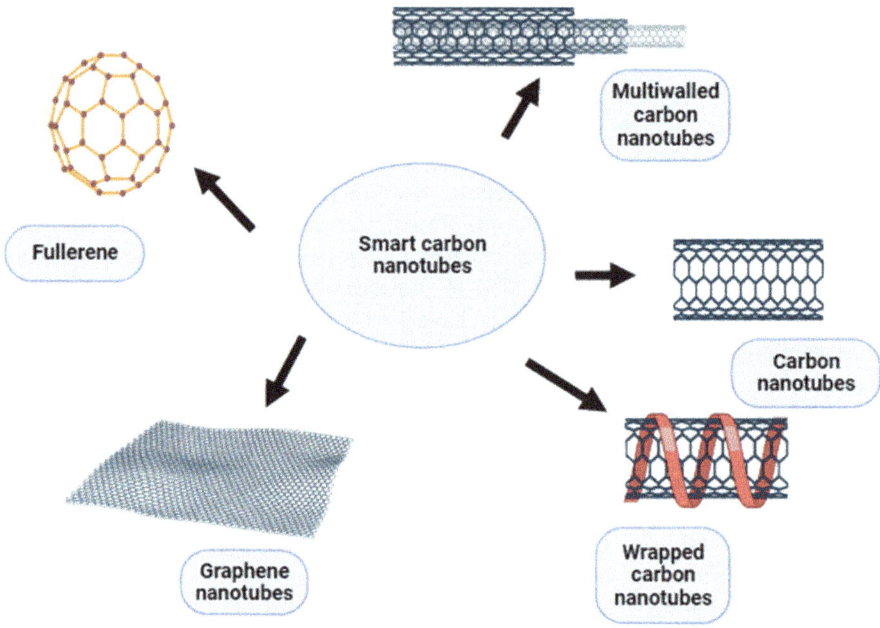

Fig. 1 Different types of smart carbon nanotubes

3 Carbon Nanotubes

Hollow, round tubes are known as carbon nanotubes (CNTs). Single-walled carbon nanotubes (SWCNTs) and multiple-walled carbon nanotubes (MWCNTs) are the two categories that they fall under based on the number of layers [28]. The SWCNT is a piece of graphite that is 1–2 nm in diameter that rolls out, while several SWCNTs are composed of central cylinder-like structures called MWCNTs with a diameter of 0.34–0.39 nm slightly larger than the value of graphite, which is 0.335 nm. The type, functionalization, and concentration of CNTs can set toxicological and physiological effects in plants. CNTs can regulate plant growth and enhance photosynthetic pigments, causing increased photosynthetic capacity, and CNTs can also enhance cell metabolism, hence enzymatic activity [29]. CNTs can enhance plant productivity in hydroponics as well as in soil conditions. According to Y. Gonzales-Garcia et al., CNTs at high levels in plants can be harmful, and their penetration into plants can cause reactive oxygen species (ROS) production. Non-enzymatic antioxidants may include glutathione, flavonoids, carotenoids, phenolic acids, alkaloids, carotenoids, non-protein amino acids, etc., and enzymatic antioxidant systems include ascorbate peroxidase (APX), glutathione reductase (GR), superoxide dismutase (SOD), mono dehydroascorbate reductase (MDHAR), dehydroascorbate reductase (DHAR), catalase (CAT), and glutathione peroxidase (GPX). To neutralize ROS, plants may become unstressed from stress [30]. During stressful conditions, extra production of ROS diminishes the cause of cellular damage, leading to programmed cell death, and decreases plant productivity. As fullerene and CNTs in Nicotiana benthamiana plants prevent TMV and tobacco mosaic virus (TMV), antioxidant enzyme activity and photosynthesis enhance [31] CNTs which can also boost disease tolerance. According to Y. Gonzales-Garcia's work on employing CNT, the GPX enzyme increased, suppressing the H_2O_2 and preventing cell wall damage. Also, GPX helped regulate cellular redox homeostasis. According to a study by Zhou et al. [32] titled "Increase in the active ingredients of traditional Chinese medicine Isatis indigotica through iron nanoparticle supplementation versus carbon nanotubes: a comparative study (2022)", CNTs did not cause significant malondialdehyde (MDA) buildup in roots at all exposure doses which suggests oxidative stress, but did so in shoots [33]. In shoots, CNTs decreased peroxidase (POD activity but not superoxide dismutase (SOD) or catalase (CAT) activity, however, in roots, CNTs decreased all enzyme activities, indicating a general decrease in the antioxidant capacity of plants.

4 The Importance of Smart Carbon Nanomaterials as an Antioxidant

Smart carbon nanomaterials also play an important role as plant antioxidants, which have several important benefits for agriculture applications and plant health (Table 1). Reactive oxygen species (ROS) from various metabolic processes and environmental stress can induce plant oxidative damage [34]. Smart carbon nanomaterials can regulate plant ROS levels by eliminating these harmful molecules, reducing oxidative stress, and preventing cell damage. Carbon nanomaterials such as carbon nanotubes (CNTs) and graphite can improve plants' resilience to biological stress factors such as salinity, drought, and exposure to heavy metals [35]. These nanomaterials can lessen the oxidative damage brought on by stressors and assist plants in surviving and thriving under challenging circumstances. Carbon nanomaterials can be designed to improve plant photosynthetic efficiency. They can act as electron carriers and catalysts for plant photosynthetic reactions by increasing carbon assimilation and plant growth. Carbon nanomaterials can be functionalized to improve plant nutrient absorption [36]. Compounds are known as antioxidants that reduce the oxidation process and have a much lower concentration than the preserved substrate. Primary antioxidants act as elimination or chain-breaking antioxidants and delay the initiation or interruption of the propagation process. The secondary antioxidants remove single oxygen, decompose non-radioactive peroxidases, chelate peroxide metal ions, and inhibit oxidation enzymes. Based on the reactiveness of antioxidants, four defenses have been described, prevention antioxidants, repair antioxidants, radical scavengers, and antioxidants based on adaptation mechanisms [37]. Antioxidants are the first line of defense to prevent the generation of red blood cells. The second line of defense mainly includes radical antioxidants. The third line of defense, consisting of antioxidant repair, intervenes after the insults caused by free radicals. The adaptation mechanism, the mechanism of action of the fourth line of defense, and the signal needed to produce free radicals are used by antioxidants, which can disturb the occurrence of free radicals that involve the intervention of the radical species [38]. Well-known endogenous antioxidants include superoxide dismutase, catalase, glutathione peroxidase, and metal-binding proteins such as lactoferrin, ferritin, Calum plasmin, uric acid, glutathione, bilirubin, and melatonin. Exogenous antioxidants are present in food and food supplements known as tocopherol, phenolic acid, carotene, and vitamin C. They slow down the endogenous antioxidants, so cell antioxidant profiles remain unchanged [39]. Gallic acid esters, synergistic hydroxy anisole hydroxy, and hydroxytoluene hydroxy, also called synthetic antioxidants, are added to foods to prevent acidity. The importance of smart carbon-based nanomaterials with antioxidation capabilities has been studied, including carbon particles, carbon sheet structures, clusters of carbon, and carbon nanotubes [40]. Smart carbon nanomaterials can be used as carriers for delivering fungicides, pesticides, and other agricultural chemicals to plants. This delivery system can reduce the chemicals required and environmental contamination, protecting plants from pests and disease (Figs. 2 and 3).

Carbon nanomaterials are generally considered more environmental friendly than traditional chemical fertilizer pesticides [41]. The use of substances reduces the environmental impact of agricultural practices and contributes to sustainable agriculture. Smart carbon nanomaterials are known for their stability and durability, ensuring long-term antioxidant effects in plants and providing permanent protection against

Table 1 Role of NMs in plant system

NMs	Specific role	Key references
CNTs	Regulate biological processes	[47]
MWCNTs	Accumulation of ROS	[48]
SWCNTs	Increased the activity of antioxidants during cryopreservation	[49]
Fullerene	Cluster cell contact was underpinned by fullerenol cytotoxicity to restrict the permeable cell walls	[50]
Graphene	Reduces the growth and increases the concentration of ROS in the cabbage and tomato plants	[51]
Fullerol	Improvement in seed germination and photosynthesis at low concentration	[1]
Oxidized MWCNTs	Plant shoot and root length increase when treated with oxidized MWCNTs	[52]
Single-walled carbon nanohorns (SWCNHs)	Plant seed exposed to SWCNHs have a higher germination rate	[53]
SWCNTs are not functionalized	Plants exposed to SWCNTs increase root length	[54]
SWCNTs and functionalized SWCNTs	SWCNTs enhance root elongation in onion plant	[55]

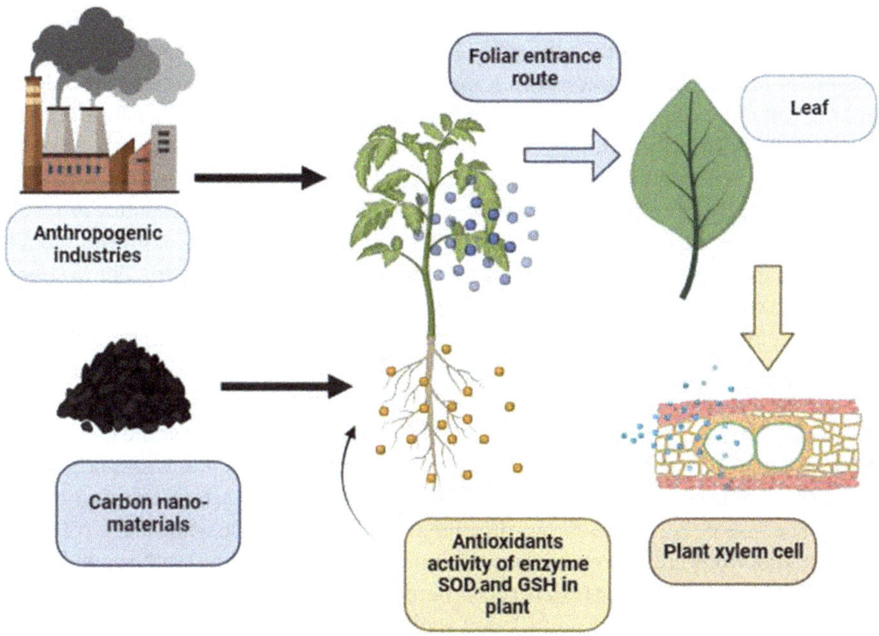

Fig. 2 Exposure of nanomaterials to enhance plant growth and their activity as antioxidants enzymes in plants

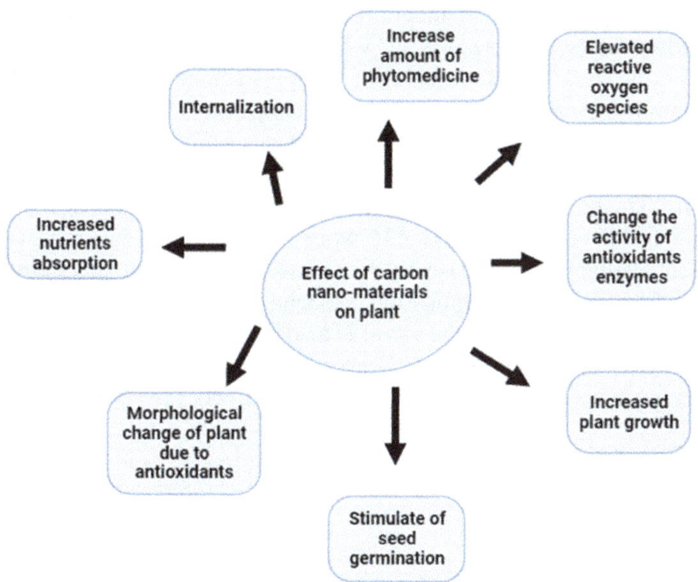

Fig. 3 Effect of carbon nanomaterials on plant growth and antioxidant activity

oxidative stress. The characteristics of carbon nanomaterials allow plants to create precise treatments specific to the site. The antioxidant properties of smart carbon nanomaterials can help reduce crop losses due to oxidative stress-related problems, such as follicles, reduced yields, or fruit destruction. Smart carbon nanomaterials can be divided into four dimensions: zero-dimension, one dimension, two dimensions, and three dimensions [42]. The zero-dimensional fullerene is an antioxidant molecule, and different fullerenes' unique structure and sorption activities are well-studied for their antioxidant activities to OH and O^2 molecules. Combining double bonds and the low-occupied molecular orbits can be responsible for the excellent antioxidation ability. Fullerene derivatives, including fullerenol, carboxy-modified fullerenol, ethylene diamine-modified fullerenol, and metal-modified fullerenol, generally exhibit good water solubility [43]. One-dimensional carbon nanotubes (CNTs) are linear cylinders of graphite sheets with distinct integral properties that are divided into single and multiple-wall CNTs of 0.4 to 2.5 nm and some nanometers to 100 nm in diameter [44]. CNTs are believed to be a fundamental waste because of their exceptional electron donation and the acceptability of carbon-splitting double bonds. The researchers have shown that intact CNTs with single walls and CNTs with phenolic functions are powerful antioxidants [45]. One of the few studies concerning fullerene-free derivatives of fullerene has demonstrated that the inhalation and assimilation of C-60 malonate tri-adduct into roots and aerial parts of test plants (wheat, barley, tomato) optimize the function of the antioxidant system and catabolic cyclic process to promote plant growth in the initial germination stage and the vegetative period [46] (Table 2).

In another study, Panova et al. mentioned that root and foliar application of amino acid derivatives of fullerene to spring wheat and barley growth and net productivity under non-stress conditions enhance plant resistance through modulating efficiency of photosynthesis and antioxidant effects against UV-B radiation stress [71]. Numerous studies have shown that carbon-based nanoparticles have an advantageous effect on plant growth, which has sparked research into fertilizers incorporating nanocarbons. The bulk of these fertilizers is built on adding nanocarbon to mineral and organic fertilizers, which typically work as a fertilizer synergist to increase plant nutrient availability, decrease nutrient losses, and promote plant development. Regardless of the method of application, especially when using CNTs, the application of CNMs positively altered the ascorbic acid level [72]. The highest levels of this molecule were produced in foliar application action at doses of 10, 50, and 100mg L1 of CNT, reaching up to 59, 63, and 57%, respectively [25] (Fig. 4).

Table 2 Inhibitory and promoter effects of pristine Fullerene and their water-soluble derivatives on plant systems

S.no	Fullerenes	Size (nm)	Plant species	Effects	References
1	Fullerene Soot	20–150	Arabidopsis	Net growth and a slight reduction in biomass, both of which were caused by the transcription of stress-related genes involved in electron transport and energy processes	[56]
2	C-60	1450–1900	Corn, soybean	decrease in biomass and net expansion	[57]
3	C-60	1450–1900	Rice	Reduced shoot height, root length and cortical cell diameter	[58]
4	C-60	30–60	Rice	rice growth and phytohormone levels are inhibited when given a high dose of medication	[59]
5	nC-60	122.4–295.3	Wheat	Roots inner walls and pore structures are blocked, and root cells and tissue structure are compressed and damaged	[60]
6	13C-label C60	122.4–295.3	Rice	regulation of some phytohormones chemical composition	[61]
7	C70(C(COOH)2)4.8	146	Tobacco cell	preventing cellular growth, viability, and mitochondrial function, generating ROS and disrupting cytoskeleton organization	[50]
8	C70(C(COOH)2)2.4	100–200	Tobacco cell	Changes in the chemical makeup of the cell wall, rupture of the cell wall, and membrane stimulation-sensitive intracellular ROS generation	[62]

(continued)

Table 2 (continued)

S.no	Fullerenes	Size (nm)	Plant species	Effects	References
9	C70-NOM	1.12–100	Onion cell	Obstructing the plant's permeable cell wall causes cell loss	[63]
10	C60(OH)20	100–700	Bitter melon	Better biomass, plant biomass, and fruit yields	[64]
11	C60(OH)22·8H2O	138	Maize	Seedlings have improved germination and drought resistance	[65]
12	C60(OH)24–26	NA	Arabidopsis	Lengths of hypocotyls of Arabidopsis	[66]
13	C60(NCH2CH2NH3 + ·CF3COO −)4·10H2O	55–155	Maize, Arabidopsis	Improve the expression of proteins related to energy and stress to accelerate growth under normal and drought stress	[67]
14	C60(OH)24	50–70	Sugar beet	Improvements intercellular water supplies to the alleviated water	[68]
15	C60(OH)24	4–11	Wheat	Improved growth, longer stem length, and fungus disease resistance	[69]
16	C60	20–40	Pharmasaat	Long leaves, healthier and deeper green leaves, the number of flowers	[70]

5 Use of Smart Carbon Nanomaterials as Antioxidants in Different Studies

- During cryopreservation, SWCNTs markedly boosted the antioxidant system's activity. SWCNTs significantly boosted the activities of enzymatic antioxidants, especially POD and CAT activities [73].
- The SOD activity did not change much. SWCNT treatment resulted in a significant increase in POD activity compared to the control method. The non-enzymatic antioxidant GSH in the SWCNTs group also nearly doubled compared to the control group when dehydrated [49].

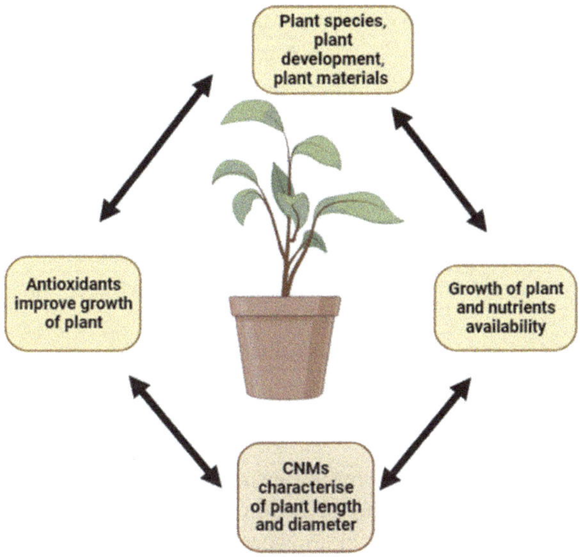

Fig. 4 Carbon nanomaterials impact on plant development, production

- In grape seedlings, MWCNT carbon nanotubes reduced salt stress and increased antioxidant activity. The application of MWCNTs to grape seedlings delayed the loss of relative water content in the roots of the plants, decreased MDA production and cell membrane damage, and increased the activity of antioxidant enzymes in the leaves of the plants, creating the ideal environment for scavenging intracellular oxygen free radicals [74].
- Due to their low toxicity and distinctive structure, graphene quantum dots (GQDs), a typical representative of CDs, are the subject of extensive research [75].
- Researchers have created GQDs with antioxidant characteristics to scavenge free radicals and protect cells from oxidative damage efficiently. They claim that surface flaws and unpaired electrons decreased ROS. In the acute kidney injury paradigm, GQDs were rationally created and demonstrated to be superior antioxidants (Tables 3 and 4).

6 Future Directions of Smart Carbon Nanomaterials as an Antioxidant

The antioxidative ability of antioxidants can be enhanced by encapsulating them into carbon nanocarriers or attaching them to the surface of nanomaterials. The future of carbon nano antioxidants should have some special characteristics, such as: -

- having a specific targeted group with the ability to selectively reach the sick organ or tissue

Table 3 A summary of the biphasic effects of MWCNTs on plant systems

Plant species	Effect on MWCNTs on plant	References
Amaranthus tricolor L	Induced plant growth reduces toxicity	[76]
Wheat and rapeseed	0.005% of MWCNT is absorbed by plant roots and transferred to the leaves	[77]
Tomato	Increasing Biomass and activating many stresses related genes	[78]
Mustard	Promoting plant germination rate	[52]
Wheat	Increase the number of cereals, Biomass, and stomatal density	[79]
Alfalfa and wheat	Increasing root elongation	[80]
Rice	Enhance the ROS activity	[58]

- The antioxidant monomer contained within nanocarriers will not be released during long-term circulation before they reach the targeted site [88].
- Has a high antioxidant capacity.
- In vitro or in vivo cytotoxicity should not exist for the nano-antioxidant.
- Developing microreactor designs that have good reaction synthesis control.
- Making nano-antioxidants more biocompatible, dependable, resilient, and stable in biological microenvironments [89]
- Streamlining the fabrication procedures for producing extremely effective nano-antioxidants.
- Low concentrations of CNMs (e.g., seed coatings) applied in the early stages of plant growth have a useful effect on the environment and edible plant segments, with a limited introduction providing this use from the point of view of nanosafety (Fig. 5).

- The exposure of nanomaterials to other toxic pollutants can increase or reduce the risks.
- Except at extremely high concentrations, which are environmentally unrealistic, NMs are not hazardous to plant safety or agronomic qualities, including harvesting and nutritional quality [90] (Fig. 6).

Table 4 A summary of the effects of SWCNTs on plant systems

Name	Plant species	Effect	References
Non functionalized SWCNTs			
	Cucumber, onion and tomato	To inhibit the expansion of the root elongation in tomato and improve the expansion of the root in onions and cucumbers	[35]
Functionalized SWCNTs	Lettuce, cabbage, carrots	The length of the lettuce roots is inhibited, and the cabbage and carrots are not affected	[35]
SWCNTs	Maize	Maize seminal root growth increase	[81]
SWCNTs and MWCNTs	Rice	Accelerate the growth of leaf and flowering rice seeds with 20 mg per liter	[82]
SWCNTs	Pea	Higher doses have been shown to adversely affect photosynthesis	[83]
SWCNTs	Arabidopsis	The trichome clusters on the surface are less than 50 mg per liter in the stimulation of plant cell	[84]
SWCNTs	Onion, tomato, turnip and radish	Improve the growth of tomatoes, onions, radish, and turnips with 10–40 mg per liter have harmful and toxic effects on the germination of onions and radish seed with 40 mg per liter increase the dry weight of tomatoes and shoots of all concentration 10–40 mg per liter	[85]
SWCNTs	Arabidopsis and rice	Show adverse effect like cell aggregation, chromatin condensation	[86]
SWCNTs	Pea	Low concentrations of 10 mg per liter do not affect leaf characteristics, but higher doses cause damage to photosynthesis	[83]
SWCNTs	Thymus digenesis	50 mg per liter provides growth, boosting fresh and dry biomass growth, increasing phenol content, and antioxidant activity	[87]

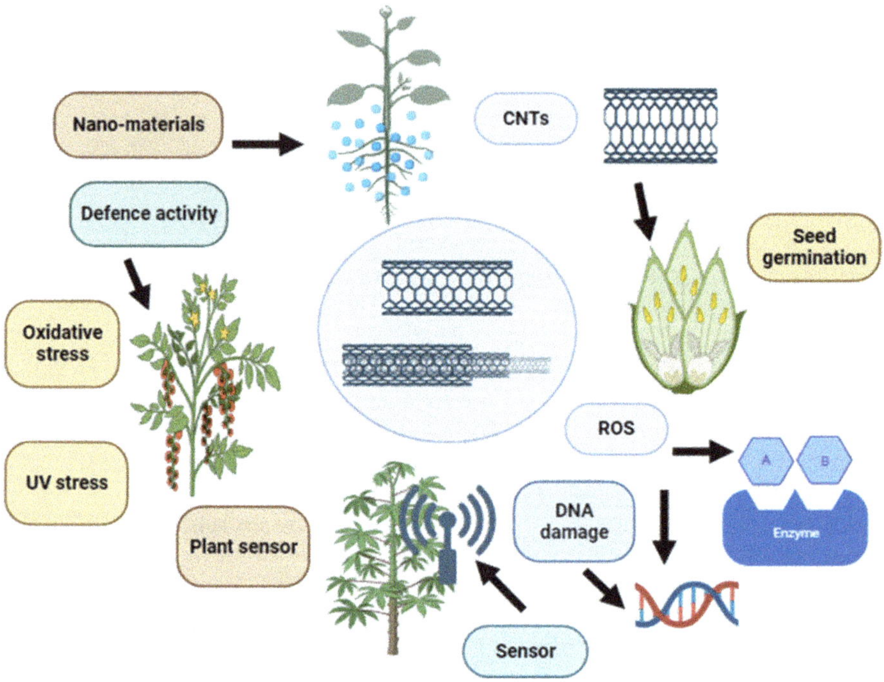

Fig. 5 CNTs enhance plant growth and are employed as biosensors, and nanomaterials' shield plants by defensive mechanisms

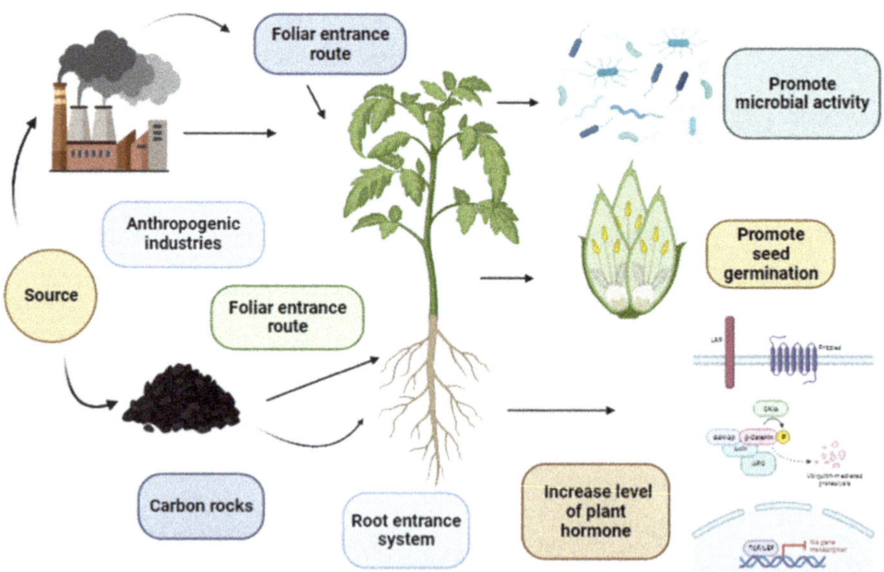

Fig. 6 Carbon nanotubes as impact on plant growth regulator

7 Conclusion

Carbon-based nanomaterials have been used as the basis for broad research efforts guided by the most important classes of nanomaterials. CBNs having multiple properties, known as chemical, mechanical, optical, and electrical, have been present in multiple areas. CBNs could be used to affect the antioxidants in the plant system. Graphene and CNTs are used in the development of nanodevices. Nanodevices are called nanometer oscillators and nano resonators. Smart carbon nanomaterials such as fullerenols, derivatives of nanosized water-soluble polyhydroxylated fullerene, are bioactive compounds used in drug development. Carbon nanomaterials have demonstrated great prospects for various plant science and agriculture applications.

Further research can lead to innovative solutions to improve plant growth and nutrient absorption. Research on the use of carbon nanomaterials, such as CNTs or graphene, acts as a carrier of nutrients and fertilizer. These materials can help control the release of nutrients from plants and improve the absorption and utilization of nutrients. Using carbon nanomaterials to improve plant stress tolerance includes salinity, drought, and temperature stress. We research the potential of carbon nanomaterials in developing new approaches to diseases and pest management in agriculture. CNMs improve soil quality and fertility; these materials can maintain humidity, improve soil structure, and promote beneficial microorganisms in the rhizosphere. The impact of CNMs on seed coatings on germination rate, early growth rate, and harvest yield is examined. Further study of CNMs has a great potential environmental impact on agriculture. Assess the long-term effects on soil health, water quality, and non-targeted organisms. To assess risk and characterize potential uses, conducting a more thorough and systematic investigation of the critical elements that influence how different carbon nanomaterials interact with living things and the environment will be crucial. More research is required to enable the safer use of CNTs in the advancement of sustainable agriculture. Further research is necessary to determine whether CNTs may affect plant gene expression. Because CNT-contaminated food products can introduce CNTs into the human body through ingestion, it is crucial to explore further the safety of eating plant parts exposed to CNT contamination. Monoculture tests must also be carried out to evaluate the cytotoxic effects of CNTs on soil microorganisms. Therefore, their toxicity must be evaluated before CNTs are commercially used for agriculture applications. Reliable and efficient methodologies should be recommended to analyze and reduce the Eco-toxicological effects of CNTs.

References

1. Husen A, Siddiqi KS (2014) Plants and microbes assisted selenium nanoparticles: characterization and application. J Nanobiotechnology 12(1):1–10
2. Zhu L, Chen L, Gu J, Ma H, Wu H (2022) Carbon-based nanomaterials for sustainable agriculture: their application as light converters, nanosensors, and delivery tools. Plants 11(4):511
3. Husen A (2020) Carbon-based nanomaterials and their interactions with agricultural crops. In: Husen A, Jawaid M (eds) Elsevier Inc. 50 Hampshire St., 5th Floor, Cambridge, MA 02139, USA, pp 199–218 https://doi.org/10.1016/B978-0-12-817852-2.00008-1
4. Husen, A (2022) Engineered nanomaterials for sustainable agricultural production, soil improvement and stress management. Elsevier Inc., 50 Hampshire St., 5th Floor, Cambridge, MA 02139, USA
5. Kim T, Sridharan I, Zhu B, Orgel J, Wang R (2015) Effect of CNT on collagen fiber structure, stiffness assembly kinetics and stem cell differentiation. Mater Sci Eng C 49:281–289
6. Mehmood A, Mubarak NM, Khalid M, Walvekar R, Abdullah EC, Siddiqui MTH, Mazari S (2020) Graphene-based nanomaterials for strain sensor application—a review. J Environ Chem Eng 8(3):103743
7. Anjum S, Ishaque S, Fatima H, Farooq W, Hano C, Abbasi BH, Anjum I (2021) Emerging applications of nanotechnology in healthcare systems: grand challenges and perspectives. Pharmaceuticals 14(8):707
8. Arya A, Husen A (2023) Smart nanomaterials in biosensing applications. In: Husen A, Siddiqi KS (eds) Advances in smart nanomaterials and their applications. Elsevier Inc., 50 Hampshire St., 5th Floor, Cambridge, MA 02139, USA, pp 207–231. ISBN: 978–0–323–99546–7
9. Arya A, Tyagi PK, Kumar S, Husen A (2023) Nanomaterials and their application in microbiology disciplines. In: Husen A, Siddiqi KS (eds) Advances in smart nanomaterials and their applications. Elsevier Inc., 50 Hampshire St., 5th Floor, Cambridge, MA 02139, USA, pp 175–194. ISBN: 978–0–323–99546–7
10. Belay T, Worku LA, Bachheti RK, Bachheti A, Husen A (2023) Nanomaterials: Introduction, synthesis, characterization and applications. In: Husen A, Siddiqi KS (eds) Advances in smart nanomaterials and their applications. Elsevier Inc., 50 Hampshire St., 5th Floor, Cambridge, MA 02139, USA, pp 1–21. ISBN: 978–0–323–99546–7
11. Chauhan AK, Singh SP, Yadav B, Khatri S, Husen A (2023) Smart nanomaterials and control of biofilms. In: Husen A, Siddiqi KS (eds) Advances in smart nanomaterials and their applications. Elsevier Inc., 50 Hampshire St., 5th Floor, Cambridge, MA 02139, USA, pp 109–122. ISBN: 978–0–323–99546–7
12. Husen A, Siddiqi KS (2023) Advances in smart nanomaterials and their applications. Elsevier Inc., 50 Hampshire St., 5th Floor, Cambridge, MA 02139, USA
13. Mondal R, Shand H, Kumar A, Sellami H, Ghorai S, Mandal AK, Husen A (2023) Lipid-based cubosome nanoparticle mediated efficient and controlled vesicular drug delivery for cancer therapy. In: Husen A, Siddiqi KS (eds) Advances in smart nanomaterials and their applications. Elsevier Inc., 50 Hampshire St., 5th Floor, Cambridge, MA 02139, USA, pp 97–107. ISBN: 978–0–323–99546–7
14. Porwal P, Taghiyari HR, Husen A (2023) Use of nanomaterials in the forest industry. In: Husen A, Siddiqi KS (eds) Advances in smart nanomaterials and their applications. Elsevier Inc., 50 Hampshire St., 5th Floor, Cambridge, MA 02139, USA, pp 469–487. ISBN: 978–0–323–99546–7
15. Raja MA, Ahmad MA, Daniyal M, Husen A (2023) Management of wastewater and other environmental issues using smart nanomaterials. In: Husen A, Siddiqi KS (eds) Advances in smart nanomaterials and their applications. Elsevier Inc., 50 Hampshire St., 5th Floor, Cambridge, MA 02139, USA, pp 489–503. ISBN: 978–0–323–99546–7

16. Singh N, Gaur S, Chawla S, Singh S, Husen A (2023) Use of smart nanomaterials in food packaging. In: Husen A, Siddiqi KS (eds) Advances in smart nanomaterials and their applications. Elsevier Inc., 50 Hampshire St., 5th Floor, Cambridge, MA 02139, USA, pp 223–245. ISBN: 978–0–323–99546–7
17. Vanti G, Belur S, Husen A (2023) Use of nanomaterials in agricultural sectors. In: Husen A, Siddiqi KS (eds) Advances in smart nanomaterials and their applications. Elsevier Inc., 50 Hampshire St., 5th Floor, Cambridge, MA 02139, USA, pp 445. ISBN: 978–0–323–99546–7
18. Worku LA, Deepti, Nigussie Y, Bachheti A, Bachheti RK, Husen A (2023) Antimicrobial activities of nanomaterials. In: Husen A, Siddiqi KS (Ed) Advances in Smart Nanomaterials and their Applications. Elsevier Inc., 50 Hampshire St., 5th Floor, Cambridge, MA 02139, USA, pp 127–148. ISBN: 978–0–323–99546–7
19. Jeevanandam J, Barhoum A, Chan YS, Dufresne A, Danquah MK (2018) Review on nanoparticles and nanostructured materials: history, sources, toxicity and regulations. Beilstein J Nanotechnol 9(1):1050–1074
20. Mishra N, Jiang C, Chen L, Paul A, Chatterjee A, Shen G (2023) Achieving abiotic stress tolerance in plants through antioxidative defense mechanisms. Front Plant Sci 14:1110622
21. Garg P, Attri P, Sharma R, Chauhan M, Chaudhary GR (2022) May Advances and perspective on antimicrobial nanomaterials for biomedical applications. Front Nanotechnol 3(4):898411
22. Baig S, Ahmed M, Batool A, Bashir A, Mumtaz S, Ikram M, Ikram M (2022). Introductory chapter: brief scientific description to carbon allotropes-technological perspective
23. Joudeh N, Linke D (2022) Nanoparticle classification, physicochemical properties, characterization, and applications: a comprehensive review for biologists. J Nanobiotechnology 20(1):262
24. Adorinni S, Rozhin P, Marchesan S (2021) Smart hydrogels meet carbon nanomaterials for new frontiers in medicine. Biomedicines 9(5):570
25. Zaytseva O, Neumann G (2016) Carbon nanomaterials: production, impact on plant development, agricultural and environmental applications. Chem Biol Technol Agric 3(1):1–26
26. Saifuddin N, Raziah AZ, Junizah AR (2013) Carbon nanotubes: a review on structure and their interaction with proteins. J Chem
27. Malhotra N, Audira G, Castillo AL, Siregar P, Ruallo JMS, Roldan MJ, Hsiao CD (2021) An update report on the biosafety and potential toxicity of fullerene-based nanomaterials toward aquatic animals. Oxid Med Cell Longev
28. Solorio-Rodriguez SA, Williams A, Poulsen SS, Knudsen KB, Jensen KA, Clausen PA, Halappanavar S (2023) Single-walled vs. multi-walled carbon nanotubes: influence of physico-chemical properties on toxicogenomics responses in mouse lungs. Nanomaterials 13(6):1059
29. Rahmani N, Radjabian T, Soltani BM (2020) Impacts of foliar exposure to multi-walled carbon nanotubes on physiological and molecular traits of Salvia verticillata L., as a medicinal plant. Plant Physiol Biochem 150:27–38
30. Irato P, Santovito G (2021) Enzymatic and non-enzymatic molecules with antioxidant function. Antioxidants 10(4):579
31. Adeel M, Farooq T, White JC, Hao Y, He Z, Rui Y (2021) Carbon-based nanomaterials suppress tobacco mosaic virus (TMV) infection and induce resistance in Nicotiana benthamiana. J Hazard Mater 404:124167
32. Zhou P, Long B, Wang R, Jiang Y, Zhao W, Li Y, Li M, Guo Z, Zhang P, Rui Y, Lynch I (2022) Increase in the active ingredients of traditional Chinese medicine Isatis indigotica through iron nanoparticles supplementation versus carbon nanotubes: a comparative study. Environ Sci 9(8):2966–2978
33. Cordiano R, Di Gioacchino M, Mangifesta R, Panzera C, Gangemi S, Minciullo PL (2023) Malondialdehyde as a potential oxidative stress marker for allergy-oriented diseases: an update. Molecules 28(16):5979
34. Hasanuzzaman M, Bhuyan MB, Zulfiqar F, Raza A, Mohsin SM, Mahmud JA, Fotopoulos V (2020) Reactive oxygen species and antioxidant defense in plants under abiotic stress: revisiting the crucial role of a universal defense regulator. Antioxidants 9(8):681

35. Safdar M, Kim W, Park S, Gwon Y, Kim YO, Kim J (2022) Engineering plants with carbon nanotubes: a sustainable agriculture approach. J. Nanobiotechnology 20(1):1–30
36. Wu H, Li Z (2022) Recent advances in nano-enabled agriculture for improving plant performance. Crop J 10(1):1–12
37. Pisoschi AM, Pop A, Iordache F, Stanca L, Bilteanu L, Serban AI (2021) Antioxidant determination with the use of carbon-based electrodes. Chemosensors 9(4):72
38. Kurutas EB (2015) The importance of antioxidants which play the role in cellular response against oxidative/nitrosative stress: current state. Nutr J 15(1):1–22
39. Bouayed J, Bohn T (2010) Exogenous antioxidants—double-edged swords in cellular redox state: health beneficial effects at physiologic doses versus deleterious effects at high doses. Oxid Med Cell Longev 3(4):228–237
40. Holmannova D, Borsky P, Svadlakova T, Borska L, Fiala Z (2022) Carbon nanoparticles and their biomedical applications. Appl Sci 12(15):7865
41. Hazarika A, Yadav M, Yadav DK, Yadav HS (2022) An overview of the role of nanoparticles in sustainable agriculture. Biocatal Agric Biotechnol 43:102399
42. Byakodi M, Shrikrishna NS, Sharma R, Bhansali S, Mishra Y, Kaushik A, Gandhi S (2022) Emerging 0D, 1D, 2D, and 3D nanostructures for efficient point-of-care biosensing. Biosens BioelectronX 12:100284
43. Sitko R, Musielak M, Serda M, Talik E, Gagor A, Zawisza B, Malecka M (2021) Graphene oxide decorated with fullerenol nanoparticles for highly efficient removal of Pb (II) ions and ultrasensitive detection by total-reflection X-ray fluorescence spectrometry. Sep Purif Technol 277:119450
44. Rahman G, Najaf Z, Mehmood A, Bilal S, Shah AUHA, Mian SA, Ali G (2019) An overview of the recent progress in the synthesis and applications of carbon nanotubes. C 5(1):3
45. Ahmadi-Majd M, Mousavi-Fard S, Rezaei Nejad A, Fanourakis D (2022) Carbon nanotubes in the holding solution stimulate flower opening and prolong vase life in carnation. Chem Biol Technol Agric 9(1):15
46. Wu Q, Fan C, Wang H, Han Y, Tai F, Wu J, He R (2023).Biphasic impacts of graphite-derived engineering carbon-based nanomaterials on plant performance: effectiveness vs. nanotoxicity. Adv Agrochem
47. Aslani F, Bagheri S, Muhd Julkapli N, Juraimi AS, Hashemi FS, Baghdadi A (2014) Effects of engineered nanomaterials on plants growth: an overview. Sci World J
48. Tan XM, Lin C, Fugetsu B (2009) Studies on toxicity of multi-walled carbon nanotubes on suspension rice cells. Carbon 47(15):3479–3487
49. Ren L, Deng S, Chu Y, Zhang Y, Zhao H, Chen H, Zhang D (2020) Single-wall carbon nanotubes improve cell survival rate and reduce oxidative injury in cryopreservation of Agapanthus praecox embryogenic callus. Plant Methods 16(1):1–12
50. Dang S, Liu Q, Zhang X, He K, Wang C, Fang X (2012) Comparative cytotoxicity study of water-soluble carbon nanoparticles on plant cells. J Nanosci Nanotechnol 12(6):4478–4484
51. Begum P, Ikhtiari R, Fugetsu B (2011) Graphene phytotoxicity in the seedling stage of cabbage, tomato, red spinach, and lettuce. Carbon 49(12):3907–3919
52. Mondal A, Basu R, Das S, Nandy P(2011) Beneficial role of carbon nanotubes on mustard plant growth: an agricultural prospect. J Nanoparticle Res 13:4519–4528
53. Lahiani MH, Chen J, Irin F, Puretzky AA, Green MJ, Khodakovskaya MV (2015) Interaction of carbon nanohorns with plants: uptake and biological effects. Carbon 81:607–619
54. Cano AM, Kohl K, Deleon S, Payton P, Irin F, Saed M, Shah SA, Green MJ, Canas-Carrel JE (2016) Determination of uptake, accumulation, and stress effects in corn (*Zea mays* L.) grown in single-wall carbon nanotube contaminated soil. 152:117–122
55. Flores D, Chacón R, Alvarado L, Schmidt A, Alvarado C, Chaves J (2014) Effect of using two different types of carbon nanotubes for blackberry (Rubus adenotrichos) in vitro plant rooting, growth and histology. Am J Plant Sci 5(24):3510
56. Landa P, Vankova R, Andrlova J, Hodek J, Marsik P, Storchova H, White JC, Vanek T (2012) Nanoparticle-specific changes in Arabidopsis thaliana gene expression after exposure to ZnO, TiO2, and fullerene soot. J Hazard MaterNov 30(241):55–62

57. De La Torre-Roche R, Hawthorne J, Deng Y, Xing B, Cai W, Newman LA, Wang Q, Ma X, Hamdi H, White JC (2013) Multiwalled carbon nanotubes and C60 fullerenes differentially impact the accumulation of weathered pesticides in four agricultural plants. Environ Sci Technol 47(21):12539–12547
58. Hao Y, Ma C, Zhang Z, Song Y, Cao W, Guo J, Zhou G, Rui Y, Liu L, Xing B (2018) Carbon nanomaterials alter plant physiology and soil bacterial community composition in a rice-soil-bacterial ecosystem. Environ Pollu 232:123–136
59. Hao Y, Ma C, White JC, Adeel M, Jiang R, Zhao Z, Rao Y, Chen G, Rui Y, Xing B (2020) Carbon-based nanomaterials alter the composition of the fungal endophyte community in rice (Oryza sativa L.). Environ Sci Nano 7(7):2047–2060
60. He A, Jiang J, Ding J, Sheng GD (2021) Blocking effect of fullerene nanoparticles (nC60) on the plant cell structure and its phytotoxicity. Chemosphere 278:130474
61. Guo KR, Adeel M, Hu F, Xiao ZZ, Wang KX, Hao Y, Rui YK, Chang XL (2021) Absorption of carbon-13 labelled fullerene (C60) on rice seedlings and effect of phytohormones on growth. J Nanosci Nanotechnol 21(6):3197–3202
62. Liu Q, Zhang X, Zhao Y, Lin J, Shu C, Wang C, Fang X (2013) Fullerene-induced increase of glycosyl residue on living plant cell wall. Environ Sci Technol 47(13):7490–7498
63. Chen R, Ratnikova TA, Stone MB, Lin S, Lard M, Huang G, Hudson JS, Ke PC (2010) Differential uptake of carbon nanoparticles by plant and mammalian cells. Small 6(5):612–617
64. Kole C, Kole P, Randunu KM, Choudhary P, Podila R, Ke PC, Marcus RK (2013) Nanobiotechnology can boost crop production and quality: first evidence from increased plant biomass, fruit yield and phytomedicine content in bitter melon (Momordica charantia). BMC Biotechnol (1):1–10
65. Liu FY, Xiong FX, Fan YK, Li J, Wang HZ, Xing GM, He R (2016) Facile and scalable fabrication engineering of fullerenol nanoparticles by improved alkaline-oxidation approach and its antioxidant potential in maize. J Nanopart Res 18:1–13
66. Gao J, Wang Y, Folta KM, Krishna V, Bai W, Indeglia P, Moudgil B (2011) Polyhydroxy fullerenes (fullers or fullerenols): beneficial effects on growth and lifespan in diverse biological models. PLoS ONE 6(5):e19976
67. Liu Y, Wang T, Cao J, Zang Z, Wu Q, Wang H, He R (2019) Quaternary ammonium salts of imino fullerenes: fabrication and effect on seed germination. J Agric Food Chem 67(49):13509–13517
68. Borišev M, Borišev I, Župunski M, Arsenov D, Pajević S, Ćurčić Ž, Djordjevic A (2016) Drought impact is alleviated in sugar beets (Beta vulgaris L.) by foliar application of fullerenol nanoparticles. PLoS One 11(11):0166248
69. Kovac T, Marcek T, Šarkanj B, Borišev I, Ižakovic M, Jukic K, Krska RF (2021) C60 (OH) 24 nanoparticles and drought impact on wheat (Triticum aestivum L.) during growth and infection with Aspergillus flavus. J Fungi 7:236
70. Ahmadi SZ, Ghorbanpour M, Aghaee A, Hadian J (2020) Deciphering morpho-physiological and phytochemical attributes of Tanacetum parthenium L. plants exposed to C60 fullerene and salicylic acid. Chemosphere 259:127406
71. Panova GG, Semenov KN, Zhuravleva AS, Khomyakov YV, Volkova EN, Mirskaya GV, Udalova OR (2023) Obtaining vegetable production enriched with minor micronutrients using fullerene derivatives. Hortic 9(7):828
72. Zhao F, Xin X, Cao Y, Su D, Ji P, Zhu Z, He Z (2021) Use of carbon nanoparticles to improve soil fertility, crop growth and nutrient uptake by corn (Zea mays L.). Nanomater 10:2717
73. Hossen S, Sukhan ZP, Cho Y, Lee WK, Kho KH (2022) Antioxidant activity and oxidative stress-oriented apoptosis pathway in saccharides supplemented cryopreserved sperm of pacific abalone, Haliotis discus hannai. Antioxidants 11(7):1303
74. Li Y, Liu M, Yang X, Zhang Y, Hui H, Zhang D, Shu J (2022) Multi-walled carbon nanotubes enhanced the antioxidative system and alleviated salt stress in grape seedlings. Sci Hortic 293:110698
75. Tian P, Tang L, Teng KS, Lau SP (2018) Graphene quantum dots from chemistry to applications. Mater. Today Chem. 10:221–258

76. Begum P, Ikhtiari R, Fugetsu B, Matsuoka M, Akasaka T, Watari F (2012) Phytotoxicity of multi-walled carbon nanotubes assessed by selected plant species in the seedling stage. Appl Surf Sci 262:120–124
77. Larue C, Pinault M, Czarny B, Georgin D, Jaillard D, Bendiab N, Mayne-L'Hermite M, Taran F, Dive V, Carrière M (2012) Quantitative evaluation of multi-walled carbon nanotube uptake in wheat and rapeseed. J Hazard MaterAug 15(227):155–163
78. Khodakovskaya MV, de Silva K, Nedosekin DA, Dervishi E, Biris AS, Shashkov EV, Galanzha EI, Zharov VP(2011) Complex genetic, photothermal, and photoacoustic analysis of nanoparticle-plant interactions. PNAS 108(3):1028–33
79. Joshi A, Sharma L, Kaur S, Dharamvir K, Nayyar H, Verma G (2020) Plant nanobionic effect of multi-walled carbon nanotubes on growth, anatomy, yield and grain composition of rice. BioNanoScienceJun 10:430–445
80. Miralles P, Johnson E, Church TL, Harris AT (2012) Multiwalled carbon nanotubes in alfalfa and wheat: toxicology and uptake. J R Soc 9(77):3514–3527
81. Yan S, Zhao L, Li H, Zhang Q, Tan J, Huang M, He S, Li L (2013) Single-walled carbon nanotubes selectively influence maize root tissue development accompanied by the change in the related gene expression. J Hazard Mater 15(246):110–118
82. Zhang H, Yue M, Zheng X, Xie C, Zhou H, Li L (2017) Physiological effects of single-and multi-walled carbon nanotubes on rice seedlings. IEEE T NANOBIOSCI 16(7):563–570
83. Velikova V, Petrova N, Kovács L, Petrova A, Koleva D, Tsonev T, Taneva S, Petrov P, Krumova S (2021) Single-walled carbon nanotubes modify leaf micromorphology, chloroplast ultrastructure and photosynthetic activity of pea plants. Int J Mol Sci 22(9):4878
84. Yuan H, Hu S, Huang P, Song H, Wang K, Ruan J, He R, Cui D (2011) Single-walled carbon nanotubes exhibit dual-phase regulation to exposed Arabidopsis mesophyll cells. Nanoscale Res Lett 6:1–9
85. Haghighi M, Teixeira da Silva JA (2014) The effect of carbon nanotubes on the seed germination and seedling growth of four vegetable species. JCSB 17:201–208
86. Shen CX, Zhang QF, Li J, Bi FC, Yao N (2010) Induction of programmed cell death in Arabidopsis and rice by single-wall carbon nanotubes. Am BotOct 97(10):1602–1609
87. Samadi S, Saharkhiz MJ, Azizi M, Samiei L, Karami A, Ghorbanpour M(2021) Single-wall carbon nano tubes (SWCNTs) penetrate Thymus daenensis Celak. Plant cells and increase secondary metabolite accumulation in vitro. Ind Crops Prod 165:113424
88. Singh AP, Biswas A, Shukla A, Maiti P (2019) Targeted therapy in chronic diseases using nanomaterial-based drug delivery vehicles. Signal Transduct Target Ther 4(1):33
89. Haghighi FH, Mercurio M, Cerra S, Salamone TA, Bianymotlagh R, Palocci C, Fratoddi I (2023) Surface modification of TiO 2 nanoparticles with organic molecules and their biological applications. J Mater Chem B
90. Mukherjee A, Majumdar S, Servin AD, Pagano L, Dhankher OP, White JC (2016) Carbon nanomaterials in agriculture: a critical review. Front Plant Sci 7:172
91. Afzal I, Javed T, Amirkhani M, Taylor AG (2020) Modern seed technology: seed coating delivery systems for enhancing seed and crop performance. Agriculture 10(11):526
92. Baun A, Sørensen SN, Rasmussen RF, Hartmann NB, Koch CB (2008) Toxicity and bioaccumulation of xenobiotic organic compounds in the presence of aqueous suspensions of aggregates of nano-C60. Aquat Toxicol 86(3):379–387
93. Cañas JE, Long M, Nations S, Vadan R, Dai L, Luo M, Ambikapathi R, Lee EH, Olszyk D (2008) Effects of functionalized and nonfunctionalized single-walled carbon nanotubes on root elongation of select crop species. Environ Toxicol Chem Int J 27(9):1922–1931
94. Long M, Nations S, Vadan R, Dai L, Luo M, Ambikapathi R, Lee EH, Olszyk D (2008) Effects of functionalized and nonfunctionalized single-walled carbon nanotubes on root elongation of select crop species. Environ Toxicol Chem Int J 27(9):1922–1931
95. Malik S, Muhammad K, Waheed Y (2023) Nanotechnology: a revolution in modern industry. Molecules 28(2):661
96. Nel A, Xia T, Madler L, Li N (2006) Toxic potential of materials at the nanolevel. Science 311(5761):622–627

97. Rentel MC, Lecourieux D, Ouaked F, Usher SL, Petersen L, Okamoto H, Knight H, Peck SC, Grierson CS, Hirt H, Knight MR (2004) OXI1 kinase is necessary for oxidative burst-mediated signalling in Arabidopsis. Nature 427(6977):858–861
98. Wu Q, Fan C, Wang H, Han Y, Tai F, Wu J, He R (2023) Biphasic impacts of graphite-derived engineering carbon-based nanomaterials on plant performance: effectiveness vs. nanotoxicity. Adv Agrochem
99. Gonzales-Garcia Y, et al.
100. Ya'Acov YL, Rapoport D, Frimer AA, Strul G, Asaf U, Felner I (1993) Buckminsterfullerene (C-60 carbon allotrope) inhibits ethylene evolution from 1-aminocyclopropane-1-carboxylic acid (ACC)-treated shoots of pea (Pisum sativum), broadbean (Vicia faba) and flowers of carnation (Dianthus caryophyllus). Ann Bot 72(5):457–461
101. Zan R, Ramasse QM, Jalil R, Bangert U (2013) Atomic structure of graphene and h-BN layers and their interactions with metals. In: Advances in graphene science. IntechOpen
102. Husen A, Siddiqi KS (2014) Carbon and fullerene nanomaterials in plant system. J Nanobiotechnology 12(1):1–16
103. Wagay JA, Singh S, Raffi MM, Rahman QI, Husen A (2019) Effect of carbon-based nanomaterials on plant functioning and rhizosphere. In: Husen A, Iqbal M (eds) N*anomaterials* and plant potential. Springer International Publishing AG, Gewerbestrasse 11, 6330 Cham, pp 553–575. https://doi.org/10.1007/978-3-030-05569-1_22

Role of Carbon Nanomaterials in the Prevention of Plant Disease

Mayur Mukut Murlidhar Sharma, Divya Kapoor, Pankaj Sharma, and Azamal Husen

Abstract Global agricultural systems are facing a lot of challenges of reduced crop yield, continuous decline in agricultural land, scarcity of irrigation water, and continuously changing climatic conditions. Agriculture, as a monoculture practice, has also to face the continuous occurrence of plant disease and increasing attack of resistant plant pathogenic microbes. The current treatments for killing plant pathogenic microbes are not environmentally friendly and sometimes fail to protect plants. The scientific and farming community is looking at nanotechnology as a potential alternative. Although many nanotechnological tools have been reported as potential weapons to render plant protection, carbon nanoparticles have recently gained attention. Their non-toxic attitude, environmental acceptability, and ability to possess biocidal properties against a wide range of plant pathogenic microbes make them superior to other nanoparticles. They kill the host cell by rupturing the cell membrane and penetrating the fungal hyphae, followed by the precipitation of genetic material. The multifarious approach reduces the chances of resistance in pathogenic microbes. In the current chapter, we have discussed different types of nanomaterials finding their origin in carbon and their plant protective role during various diseases. We have also tried to highlight the potential mechanism governing their biocidal nature.

M. M. M. Sharma
Department of Agriculture and Industry, Kangwon National University, Chuncheon, Republic of Korea

D. Kapoor · P. Sharma (✉)
Department of Microbiology, CCS Haryana Agricultural University, Hisar, Haryana 125004, India
e-mail: sharmap@uic.edu; psk14495@gmail.com

P. Sharma
Department of Ophthalmology and Visual Sciences, University of Illinois Chicago, Chicago, IL 60612, USA

A. Husen
Department of Biotechnology, Smt. S. S. Patel Nootan Science and Commerce College, Sankalchand Patel University, Visnagar, Gujarat 384315, India

Department of Biotechnology, Graphic Era (Deemed to be University), Dehradun, Uttarakhand 248002, India

Wolaita Sodo University, Wolaita, Ethiopia

A. Bachheti et al. (eds.), *Carbon-Based Nanomaterials*, Smart Nanomaterials Technology, https://doi.org/10.1007/978-981-97-0240-4_6

Keywords Carbon nanomaterials · Phyto-pathogenic fungi · Plant-parasitic nematode · Fullerene nanomaterials · Graphene nanomaterials

1 Introduction

The Global Population is increasing at an alarming rate and is expected to rise at even higher rates in the future. The land mass is, however, diminishing, making the situation even worse, and it is obligatory to deduce higher food production in the limited land. Numerous ways have been tried and adapted, including excessive fertilizers and pesticides, which indeed provided a boom to the productivity of crops but have also degraded the quality of the produced, accompanied by detrimental health consequences. The emergence of nanotechnology, however, has opened doors to their potential application to challenge agriculture-oriented problems [1–9]. The use of nanomaterial in the detection of various plant pathogens and their use as adjuvants of pesticides and fertilizers and also the practice of using fertilizers and other micronutrients in nano form has yielded prominent results owing to the small size of particles, which enables them to act more effectively as compared to their conventional counterparts [10, 11]. Application of nanomaterials in optimum conditions yields a significantly positive impact on the growth and development of plants and renders minimum toxicity to the environment and the plant tissues [12–18]. The literature is flooded with numerous such studies where the use of nanoparticles as a substitute to or along with the conventional counterpart has resulted in better growth and improved productivity [19, 20]. Different kinds of nanoparticles have been examined in the greenhouse and field studies for the improvement in plant growth and health, such as iron oxide nanoparticles [21], zinc oxide nanoparticles [22], titanium dioxide nanoparticles [23], Silver nanoparticles [24], and Copper nanoparticles [25]. Apart from the direct use of nanoparticles for plant growth and development, nanotechnology is also used to detect plant problems. For instance, nano-based sensors have been in practice to see different kinds of plant pathogens and the emergence of diseases associated with them, which could be detrimental to plant health [11]. For example, a Graphene-based sensor has been used to detect false smut of rice [26]. In the latest advancements in nanotechnology from the agricultural perspective, the spotlight is on carbon-based nanomaterials. The extra emphasis on carbon nanoparticles in recent years owes to their unique physical properties as well as reduced toxicity as compared to other nano counterparts and their immaculate biocompatibility, which makes them an excellent subject to study for plants and their practical use in the agriculture sector as well [27]. Carbon nanomaterials have several applications in the agriculture systems, such as their direct use as plant growth promoters and using them as a carrier to deliver different agrochemicals, including fertilizers and pesticides (Fig. 1). For instance, the foliar spray and complete soil drench using carbon nanoparticles in maize increased the plant height and improved different growth parameters such as biomass of roots and shoots and nutrient total nutrient uptake [28, 29]. The carbon nanomaterials, owing to their low toxicity and biodegradability, are also being looked upon as potential agents to ameliorate plant diseases. Although the anti-pathogenic properties of carbon nanoparticles have not been studied intensely, there are carbon nanomaterials such as single- and multi-walled carbon nanotubes, graphene oxides, and fullerenes which are used as carriers for introducing the conventional form of chemical pesticides and also as antimicrobial agents [30]. This chapter gathers all the information available in the literature regarding past studies on the anti-pathogenic activities of carbon-based nanoparticles and nanomaterials and further prospects for their potential application.

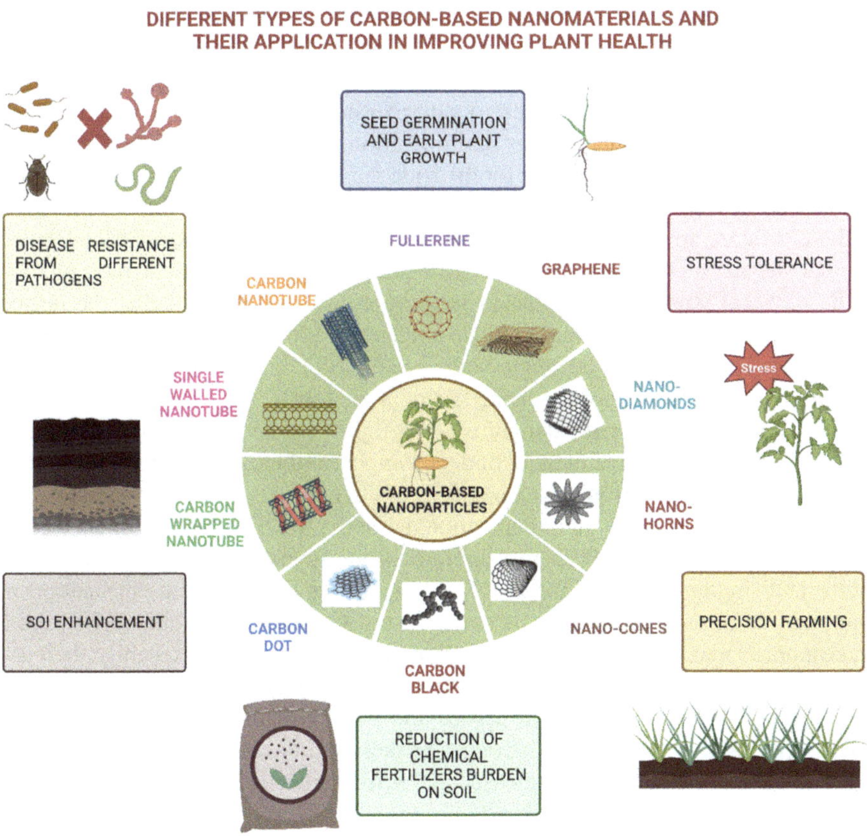

Fig. 1 A portrayal depicting different types of carbon nanomaterials and their application in plant disease protection

2 Types of Carbon Nanomaterials

Carbon nanomaterials are majorly comprised of carbon nanotubes (CNTs) which can be single-walled or multi-walled; Graphenes mainly constitute graphene oxide sheets and fullerenes. Other carbon nanomaterials include nano-onions, nano-cones, nano-horns, carbon dots, and nano-diamonds [31, 32].

2.1 *Carbon Nanotubes*

Carbon nanotubes are the widely used carbon-based nanomaterials. Since their discovery, they have been the most popular form of carbon nanomaterials. They have a cylindrical structure and sp^2 hybridization. The sp^2 hybridization and their physical properties are responsible for the wide range of optical and electrical properties they possess [33]. In the carbon nanotubes, in an ideal scenario, the carbon bonds in the CNTs are arranged in a hexagonal lattice except for the ends. CNTs formed in other shapes, such as heptagons and pentagon, considerably degrade their quality. If the physical strength of the CNTs is to be considered numerically, the elastic modulus approaching 1 TPa and a tensile strength of 100 GPa has been found for the walls [34]. The CNTs can be single-walled (SWCNT) or multi-walled (MWCNT), depending on the number of graphene layers in the tube.

SWCNT has 0.4 and 2 nm diameters and is tightly packed in a hexagonal bundle. MWCNT comprises two or more cylinders composed of individual graphene sheets, and the diameter varies from 1 to 3 nm [35]. CNTs can be synthesized by laser ablation method and chemical vapor deposition methods. Arc discharge signifies the use of high temperatures (<3000 °C) to evaporate carbon atoms into plasma, forming both multi- and single-walled CNTs [36]. The tremendous chemical and physical properties of CNTs, such as high tensile strength, great elasticity, and ultra-lightweight, make them an ideal candidate to study their use in different sectors. Recently, their use in the agriculture sector is also well looked upon. They are considered novel fertilizers due to their slow-release properties and plant growth-boosting effect. CNTs can also be used in agricultural practices to deliver pesticides and other necessary agrochemicals, which are slow-released and self-regulated [37].

2.2 *Graphene*

Graphene is formed by a hexagonal structure that consists of a sp^2 hybridized carbon atom and is referred to as the "mother" of all graphite carbon materials because of its building block features. Graphene was initially extracted from graphite via mechanical cleavage by a scotch tape to attach flakes of graphene layers [38]. The mechanically exfoliated graphene is defect-free and pure, so studies are going on to synthesize it on a large scale, including graphene oxide, reduced graphene oxide, and thermally reduced graphene oxide. The remarkable properties of graphene sheets make them attractive materials for use in mechanical, electrical, and aerospace engineering. They are an emerging material for their use in biomedical sciences and plant sciences as biosensors, bioimaging, and tissue culture [39]. In agriculture, graphene is being studied for its implications, such as in tissue culture, gene delivery, and seed germination. Its role is also being examined in seed germination and its reaction and impact on plant physiology. For instance, the existing tools and techniques available for analysis are labor intensive such as PCR, ELISA, and DNA hybridization. The graphene-based nanosensors can be used owing to their accuracy and quickness [40]. The different graphene-based sensors have already been reported and used for biomolecular, immunosensing, and fluorescence labels. Another important use of graphene in the agriculture world is the improvement of water use efficiency. Graphene is hydrophilic in nature, and due to its properties of adsorbed water, the water can be held to the field surface and can be prevented from evaporation [41]. This property has been successfully exploited in enhancing seed germination, as reported in tomatoes and spinach. Alongside increasing water use efficiency, graphene nanomaterials can be used as carriers to deliver fertilizers and pesticides efficiently. Conventional fertilizers are associated with the problem of leaching and run-off, leading to the problem of eutrophication. Hence, nano-substitutes are being used and studied to deliver chemical fertilizers efficiently. For instance, the graphene sheets are used to deliver KNO_3, which extends the release process to 8 h [42].

2.3 Fullerene

Fullerene is the first member of the nanocarbon family with a close-caged structure, and its discovery in the 1960s has opened doors to a new era of nanoscience. From all the available allotropes of the elemental carbon, the fullerene has a specified spherical structure containing only the sp^2 hybridization in its atoms. Their unique system in 0-D makes them unique and special for designing nanomaterials from them which can be used in different fields such as optoelectronics, biomedical sciences, and other allied fields of science. The properties of fullerene make it oxygen deficient and thus increase its reactivity, opening further doors to explore its functionalities [43]. The fullerene-based nanomaterials can be categorized into domains, such as fullerene-based nanoparticles, fullerene nanowires, fullerene nanocomposites, and fullerene derivatives. All of them are used in multiple fields such as electronics, physical, and medical sciences. For instance, NIR light-harvesting fullerene-based nanoparticles were used in cancer therapy and exhibited anti-tumor properties in vivo and in vitro [44]. The Fullerene nanowires are also known for their extraordinary physical and chemical properties. Their cage-like structure contains 60 carbon atoms in 20 hexagons and 12 pentagons, making them unique and strong. They have been widely used in optical, electrical, mechanical, electrochemical, and biological domains [45]. They have been used in biology for drug delivery, targeted imaging, and treating

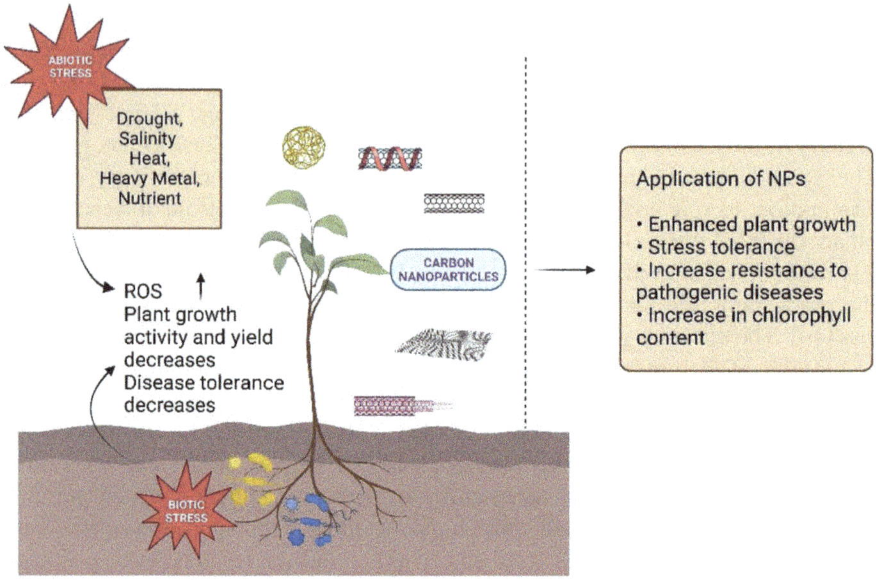

Fig. 2 A portrayal depicting the role of carbon nanoparticles in biotic and abiotic stress management in plant systems

multi-drug resistance pathogens [46]. The "bulkyball" structure of the fullerenes is considered highly favorable for making nanocomposites and other derivatives. The leading derivatives formed using fullerene are diphosphonate, phosphonate, and organophosphorus compounds. Their derivatives have been used vastly as antioxidants, pharmaceuticals, biopolymers, and catalysts [47]. Among all the fullerene nanomaterials, the fullerene derivatives are commonly used in agriculture to enhance productivity and plant growth and development. For instance, Panova et al. [48] used C60 adduct with the three essential amino acids, threonine, lysine, and arginine, for their potential effect on wheat and barley. These conjugates stimulated the growth of plants, where their biomass was increased by 20% compared to the control plants (Fig. 2).

2.4 Other Carbon Nanomaterials

Other uncategorized miscellaneous carbon nanomaterials include nano-onions, nanocones, nano-horns, carbon dots, nanoi-dots, and nano-diamonds. Nano-onions were discovered in 1992 and found as concentric shells of carbon atoms. The characteristics and features of CNOs make them good candidates for application in various fields. Their size generally depends upon the method used for their synthesis, although most exhibit a higher area-to-volume ratio [49]. Nano-onions are highly used in optics and electronics such as energy storage devices and hyperlubricants. Recently, nano-onions have been used for cellular imaging in biological methods. The plant-based applications of nano-onions are limited, hence, only constrained reports are available regarding their use in plant sciences [50].

For instance, Sonkar et al. [51] used nano-onion as growth stimulants for gram plants, effectively increasing the plant biomass. Carbon nano-horns (CNHs), sometimes known as nano-cones, are closed cages of sp^2-bonded carbon atoms, typically 2–5 nm in diameter and 40–50 nm in length. In one sense, they can be considered a high-aspect ratio subset of fullerenes due to their closed cage structure. However, they may be opened via oxidation to increase their surface area and provide access to their interior cavity. However, their elongated shape means that a closer structural analog is short single-walled carbon nanotubes, with which they share much common chemistry. Indeed, nano-horns are under investigation as replacement candidates for nanotubes in various fields, notably energy conversion, drug delivery, gas storage, and supercapacitors [52]. The applications associated with carbon nanohorns are drug carriers, reproducible sensors, and photothermal cancer treatment [52]. The carbon dots are carbon-based nanomaterials ranging less than 10 nm in size. They are unique carbon nanomaterials having photoluminescence and photoelectrochemical properties. They have unique optical properties, good biocompatibility, and less toxicity which makes them a good candidate for use in place of conventional fluorescent materials. Their use as fluorescent material for sensory purposes has been under study, which can cover a broad range of things to analyze, such as micro and macromolecules, living cells such as bacteria, and ions [53]. Their use in agricultural systems has been studied less, and few reports have been published. For instance, carbon dots have been used in transfecting foreign DNA into the plant system [54]. Other applications of carbon dots in plants include bioimaging, biosensing, and photocatalysis [55]. Carbon dots' ability in light conversion is also being studied to enhance the photosynthetic capabilities of plants [56]. Other nanomaterials which are being explored among all others are nano-diamonds. The nano-diamonds are highly biocompatible and are among the least toxic. Their properties to imbibe various bioactive compounds to their surface enables them to be used in biomedical sciences. They are highly efficient in bioimaging, drug release, and enhanced therapeutic efficacy [57]. In plant systems, their application is minimal due to the high cost, and they are predominantly used in the bioimaging of plant cells [58].

3 Application of Carbon Nanomaterials in the Prevention of Plant Disease

3.1 Carbon Nanotubes

Nanoscience and nanoengineering technologies are recently dissipated mainly to the agricultural sector to enhance crop productivity. Among a gamut of nanomaterials used for controlling plant diseases, carbon nanotubes have emerged as a potential strategy for improving plant productivity and controlling disease severity [59]. Carbon Nanotubes (CNTs) have been reported to exhibit lesser phytotoxicity as compared to other nanomaterials, further advocating its superiority over other nanomaterials [60]. Although concentration-dependent phytotoxicity of carbon nanotubes has also been reported, they can work at sub-lethal concentrations when assessed in terms of killing plant pathogens. In addition, CNTs can also increase plant productivity by enhancing the plant's photosynthetic capacity, stimulating plant metabolism, and increasing the levels of plant photosynthetic components. They possess antimicrobial activity as they can generate membrane damage in pathogenic microbes and induce oxidative damage upon entry into the microbial cell. The application of carbon nanotubes for controlling plant pathogens has been explored in diverse crops. For instance, applying CNTs in tomato crops significantly decreased the incidence and severity of phytopathogenic fungi *Alterneria solani*. In addition, the flavonoid content, enzymatic activity, and water use efficiency were also found to be increased [61]. The antifungal activity of single-walled and multiwalled CNT has also been explored against *Fusarium graminearum* and *Fusarium poae*. Single-walled CNTs were found to exhibit the greatest antifungal activity. They inhibited the growth of fungus by depositing in the surface of spores and by inhibiting the water uptake. Further, they also induced plasmolysis in the fungal spores [62]. Further, the surface modification of CNTs can also enhance their antimicrobial activity. For instance, Wang et al. [63] explored the antifungal activities of multiwalled CNTs having OH–, COOH–, and NH_2– as functional groups on their surface. The surface-modified CNTs exhibited significantly higher antifungal activity. Similarly, multiwalled CNTs have also been reported to show antifungal activity against Bort *cinerea* at a concentration of 200 mg/L [64]. In a similar way, multiwalled CNTs exhibited decent biocidal activity against the fungi *Podosphaera pannosa*, which is known to cause powdery mildew on the plants in the rose family [65].

Interestingly, CNTs have also been reported to inhibit the growth and activity of viral pathogens infecting plant systems. The foliar application of CNTs induced the upregulation of plant defense-related and phytohormone-related genes. The treatment also resulted in the inhibition of viral replication along with the restriction of the viral movement [66]. Similarly, the foliar application of CNTs in tobacco plants infected with the Turnip mosaic virus significantly reduced the disease severity and increased plant growth. The foliar treatment increased the plant fresh biomass by more than 55%, which further advocates the non-toxic nature of CNTs toward plant systems [67].

CNTs have also been reported to possess antimicrobial activity against bacterial plant pathogens. Single-walled and multiwalled CNTs have been reported to significantly inhibit *Ralstonia solanacearum*, a bacterial pathogen that is known for causing bacterial wilt in a diverse array of agriculturally important plants [68]. The antibacterial activity results from the cell membrane's destruction, followed by cytoplasm leakage.

3.2 Graphene Nanomaterials

Graphene nanomaterials have successfully been used in mammalian cells as nanocarriers for gene delivery systems with higher efficacy than traditional methods. Although their agricultural applications are still restricted, they are being projected as potential gene delivery vehicles in plant systems that can be used in the genetic engineering of crops. Currently, they find numerous applications in the agricultural sector in the form of biosensors as agents for successfully altering plant physiology and improving seed germination, plant biomass, and water use efficiency. Further, graphene nanomaterials are also being used as pesticide delivery vectors, thus providing an efficient tool for plant pest management [69]. Recently, it has also been reported that graphene nanomaterials possess synergistic activity with some insecticides. For instance, Wang et al. [70] reported that the addition of graphene nanomaterials could significantly enhance the activities of commercially available acaricides, namely pyridazine, chlorpyrifos, and beta-cyfluthrin correspondingly against two economically important spider mites, *Tetranychus*, *truncatus*, and *T. urticae*. The nanomaterial-acaricide complex was found to possess increased adhesion over the surface of pests, thus advocating a reduced loss of pesticide.

Graphene nanomaterials are also known to indirectly protect against plant disease by increasing plant growth and health, thus rendering them more resilient toward different diseases and attacks by pathogenic microbes. For instance, Zhang et al. [71] reported that treating *Aloe vera* with graphene oxide nanoparticles at the concentration of 50 mg/L enhanced the photosynthetic efficiency of leaves, along with a significant increase in the plant growth parameters. Similarly, Chen et al. [72] reported that treating maize seeds with 50 mg/L graphene significantly improved plant growth and health. Graphene oxide nanomaterials are also known to improve plant health by acting as water transporter. This ability enhances seed germination, thus ultimately improving the plant yield [73]. Similarly, Zhang et al. [74] also reported the positive effect of graphene nanoparticles on tomato seed germination. The enhanced seed germination was attributed to enhanced water uptake.

Graphene nanoparticles protect agriculturally important plants from different diseases owing to their possession of biocidal activities against different plant pathogenic microbes. For instance, Siddiqui et al. [75] reported that the treatment of carrots with graphene oxide nanoparticles significantly reduced the disease severity caused by *Pectobacterium carotovorum*, *Xanthomonas campestris* pv. *carotae*, *Meloidogyne javanica*, *Alternaria dauci* and *Fusarium solani*. In addition to its possession of broad-spectrum biocidal activity, the treatment also significantly upregulated plant growth parameters as assessed by a significant increase in plant growth, chlorophyll, carotenoid, and proline contents. Similarly, El-Abeid et al. [76] reported that Graphene Oxide Nanosheet-Decorated Copper Oxide Nanoparticles possessed robust antifungal activity against wild strains of *Fusarium oxysporum* affecting tomato and pepper plants. The treatment of plants with nanoparticles at the concentration of 1 mg/L reduced the disease severity below 5% without any phytotoxicity. Likewise, Graphene Oxide-Silver Nanocomposite has also been reported to inhibit the phytopathogenic fungi *Fusarium graminearum significantly* [77]. Alongside its possession of fungicidal activity, these nanomaterials are also known for their robust bactericidal activity against different plant pathogenic bacteria. For instance, Chen et al. [78] reported that applying graphene oxide at the concentration of 250 μg/ml killed more than 90% population of *Xanthomonas oryzae pv. Oryza* (*Xoo*) is a key pathogenic bacteria that causes infections in rice.

3.3 Fullerene

Fullerene and its derivatives are not much explored in the agricultural sector. However, a few studies have been done that suggest its potential as a plant growth-promoting agent. For instance, Kole et al. [79] reported that the yield of bitter melon seeds was increased by 112–128% on treatment with fullerene. Alongside, plant biomass, fruit yield, and phytomedicine content were also found to be significantly higher in the treated plants. Further, the authors reported a 54% increment in biomass and a 24% hike in water content. Surprisingly, the fruit length was increased by 20%, whereas the fruit count was increased by 59%. The net fruit yield was found to be raised by approximately 128%. Although, there are numerous reports that highlight the phytotoxic attributes of fullerene nanoparticles. For instance, He et al. [80] reported the potential of fullerene to block transpiration and damage the root structure of plants. The phytotoxic attribute of fullerene nanoparticles can be exploited for their application as total killer herbicides. Although no report regarding their application as an adjuvant to herbicides exists, they hold a robust potential and need further scientific exploration. Several reports have also highlighted that plant treatment with fullerene nanoparticles stimulates the phytoremediation of several pesticides (Table 1).

4 Conclusion

Plants, being sessile, must face many environmental challenges, including the increased attacks by newer and resistant varieties of plant pathogenic microbes. A gamut of nanoparticles has been reported for the possession of biocidal activities against different plant pathogenic microorganisms, including bacteria, viruses,

Table 1 List of carbon nanomaterials reported for protecting plants from different kinds of diseases

S. No.	Type of carbon nanomaterial	Effect on plant	Mechanism of action	Remarks	Key references
1	Carbon dots (CDs)	Increased seed germination, root elongation, seedling length, enzyme (RuBisCO) activity, and carbohydrate generation A total of 14.8% enhancement in rice yield	CDs enter the nucleus, loosening the DNA structure, and increasing the thionin (*Os06g32600*) gene expression, which ultimately improves the rice-plant disease-resistance ability	~5 nm size	[81]
2	Lignin-loaded carbon nanoparticles (L-CNPs)	Showed beneficial effects in the earliest stages of maize development and substantial augmentation in the level of carotenoid, anthocyanin, and chlorophyll pigments Reduced stalk rot disease by 86% and 81% at different concentrations	Treatment improved the activity of all the defense-related enzymes in the infected plants	100 and 500 mg/L	[82]
3	Carbon nanoparticles (CNPs)	NPs at an intermediate concentration (100–150 μM) favor the growth of crops by increasing the total chlorophyll content (1.9-fold), protein content (1.14-fold), and plant biomass (fresh weight: 1.2-fold, dry weight: 1.14-fold)	Enhance the stress-tolerant and phytoremediation efficiency of plants in a polluted environment due to their enormous absorption potential	25–200 μM	[83]
4	Flue-cured tobacco cultivar Zhongyan100	The height increased by 6.33, 10.56, and 10.00, whereas the leaf area increased by 6.64, 19.51, and 21.58% at the maturity stage	Increased the contents of chlorophyll and soluble protein of leaves, which are important factors of growth improvement in tobacco plant	0, 25, 75, and 125 mg pot^{-1}	[84]

(continued)

Table 1 (continued)

S. No.	Type of carbon nanomaterial	Effect on plant	Mechanism of action	Remarks	Key references
5	CNPs (graphite nanopowder, type-1, APS: 400 nm)	The application of CNPs enhanced the timing to flower	It down-regulated phytochrome B in a CONSTANS-independent but photoperiod-dependent manner in *Arabidopsis thaliana*	10, 100, and 500 μg/ml	[85]
6	CuNPs using C-TAB-IPA (cetyl trimethylammonium bromide)	The activity of copper nanoparticles was found to be superior to that of the commercially available fungicide bavistin against plant pathogenic fungi *F. oxysporum*, *C. lunata*, *A. alternata*, and *P. destructiva*	It inhibited the growth of pathogen	3–10 nm	[86]
7	CuO, Cu_2O nanoparticles	It showed antifungal activity when tested in the field on tomato (*Lycopersicon esculentum*) plants against *Phytophthora infestans*	It inhibited the growth of pathogen	11–55 nm	[87]
8	Multi-walled carbon nanotubes	It showed antifungal activity against *Alternaria solani* on tomato	Enzymatic degradation	30–50 nm	[61]
9	Single-walled carbon nanotubes (SWCNTs)	It showed antifungal activity against *Ralstonia Solanocerum*	Damages the cell membrane of the fungi causing the release of cytoplasmic contents	1 mg/ml	[68]
10	Single-walled carbon nanotubes (SWCNTs) and multi-walled carbon nanotubes (MWCNTs)	Accelerate leaf growth and development of rice seedlings accompanied with the increased chlorophyll content and net photosynthetic rate (PN)	A possible link between reactive oxygen species and plant hormones under CNTs treatment to promote rice seedlings growth	20 mg/L	[88]

fungi, and nematodes. However, they do possess toxicity against the plant cell as well. Different carbon nanomaterials' biocompatible and ecologically acceptable nature makes them superior to other nanoparticles. Their exertion of broad-spectrum biocidal activity also suggests the possibility of using a single biocide as a potential replacement for pesticides, fungicides, and nematicides. This practice can also prove essential for annually reducing xenobiotic compounds entering the environment. Although carbon nanomaterials possess immense potential, some, like fullerene, still need to be explored. Further research involving such nanomaterials can render them novel agricultural applications that can be a boon for the agricultural community.

References

1. Godeto YG, Ayele A, Ahmed IN, Husen A, Bachheti RK (2023) Medicinal plant-based metabolites in nanoparticles synthesis and their cutting-edge applications: an overview. In: Bachheti RK, Bachheti A (eds) Secondary metabolites from medicinal plants. CRC Press, Boca Raton, USA, pp 1–34
2. Husen A, Siddiqi KS (2014) Phytosynthesis of nanoparticles: concept, controversy and application. Nanoscale Res Lett 9(229):1–24
3. Husen A, Siddiqi KS (2014) Carbon and fullerene nanomaterials in plant system. J Nanobiotechnol 12(1):1–16
4. Husen A, Siddiqi KS (2023) Advances in smart nanomaterials and their applications. Elsevier Inc., Cambridge, MA, USA
5. Husen A (2020) Interactions of metal and metal-oxide nanomaterials with agricultural crops: an overview. In: Husen A, Jawaid M (eds) Nanomaterials for agriculture and forestry applications. Elsevier Inc., Cambridge, MA, USA, pp 167–197. https://doi.org/10.1016/B978-0-12-817852-2.00007-X
6. Husen A (2020) Carbon-based nanomaterials and their interactions with agricultural crops. In: Husen A, Jawaid M (eds) Nanomaterials for agriculture and forestry applications. Elsevier Inc., Cambridge, MA, USA, pp 199–218. https://doi.org/10.1016/B978-0-12-817852-2.00008-1
7. Husen A (2022) Engineered nanomaterials for sustainable agricultural production, soil improvement and stress management. Elsevier Inc., Cambridge, MA, USA
8. Husen A (2023) Nanomaterials and nanocomposites exposures to plants (response, interaction, phytotoxicity and defense mechanisms). Springer Nature Singapore Pte Ltd., Singapore, Singapore. https://doi.org/10.1007/978-981-99-2419-6
9. Kaphle A, Navya PN, Umapathi A, Daima HK (2018) Nanomaterials for agriculture, food and environment: applications, toxicity and regulation. Environ Chem Lett 16:43–58
10. Hazarika A, Yadav M, Yadav DK, Yadav HS (2022) An overview of the role of nanoparticles in sustainable agriculture. Biocatal Agric Biotechnol 43:102399
11. Sharma P, Pandey V, Sharma MM, Patra A, Singh B, Mehta S, Husen A (2021) A review on biosensors and nanosensors application in agroecosystems. Nanoscale Res Lett 16:1–24
12. Khot LR, Sankaran S, Maja JM, Ehsani R, Schuster EW (2012) Applications of nanomaterials in agricultural production and crop protection: a review. Crop Prot 35:64–70
13. Loyal A, Pahuja SK, Rani N, Pooja, Srivastava RK, Sharma P (2023) Titanium oxide nanoparticles: plant response, interaction, phytotoxicity, and defence mechanisms. In: Nanomaterials and nanocomposites exposures to plants: response, interaction, phytotoxicity and defense mechanisms. Springer Nature Singapore, Singapore, pp 263–284
14. Loyal A, Pahuja SK, Sharma P, Malik A, Srivastava RK, Mehta S (2023) Potential environmental and human health implications of nanomaterials used in sustainable agriculture and soil improvement. In: Engineered nanomaterials for sustainable agricultural production, soil improvement and stress management. Academic Press, pp 387–412
15. Kapoor D, Yadav S, Sharma MM, Sharma P (2023) Interaction between metal nanoparticles and PGPR on the plant growth and development. In: Nanomaterials and nanocomposites exposures to plants: response, interaction, phytotoxicity and defense mechanisms. Springer Nature Singapore, Singapore, pp 327–351

16. Sharma MM, Kapoor D, Rohilla R, Sharma P (2023) Nanomaterials and their toxicity to beneficial soil microbiota and fungi associated plants rhizosphere. In: Nanomaterials and nanocomposites exposures to plants: response, interaction, phytotoxicity and defense mechanisms. Springer Nature Singapore, Singapore, pp 353–380
17. Sharma P, Sangwan S, Mehta S (2023) Emerging role of phosphate nanoparticles in agriculture practices. In: Engineered nanomaterials for sustainable agricultural production, soil improvement and stress management. Academic Press, pp 71–97
18. Singh S, Sangwan S, Sharma P, Devi P, Moond M (2021) Nanotechnology for sustainable agriculture: an emerging perspective. J Nanosci Nanotechnol 21(6):3453–3465
19. Bachheti RK, Bachheti A, Husen A (eds) (2023) Nanomaterials for environmental and agricultural sectors. Springer Nature, Singapore. https://doi.org/10.1007/978-981-99-2874-3
20. Liu R, Lal R (2014) Synthetic apatite nanoparticles as a phosphorus fertilizer for soybean (Glycine max). Sci Rep 4:5686
21. Rui M, Ma C, Hao Y, Guo J, Rui Y, Tang X, Zhao Q, Fan X, Zhang Z, Hou T, Zhu S (2016) Iron oxide nanoparticles as a potential iron fertilizer for peanut (Arachis hypogaea). Front Plant Sci 7:815
22. Milani N, McLaughlin MJ, Stacey SP, Kirby JK, Hettiarachchi GM, Beak DG, Cornelis G (2012) Dissolution kinetics of macronutrient fertilizers coated with manufactured zinc oxide nanoparticles. J Agric Food Chem 60(16):3991–3998
23. Irshad MA, Nawaz R, ur Rehman MZ, Adrees M, Rizwan M, Ali S, Ahmad S, Tasleem S (2021) Synthesis, characterization and advanced sustainable applications of titanium dioxide nanoparticles: a review. Ecotoxicol Environ Saf 212:111978
24. Amna AB, Azeem MA, Qayyum A, Mustafa G, Ahmad MA, Javed MT, Chaudhary H (2022) Bio-fabricated silver nanoparticles: a sustainable approach for augmentation of plant growth and pathogen control. In: Sustainable agriculture reviews 53: nanoparticles: a new tool to enhance stress tolerance 2022 Jan 28. Springer International Publishing, Cham, pp 345–371
25. Kasana RC, Panwar NR, Kaul RK, Kumar P (2016) Copper nanoparticles in agriculture: biological synthesis and antimicrobial activity. Nanosci Food Agric 3(2016):129–143
26. Rana K, Mittal J, Narang J, Mishra A, Pudake RN (2021) Graphene based electrochemical DNA biosensor for detection of false smut of rice (Ustilaginoidea virens). Plant Pathol J 37(3):291
27. Shojaei TR, Salleh MA, Tabatabaei M, Mobli H, Aghbashlo M, Rashid SA, Tan T (2019) Applications of nanotechnology and carbon nanoparticles in agriculture. In: Synthesis, technology and applications of carbon nanomaterials. Elsevier, pp 247–277
28. Sangwan S, Sharma P, Wati L, Mehta S (2023) Effect of chitosan nanoparticles on growth and physiology of crop plants. In: Engineered nanomaterials for sustainable agricultural production, soil improvement and stress management. Academic Press, pp 99–123
29. Xin X, Nepal J, Wright AL, Yang X, He Z (2022) Carbon nanoparticles improve corn (Zea mays L.) growth and soil quality: comparison of foliar spray and soil drench application. J Clean Prod 363:132630
30. Elmer W, White JC (2018) The future of nanotechnology in plant pathology. Annu Rev Phytopathol 56:111–133
31. Belay T, Worku LA, Bachheti RK, Bachheti A, Husen A (2023) Nanomaterials: introduction, synthesis, characterization, and applications. In: Husen A, Siddiqi KS (eds) Advances in smart nanomaterials and their applications. Elsevier, Amsterdam, Netherlands, pp 1–21
32. Mukherjee A, Majumdar S, Servin AD, Pagano L, Dhankher OP, White JC (2016) Carbon nanomaterials in agriculture: a critical review. Front Plant Sci 7:172
33. Cha C, Shin SR, Annabi N, Dokmeci MR, Khademhosseini A (2013) Carbon-based nanomaterials: multifunctional materials for biomedical engineering. ACS Nano 7(4):2891–2897
34. De Volder MF, Tawfick SH, Baughman RH, Hart AJ (2013) Carbon nanotubes: present and future commercial applications. Science 339(6119):535–539
35. He H, Pham-Huy LA, Dramou P, Xiao D, Zuo P, Pham-Huy C (2013) Carbon nanotubes: applications in pharmacy and medicine. Biomed Res Int. https://doi.org/10.1155/2013/578290
36. Anzar N, Hasan R, Tyagi M, Yadav N, Narang J (2020) Carbon nanotube—a review on synthesis, properties and plethora of applications in the field of biomedical science. Sens Int 1:100003

37. Safdar M, Kim W, Park S, Gwon Y, Kim YO, Kim J (2022) Engineering plants with carbon nanotubes: a sustainable agriculture approach. J Nanobiotechnol 20(1):1–30
38. Novoselov KS, Geim AK, Morozov SV, Jiang DE, Zhang Y, Dubonos SV, Grigorieva IV, Firsov AA (2004) Electric field effect in atomically thin carbon films. Science 306(5696):666–669
39. Liao C, Li Y, Tjong SC (2018) Graphene nanomaterials: synthesis, biocompatibility, and cytotoxicity. Int J Mol Sci 19(11):3564
40. Pandey RR, Chusuei CC (2021) Carbon nanotubes, graphene, and carbon dots as electrochemical biosensing composites. Molecules 26(21):6674
41. Zhao D, Fang Z, Tang Y, Tao J (2020) Graphene oxide as an effective soil water retention agent can confer drought stress tolerance to Paeonia ostii without toxicity. Environ Sci Technol 54(13):8269–8279
42. Zhang M, Gao B, Chen J, Li Y, Creamer AE, Chen H (2014) Slow-release fertilizer encapsulated by graphene oxide films. Chem Eng J 255:107–113
43. Kausar A (2021) Advances in condensation polymer containing zero-dimensional nanocarbon reinforcement—fullerene, carbon nano-onion, and nanodiamond. Polym-Plast Technol Mater 60(7):695–713
44. Shi H, Gu R, Xu W, Huang H, Xue L, Wang W, Zhang Y, Si W, Dong X (2019) Near-infrared light-harvesting fullerene-based nanoparticles for promoted synergetic tumor phototheranostics. ACS Appl Mater Interfaces 11(48):44970–44977
45. Fan X, Soin N, Li H, Li H, Xia X, Geng J (2020) Fullerene (C60) nanowires: the preparation, characterization, and potential applications. Energy Environ Mater 3(4):469–491
46. Stoilova O, Jérôme C, Detrembleur C, Mouithys-Mickalad A, Manolova N, Rashkov I, Jérôme R (2007) C60-containing nanostructured polymeric materials with potential biomedical applications. Polymer 48(7):1835–1843
47. Azizi-Lalabadi M, Hashemi H, Feng J, Jafari SM (2020) Carbon nanomaterials against pathogens; the antimicrobial activity of carbon nanotubes, graphene/graphene oxide, fullerenes, and their nanocomposites. Adv Coll Interface Sci 284:102250
48. Panova GG, Kanash EV, Semenov KN, Charykov NA, Khomyakov YV, Anikina LM, Artem'eva AM, Kornyukhin DL, Vertebnyi VE, Sinyavina NG, Udalova OR (2018) Fullerene derivatives influence production process, growth and resistance to oxidative stress in barley and wheat plants. Sel'skokhozyaĭstvennaya Biol 53(1):38–49
49. Bartelmess J, Giordani S (2014) Carbon nano-onions (multi-layer fullerenes): chemistry and applications. Beilstein J Nanotechnol 5(1):1980–1998
50. Mykhailiv O, Zubyk H, Plonska-Brzezinska ME (2017) Carbon nano-onions: unique carbon nanostructures with fascinating properties and their potential applications. Inorg Chim Acta 468:49–66
51. Sonkar SK, Roy M, Babar DG, Sarkar S (2012) Water soluble carbon nano-onions from wood wool as growth promoters for gram plants. Nanoscale 4(24):7670–7675
52. Karousis N, Suarez-Martinez I, Ewels CP, Tagmatarchis N (2016) Structure, properties, functionalization, and applications of carbon nanohorns. Chem Rev 116(8):4850–4883
53. Sun X, Lei Y (2017) Fluorescent carbon dots and their sensing applications. TrAC, Trends Anal Chem 89:163–180
54. Wang B, Huang J, Zhang M, Wang Y, Wang H, Ma Y, Zhao X, Wang X, Liu C, Huang H, Liu Y (2020) Carbon dots enable efficient delivery of functional DNA in plants. ACS Appl Bio Mater 3(12):8857–8864
55. Li Y, Xu X, Wu Y, Zhuang J, Zhang X, Zhang H, Lei B, Hu C, Liu Y (2020) A review on the effects of carbon dots in plant systems. Mater Chem Front 4(2):437–448
56. Guirguis A, Yang W, Conlan XA, Kong L, Cahill DM, Wang Y (2023) Boosting plant photosynthesis with carbon dots: a critical review of performance and prospects. Small 28:2300671
57. Gupta C, Prakash D, Gupta S (2017) Cancer treatment with nano-diamonds. Front Biosci (Schol Ed) 9:62–70
58. Su LX, Lou Q, Jiao Z, Shan CX (2016) Plant cell imaging based on nanodiamonds with excitation-dependent fluorescence. Nanoscale Res Lett 11:1–7

59. González-Grandío E, Demirer GS, Jackson CT, Yang D, Ebert S, Molawi K, Keller H, Landry MP (2021) Carbon nanotube biocompatibility in plants is determined by their surface chemistry. J Nanobiotechnol 19(1):1–5
60. Smirnova E, Gusev A, Zaytseva O, Sheina O, Tkachev A, Kuznetsova E, Lazareva E, Onishchenko G, Feofanov A, Kirpichnikov M (2012) Uptake and accumulation of multiwalled carbon nanotubes change the morphometric and biochemical characteristics of Onobrychis arenaria seedlings. Front Chem Sci Eng 6:132–138
61. González-García Y, Cadenas-Pliego G, Alpuche-Solís ÁG, Cabrera RI, Juárez-Maldonado A (2021) Carbon nanotubes decrease the negative impact of Alternaria solani in tomato crop. Nanomaterials 11(5):1080
62. Wang X, Liu X, Chen J, Han H, Yuan Z (2014) Evaluation and mechanism of antifungal effects of carbon nanomaterials in controlling plant fungal pathogen. Carbon 68:798–806
63. Wang X, Zhou Z, Chen F (2017) Surface modification of carbon nanotubes with an enhanced antifungal activity for the control of plant fungal pathogen. Materials 10(12):1375
64. Hao Y, Cao X, Ma C, Zhang Z, Zhao N, Ali A, Hou T, Xiang Z, Zhuang J, Wu S, Xing B (2017) Potential applications and antifungal activities of engineered nanomaterials against gray mold disease agent Botrytis cinerea on rose petals. Front Plant Sci 8:1332
65. Hao Y, Fang P, Ma C, White JC, Xiang Z, Wang H, Zhang Z, Rui Y, Xing B (2019) Engineered nanomaterials inhibit Podosphaera pannosa infection on rose leaves by regulating phytohormones. Environ Res 170:1–6
66. Adeel M, Farooq T, White JC, Hao Y, He Z, Rui Y (2021) Carbon-based nanomaterials suppress tobacco mosaic virus (TMV) infection and induce resistance in *Nicotiana benthamiana.* J Hazard Mater 404:124167
67. Hao Y, Yuan W, Ma C, White JC, Zhang Z, Adeel M, Zhou T, Rui Y, Xing B (2018) Engineered nanomaterials suppress Turnip mosaic virus infection in tobacco (Nicotiana benthamiana). Environ Sci Nano 5(7):1685–1693
68. Wang X, Liu X, Han H (2013) Evaluation of antibacterial effects of carbon nanomaterials against copper-resistant *Ralstonia solanacearum.* Colloids Surf, B 103:136–142
69. Chen Z, Zhao J, Cao J, Zhao Y, Huang J, Zheng Z, Li W, Jiang S, Qiao J, Xing B, Zhang J (2022) Opportunities for graphene, single-walled and multi-walled carbon nanotube applications in agriculture: a review. Crop Des 1(1):100006
70. Wang X, Xie H, Wang Z, He K (2019) Graphene oxide as a pesticide delivery vector for enhancing acaricidal activity against spider mites. Colloids Surf, B 173:632–638
71. Zhang X, Cao H, Zhao J, Wang H, Xing B, Chen Z, Li X, Zhang J (2021) Graphene oxide exhibited positive effects on the growth of Aloe vera L. Physiol Mol Biol Plants 27:815–824
72. Chen Z, Zhao J, Song J, Han S, Du Y, Qiao Y, Liu Z, Qiao J, Li W, Li J, Wang H (2021) Influence of graphene on the multiple metabolic pathways of Zea mays roots based on transcriptome analysis. PLoS ONE 16(1):e0244856
73. He Y, Hu R, Zhong Y, Zhao X, Chen Q, Zhu H (2018) Graphene oxide as a water transporter promoting germination of plants in soil. Nano Res 11:1928–1937
74. Zhang M, Gao B, Chen J, Li Y (2015) Effects of graphene on seed germination and seedling growth. J Nanopart Res 17:1–8
75. Siddiqui ZA, Parveen A, Ahmad L, Hashem A (2019) Effects of graphene oxide and zinc oxide nanoparticles on growth, chlorophyll, carotenoids, proline contents and diseases of carrot. Sci Hortic 249:374–382
76. El-Abeid SE, Ahmed Y, Daròs JA, Mohamed MA (2020) Reduced graphene oxide nanosheet-decorated copper oxide nanoparticles: a potent antifungal nanocomposite against fusarium root rot and wilt diseases of tomato and pepper plants. Nanomaterials 10(5):1001
77. Chen J, Sun L, Cheng Y, Lu Z, Shao K, Li T, Hu C, Han H (2016) Graphene oxide-silver nanocomposite: novel agricultural antifungal agent against *Fusarium graminearum* for crop disease prevention. ACS Appl Mater Interfaces 8(36):24057–24070
78. Chen J, Wang X, Han H (2013) A new function of graphene oxide emerges: inactivating phytopathogenic bacterium *Xanthomonas oryzae* pv. Oryzae. J Nanoparticle Res 15:1–4

79. Kole C, Kole P, Randunu KM, Choudhary P, Podila R, Ke PC, Rao AM, Marcus RK (2013) Nanobiotechnology can boost crop production and quality: first evidence from increased plant biomass, fruit yield and phytomedicine content in bitter melon (Momordica charantia). BMC Biotechnol 13(1):1
80. He A, Jiang J, Ding J, Sheng GD (2021) Blocking effect of fullerene nanoparticles (nC60) on the plant cell structure and its phytotoxicity. Chemosphere 278:130474
81. Li H, Huang J, Lu F, Liu Y, Song Y, Sun Y, Zhong J, Huang H, Wang Y, Li S, Lifshitz Y (2018) Impacts of carbon dots on rice plants: boosting the growth and improving the disease resistance. ACS Appl Bio Mater 1(3):663–672
82. El-Ganainy SM, Mosa MA, Ismail AM, Khalil AE (2023) Lignin-loaded carbon nanoparticles as a promising control agent against Fusarium verticillioides in maize: physiological and biochemical analyses. Polymers 15(5):1193
83. Shekhawat GS, Mahawar L, Rajput P, Rajput VD, Minkina T, Singh RK (2021) Role of engineered carbon nanoparticles (CNPs) in promoting growth and metabolism of Vigna radiata (L.) Wilczek: insights into the biochemical and physiological responses. Plants 10(7):1317
84. Liang T, Yin Q, Zhang Y, Wang B, Guo W, Wang J, Xie J (2013) Effects of carbon nanoparticles application on the growth, physiological characteristics and nutrient accumulation in tobacco plants. J Food Agric Environ 11(3/4):954–958
85. Kumar A, Singh A, Panigrahy M, Sahoo PK, Panigrahi KC (2018) Carbon nanoparticles influence photomorphogenesis and flowering time in Arabidopsis thaliana. Plant Cell Rep 37:901–912
86. Kanhed P, Birla S, Gaikwad S, Gade A, Seabra AB, Rubilar O, Duran N, Rai M (2014) In vitro antifungal efficacy of copper nanoparticles against selected crop pathogenic fungi. Mater Lett 115:13–17
87. Giannousi K, Avramidis I, Dendrinou-Samara C (2013) Synthesis, characterization and evaluation of copper based nanoparticles as agrochemicals against Phytophthora infestans. RSC Adv 3(44):21743–21752
88. Zhang H, Yue M, Zheng X, Xie C, Zhou H, Li L (2017) Physiological effects of single- and multi-walled carbon nanotubes on rice seedlings. IEEE Trans Nanobiosci 16(7):563–570

Recent Research on the Use of Carbon Nanomaterials in Plant Growth and Development

Kalash Aggrawal, Versha Dixit, Anish Kumar Pal, Kundan Kumar Chaubey, Saswati Barman, Shreya Pandey, Sanjana Rajawat, Teeya Khandelwal, and Megha Gangwar

Abstract Carbon nanomaterials are new and adaptable tools that can be used to improve many areas of plant growth and development, such as plant physiology, morphology, and the generation of primary and secondary metabolites. Nanotechnology has advanced and pioneered in science and technology over the past few years with the potential to change technologies in business, medicine, and agriculture. Carbon-based nanomaterials include semiconductors, zero-valent metals, quantum dots, nanopolymers, and dendrimers, with different properties such as nanofiber, nanowire, and nanosheets. Carbon-based nanomaterials are beneficial for plant growth, development, and protection against abiotic stress such as drought, high soil salinity, heat, cold, and other oxidative stress. The potential application of carbon nanomaterials, such as carbon nanotubes, graphene, and carbon nanodots, is increasing in plant development. These materials with unique properties have drawn much attention, which has led to the current large-scale expansion of industrial production. Regardless of the application area, this is also linked to an upward trend in the purposeful or accidental release of carbon nanoparticles into the environment, where it is still difficult to forecast how these materials may affect living things. The various kinds of carbon-based nanomaterials, the production methods, and the current developments in agricultural and environmental approaches have been covered in this chapter. The important information on the use of carbon nanoparticles

---

K. Aggrawal
School of Biotechnology, Gautam Buddha University, Greater Noida, Uttar Pradesh 201312, India

V. Dixit
Department of Life Sciences, Christ University, Bangalore, Karnataka 560029, India

A. K. Pal
Division of Research and Innovation, School of Applied and Life Sciences, Uttaranchal University, Arcadia Grant, P.O. Chandanwari, Premnagar, Dehradun, Uttarakhand 248007, India

K. K. Chaubey (✉) · S. Barman · S. Pandey · S. Rajawat · T. Khandelwal · M. Gangwar
Department of Biotechnology, School of Basic and Applied Sciences, Sanskriti University, Mathura, Uttar Pradesh 281401, India
e-mail: kundan2006chaubey@gmail.com

A. Bachheti et al. (eds.), *Carbon-Based Nanomaterials*, Smart Nanomaterials Technology, https://doi.org/10.1007/978-981-97-0240-4_7

is summarized, as is the current state of research pertaining to the effect of carbon nanomaterials on plant growth and development.

Keywords Carbon · Nanotechnology · Graphene · Plant development · Fullerenes · Carbon-based nanomaterials

1 Introduction

Chemical fertilizers are frequently employed in conventional agricultural systems to increase output. Chemical fertilizer use, both excessive and continuous, contributes significantly, either directly or indirectly, to altering environmental circumstances. A new sustainable farming method is needed to address this problem and ensure food security for the world's expanding population. Pesticides and chemical fertilizers increased agricultural output, mostly due to the growing global population and lack of land resources. The increased usage of pesticides and chemical fertilizers to boost agricultural productivity was primarily driven by the world's expanding population and scarcity of land resources. The following are some of the negative consequences and side effects of careless and excessive usage of chemical fertilizers: Fertilizers are more expensive than organically generated manure since they are manufactured in factories. However, because it is essential for plant nutrition, it is expensive due to its high demand. Fertilizers should be used cautiously, or they risk harming the roots and other plant tissues and maybe killing the plant. Fertilizers are administered according to the needs of the plant. Unnecessary plant use may damage their health, particularly if cultivated in rich, fertile soil [1]. These dangerous fertilizers affect both humans and plants. It may introduce potentially dangerous substances into our food, affecting Fertilizers that influence the ecosystem, either directly or indirectly, through soil and groundwater pollution. Fertilizers can impact both soil quality and soil microbial health. Long-term use affects soil pH, limiting naturally beneficial microbial activity for plants [2].

2 Overview of Carbon Nanotubes and Graphene

Because of their remarkable optical, electrical, mechanical, and thermal capabilities, carbon nanomaterials (CNMs), a form of engineered nanomaterial (ENM), are becoming more and more popular. CNM When it comes to progressed dramatically during the previous twenty years. Heat depends on the production and pressure circumstances, carbon adopts thermodynamically advantageous configurations ranging from sp^3 to sp^2, resulting in structures like nanodiamonds and sheets of graphene. According to Mauerer and Elimelech (2008), CNMs combine nanoscale quantum phenomena with resonant structural stability and customizable unique physicochemical features. Allotropy refers to a chemical element's ability to build distinct molecule structures with the same type of atoms. Carbon is an interesting element because it can create diverse structures with fundamentally different properties. A few difficulties facing agriculture today include climate change, plant pathogenic organisms, contagious crop diseases, a lack of nutrients in the soil, decreased crop yields, ignorance of genetically modified crops, and a labor shortage [3]. These difficulties endanger agricultural sustainability. Salt, drought, high temperatures, and other extreme climatic conditions are becoming more common as a result of climate change, and they may have a significant yearly influence on world food output, yield, and quality. Climate change has increased the frequency and severity of abiotic stressors, notably drought and high temperatures, causing substantial losses in essential cereal crops, including wheat, maize, and barely [4]. Furthermore, biotic pressures such as pests and diseases lowered agricultural productivity by 20–30% yearly, regarded as the most important discouraging barrier to achieving global food security [4]. Several different tactics for increasing agricultural sustainability have been utilized in previous decades, including crop yield rotation, satellite agriculture, metropolitan agriculture, commercial fertilizers and pesticides, genetic modification, selective plant breeding, and gene editing techniques. Crop breeding approaches have also included whole-genome sequence-based methodologies, physical mapping, soma clonal variations, genetic selection, and mutagenesis breeding. In actuality, not many food crops, including maize, soybean, canola, rice, potato, papaya, squash, and apple, have genetic engineering variations that utilize cutting-edge biotechnology to add, delete, or rearrange certain genes [5]. In contrast, typical commercial breeding produces hundreds of crop types yearly to boost agricultural production, food security, nutritional diversity, and consumer choice. Choosing good parent plants helps traditional plant breeding create new combinations in the following generation. These strategies have been created to enhance commercially beneficial properties in plants, such as agriculture production, by boosting stress tolerance and breeding such crops with varied stress-tolerant features [6]. However, there are a number of important disadvantages to current techniques, such as the insufficient availability of certain crop genes, substantial ecological repercussions, very low success rate, lengthy process, and public worries about GM crops. Another method is chemical priming, a novel alternate method. It can help crops endure environmental difficulties by establishing defense systems without applying genetic alterations. In order to address these difficulties, farmers have begun to use agrochemicals excessively to decrease loss and increase crop yields. However, extreme use of conventional fertilizers produces major environmental difficulties such as groundwater contamination, eutrophication of the water supply, air pollution, and soil degradation [7]. Furthermore, chemical fertilizers are useless due to volatilization and leaching, both harming the environment and raising production costs. This makes achieving satisfactory agricultural sustainability challenging [8].

With regard to production and prospective applications, an emphasis on the agricultural and environmental sectors, as well as the significance of developing new useful products and technologies [9]. The potential effects on living things, particularly plants, are essential to food chains in both agricultural and natural environments.

In these environments, it is possible to expect an increase in the carbon nanomaterials added as a result of either deliberate use in agricultural and environmental applications or unintentional contaminants released accidentally [10].

3 Carbon Nanotube Classifications

There are two forms of carbon nanotubes:

(1) Single-layer graphene is used to create single-walled carbon nanotubes. For its synthesis, a catalyst is needed. SWCTs lack a complicated structure and have low levels of purity. It is easily twistable [11].

(2) Numerous-walled carbon Nanotubes are made up of many graphene layers. For its synthesis, no catalyst is necessary. High purity and complex structure characterize MWCTs. It is difficult to twist. Since the first graphene manufacturing was disclosed in 2004, it has been widely explored by the scientific and engineering community for over fifteen years. Graphene, a single layer of two-dimensional carbon atoms in a hexagonal lattice structure, has been used extensively in various applications, including electronics, energy-storing batteries, supercapacitors, fuel cells, and solar cells. This is because of its high surface area, thermal conductivity, electrical conductivity, and mechanical strength. Graphene is the name of an allotrope of carbon made of hexagons. Fullerenes (0D), nanotubes (1D), and graphite (3D) are some other allotropes. Graphene-based nanomaterials include reduced graphene oxide, graphene quantum dots, and graphene oxides. Sp2 is the optimum component of graphene. The scientific and technical communities have been studying graphene extensively for more than 15 years because it is a two-dimensional sheet of carbon atoms with sp2 hybridization resulting from the mixing of S, Px, and Py, orbitals. Each carbon atom's remaining pz orbital establishes bonds with three of its neighbors, forming the valence band of vacant orbitals, or the conduct band. The sigma bonds that make up the core of the hexagonal structure are formed by three of the four valence electrons that carbon has. The last electron makes a third of a connection with the carbon atom next to it. Because of these weak out-of-plane interactions, out-of-plane electrical and thermal conductivities are around 103 times lower than those of in-plane interactions. A single layer of graphene comprises Carbon atoms organized in a hexagonal pattern. Bi-layer graphene has two layers, while few-layer graphene has three to ten sheets in two dimensions. Thick graphene has less scientific relevance and is defined as having more than 10 layers. Carbon atoms can be packed in bilayer and few-layer graphene in various ways, such as Bernal or AB stacking, rhombohedral or ABD stacking, and hexagonal or AA stacking. Layers are wrapped at a little angle in twisted bilayer graphene.

4 Importance of Nanomaterials in Agriculture

Nanotechnology has enormous potential for improving people's life quality applications in agricultural and food systems. The development of novel agro-products based on nanotechnology, such as nanosensors, nanofertilizers, nanopesticides, and nanoformulations of biocontrol agents, is now under consideration [12–29]. A number of nanomaterials have been advocated for use in farming to aid in reducing pesticide consumption through smart delivery systems and limit nutrient management. Plant breeding and genetics have also been investigated using nanotechnology-derived devices [30]. Additionally, agricultural products and by-products can be used to generate bio-nanotech companies. Nonetheless, the potential advantages of nanotechnology applications in agriculture still need to be expanded and have not been exploited to the same level as in other industrial sectors (Fig. 1). Agricultural nanotechnology research is being heavily pursued in the search for by using cutting-edge methods for tracking, evaluating, and managing agricultural practices, it is possible to find answers to issues with sustainability, higher production, disease management, and crop protection in the agricultural and environmental fields [31]. Utilizing biosensors with enhanced technology in molecular biology, nanomaterials, and microfluidics has huge possibilities for crop productivity. These are also used to track the soil's microorganisms' activity and ability to forecast the likelihood of soil illness. The primary idea behind soil testing with a biosensor is to compare positive and negative effects of bacteria in soil based on variations in oxygen utilization during their breathing [32]. They also provide numerous chances for sensing pollutants and their effects by utilizing new nanomaterial characteristics [33, 34]. For methyl salicylate and azelaic acid, Biosensors for measuring nitrate concentrations in plants and identifying unhealthy plant markers have been examined for methyl salicylate and azelaic acid [35]. Biosensors are used to monitor citrus fruit with *penicillium digitatum* infection sensors are used to aid in the efficient use of natural agricultural resources such as nutrients, water, and chemicals through precision farming techniques such as satellite monitoring, geographic systems, and remote detecting tools that can detect pests on crop [36]. Smart delivery technologies and nanosensors indicators of strain such as drought. Using separate sensors linked to real-time GPS tracking plays an important role in tools with nanotechnology assistance. Due to their effective response devices for detecting gases and microorganisms in food nanosensors are increasingly popular and playing a significant role in the growth of the food industry [37]. Allergens have also been discovered. The study is about to be commercialized using biosensors and nanoparticles. Additionally, these devices can monitor temperature, time, and expiration dates (Table 1).

Agricultural products enter our lives in a variety of ways, including food, fuel, furniture, textiles, and feedstock. Agricultural productivity, on the other hand, is severely hampered by a lack of space, illness, and changes in agro-climatic conditions [43]. It has been established that some practices, including the use of pesticides and fertilizers to boost crop yield, are grave and lethal acts. As a result, there is an urgent need to modernize agricultural practices and processes using next-generation

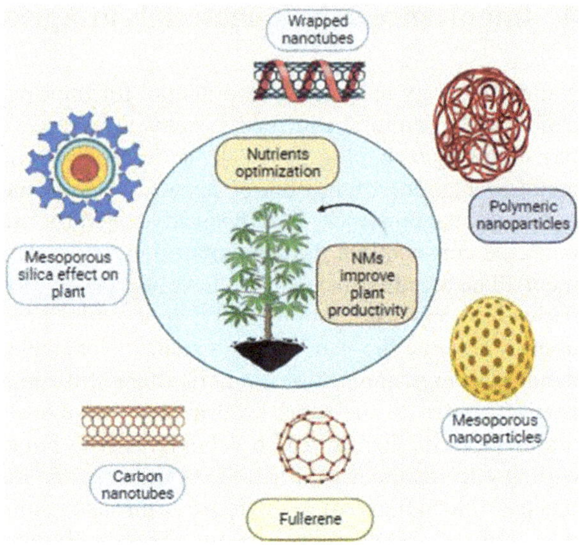

Fig. 1 Industrial use of CNTs

Table 1 Benefits and challenges of carbon nanoparticles in agricultural

S. no.	Prospects for agricultural	Potential benefits and challenges	References
1	Enhanced nutrient delivery	Improved nutrient uptake and utilization by plants Reduced fertilizer use and environmental impact	[38]
2	Enhanced pest control	Potential for controlled release of pesticides Reduced pesticide runoff and environmental contamination	[39]
3	Soil remediation and improvement	Enhanced soil structure and water retention Removal of contaminants from soil	[40]
4	Disease resistance enhancement	Increased plant immunity and resistance to pathogens Reduced need for chemical fungicides	[3]
5	Increased crop yield	Improved photosynthesis and plant growth Higher crop yields and quality	[41]
6	Environmental concerns	Potential ecological impacts and unintended consequences Need for rigorous safety assessment	[42]
7	Regulatory and ethical considerations	Ongoing regulatory scrutiny and public perception Ethical concerns regarding nanomaterial use in agriculture	[28]

technologies. This is where the application of nanotechnology in agriculture comes into play. Agriculture uses nanotechnology to increase agricultural output while guaranteeing food safety and nutritional value. To increase crop output, it is necessary to use nanocarriers, which enable the delayed release and administration of these nano-agrochemicals, to administer fertilizers, insecticides, herbicides, and plant growth regulators (Figs. 2 and 4) [44].

Nanosensors can detect nano-agrochemicals and pathogens in food and agricultural systems at trace levels. According to a recent study, copper-doped montmorillonite monitors propine b fungicide in aquatic environments (both fresh and saline water). Graphene nanoparticles can detect pathogens in wastewater and cleanse, making them useful in aquaculture. Gold, silver, copper, and carbon nanotubes are also utilized as nanosensors for real-time monitoring in agriculture [30]. The seed determines the production of any crop. Before delivering the seeds to farmers for planting, they were routinely checked for germination. Despite being examined in well-equipped laboratories, the seeds do not replicate owing to moisture conditions in rain-fed locations. Developing methods for rain-fed agriculture is critical, particularly in India, where rain-fed agriculture accounts for more than 60% of the net planted area. Researchers are focusing on producing metal oxide nanoparticles and carbon nanotubes to boost the fertilization of rain-fed crops [45]. Carbon nanotubes pierce the seed coat, creating a new hole for water permeation and acting as a conduit to transfer water from the substrate to the seeds, which is used in rain-fed agricultural systems to aid fertilization.

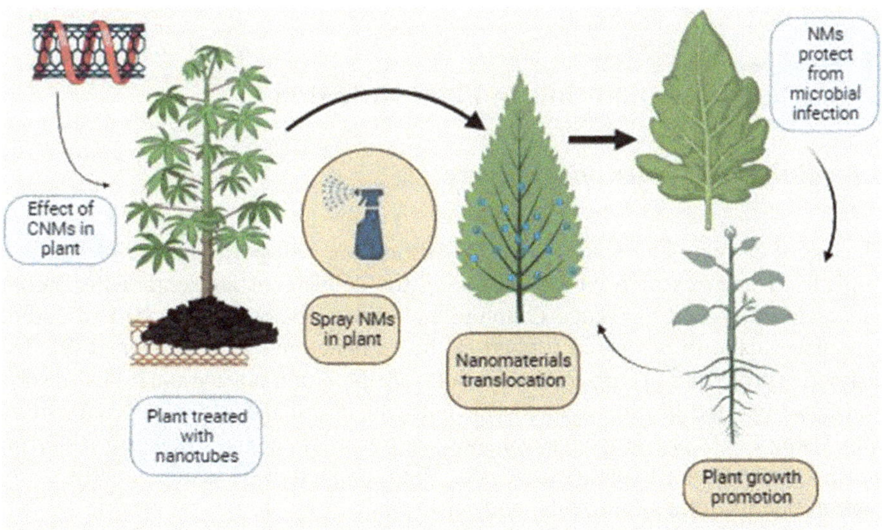

Fig. 2 Carbon nanomaterials enhanced plant growth and protect from microbial infection

5 Synthesis of Carbon Nanotubes

Commercial Production of Nanomaterials Based on Carbon. Nanotubes are made up of carbon. The fundamental methods of CNT production include laser ablation, chemical vapor deposition (CVD), and arc discharge. CVD is currently one of the most researched and widely utilized processes for CNT synthesis. Other techniques, such as Arc discharge and laser ablation, require more complicated machines and more severe temperature and pressure conditions. CVD synthesis is more suitable for the production of large-scale CNT. The foundations of CVD synthesis are hydrocarbon decomposition into the carbon, followed by the formation of carbon nanostructures on diverse substrates, including growth catalysts for nanotubes. Single-walled carbon nanotubes (SWCNTs) diameter is 0.5–5 nm, while multi-walled carbon nanotubes (MWCNTs) is 8–100 nm. Metal nanoparticles are often used as catalysts and their size is significantly related to the diameter of the nanotubes formed. SWCNT and MWCNT typically use nickel, cobalt, or iron nanoparticles as catalysts [46]. CVD reactors typically comprise inert gas tubes, a reaction chamber, and hydrocarbons. Methane is commonly employed in creating SWCNTs, while ethylene or acetylene is used in producing MWCNTs. To summarize the method, the substrate is heated to 850–1000 °C for SWCNT manufacturing and 550–700 °C for MWCNT manufacturing. By heating hydrocarbons, carbon is created, which dissolves into metal nanoparticle catalysts. When a certain carbon concentration is obtained due to the continuous supply of hydrocarbon carbon, it produces a semi-fullerene cap as the starting structure for producing a nanotube of a cylindrical shell [47].

6 Carbon Nanomaterials in Plant Nutrition

6.1 Enhanced Nutrient Delivery

The use of carbon nanomaterials for the absorption, transport, and use of nutrients in plants is new research called enhancing the delivery of nutrients using carbon nanomaterials based in plants. Graphite and carbon nanotubes are part of carbon nanomaterials with advanced properties that enhance nutrition delivery [48]. The distinctive properties of carbon nanoparticles are their high surface and robust absorption capacity. Modification of nanoscale nutrients in soil and attract and retain nutrients. Using CNTs and graphene as nanoscale transport channels makes transferring nutrients from the soil to plant roots easier. High conductivity ions transfer effectively. Carbon nanomaterials, nutrients, and fertilizers can be stored, preventing leakage and volatilization. Overuse of fertilizers causes environmental pollution caused by nutrient leakage, soil erosion, surface water, eutrophication, and nitrate contamination of groundwater. Nano-based fertilizers often contain nutrient stimulates encapsulated in nanoscale polymers, chelators, or emulsifiers that allow for a slow, targeted, and efficient release and thus release of potential for fertilizer overuse. Crops can

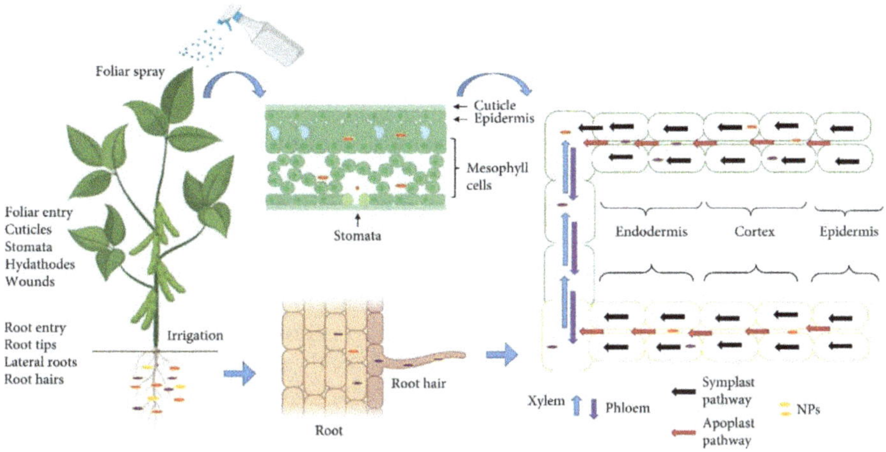

Fig. 3 Uptake, translocation and consequences of nanomaterials on plant growth and stress adaptation

get nutrients from nanofertilizers in three different ways: within nanomaterials like nanotubes or nanoporous materials, inside nanomaterials wrapped in thin protective polymer films, or as particles or nanoscale emulsions [49]. For instance, biodegradable polymer chitosan nanoparticles (78 nm) can slowly release encapsulated fertilizers, and in some situations, nanomaterials can promote plant growth (Fig. 3). Nano fertilizers have a significant potential to provide nutrients to plant roots more effectively because of the large surface area compared to the amount of nanomaterials. Because they remain in the pore water, collect nutrients from the roots, may enter through stems, and are not easily linked to soil, nanoscale fertilizers are more effective at supplying plants with nutrients. Reducing pollution and enhancing the value of agricultural practices help advance precision agriculture [50].

6.2 Role in Micronutrient Uptake and Stimulating Photosynthesis with Nanomaterials

For plants to tolerate drought, nanoparticles (NPs) are essential (Fig. 4). There are two primary ways (physical or chemical) to make NPs. An environmentally friendly procedure is the green synthesis of NPs. There are various types of NPs, including carbon-, metal-, and metal-oxide-based NPs. Under drought stress (DS), NPs boost gene expression and shield plants from harmful impacts [51]. Drought stress (DS) severely threatens worldwide agricultural output and food security. Under the current conditions of fast climate change and increasing drought intensity, nanoparticles (NPs) have emerged as a viable tool for increasing agricultural yield. DS has a deleterious impact on plant development and physiological and metabolic processes, as well as disrupting cellular membranes, nutrition and water intake, photosynthesis tools, and antioxidant functions (Fig. 5). The use of NPs preserves the membranes, maintains the water connection, and improves nutrient and water absorption, improving plant growth under DS. NPs protect the photosynthetic machinery and boost photosynthetic efficiency, osmolyte, hormone, phenolic accumulation, antioxidant activities, and gene expression, enabling plants greater resilience to DS. DS has a significant impact on crop root shape and geographical dispersion [52] found that DS lowered seed yield by decreasing photosynthesis, transpiration, and chlorophyll components [53] investigated the effect of DS on the production and quality of maize (Zea mays) and sorghum (Sorghum bicolor) and determined that DS had a significant impact on the substrate composition in these crops [53]. Nanotechnology (NT) has developed as a promising topic, with applications in agriculture, food, and medicine [54]. Titanium dioxide (TiO_2), iron oxide (Fe_3O_4), zinc oxide (ZnO), and other nanoparticles (NPs). Because of their non-threatening application in agriculture, silicon oxide (SiO_2), copper (Cu-NPs), and selenium (Se-NPs) have attracted substantial attention recently [55]. NPs may be produced via a variety of methods, including chemical, green, and physical procedures [56]. NPs favorably influence plant growth and development, however, these effects vary depending on the origin, size, concentration, and timing of crop administration [57]. NPs have recently increased plant tolerance to biotic and abiotic stressors. NPs protect plants from oxidative stress by boosting antioxidant activity [58]. NPs can mitigate drought-induced harmful effects by lowering hydrogen peroxide (H_2O_2) and malondialdehyde (MDA) buildup and preserving

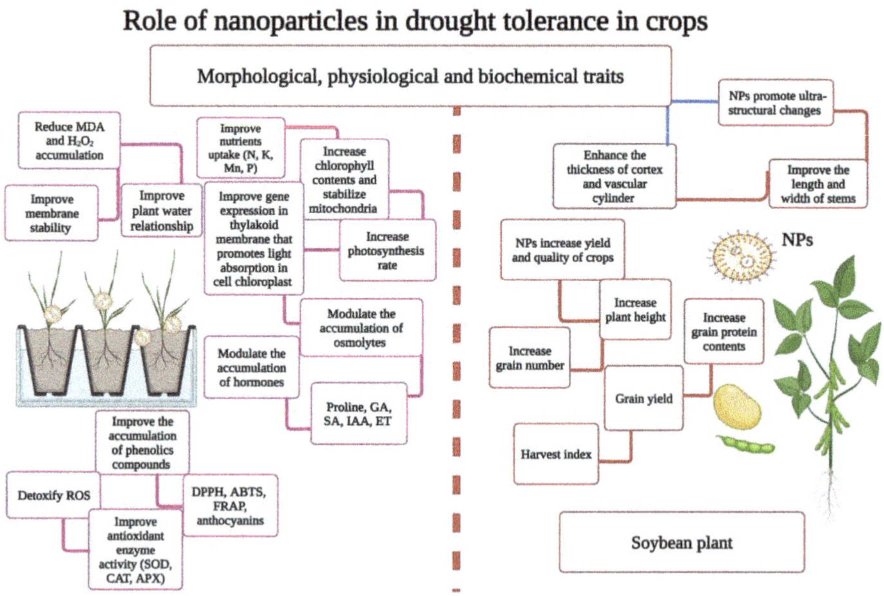

Fig. 4 Illustrate factors affecting how nanoparticles are absorbed, taken up, transported, and absorbed by plants

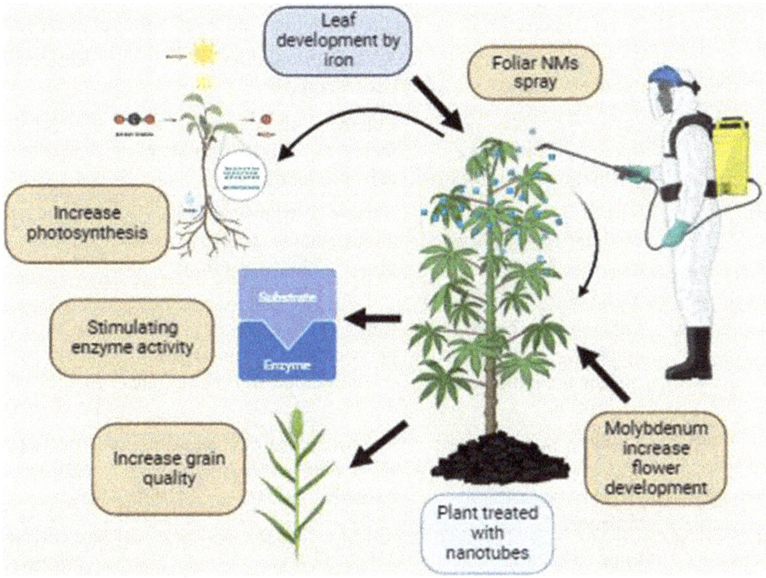

Fig. 5 Micronutrients and nanomaterials increase plant activity and stimulating photosynthesis

photosynthetic apparatus efficiency [58]. NPs may also permeate plant chloroplasts and reach the photo system-II (PS-II) reaction center, increasing electron transmission and light absorption in chloroplasts under DS and, therefore, boosting photosynthetic efficiency and plant development. The usage of NPs (TiO_2) also boosts nitrate reductase activity and osmolyte buildup. Increases in NR activity caused by NP cause NO production, which boosts proline and glycine betaine (GB) synthesis and protects plants from oxidative damage [59]. NPs also greatly enhance the accumulation of glucose, fructose, trehalose, and sucrose, which improves drought tolerance by increasing antioxidant activities. By scavenging ROS, NP-mediated increases in antioxidant activity reduce the negative effects of DS on plants (Table 2).

6.3 Improved CO_2 Fixation

Carbon nanoparticles can potentially play a significant role in improving CO_2 fixation and addressing climate change. CO_2 fixation refers to capturing and converting carbon dioxide (CO_2) from the atmosphere into a stable form, thereby reducing its concentration in the atmosphere and mitigating the greenhouse effect. Here's how carbon nanoparticles can contribute to this process.

1. Carbon nanoparticles can act as catalysts in chemical reactions involving CO_2 fixation. They can enhance the efficiency of CO_2 conversion processes, such as converting CO_2 into valuable chemicals or fuels through electrochemical or

Table 2 Role of carbon nanoparticles in micronutrient uptake and stimulating photosynthesis

Carbon nanomaterial	Role in micronutrient uptake	Role in stimulating photosynthesis	References
Carbon nanotubes (CNTs)	Serve as carriers for micronutrients and facilitate their transport	Enhance light absorption by, increasing energy into plant cells production	[60]
Graphene	Improves nutrient availability in soil by increasing cation exchange capacity	Enhance electron transfer photosystems, accelerating photosynthetic reactions	[61]
Carbon quantum dots (CQDs)	Enhance micronutrient uptake by increasing root surface area and nutrient adsorption	Improve the efficiency of photosystem II, leading to increased photosynthetic rates	[62]
Fullerene (C60)	Assist in micronutrient uptake by enhancing root growth and nutrient transport	Protect chlorophyll and other photosynthetic pigments from oxidative stress	Husen et al. (2014)
MWCNTs	Improving crop production through improved efficiency during propagation	MWCNT reduced photosynthesis pigments and increased oxidative stress	Rahmani et al. (2020)
SWCNTs	Micronutrients are needed in very small amounts for the development of plant	SWCNTs protect the photosynthesis of algae from photoinhibition	[63]
Graphene dots	Toxic effect on root and shoot elongation is the exposure of graphene dots inhibited the expansion of the roots	Graphene-based nanomaterials can damage the photosystem by destroying the structure of the chloroplast and decline Rubisco enzymes activity	[64]

photochemical reactions. These nanoparticles can facilitate the activation of CO_2 molecules and promote their transformation into more useful forms.

2. Carbon nanoparticles have a high surface area, which makes them effective adsorbents for capturing CO_2 from the atmosphere or industrial emissions. Their porous structure allows for the physical adsorption of CO_2 molecules, helping to reduce atmospheric CO_2 levels. Functionalizing carbon nanoparticles with specific groups or chemicals can enhance their CO_2 adsorption capacity.
3. Carbon nanoparticles can be used as carriers or sorbents for CO_2 capture and storage technologies. They can help absorb CO_2 from power plants, industrial processes, or even directly from the air and store it in geological formations or other secure storage sites.
4. Researchers are developing carbon-based nanocomposites that incorporate carbon nanoparticles into materials like polymers or concrete. These nanocomposites can provide structural support while simultaneously facilitating CO_2

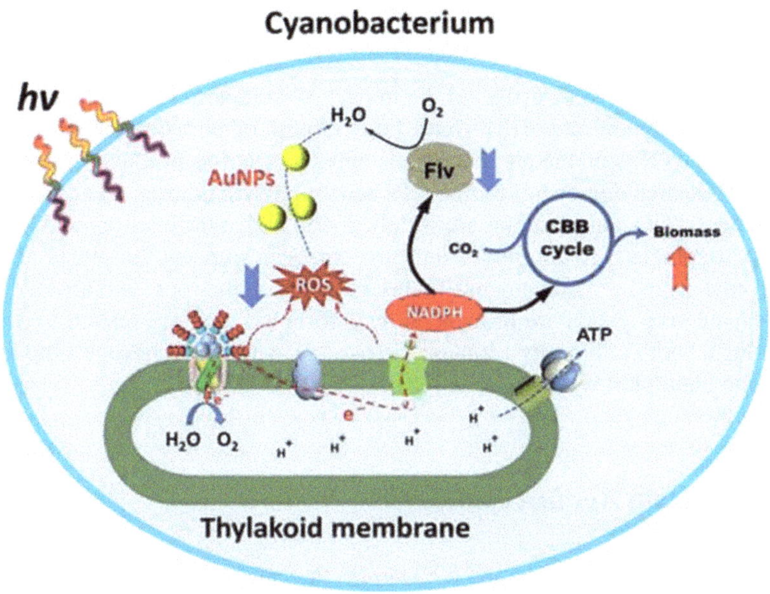

Fig. 6 Falling of light on thylakoid membrane stimulating photosynthesis

adsorption or conversion. They can be used in construction materials that could passively capture and store CO_2 over time [65].

5. Carbon nanoparticles can be used to enhance photosynthesis in plants. By introducing nanoparticles with specific properties into plant leaves, researchers aim to increase the efficiency of CO_2 uptake and fixation during photosynthesis. This can lead to higher crop yields and enhanced carbon sequestration in forests and agriculture (Fig. 6).
6. Carbon nanoparticles can also be incorporated into energy storage systems such as supercapacitors and batteries. These systems can help integrate renewable energy sources more effectively, reducing the need for fossil fuel-based power generation and, in turn, reducing CO_2 emissions.
7. Carbon-based nanoparticles can be used as co-catalysts in photoelectrochemical water-splitting processes, generating hydrogen fuel while simultaneously reducing CO_2 to form other valuable products [66].

It's important to note that while carbon nanoparticles are promising to improve CO_2 fixation, there are challenges to address, including scalability, cost-effectiveness, and potential environmental impacts. Additionally, research in this field is ongoing, and new developments are continually emerging. The successful implementation of carbon nanoparticles in CO_2 fixation strategies will likely require interdisciplinary collaboration among scientists, engineers, and policymakers [67].

6.4 Increased Chlorophyll Production

Numerous writers have investigated the impact of nanoparticles on plants and their ability to photosynthesize. There have been reports of both long-term and short-term impacts of NPs on the photosynthetic apparatus and its productivity. However, numerous research demonstrates that NPs benefit photosynthesis, plant growth, and development. CuO nanoparticles are widely distributed in the environment and must impact how plants perform photosynthesis. Nitrogen impacts the plants ability to photosynthesis and the movement of nutrients through the roots and above ground. The ultimate crop yields are influenced by chlorophyll and are related to nitrogen levels. Sugar beet's chlorophyll content can rise due to increased nitrogen application, raising the plant photosynthetic rate [68].

6.5 Protection Against Pathogens

Carbon nanoparticles have distinctive features that successfully impact CNT application in plant growth and diagnostic techniques of plant pathogens. The distinctive electrochemical properties of carbon nanoparticles facilitate using CNTs as constituents in biosensors. Carbon-based nanomaterials can successfully control, diagnose, and decrease fungal aflatoxins [69].

6.6 Biochemical and Physiological Effects

6.6.1 Influence on Enzymatic Activity

One requirement for the biomedical application of carbon-based nano enzymes is high catalytic activity. Through approaches including morphological control, surface modification, and heteroatom doping, researchers have recently achieved the rational design of some carbon-based nano enzymes using experimental methods and theoretical calculations. Carbon-based nano enzyme's design and activity regulation methods will be examined (Fig. 9) [70].

6.6.2 Changes in Plant Metabolism

Significant technological advances and improvements in agriculture have been developed in recent years to tackle the mounting difficulties of sustainable agricultural production and food security [71]. To fulfill the demands of 9 billion people, the globe must produce 50% more food by 2050. Because land and water supplies are limited, this aim can only be achieved by technical interventions to boost production.

The primary challenges that have brought academics' attention to adopting alternative approaches include a need for more arable land, irrigation, and dependency on traditional crops [72]. Because of the evident exceptional and extraordinary capabilities of nanoparticles, it is not unexpected that attempts have been made to enhance the agriculture sector utilizing nanotechnology and nanomaterials. Among the nanomaterials, metal oxides, silicates, ceramics, magnetic materials, semiconductors of quantum dots (QD), polymers, lipids, dendrimers, and emulsions. The objectives are to reduce the amount of plant protection products (PPP) used, reduce nutrient losses during fertilization, and increase income in the agriculture industry through better nutrient management. Many elements, including well-known effects, possible toxicity, monitored destiny, and overdosing levels, depend on the rising usage of nanoparticles (NPs) in agriculture [73]. Plants are important to all ecosystems, and NPs may communicate with them. As a result, it is possible to deduce that NPs may communicate with plants and that these interactions, such as NP ingestion and accumulation in plant biomass, influence their environmental destiny and transit. However, nanomaterials may enter live plant tissues, their aggregation in the environment and efficacy as smart delivery systems in living plants have consequences [74]. It is critical to determine if plants can consume intact NPs and transport them to other plant tissues. NPs can enter plant tissue through root tissues, above-ground tissues (e.g., cuticles, stomata, hydathodes, and trichrome), root junctions, and wound areas (Fig. 7).

Plant scientists have reported that nanomaterials improve agronomic capacities. NPs are employed as sensing materials, herbicides, insecticides, and nanofertilizers

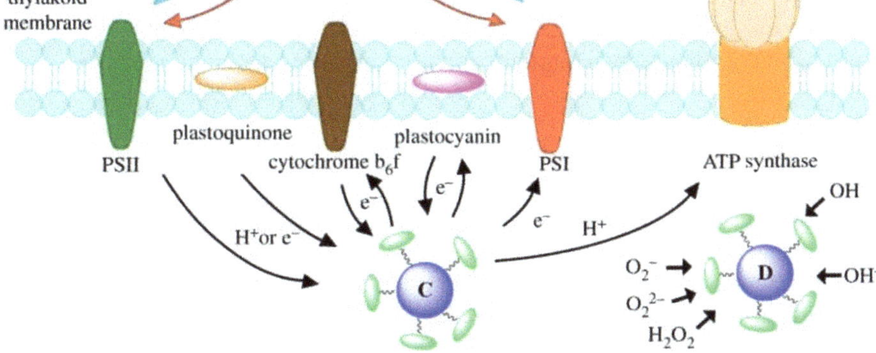

Fig. 7 Potential advantageous interactions between NPs and photosynthetic infrastructure

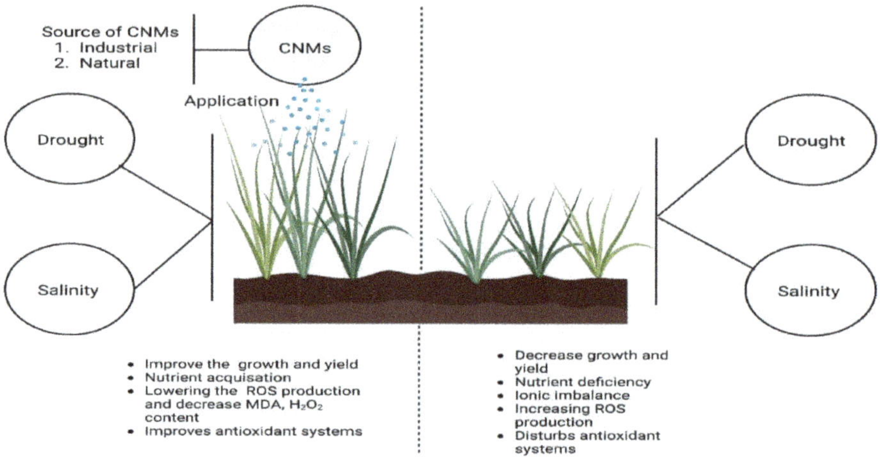

Fig. 8 Carbon nanomaterials and stress tolerance

to increase plant nutrition and as agrochemical carriers for controlled release. Nanomaterials also hasten plant adaptability to environmental changes and improve plant tolerance [75]. However, rather than serving as nanocarriers, some nanoparticles (NPs) with specific physicochemical features naturally boost plant growth and stress tolerance (Fig. 8). The biological activity of nanoparticles is determined by their physicochemical qualities, the mode of administration (soil, foliar, and hydroponics), and the concentration utilized. Nanomaterials have also been shown to play an important role in plant tissue in vitro cultures, including callus induction, organogenesis, somatic embryogenesis, and secondary metabolite synthesis [76]. Such research indicates that plant cells may acquire nanoparticles through their cell walls, although the process is yet unknown. Although the use of nanoparticles for DNA transport in plant cells has previously been investigated, significant progress has been achieved in recent years. This research will look at the function of nanotechnology in the transport of NPs and the usage of NPs in increasing seed germination and plant growth under natural and environmental stress conditions [77]. Furthermore, we discuss recent nanotechnology breakthroughs and accompanying problems (Fig. 9).

6.6.3 Potential Ecological Impacts

Environmental emissions will inevitably result from production, usage, and disposal. They can enter the environment through wastewater treatment streams, landfills, and burning products containing nanomaterials, though most likely in altered forms from their original counterpart [78]. Additionally, some nanoparticles are used in environmental cleanup processes, and as a result, they are dispersed in the environment as key nanoscale dissolving agents.

1. Speciation (i.e., joining forces with other chemically dissolved ions or molecules)

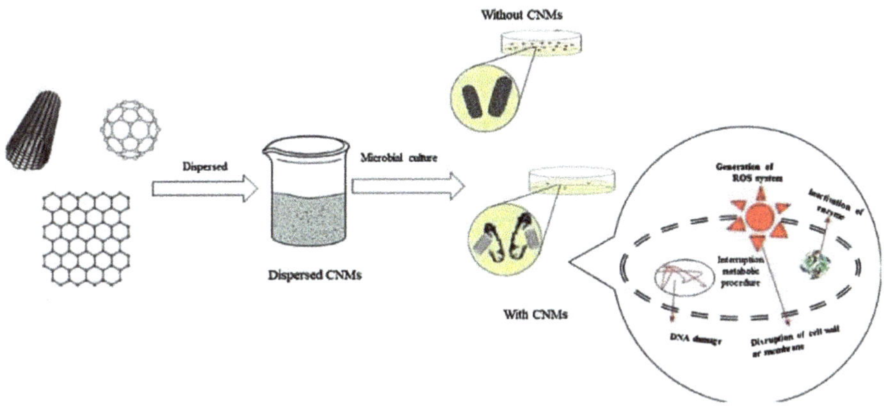

Fig. 9 Carbon nanomaterials influence the enzymatic activity and prevent DNA damage

2. Complete mineralization (to carbon dioxide and water) and/or biological or chemical modification into other substances
3. Aggregation and disaggregation
4. Complacent

When evaluating the potential effects of nanomaterials on the environment, exposure concentrations or dosages should be taken seriously. It is important to analyze the nanomaterial in all of its manifestations, including all physical and chemical species and the free nanoparticulate form [79]. It is critical to understand that because of their propensity to collect, agglomerate, and possibly adsorb to or interact with biological matter, nanomaterials may congregate in some "hot spots." Additionally, it is anticipated that some of the nanomaterials going through the stream of traditional waste treatment may get attached to the solid phase and deposit in areas of the environment where they may be subjected to higher loads. According to Lead and Wilkinson, the main determinants of colloid stability in natural settings are pH, ionic strength, and the presence of organic matter. In seawater with high pH and ionic strength compared to freshwater, electric double layers of colloid particles are substantially thinner, allowing for a closer interparticle approach, often resulting in increased aggregation [80]. Additionally, the fate and behavior of the materials will be influenced by their intrinsic qualities and characteristics, particularly their special chemistry. Nanomaterials' surface properties play a crucial role in their aggregation behavior, which affects their mobility in terrestrial and aquatic systems and their interactions with and general bioavailability to organisms.

The "brown waters" humic and fulvic acids will coat the nanoparticles, potentially keeping them around longer and dispersing them more evenly. For instance, natural organic matter (NOM) and iron oxide in freshwater systems have a stabilizing effect on fullerene and carbon nanotube suspensions in the water. Contrarily, Baalousha et al. showed that the effects of humic acids and changing pH can have combined effects on the fate of iron oxide nanoparticles, with rising pH leading to a higher level of aggregation. A similar effect has recently been shown with CeO_2 nanoparticles. Nanomaterials' environmental persistence (i.e., resistance to transformation and degradation) is influenced by the chemical composition of both the core and the surface material. The majority of nanomaterials can persist in their initial particle state, although this cannot be expected in all cases. Although there is a paucity of data in this area, the organic coats of nanoparticles seem to be easily altered or removed.

As was already mentioned, it is feasible for at least some metal nanoparticles to disintegrate. Whether the substance is destroyed during the dissolution process, nanoparticles disappear and lose their persistent properties. In standard tests for a chemical compound's readiness for biodegradation, the amount of dissolved organic carbon lost or CO_2 produced is considered. Therefore, it is crucial to establish whether the nanomaterial can be used to feed or supply energy to microorganisms. Second, the nanomaterial must be available to bacteria to be degradable. Testing in surface water may not be essential since biodegradation is unlikely if the substance is not soluble in water or unlikely to survive in the water column. The biodegradation screening techniques (such as for ready biodegradability) that measure dissolved carbon do not apply to nanomaterials containing carbon. Although they need a lot of test material, the techniques for measuring carbon dioxide generation or oxygen intake are theoretically practical. It is crucial to consider whether any carbon-based nanomaterials, including fullerenes and nanotubes, can degrade and under what circumstances. However, some information suggested that wood decay fungus might absorb fullerenes, indicating that the carbon in fullerenes might be digested.

7 Conclusion

CNPs were crucial in improving plant growth and biomass production in several crop plants. It also increases nutrient absorption, i.e., N, P, and K, which ultimately increases soil nutrient retention and enzyme activity. Despite all these positive effects, if CNP application is increased at a larger rate (>400 mg kg1), it might inhibit the growth of the plants. Rapid developments in nanotechnology promise advancements in various fields of science and technology. For industrial-scale production, nanomaterials, especially carbon ones, are developed for various implementations, including those in the agriculture and environmental fields. The actual concentrations and behavior of these elements in natural ecosystems and their interconnection with living organisms are required for safety assessments, yet more data is needed. This is made even more challenging by the large variety of nanomaterials, each of which has distinct properties, as well as the conformational changes that take place in carbon nanomaterials when they interact with the different elements of various incubation media (such as agglomeration), which may have an impact on bioavailability and toxicity. Despite these issues, at least a few essential traits of carbon nanoparticles that are crucial for comprehending how they interact with plants have been identified. It will be essential to perform a more thorough and systematic examination of the fundamental factors affecting how several carbon nanomaterials interact with living things and the environment to assess danger and characterize potential uses.

References

1. Pathak VM, Verma VK, Rawat BS, Kaur B, Babu N, Sharma A, Dewali S, Yadav M, Kumari R, Singh S, Mohapatra A, Pandey V, Rana N, Cunill JM (2022) Current status of pesticide effects on the environment and human health their it's eco-friendly management as bioremediation: a comprehensive review. Front Microbiol 13:962619
2. Bai YC, Chang YY, Hussain M, Lu B, Zhang JP, Song XB, Lei XS, Pei D (2020) Soil chemical and microbiological properties are changed by long-term chemical fertilizers that limit ecosystem functioning. Microorganisms 8(5):694
3. Singh BK, Delgado-Baquerizo M, Egidi E, Guirado E, Leach JE, Liu H, Trivedi P (2023) Climate change impacts on plant pathogens, food security, and paths forward. Nat Rev Microbiol 1–17
4. Sánchez-Bermúdez et al (2022) Effects of combined abiotic stresses related to climate change on root growth in crops. Front Plant Sci 13:918537
5. Animasaun DA et al (2023) Modern plant biotechnology: an antidote against global food insecurity. J Agron 13(8):2038
6. Neme K, Nafady A, Uddin S, Tola YB (2021) Application of nanotechnology in agriculture, postharvest loss reduction and food processing: food security implication and challenges. Heliyon 7(12)
7. Ayiti OE, Babalola OO (2022) Factors influencing soil nitrification process and the effect on environment and health. Front Sustain Food Syst 6:821994
8. Krasilnikov P, Taboada MA, Amanullah (2022) Fertilizer use, soil health, and agricultural sustainability. Agriculture 12(4):462
9. Sharma S, Verma K, Hardaha P (2023) Implementation of artificial intelligence in agriculture. JCCE 2(2):155–162
10. Aslani F, Bagheri S, Muhd Julkapli N, Juraimi AS, Hashemi FSG, Baghdadi A (2014) Effects of engineered nanomaterials on plant growth: an overview. Sci World J 2014
11. Rouf SA, Usman Z, Masood HT, Majeed AM, Sarwar M, Abbas W (2021) Synthesis and purification of carbon nanotubes. In: Carbon nanotubes-redefining the world of electronics. IntechOpen, London
12. Bachheti RK, Bachheti A, Husen A (eds) (2023) Nanomaterials for environmental and agricultural sectors. Springer, Singapore. https://doi.org/10.1007/978-981-99-2874-3
13. Husen A, Siddiqi KS (2014) Phytosynthesis of nanoparticles: concept, controversy and application. Nanoscale Res Lett 9(229):1–24
14. Husen A, Siddiqi KS (2014) Carbon and fullerene nanomaterials in plant system. J Nanobiotechnol 12(1):1–16
15. Husen A, Siddiqi KS (2023) Advances in smart nanomaterials and their applications. Elsevier Inc., Cambridge, MA, USA
16. Husen A (2020) Interactions of metal and metal-oxide nanomaterials with agricultural crops: an overview. In: Husen A, Jawaid M (eds) Nanomaterials for agriculture and forestry applications. Elsevier Inc., Cambridge, MA, USA, pp 167–197. https://doi.org/10.1016/B978-0-12-817852-2.00007-X
17. Husen A (2020) Carbon-based nanomaterials and their interactions with agricultural crops. In: Husen A, Jawaid M (eds) Nanomaterials for agriculture and forestry applications. Elsevier Inc., Cambridge, MA, USA, pp 199–218. https://doi.org/10.1016/B978-0-12-817852-2.00008-1
18. Husen A (2022) Engineered nanomaterials for sustainable agricultural production, soil improvement and stress management. Elsevier Inc., Cambridge, MA, USA

19. Husen A (2023) Nanomaterials and nanocomposites exposures to plants (response, interaction, phytotoxicity and defense mechanisms). Springer Nature Singapore Pte Ltd., Singapore, Singapore. https://doi.org/10.1007/978-981-99-2419-6
20. Husen A, Iqbal M (2019) Nanomaterials and plant potential. Springer International Publishing AG, Cham, Switzerland. https://doi.org/10.1007/978-3-030-05569-1
21. Kumar A, Choudhary A, Kaur H, Guha S, Mehta S, Husen A (2022) Potential applications of engineered nanoparticles in plant disease management: a critical update. Chemosphere 295:133798
22. Kumar A, Choudhary A, Kaur H, Mehta S, Husen A (2021) Metal-based nanoparticles, sensors and their multifaceted application in food packaging. J Nanobiotechnol 19(256):1–25. https://doi.org/10.1186/s12951-021-00996-0
23. Kumar A, Choudhary A, Kaur H, Mehta S, Husen A (2021) Smart nanomaterial and nanocomposite with advanced agrochemical activities. Nanoscale Res Lett 16(156):1–26
24. Sharma P, Pandey V, Sharma MMM, Patra A, Singh B, Mehta S, Husen A (2021) A review on biosensors and nanosensors application in agroecosystems. Nano Res Lett 16:136. https://doi.org/10.1186/s11671-021-03593-0
25. Siddiqi KS, Husen A (2022) Plant response to silver nanoparticles: a critical review. Crit Rev Biotechnol 42(7):773–990. https://doi.org/10.1080/07388551.2021.1975091
26. Vanti G, Belur S, Husen A (2023) Use of nanomaterials in agricultural sectors. In: Husen A, Siddiqi KS (eds) Advances in smart nanomaterials and their applications. Elsevier Inc., Cambridge, MA, USA
27. Wagay JA, Singh S, Raffi MM, Rahman QI, Husen A (2019) Effect of carbon-based nanomaterials on plant functioning and rhizosphere. In: Husen A, Iqbal M (eds) Nanomaterials and plant potential. Springer International Publishing AG, Cham, pp 553–575. https://doi.org/10.1007/978-3-030-05569-1_22
28. Yadav A, Yadav K, Ahmad R, Abd-Elsalam KA (2023) Emerging frontiers in nanotechnology for precision agriculture: advancements, hurdles and prospects. Agrochemicals 2(2):220–256
29. Husen A, Iqbal M (2019) Nanomaterials and plant potential: an overview. Springer International Publishing, pp 3–29
30. Prabha AS, Thangakani JA, Devi NR, Dorothy R, Nguyen TA, Kumaran SS, Rajendran S (2022) Nanotechnology and sustainable agriculture. In: Nanosensors for smart agriculture. Elsevier, pp 25–39
31. Behl T, Kaur I, Sehgal A, Singh S, Sharma N, Bhatia S, Bungau S (2022) The dichotomy of nanotechnology as the cutting edge of agriculture: nano-farming as an asset versus nanotoxicity. Chemosphere 288:132533
32. Nadporozhskaya M, Kovsh N, Paolesse R, Lvova L (2022) Recent advances in chemical sensors for soil analysis: a review. Chemosensors 10(1):35
33. Baruah S, Dutta J (2009) Nanotechnology applications in pollution sensing and degradation in agriculture: a review. Environ Chem Lett 17:191–204
34. Baruah S, Dutta J (2009) Nanotechnology applications in pollution sensing and degradation in agriculture: a review. Environ Chem Lett 7:191–204
35. Griesche C, Baeumner AJ (2020) Biosensors to support sustainable agriculture and food safety. TrAC 128:115906
36. Chalupowicz D, Veltman B, Droby S, Eltzov E (2020) Evaluating the use of biosensors for monitoring of Penicillium digitatum infection in citrus fruit. Sens Actuators B: Chem 311:127896
37. Warriner K, Reddy SM, Namvar A, Neethirajan S (2014) Developments in nanoparticles for use in biosensors to assess food safety and quality. Trends Food Sci Technol 40(2):183–199
38. Babu S, Singh R, Yadav D, Rathore SS, Raj R, Avasthe R, Singh VK (2022) Nanofertilizers for agricultural and environmental sustainability. Chemosphere 292:133451
39. Tudi M, Daniel Ruan H, Wang L, Lyu J, Sadler R, Connell D, Phung DT (2021) Agriculture development, pesticide application, and its impact on the environment. Int J Environ Res Public Health 18(3):1112

40. Tiwari M, Tripathy DB (2023) Soil contaminants and their removal through surfactant-enhanced soil remediation: a comprehensive review. Sustainability 15(17):13161
41. Muhie SH (2022) Optimization of photosynthesis for sustainable crop production. CABI Agric Biosci 3(1):1–8
42. Chu EW, Karr JR (2017) Environmental impact: concept, consequences, measurement. In: Reference module in life sciences
43. Tun Oo A, Boughton D, Aung N (2023) Climate change adaptation and the agriculture-food system in Myanmar. Climate 11(6):124
44. Fraceto LF, Grillo R, de Medeiros GA, Scognamiglio V, Rea G, Bartolucci C (2016) Nanotechnology in agriculture: which innovation potential does it have. Front Environ Sci 4:20
45. Mittal D, Kaur G, Singh P, Yadav K, Ali SA (2020) Nanoparticle-based sustainable agriculture and food science: recent advances and future outlook. Front Nanotechnol 2:579954
46. Moses JC, Gangrade A, Mandal BB (2019) Carbon nanotubes and their polymer nanocomposites. In: Nanomaterials and polymer nanocomposites. Elsevier, pp 145–175
47. Zaytseva O, Neumann G (2016) Carbon nanomaterials: production, impact on plant development, agricultural and environmental applications. Chem Biol Technol Agric 3(1):1–2
48. Mahor A, Singh PP, Bharadwaj P, Sharma N, Yadav S, Rosenholm JM, Bansal KK (2021) Carbon-based nanomaterials for delivery of biologicals and therapeutics: a cutting-edge technology. C 7(1):19
49. Yadav N, Garg VK, Chhillar AK, Rana JS (2023) Recent advances in nanotechnology for the improvement of conventional agricultural systems: a review. Plant Nano Biol 100032
50. Subramanian KS, Thirunavukkarasu M (2017) Nano-fertilizers and nutrient transformations in soil. In: Nanoscience and plant–soil systems, pp 305–319
51. Sharaya R, Nehra A, Kalwan G, Agarwala N, Khan NA, Tuteja N, Gill R, Gill SS (2023) Piriformospora indica (Serendipity indica): a potential tool for alleviation of heavy metal toxicity in plants. In: Biostimulants in alleviation of metal toxicity in plants. Academic Press, pp 401–422
52. Mafakheri A, Siosemardeh AF, Bahramnejad B, Struik PC, Sohrabi Y (2010) Effect of drought stress on yield, proline, and chlorophyll contents in three chickpea cultivars. Aust J Crop Sci 4(8):580–585
53. Schittenhelm S (2010) Effect of drought stress on yield and quality of maize sunflower and maize/sorghum intercrops for biogas production. J Agron Crop Sci 196(4):253–261
54. Alabdallah NM, Hasan MM (2021) Plant-based green synthesis of silver nanoparticles and its effective role in abiotic stress tolerance in crop plants. Saudi J Biol Sci 28(10):5631–5639
55. Hafeez U, Akbar W (2015) Impact of training on employees performance (evidence from pharmaceutical companies in Karachi, Pakistan). Bus Manag Strat 6(1):49–64
56. Akhtar A, Bakhtawar B, Akhtar S (2022) Extreme Programming vs Scrum: a comparison of agile models. Technol Innov Manag Rev 2(2):80–96
57. Rubilar G (2013) Repertorios y aproximaciones biográfico-narrativas. Testimonios y análisis de prácticas investigativas en trabajadores sociales. Forum Qual Sozialforschung/Forum: Qual Soc Res 14(2)
58. Ahmed Z, Le HP (2021) Linking Information Communication Technology, trade globalization index, and CO_2 emissions: evidence from advanced panel techniques. Environ Sci Pollut Res 28:8770–8781
59. Khan ZS, Rizwan M, Hafeez M, Ali S, Adrees M, Qayyum MF, Khalid S, ur Rehman MZ, Sarwar MA (2020) Effects of silicon nanoparticles on growth and physiology of wheat in cadmium contaminated soil under different soil moisture levels. Environ Sci Pollut Res
60. Safdar M, Kim W, Park S, Gwon Y, Kim YO, Kim J (2022) Engineering plants with carbon nanotubes: a sustainable agriculture approach. J Nanobiotechnol 20(1):1–30
61. Zhao Y, Hua X, Rui Q, Wang D (2023) Exposure to multi-walled carbon nanotubes causes suppression in octopamine signal associated with transgenerational toxicity induction in C. elegans. Chemosphere 318:137986

62. Tan TL, Nulit R, Jusoh M, Rashid SA (2023) Recent developments, applications and challenges for carbon quantum dots as a photosynthesis enhancer in agriculture. RSC Adv 13(36):25093–25117
63. Antal TK, Volgusheva AA, Kukarskikh GP, Lukashev EP, Bulychev AA, Margonelli A, Lambreva MDM (2022) Single-walled carbon nanotubes protect photosynthetic reactions in Chlamydomonas reinhardtii against photoinhibition. Plant Physiol Biochem 192:298–307
64. Xiao X, Wang X, Liu L, Chen C, Sha A, Li J (2022) Effects of three graphene-based materials on the growth and photosynthesis of Brassica napus L. Ecotoxicol Environ Saf 234:113383
65. Araby S et al (2021) Recent advances in carbon-based nanomaterials for flame retardant polymers and composites. Composites 212:108675
66. Maeda K et al (2013) Direct water splitting into hydrogen and oxygen under visible light by using modified TaON photocatalysts with d0 electronic configuration. Chem Eur J 19(16):4986–4991
67. Machín A, Cotto M, Ducongé J, Márquez F (2023) Artificial photosynthesis: current advancements and future prospects. Biomimetics 8(3):298
68. Noor H, Ding P, Ren A, Sun M, Gao Z (2023) Effects of Nitrogen Fertilizer on Photosynthetic Characteristics and Yield. Agronomy 13(6):1550
69. Ma X, Li X, Zhang W, Meng F, Wang X, Qin Y, Zhang M (2021) Carbon-based nanocomposite smart sensors for the rapid detection of mycotoxins. Nanomaterials 11(11):2851
70. Sun Y et al (2023) Slippery graphene-bridging liquid metal layered heterostructure nanocomposite for stable high-performance electromagnetic interference shielding. ACS Nano 17(13):12616–12628
71. Arora NK (2018) Agricultural sustainability and food security. Environ Sustain 1(3):217–219
72. Gomiero T (2016) Soil degradation, land scarcity and food security: reviewing a complex challenge. Sustainability 8(3):281
73. Shang Y, Hasan MK, Ahammed GJ, Li M, Yin H, Zhou J (2019) Applications of nanotechnology in plant growth and crop protection: a review. Molecules 24(14):2558
74. Pérez-de-Luque A (2017) Interaction of nanomaterials with plants: what do we need for real applications in agriculture. Front Environ Sci 5:12
75. Gupta A, Rayeen F, Mishra R, Tripathi M, Pathak N (2023) Nanotechnology applications in sustainable agriculture: an emerging ecofriendly approach. Plant Nano Biol 100033
76. Ali S, Mehmood A, Khan N (2021) Uptake, translocation, and consequences of nanomaterials on plant growth and stress adaptation. J Nanomater 2021:1–17
77. Wohlmuth J, Tekielska D, Čechová J, Baránek M (2022) Interaction of the nanoparticles and plants in selective growth stages—usual effects and resulting impact on usage perspectives. Plants 11(18):2405
78. Zahra Z, Habib Z, Hyun S, Sajid M (2022) Nanowaste: another future waste, its sources, release mechanism, and removal strategies in the environment. Sustainability 14(4):2041
79. Harish V, Ansari MM, Tewari D, Gaur M, Yadav AB, García-Betancourt ML, Barhoum A (2022) Nanoparticle and nanostructure synthesis and controlled growth methods. Nanomaterials 12(18):3226
80. Geitner NK, Hendren CO, Cornelis G, Kaegi R, Lead JR, Lowry GV, Wiesner MR (2020) Harmonizing across environmental nanomaterial testing media for increased comparability of nanomaterial datasets. Environ Sci Nano 7(1):13–36

Role of Carbon Nanotubes for Herbicide Detection and Remediation

Noureddine Chaachouay and Lahcen Zidane

Abstract Carbon nanotubes are cylindrical structures made of carbon that have diameters as small as nanometers and lengths that range from micrometers to millimeters. They are widely sought after in a variety of sectors, including herbicide detection and cleanup, because of their exceptional qualities, which include excellent strength, stiffness, and outstanding thermal and electrical conductivity. Carbon nanotubes may be produced using arc discharge, template synthesis, laser ablation, and chemical vapor deposition. Carbon nanotubes have distinctive characteristics that benefit them in several applications, such as identifying herbicides. Due to their heightened sensitivity, these detectors are capable of effectively detecting low levels of substances in water and soil. Carbon nanotubes (CNTs) have versatile applications in several sensing systems, including electrochemical and optical sensors. They possess a relatively low cost and are easily manufactured, making them a desirable choice for sensor applications. Carbon nanotubes can remove herbicides from soil or water by adsorption, effectively minimizing environmental pollution. Carbon nanotubes (CNTs) may serve as a photocatalyst for the degradation of herbicides. This process involves the generation of electrons and holes, which then interact with herbicide molecules, breaking them down into non-toxic substances. In addition, carbon nanotubes (CNTs) may serve as electrodes in the electrochemical breakdown of herbicides. To summarize, using carbon nanotubes to detect and remove herbicides is a lucrative field of study that has promise for its possible applications in agricultural and environmental conservation. By investigating the properties of carbon nanotubes and evaluating different detection and remediation strategies, researchers can develop effective and sustainable solutions for herbicide contamination.

Keywords Agriculture · Carbon nanotubes · Detection · Herbicide · Nanotechnology · Remediation

---

N. Chaachouay (✉)
Agri-Food and Health Laboratory (AFHL), Faculty of Sciences and Techniques of Settat, Hassan First University, PoBox 382, 26000 Settat, Morocco
e-mail: noureddine.chaachouay@uhp.ac.ma

L. Zidane
Animal Productions and Agro-Industry Laboratory, Department of Biology, Faculty of Sciences, Ibn Tofail University, BP 133, 14000 Kenitra, Morocco

A. Bachheti et al. (eds.), *Carbon-Based Nanomaterials*, Smart Nanomaterials Technology, https://doi.org/10.1007/978-981-97-0240-4_8

List of Abbreviations

CVD	Chemical Vapor Deposition
PVD	Physical Vapor Deposition
CNTs	Carbon nanotubes
MWCNTs	Multi-Walled Carbon Nanotubes
SERS	Surface-Enhanced Raman Scattering
SWCNTs	Single-Walled Carbon Nanotubes

1 Introduction

Nanotechnology can serve this purpose by using atoms and molecules to make materials with nano-scale dimensions [1–3]. Nanotechnology-based techniques have worldwide attention for detecting, degrading, and removing hazardous pesticides, offering high specificity during detection and degradation. Various nanomaterials, including nanoparticles, nanotubes, and nano-composites, have led to an extensive investigation in this field through multiple procedures [4–9].

Carbon nanotubes (CNTs) are cylindrical structures of carbon atoms with a diameter on the order of nanometers and a length ranging from micrometers to millimeters. The Japanese physicist Iijima Sumio made the first discovery of these entities in 1991 [10–12]. Carbon nanotubes possess extraordinary characteristics, including increased heat conductivity, rigidity, durability, and exceptional electrical conductivity. These characteristics make them appealing for various applications, such as electronics, materials research, energy storage, and herbicide detection and cleanup.

Several research investigations have explored the use of carbon nanotube based nano-composites to improve the detection and treatment of herbicides [13–17]. The nano-composites use CNTs in combination with other substances, such as iron oxide nanoparticles or graphene oxide, to enhance their ability to adsorb and increase their overall effectiveness. Carbon nanotubes may be modified with particular molecules that have a preference for various herbicides, enabling the identification and measurement of herbicides in samples [18–20]. These technologies include fluorescence, electrochemical sensing, and Surface-Enhanced Raman Scattering (SERS) as viable approaches.

Carbon nanotubes may serve as adsorbents for herbicide cleanup, effectively extracting herbicides from polluted water or soil. Carbon nanotubes possess a large surface area, which offers several active sites for adsorption. Simultaneously, their electrical conductivity facilitates the effective electron transfer during the oxidation or reduction of the adsorbed herbicides [21–24]. Moreover, carbon nanotubes may be modified with certain chemicals to increase their specificity towards specific herbicides. The next chapter will explore the possible uses of carbon nanotubes in the detection and treatment of herbicides.

2 Structure and Type of Carbon Nanotubes

Carbon nanotubes are carbon structures that have a cylindrical shape with diameters that usually vary from a few nanometers to a few micrometres. They may also reach lengths of several centimeters [25, 26]. Carbon nanotubes may be conceptualized as cylindrical structures formed by rolling up sheets of graphene, a two-dimensional lattice composed of hexagonal arrangements of carbon atoms. [27, 28]. There are two main types of carbon nanotubes: multi-walled carbon nanotubes (MWCNTs) and single-walled carbon nanotubes (SWCNTs) [29–31]. Single-walled carbon nanotubes (SWCNTs) are cylindrical nanostructures composed of carbon atoms [32, 33]. Carbon nanotubes are cylindrical structures composed of carbon atoms organized in hexagons. They may have diameters as tiny as a few nanometers and lengths that extend up to several millimeters. The characteristics of these substances differ based on their chirality, which refers to the arrangement of carbon atoms at different angles. They may exhibit either semiconducting or metallic behavior. MWCNTs, in contrast to SWCNTs, consist of several layers of graphene sheets that are rolled into concentric cylinders. These cylinders have diameters that vary from a few to tens of nanometers [34, 35]. The number of layers might range from a small quantity to a large amount, depending on the synthesis process used.

3 Properties of Carbon Nanotubes

Cylindrical carbon structures, consisting of carbon atoms and carbon nanotubes, exhibit remarkable mechanical, electrical, thermal, and chemical capabilities, making them very suitable for a wide range of applications. Carbon nanotubes possess many qualities [1, 20, 25, 36–39] (Fig. 1).

- **Mechanical properties:** Carbon nanotubes are acknowledged for their exceptional strength and stiffness. They possess an impressive tensile strength of up to 63 GPa and Young's modulus of up to 1 TPa. Although they have excellent resilience, they exhibit remarkable flexibility, allowing them to bend and twist without harm effortlessly.
- **Electrical properties:** Carbon nanotubes exhibit exceptional electrical conductivity, reaching values as high as 108 S/m. Additionally, they have distinctive field emission characteristics, making them valuable in electrical devices, including transistors, sensors, and solar cells.
- **Thermal properties:** CNTs exhibit exceptional thermal conductivity, reaching values as high as 3500 W/mK, surpassing all known materials. These characteristics make them appropriate for applications involving thermal management and heat dispersion.
- **Chemical properties:** CNTs exhibit exceptional stability and resistance to chemical reactions, making them very useful in many chemical applications, including catalysis, drug delivery, and water purification.

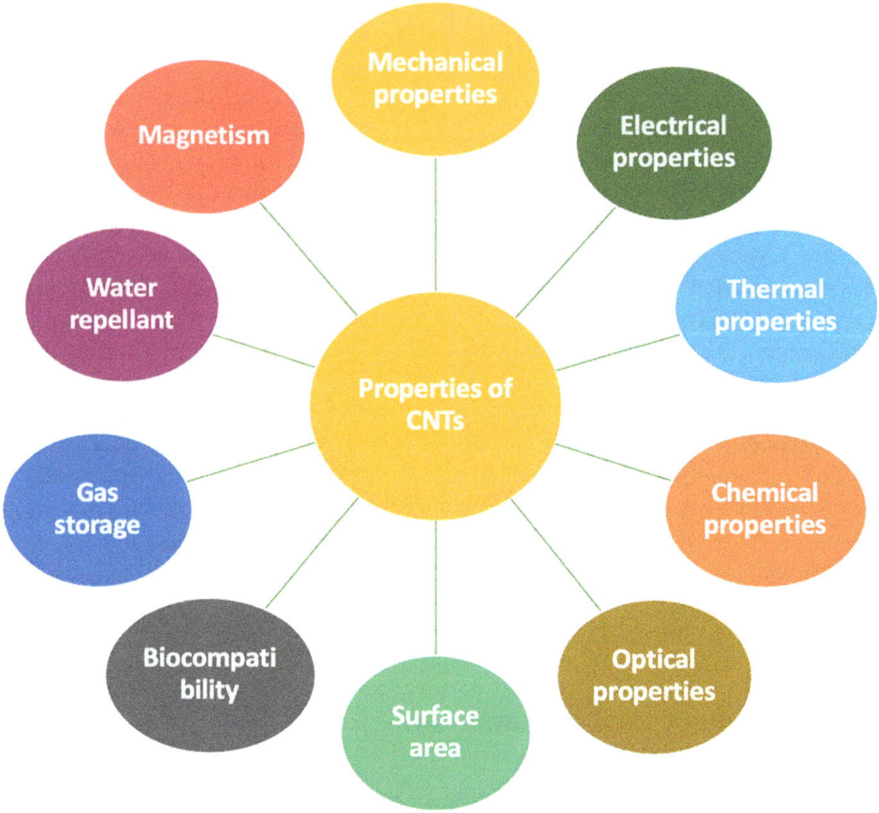

Fig. 1 Different properties of carbon nanotubes

- **Optical properties:** CNTs possess distinctive optical characteristics, such as efficient absorption and emission of light, as well as an adjustable bandgap. These characteristics make them valuable in optoelectronic applications, such as photovoltaics and light-emitting devices.
- **Surface area:** CNTs possess an exceptionally high surface area to volume ratio, making them valuable in several fields, such as energy storage and catalysis.
- **Biocompatibility:** Certain carbon nanotubes exhibit biocompatibility, indicating their lack of harmful effects on living cells. These characteristics make them favorable for biological applications like medication delivery and tissue engineering.
- **Gas storage:** CNTs can efficiently store significant quantities of gases, such as hydrogen and methane. This characteristic makes them valuable for many energy storage purposes.

- **Water repellant:** Certain carbon nanotubes have superhydrophobic properties, indicating that they can repel water and are resistant to becoming wet. This characteristic makes them valuable in applications involving self-cleaning surfaces and anti-corrosion coatings.
- **Magnetism:** Carbon nanotubes possess magnetic characteristics, making them valuable in several applications, including magnetic data storage and sensing.

4 Carbon Nanotubes Synthesis

CNTs are cylindrical carbon structures with distinctive thermal, electrical, and mechanical characteristics. The following are the sequential procedures for synthesizing carbon nanotubes [36, 40–43] (Fig. 2):

- **Carbon source selection:** To start the process of synthesizing carbon nanotubes, it is necessary to choose an appropriate carbon source carefully. Carbon sources often used include hydrocarbons like acetylene, ethylene, and methane.
- **Catalyst preparation:** The formation of carbon nanotubes necessitates the presence of a catalyst. Transition metals, such as nickel, cobalt, and iron, are the most often used motivators. The catalyst is often synthesized by applying a thin metal

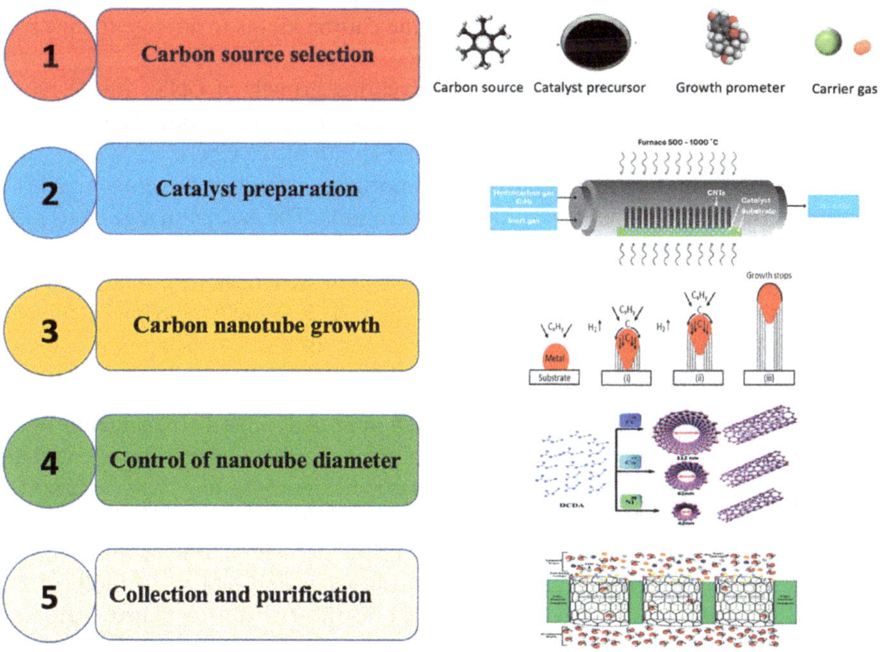

Fig. 2 Steps in the standard CVD Carbon nanotubes synthesis process

coating onto a substrate by physical vapor deposition (PVD) or chemical vapor deposition (CVD) processes.
- **Carbon nanotube growth:** The carbon source is placed into the growth chamber once the catalyst has been created. The carbon source is often subjected to elevated temperatures (800–1000 °C) in the presence of the catalyst. This results in the carbon atoms undergoing nucleation and arranging themselves into a cylindrical shape around the catalyst particle.
- **Control of nanotube diameter:** The nanotube diameter may be controlled by altering the size of the catalyst particle. Decreasing the catalyst particle size will result in the production of nanotubes with a reduced diameter.
- **Collection and purification:** After the growth process, it is necessary to gather and refine the carbon nanotubes. The predominant sample collection approach involves using a substrate coated with a thin catalyst coating. Subsequently, the substrate may be detached, and the nanotubes can be gathered from the catalyst layer using a solvent such as ethanol. To accomplish purification, the nanotubes may be washed with acid to eliminate any remaining catalyst particles.

Various methods may be used to synthesize carbon nanotubes, such as laser ablation, template synthesis, chemical vapor deposition, and arc discharge. The strategies included in this list are as follows [44–47]:

- **Arc Discharge:** This method uses an electric arc between two graphite electrodes inside a gas environment that lacks reactivity. The elevated temperature and pressure the angle generates prompt the carbon atoms to arrange themselves into carbon nanotubes.
- **Template Synthesis:** This method involves the growth of CNTs inside nanoporous templates made of either aluminum oxide or silicon dioxide. The templates regulate the dimensions of the diameter and length of the CNTs.
- **Laser Ablation:** This method involves vaporizing a graphite target using a laser inside a furnace that operates at high temperatures. The carbon in a gaseous state is then cooled and solidified into CNTs.
- **Chemical Vapor Deposition:** In this method, a furnace operating at high temperatures is used to inject a gas containing carbon, such as methane or ethylene, into a material called a substrate. The carbon atoms in the gas then settle onto the substrate, forming carbon nanotubes.

The advantages and disadvantages of each approach must be considered when choosing the appropriate procedure for producing nanotubes, taking into account the desired qualities of the nanotubes and the particular purpose they will be used for [30, 40]. For instance, the arc discharge process is relatively uncomplicated and can generate substantial nanotubes. However, the resultant material often exhibits impurities and may possess imperfections. Chemical vapor deposition is a method that may produce nanotubes of superior quality while allowing for regulated attributes. However, it requires more meticulous regulation of the reaction conditions and can incur higher costs.

5 Applications of Carbon Nanotubes for Herbicide Detection

Carbon nanotubes possess distinctive physical and chemical characteristics, making them valuable in several domains, such as sensing and detection [48, 1, 49–51. Using carbon nanotubes for herbicide detection has gained significant attention in recent years. Herbicides are chemical substances used in agriculture to manage and suppress the growth of unwanted plants, often known as weeds. The utilization of herbicides has substantially increased in the last several decades. Nevertheless, herbicides may adversely affect human health and the environment, underscoring the need to detect their presence. CNTs have high efficacy in seeing herbicides in aqueous and terrestrial environments. They are applicable in many sensing systems, including electrochemical and optical sensors. Electrochemical sensors use carbon nanotubes as the sensing component, using variations in CNT conductivity to identify the presence of herbicides. Carbon nanotubes can increase the sensitivity of optical sensors, enabling the detection of trace amounts of herbicides. Using CNTs for herbicide detection has the advantage of their exceptional sensitivity. CNT-based sensors have superior sensitivity, enabling the detection of herbicides at concentrations as minute as parts per trillion. Consequently, these sensors are very efficient in identifying even minuscule quantities of herbicides in the surrounding environment. Furthermore, carbon nanotubes are cost-effective and easily manufacturable, which enhances their appeal as a sensor option. Here are many ways in which carbon nanotubes may be used in this particular sector [1, 21, 31, 52–54]:

- **Field monitoring:** CNTs can monitor the presence of herbicides in agricultural fields in real time. The sensors may be positioned inside the soil or affixed to plants to ascertain the presence and intensity of herbicides. This data may be used to enhance the efficiency of herbicide application, minimize unnecessary usage, and enhance agricultural productivity.
- **Water quality monitoring:** Herbicides have the potential to infiltrate water bodies and pollute them, leading to negative impacts on both aquatic organisms and human well-being. Carbon nanotubes have the potential to be used in the creation of water quality sensors that can accurately identify and measure small amounts of herbicides present in bodies of water.
- **Food safety:** The presence of herbicide residues in food items may constitute a substantial health hazard to people. Carbon nanotube sensors can identify traces of herbicide residues in food items, namely vegetables and fruits, guaranteeing their suitability for ingestion.
- **Environmental monitoring:** Carbon nanotubes have the potential to be used in the creation of sensors capable of detecting traces of herbicide in various ecological components, such as soil and air. This data may be used to evaluate the repercussions of herbicide utilization on the environment and to enforce suitable steps to mitigate their consequences.

Carbon nanotubes possess distinctive physical and chemical characteristics, making them very efficient in sensing and detecting applications. The use of carbon nanotubes for the detection of herbicides has attracted considerable interest as a result of the growing prevalence of herbicide usage in agriculture and the potential negative impacts on human health and the environment [31, 48]. Carbon nanotube sensors provide a sympathetic and economical method for detecting herbicides in the environment, making them indispensable for monitoring and alleviating the effects of herbicides.

6 Applications of Carbon Nanotubes for Herbicide Remediation

CNTs have undergone much research due to their significant promise in environmental remediation. An area with great potential for application is the cleanup of herbicides in soil and water [21, 22]. Herbicides are extensively used to manage undesirable vegetation in the fields of agriculture and forestry. Nevertheless, their entry into water bodies or soil may pose a substantial risk to human health and the environment. Herbicides have the potential to pollute groundwater, surface water, and soil, leading to negative impacts on ecosystems, human health, and animals. Hence, it is essential to devise efficient techniques for remediating soil and water polluted by herbicides. Adsorption is a method that may be used to utilize CNTs for herbicide cleanup [23, 31]. Carbon nanotubes possess a substantial surface area, enabling effective adsorption of herbicides via both physical and chemical interactions. The large surface area of CNTs may enhance the interaction between the herbicide and the adsorbent, hence enhancing the efficacy of adsorption. Adsorption may effectively eliminate the herbicide from soil or water, thus mitigating the potential for environmental pollution [55]. Photocatalysis is a further method by which carbon nanotubes may be used for herbicide cleanup. Carbon nanotubes can function as a photocatalyst for the degradation of herbicides. Exposure of carbon nanotubes to light results in the production of electrons and holes, which may undergo reactions with herbicide molecules. This ultimately leads to the breakdown of herbicides into non-toxic substances. The increased surface area of carbon nanotubes may augment the efficacy of photocatalytic degradation by facilitating more interaction between the herbicide and the photocatalyst. Carbon nanotubes may serve as electrodes in the electrochemical breakdown of herbicides.

The exceptional conductivity of CNTs may significantly augment the electrochemical process, hence leading to the degradation of the herbicide. The electrochemical technique may also generate hydrogen peroxide, enhancing the breakdown of herbicides [56, 57]. Ultimately, CNTs may be an effective barrier in nanofiltration processes, effectively eliminating herbicides from water. CNTs' diminutive dimensions enable them to remove pesticides while permitting the passage of water selectively. Overall, using carbon nanotubes for herbicide remediation has promised a robust and streamlined approach to eliminate herbicides from soil and water. Nevertheless, it is crucial to conduct a comprehensive analysis to fully understand the potential environmental consequences of using CNTs and to enhance the technology for efficient application. [23, 24]. Furthermore, there are still challenges that need to be tackled, such as the cost-effectiveness and scalability of the technology. However, the possible advantages of using CNTs for herbicide remediation make it a captivating field of study and advancement in environmental remediation.

7 Potential Toxicity of Carbon Nanotubes

CNTs have extraordinary physical characteristics such as great mechanical strength, electrical conductivity, and thermal conductivity, which make them highly sought after for a wide range of applications [44, 46, 48]. Nevertheless, there are apprehensions over their possible toxicity, especially when they are breathed or come into touch with live cells. Multiple inquiries have shown that inhalation of CNTs may result in pulmonary harm and inflammation, akin to the consequences of asbestos exposure. This is because carbon nanotubes have the same form as asbestos fibers, which allows them to become stuck in the lungs. This may result in the development of illnesses, including mesothelioma and lung cancer [58–60].

Furthermore, carbon nanotubes can permeate cell membranes and amass in different organs, resulting in oxidative stress and DNA harm. This has the potential to elevate the susceptibility to cancer and other ailments [61]. The toxicity of CNTs is heavily dependent on their surface area, shape, size, and other features. Hence, it is essential to exercise care while dealing with them and properly evaluate their possible hazards before their use in any application. To reduce the potential toxicity of carbon nanotubes, scientists are investigating several approaches, such as altering their surface characteristics or enclosing them in biocompatible substances. It is crucial to ensure the appropriate disposal of carbon nanotubes to avoid potential environmental pollution.

8 Limitations of the Application of Carbon Nanotubes

Carbon nanotubes have unique physical and chemical characteristics that make them very promising for the detection and remediation of herbicides. Nevertheless, it is essential to consider specific constraints before using CNTs for such purposes [62, 63]. An inherent shortcoming of CNTs is their limited ability to distinguish between various kinds of herbicides, which may result in inaccurate detection and subsequent treatment. Furthermore, a range of environmental parameters such as temperature, pH, and ionic strength may impact the efficacy of carbon nanotubes in the detection and remediation of herbicides. The exorbitant expenses associated with the production and functionalization of CNTs might likewise impose constraints on their extensive use. In addition, there may be regulatory issues arising from possible effects on human health and the environment [62, 64, 65]. It is vital to consider practical factors, such as the challenges of increasing the process size from a laboratory to an industrial scale and incorporating carbon nanotubes into current treatment systems. In conclusion, while CNTs have the potential for detecting and removing herbicides, it is essential to do more studies to overcome these limitations and enhance their effectiveness in these specific uses.

9 Conclusions

Carbon nanotubes have lately attracted considerable interest because of their potential in detecting and remedying herbicides. These features, which are exclusive to them, make them well-suited for applications involving sensing and repair. Multiple studies have shown evidence that carbon nanotubes can detect a wide range of herbicides with exceptional sensitivity and selectivity. The enhanced surface area of CNTs facilitates the adequate adhesion of herbicides, resulting in precise and dependable detection. CNTs show great potential as a viable option for creating herbicide sensors capable of detecting minute quantities of herbicides in polluted water and soil. CNTs may also function as efficient adsorbents for herbicides in polluted water and soil, resulting in their elimination and remediation. Due to their high adsorption capacities, they may effectively catch herbicides, therefore avoiding their infiltration into groundwater or absorption by plants. Implementing this measure may effectively mitigate environmental pollution and reduce the adverse impacts of herbicides on human health and ecology. Utilizing CNTs for herbicide detection and cleanup can enhance the effectiveness and precision of existing techniques, which often incur high expenses and consume significant amounts of time. This is especially crucial in agricultural environments, where the rapid occurrence of herbicide use and contamination might be prevalent. CNT systems provide a quick and economical method for identifying and eliminating herbicides from polluted environments. Nevertheless, there are several difficulties linked to the implementation of carbon nanotube-based herbicide detection and cleanup. An obstacle lies in optimizing CNT production and functionalization to guarantee efficient detection and elimination of herbicides. Another obstacle lies in the possibility of CNT toxicity and its possible environmental consequences, necessitating more investigation and assessment. Notwithstanding these difficulties, persistent analysis in this field offers potential for the development of efficient and enduring methods for the detection and cleanup of herbicides. CNTs have unique characteristics that position them as a viable option for tackling the issues related to herbicide pollution. They offer a more environmentally friendly and practical approach to detect and eliminate herbicides from the surroundings.

References

1. Chaachouay N, Azeroual A, Bencharki B, Douira A, Zidane L (2023) Chapter 3—Hormonal interactions during fruit development and ripening. In: Husen A, Zhang W (eds) Hormonal cross-talk, plant defense and development. Academic Press, pp 37–46 https://doi.org/10.1016/B978-0-323-95375-7.00014-8
2. Lee Y-C, Moon J-Y, Lee Y-C, Moon J-Y (2020) Introduction to nanotechnology and bionanotechnology. Introd Bionanotechnol 1–14
3. Meetoo DD (2011) Nanotechnology and the food sector : from the farm to the table. Emir J Food Agric 387–403
4. Husen A, Siddiqi KS (2023) Advances in smart nanomaterials and their applications. Elsevier Inc., 50 Hampshire St., 5th Floor, Cambridge, MA 02139, USA
5. Husen A (2022) Engineered nanomaterials for sustainable agricultural production, soil improvement and stress management. Elsevier Inc., 50 Hampshire St., 5th Floor, Cambridge, MA 02139, USA
6. Husen A (2023) Nanomaterials and nanocomposites exposures to plants (Response, Interaction, Phytotoxicity and Defense Mechanisms). Springer Nature Singapore Pte Ltd., 152 Beach Road, #21-01/04 Gateway East, Singapore 189721, Singapore. https://doi.org/10.1007/978-981-99-2419-6
7. Husen A, Iqbal M (2019). Nanomaterials and plant potential. Springer International Publishing AG, Gewerbestrasse 11, 6330 Cham, Switzerland. https://doi.org/10.1007/978-3-030-05569-1
8. Kassem A, Ayoub GM, Malaeb L (2019) Antibacterial activity of chitosan nano-composites and carbon nanotubes: a review. Sci Total Environ 668:566–576
9. Vallinayagam S, Rajendran K, Lakkaboyana SK, Soontarapa K, Remya RR, Sharma VK, Kumar V, Venkateswarlu K, Koduru JR (2021) Recent developments in magnetic nanoparticles and nano-composites for wastewater treatment. J Environ Chem Eng 9(6):106553
10. Aqel A, Abou El-Nour KM, Ammar RA, Al-Warthan A (2012) Carbon nanotubes, science and technology part (I) structure, synthesis and characterisation. Arab J Chem 5(1):1–23
11. Burstein E (2003) A major milestone in nanoscale material science: the 2002 Benjamin Franklin Medal in Physics presented to Sumio Iijima. J Franklin Inst 340(3–4):221–242
12. Hirsch A, Vostrowsky O (2005) Functionalization of carbon nanotubes. Funct Mol Nanostructures 193–237
13. Ali A, Ahmed Z, Maqbool R, Shahzad K, Shah ZH, Ali MZ, Alsamadany H, Bilal M (2022) Nanobioremediation of insecticides and herbicides. In: Biodegradation and biodeterioration at the nanoscale. Elsevier, pp 655–674
14. Bhalla G, Bhalla B, Kumar V, Sharma A, Upadhye VJ, Pathak D (2022) Remediation of pesticide residues from contaminated water using various nanomaterials and nanocomposites. In: Pesticides remediation technologies from water and wastewater. Elsevier, pp 229–251
15. Francis S, Vijayan R, Koshy EP, Mathew B (2022) Sensing of environmentally hazardous chemicals by carbon nanotubes. In: Carbon nanotubes for energy and environmental applications. Apple Academic Press, pp 211–225
16. Rawtani D, Khatri N, Tyagi S, Pandey G (2018) Nanotechnology-based recent approaches for sensing and remediation of pesticides. J Environ Manage 206:749–762
17. Thakur D, Jain A, Ghoshal G (2016) Evaluation of phytochemical, antioxidant and antimicrobial properties of glycyrrhizin extracted from roots of Glycyrrhiza Glabra.
18. Azizi-Lalabadi M, Hashemi H, Feng J, Jafari SM (2020) Carbon nanomaterials against pathogens; the antimicrobial activity of carbon nanotubes, graphene/graphene oxide, fullerenes, and their nanocomposites. Adv Coll Interface Sci 284:102250
19. Bilal M, Nguyen TA, Iqbal HM (2020) Multifunctional carbon nanotubes and their derived nano-constructs for enzyme immobilization–a paradigm shift in biocatalyst design. Coord Chem Rev 422:213475
20. Ruoff RS, Lorents DC (1995) Mechanical and thermal properties of carbon nanotubes. Carbon 33(7):925–930

21. Alghuthaymi M, Mostafa M, Abd-Elsalam KA (2020) Carbon nanotubes : an efficient sorbent for herbicide sensing and remediation. In: Carbon nanomaterials for agri-food and environmental applications. Elsevier, pp 429–457
22. Deng J, Shao Y, Gao N, Deng Y, Tan C, Zhou S, Hu X (2012) Multiwalled carbon nanotubes as adsorbents for removal of herbicide diuron from aqueous solution. Chem Eng J 193:339–347
23. Deokar SK, Bajad GS, Bhonde P, Vijayakumar RP, Mandavgane SA (2017) Adsorptive removal of diuron herbicide on carbon nanotubes synthesized from plastic waste. J Polym Environ 25:165–175
24. Duman O, Özcan C, Polat TG, Tunc S (2019) Carbon nanotube-based magnetic and non-magnetic adsorbents for the high-efficiency removal of diquat dibromide herbicide from water: OMWCNT, OMWCNT-Fe3O4 and OMWCNT-κ-carrageenan-Fe_3O_4 nanocomposites. Environ Pollut 244:723–732
25. Dresselhaus MS, Dresselhaus G, Charlier J-C, Hernandez E (2004) Electronic, thermal and mechanical properties of carbon nanotubes. Philos Trans R Soc London Ser A: Math, Phys Eng Sci 362(1823):2065–2098
26. Kane CL, Mele EJ (1997) Size, shape, and low energy electronic structure of carbon nanotubes. Phys Rev Lett 78(10):1932
27. Lambin P, Loiseau A, Culot C, Biro LP (2002) Structure of carbon nanotubes probed by local and global probes. Carbon 40(10):1635–1648
28. Zhou O, Shimoda H, Gao B, Oh S, Fleming L, Yue G (2002) Materials science of carbon nanotubes: fabrication, integration, and properties of macroscopic structures of carbon nanotubes. Acc Chem Res 35(12):1045–1053
29. Merkoçi A (2006) Carbon nanotubes in analytical sciences. Microchim Acta 152:157–174
30. Prasek J, Drbohlavova J, Chomoucka J, Hubalek J, Jasek O, Adam V, Kizek R (2011) Methods for carbon nanotubes synthesis. J Mater Chem 21(40):15872–15884
31. Wang J, Shen B, Lan M, Kang D, Wu C (2020) Carbon nanotubes (CNTs) production from catalytic pyrolysis of waste plastics: the influence of catalyst and reaction pressure. Catal Today 351:50–57
32. Moisala A, Nasibulin AG, Kauppinen EI (2003) The role of metal nanoparticles in the catalytic production of single-walled carbon nanotubes—a review. J Phys: Condens Matter 15(42):S3011
33. Singh P, Campidelli S, Giordani S, Bonifazi D, Bianco A, Prato M (2009) Organic functionalisation and characterisation of single-walled carbon nanotubes. Chem Soc Rev 38(8):2214–2230
34. Majeed S, Fierro D, Buhr K, Wind J, Du B, Boschetti-de-Fierro A, Abetz V (2012) Multi-walled carbon nanotubes (MWCNTs) mixed polyacrylonitrile (PAN) ultrafiltration membranes. J Membr Sci 403:101–109
35. Monteiro-Riviere NA, Nemanich RJ, Inman AO, Wang YY, Riviere JE (2005) Multi-walled carbon nanotube interactions with human epidermal keratinocytes. Toxicol Lett 155(3):377–384
36. Hynes NRJ, Sankaranarayanan R, Kathiresan M, Senthamaraikannan P, Khan A, Asiri AM, Khan I (2019) Synthesis, properties, and characterization of carbon nanotube-reinforced metal matrix composites. In: Nanocarbon and its Composites. Elsevier, pp 805–830
37. Odom TW, Huang J-L, Kim P, Lieber CM (2000) Structure and electronic properties of carbon nanotubes. J Phys Chem B 104(13):2794–2809). ACS Publications
38. Salvetat J-P, Bonard J-M, Thomson NH, Kulik AJ, Forro L, Benoit W, Zuppiroli L (1999) Mechanical properties of carbon nanotubes. Appl Phys A 69:255–260
39. Yakobson BI, Avouris P (2001) Mechanical properties of carbon nanotubes. In: Carbon nanotubes : synthesis, structure, properties, and applications. Springer, pp 287–327
40. Andrews R, Jacques D, Qian D, Rantell T (2002) Multiwall carbon nanotubes : synthesis and application. Acc Chem Res 35(12):1008–1017
41. Baddour CE, Briens C (2005) Carbon nanotube synthesis : a review. Inter J Chem Reactor Eng 3(1)
42. Joselevich E, Dai H, Liu J, Hata K, Windle HA (2008) Carbon nanotube synthesis and organization. Carbon nanotubes: advanced topics in the synthesis, structure, properties and applications. pp 101–165

43. Kumar M (2011) Carbon nanotube synthesis and growth mechanism. Carbon nanotubes-synthesis, characterization, applications. pp 147–170
44. Dai H (2002) Carbon nanotubes: synthesis, integration, and properties. Acc Chem Res 35(12):1035–1044
45. Sinnott SB, Andrews R (2001) Carbon nanotubes : synthesis, properties, and applications. Crit Rev Solid State Mater Sci 26(3):145–249
46. Smalley RE (2003) Carbon nanotubes : synthesis, structure, properties, and applications
47. Yellampalli S (2011) Carbon nanotubes : synthesis, characterization, applications. BoD–Books on Demand
48. Anzar N, Hasan R, Tyagi M, Yadav N, Narang J (2020) Carbon nanotube-a review on synthesis, properties and plethora of applications in the field of biomedical science. Sensors International 1:100003
49. Fam DWH, Palaniappan A, Tok AIY, Liedberg B, Moochhala SM (2011) A review on technological aspects influencing commercialization of carbon nanotube sensors. Sens Actuators B Chem 157(1):1–7
50. Tran T-T, Mulchandani A (2016) Carbon nanotubes and graphene nano field-effect transistor-based biosensors. TrAC, Trends Anal Chem 79:222–232
51. Zhang T, Mubeen S, Myung NV, Deshusses MA (2008) Recent progress in carbon nanotube-based gas sensors. Nanotechnology 19(33):332001
52. Chen H, Zuo X, Su S, Tang Z, Wu A, Song S, Zhang D, Fan C (2008) An electrochemical sensor for pesticide assays based on carbon nanotube-enhanced acetycholinesterase activity. Analyst 133(9):1182–1186
53. López-Feria S, Cárdenas S, Valcárcel M (2009) One step carbon nanotubes-based solid-phase extraction for the gas chromatographic–mass spectrometric multiclass pesticide control in virgin olive oils. J Chromatogr A 1216(43):7346–7350
54. Thakkar KN, Mhatre SS, Parikh RY (2010) Biological synthesis of metallic nanoparticles. Nanomedicine Nanotechnol Biol Med 6(2):257–262
55. Dehghani MH, Kamalian S, Shayeghi M, Yousefi M, Heidarinejad Z, Agarwal S, Gupta VK (2019) High-performance removal of diazinon pesticide from water using multi-walled carbon nanotubes. Microchem J 145:486–491
56. Diel JC, Franco DS, Nunes I dos S, Pereira HA, Moreira KS, Thiago A de L, Foletto EL, Dotto GL (2021) Carbon nanotubes impregnated with metallic nanoparticles and their application as an adsorbent for the glyphosate removal in an aqueous matrix. J Environ Chem Eng9(2):105178
57. Thakkar JB, Gupta S, Prabha CR (2019) Acetylcholine esterase enzyme doped multiwalled carbon nanotubes for the detection of organophosphorus pesticide using cyclic voltammetry. Int J Biol Macromol 137:895–903
58. Chaachouay N, Benkhnigue O, Douira A, Zidane L (2021) Poisonous medicinal plants used in the popular pharmacopoeia of the Rif, northern Morocco. Toxicon 189:24–32
59. Francis AP, Devasena T (2018) Toxicity of carbon nanotubes: a review. Toxicol Ind Health 34(3):200–210
60. Luongo LA, Zhang XJ (2010) Toxicity of carbon nanotubes to the activated sludge process. J Hazard Mater 178(1–3):356–362
61. Wang J, Xu Y, Yang Z, Huang R, Chen J, Wang R, Lin Y (2013) Toxicity of carbon nanotubes. Curr Drug Metab 14(8):891–899
62. Gu W, Yushin G (2014) Review of nanostructured carbon materials for electrochemical capacitor applications: advantages and limitations of activated carbon, carbide-derived carbon, zeolite-templated carbon, carbon aerogels, carbon nanotubes, onion-like carbon, and graphene. Wiley Interdiscip Rev: Energy Environ 3(5):424–473
63. Shao W, Arghya P, Yiyong M, Rodes L, Prakash S (2013) Carbon nanotubes for use in medicine : potentials and limitations. Synth Appl Carbon Nanotub Compos 13:285–311
64. Merum S, Veluru JB, Seeram R (2017) Functionalized carbon nanotubes in bio-world: applications, limitations and future directions. Mater Sci Eng, B 223:43–63
65. Wang CY, Ru CQ, Mioduchowski A (2004) Applicability and limitations of simplified elastic shell equations for carbon nanotubes. J Appl Mech 71(5):622–631

66. Chaachouay N, Azeroual A, Bencharki B, Zidane L (2023) Chapter 17—Food safety and biotechnological products. In: Husen A, Ahmad A (eds) Genomics, transcriptomics, proteomics and metabolomics of crop plants. Academic Press, pp 353–363. https://doi.org/10.1016/B978-0-323-95989-6.00017-6
67. Chaachouay N, Azeroual A, Bencherki B, Douira A, Zidane L (2023) Various metabolites and or bioactive compounds from vegetables, and their use nanoparticles synthesis, and applications. In: Husen A (ed) Nanomaterials from agricultural and horticultural products. Springer Nature, pp 187–209. https://doi.org/10.1007/978-981-99-3435-5_10
68. Chaachouay N, Azeroual A, Benkhnigue O, Zidane L (2023) Alkaloids : a suitable precursor for nanomaterials synthesis, and their various applications. In: Husen A (ed) Secondary metabolites based green synthesis of nanomaterials and their applications. Springer Nature Singapore, pp 23–48. https://doi.org/10.1007/978-981-99-0927-8_2

Drug Delivery Using Carbon Nanomaterials

Maharudra Pratap Singh, Akhilesh Kumar Singh, Prakash Kumar Sarangi, Brijesh Pandey, Anand Prakash, and Rishi Kumar Singh

Abstract Carbon nanomaterials (CNs) possess unique physical, chemical, and physiological properties that make them highly valuable as carriers for biologically important molecules. However, their use is limited due to potential health and environmental effects. Nevertheless, CNs exhibit great potential in drug delivery owing to their high biocompatibility and surface area. Fullerenes, carbon nanotubes, graphenes, carbon nanodiamonds, porous carbon, and carbon dots are all CNs with broad applicability in drug delivery due to their small size and biological activity. Additionally, they can be customized with targeting ligands to deliver drugs to specific diseased cells or tissues and improve drug stability and solubility, resulting in better therapeutic outcomes. Recently, functionalized CNs with efficient drug-loading capacity, biocompatibility, and immunogenicity have been developed. These CNs have a high surface area, multifunctional surface chemistry, and excellent optical activity. Enzymatic degradation, surface modification, biological interactions, and biocorona play critical roles in achieving effective drug delivery. Recent advancements in carbon-based drug delivery hold the potential to improve disease treatment. CNs can be customized with bioactive peptides, proteins, nucleic acids, and drugs to create composites with low toxicity and high pharmaceutical efficacy. This chapter delves into the potential of CNs in drug delivery for various human diseases, including cancer, neurodegenerative conditions, tuberculosis, fungal infections, and other issues.

M. P. Singh · A. K. Singh (✉) · B. Pandey · A. Prakash (✉)
Department of Biotechnology, School of Life Sciences, Mahatma Gandhi Central University, Motihari, Bihar 845401, India
e-mail: akhiliit@gmail.com; akhileshsingh@mgcub.ac.in

A. Prakash
e-mail: anandprakash@mgcub.ac.in

P. K. Sarangi
College of Agriculture, Central Agricultural University, Imphal 795004, India

R. K. Singh
School of Biosciences, IMS Institute University Courses Campus, Uttar Pradesh, Ghaziabad, India

A. Bachheti et al. (eds.), *Carbon-Based Nanomaterials*, Smart Nanomaterials Technology, https://doi.org/10.1007/978-981-97-0240-4_9

Keywords Carbon nanomaterial · SWCNTs · MWCNTs · Nanodiamonds · Fullerenes · Bio-corona · Toxicity · Drug delivery systems

1 Introduction

Advancements in biomedical and pharmaceutical fields have improved life expectancy and health conditions, leading to increased interest in nano-drug delivery systems (NDDs) for enhanced drug targeting, reduced side effects, and patient compliance, offering precise drug release control [1–10]. With NDDS that can pass through physiological and biological barriers due to their small size—between 1 and 100 nm—nanotechnology has completely transformed the healthcare sector. Their pharmacokinetics and pharmacodynamic properties can be easily modified utilizing chemical methods, making them extremely versatile and flexible for a wide range of drug delivery applications. As a result, pharmacological therapy is more efficient and effective, and it allows for personalized drug delivery by adhering to specific cells or tissues. Moreover, because of their small size, they aren't eliminated by the immune system, which prolongs their half-life and improves the bioavailability of drugs [11]. A growing number of studies are being conducted to investigate the potential of carbon nanomaterials (CNs), including fullerenes, graphene, carbon nanotubes, mesoporous carbon, and nanodiamonds, in drug delivery applications. Carbon, the building block of life and the main component of DNA, has a special electron arrangement that allows it to exist in a variety of forms. Carbon atom hybridization and bonding dictate its diverse allotropes and configurations, which makes it a perfect material for drug delivery. Furthermore, carbon's propensity to bind with practically all elements makes it a good contender for a wide range of applications [12, 13]. Significant interest in advanced nanomedicine has been sparked by the unique physical properties and composition of CNs. With the capacity to be tailored with a variety of biomolecules, these unique structures have improved therapeutic efficacy and a decreased frequency of adverse effects. Additionally, they exhibit remarkable visual qualities and are compatible with living things, making them an attractive choice for drug delivery systems [14, 15]. The use of functionalized CNs as biocompatible scaffolds and nanomedicines has been studied. Chemical modification of certain moieties, such as polymers, compounds, and functional groups, has shown promise for biomedical applications [16]. CNs have been explored for cancer treatment because of their ability to supramolecular π-π stack, adsorb, and transform light into heat. Additionally, they are biocompatible and have the ability to cross cellular barriers, which makes them valuable in a variety of medical applications like drug delivery, tissue engineering, disease diagnosis, and biological molecule detection [17]. CNs have drawn a lot of interest in the field of drug delivery because of their distinctive characteristics, including their large surface area and flexible surface chemistry. Enhancing drug solubility, stability, and targeted distribution to specific tissues are all potential benefits of these compounds. Moreover, it is simple to functionalize a wide range of substances into carbon nanomaterials, which permits controlled drug release and minimizes side effects [18].

2 Classification of Carbon Nanomaterial

Carbon nanomaterials are classified according to their shape and structure, and these include porous carbon, carbon dots, nanodiamonds, fullerenes, nanotubes, and graphene (Fig. 1). The components of these nanomaterials are all carbon atoms.

2.1 *Porous Carbon*

Porous carbon compounds, often known as activated carbons, are gaining popularity due to their significant surface area and excellent physicochemical properties. These nanomaterials are 50 nm in size and fall into three different sizes: macropores, mesopores, and micropores [19]. They are usually produced at high temperatures by pyrolyzing and chemically activating organic precursors. Of all the carbon varieties, mixed porous carbon—a blend of all three—has a large surface area. Because of their resistance to mass transfer in 3D bulk forms, mesoporous carbon nanomaterials (MCNs) are frequently used in adsorption, separation, and drug delivery. Their distinct shape enables efficient adsorption and a substantial loading capacity for compounds, making them intriguing candidates for controlled drug delivery systems [20].

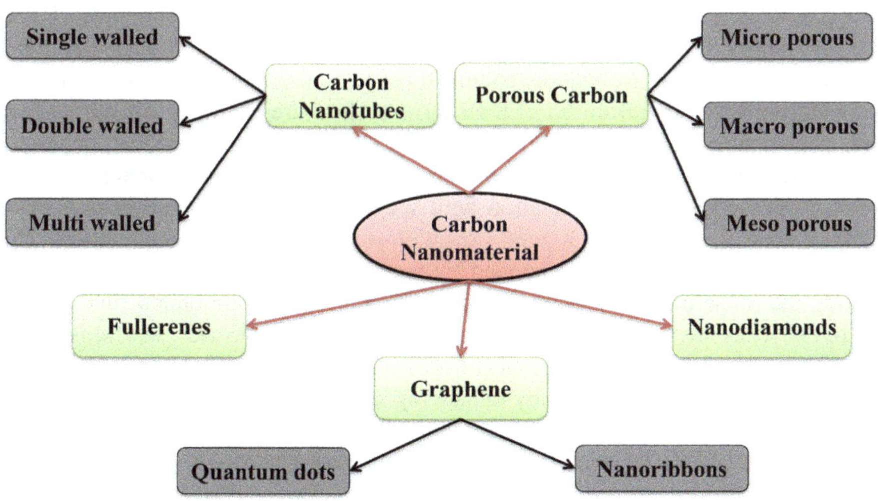

Fig. 1 Classification of carbon nanomaterials

2.2 Carbon Dots

Carbon dots (CDs) are microscopic, quasi-spherical carbon nanomaterials that remove pathogens through the emission of reactive oxygen species (ROS) when exposed to light. Because of their low level of toxicity and biocompatibility, they have received attention in biomedical research. In comparison to conventional antibiotics and antiviral medications, CDs have the ability to fight bacteria and viruses that have become resistant to drugs. CDs are able to be functionalized with targeting ligands, improving their therapeutic efficiency via targeted drug delivery to infected cells or tissues [21]. CDs are materials that imitate enzyme form and functionalities and are employed in biosensing because of their water solubility, eco-friendliness, conductivity, optical characteristics, low toxicity, biocompatibility, and theranostics. These substances present a viable option for drug delivery systems due to their advantageous properties, such as changeable surface chemistry, which allows for tailored distribution and better drug efficacy [22]. The bottom-up method for synthesizing CDs involves hydrothermal treatment of molecular precursors, while template methods are used for spherical polymer synthesis due to their ease of operation and accessibility, allowing precise control over size and shape [23].

2.3 Nanodiamonds

Nanodiamonds (NDs) are diamond-like 3D cubic lattices containing tetrahedrally linked carbon atoms. They are divided into two types: detonation nanodiamonds (DNDs) and fluorescence nanodiamonds (FNDs). DNDs have a monodispersity close to 5 nm. They are commonly produced using trinitrotoluene-like explosives and hexogen, while FNDs are made at high pressures and temperatures and have a larger size distribution [24]. NDs, characterized by sp2 and sp3 bonds, are versatile materials that easily adhere to ligands, chemical compounds, and drug molecules. NDs are widely investigated because of their large surface area and ease of fabrication. Nitrogen-vacancy defect centers glow at 550–800 nm, making them ideal for single-particle tracking and bio-imaging. Their unique spectroscopic and optical properties, stability, as well as functionalization with specific ligands or targeting molecules increase their potential for targeted drug delivery and precision medication [25]. By conjugation, NDs provide enhanced biocompatibility and pH-responsive release, making them useful for the drug delivery of low-soluble compounds such as G9a inhibitors. They are being employed more and more in theranostics, scaffolds, tissue engineering, bone surgery, and genetic material delivery because of their great surface area and stability. Furthermore, by enhancing solubility and absorption, NDs may increase the effectiveness of chemotherapeutic drugs [26]. Carbon nanodiamond particles give rise to unique compounds known as carbon nano-onions (CNOs), which have distinct chemical and physical properties. Due to their large surface area, biocompatibility, and capacity to encapsulate a variety of compounds, these quasi-spherical and polyhedral-shaped graphitic layers hold promise as drug carriers. With the help of particular ligands, CNOs can be functionalized to improve cellular uptake and regulate drug release [27]. Numerous studies have validated the biocompatibility of CNOs, which have benefits such as high purity, stability, and ease of fabrication. It can be customized to obtain specific applications, such as targeted drug delivery as well as bioimaging. Their adaptable surface chemistry enables incorporation with biomedical devices, hence extending their potential in nanomedicine [28].

2.4 *Fullerenes*

Fullerenes are spherical carbon compounds with numerous atomic Cn clusters that form pentagons and hexagons at their vertices. The most extensively studied fullerene, fullerene C60, has a distinct structure with 60 carbon atoms organized in the form of a soccer ball-like shape. Its discovery opens up new avenues of research in sectors like nanotechnology, materials science, and medicine. Novel materials and bioactive redox drugs can be produced through the chemical functionalization of fullerenes. In the biomedical domains, pure C60—which is non-toxic and advantageous to human health—is being researched for drug administration, targeted imaging, and quenching reactive oxygen species. Their special qualities allow for the accurate targeting of particular cells or tissues, which results in more potent therapies. Additionally, non-invasive imaging approaches utilizing C60-based nanomaterials are being investigated to offer insights into the course of disease and the response to therapy [29].

2.5 *Carbon Nanotubes*

Carbon nanotubes (CNTs) are cylindrical tubes with different lengths, diameters, layers, and chirality that provide flexibility, strength, and rigidity. They can be single, double, multi-walled, or functionalized. Synthesis techniques include arc discharge, chemical vapour deposition, laser ablation, and high-pressure carbon monoxide disproportionation.

2.5.1 Single-Walled Carbon Nanotubes

Single-walled carbon nanotubes (SWCNTs) are 1,000:1 diameter-to-diameter graphene sheets [30]. With sp2 hybridization, they have hollow and cylindrical one-dimensional forms. SWCNTs are widely employed in nanotechnological applications and have a diameter of 1 nm. However, their principal drawback is expensive manufacturing costs. Nanohorns, modified SWCNTs with a single hexagonal carbon ring, are used for drug delivery and targeting malignant cancer molecules due to their unique feature and increased diameter [31]. SWCNTs are widely used in biomedical fields as cell culture growth scaffolds and diagnostic devices. However, their low solubility and difficulty dispersing in aqueous phases make them challenging to disperse. Chemical modification suggests solubilizing and dispersing SWCNTs in water to increase their biomedical and pharmaceutical interest, as well as their potential for targeted drug delivery. Functionalizing their surface with hydrophilic molecules allows better dispersion in water, preserving their unique properties [32].

2.5.2 Double-Wall Carbon Nanotubes

Double-wall carbon nanotubes (DWCNTs) are a new class of carbon nanotubes with a double-layer structure consisting of two graphene sheets and two carbon nanotubes. Compared to SWNTs, this structure improves mechanical strength and electrical conductivity, and because it can encapsulate different molecules or nanoparticles, it is perfect for drug delivery applications [33]. DWCNTs, known for their strength, thermal stability, chemical resistance, and optical and electrical properties, are crucial for drug delivery systems and tissue engineering due to their efficient charge transport and enhanced conductivity [34]. Due to their superior conductivity and aspect ratio, DWCNTs find application in tissue engineering, biosensors, and nanotechnology. Utilizing tissue engineering, they have been utilized to create neurons, immunosensors for adiponectin, and nanoelectronic devices that improve performance [35]. The ability of DWCNTs to encapsulate and release pharmacological molecules with regulated release kinetics makes them promising for use in drug delivery systems. Their unique characteristics make them valuable for a range of applications in nanotechnology.

2.5.3 Multi-walled Carbon Nanotubes

Multi-walled carbon nanotubes (MWCNTs) are a form of nanotube composed of a multitude of graphene sheets with complex electrical properties. They are made by stacking layers with a 3.4 Å interlayer dispersion, and they vary in size from 5 to 50 nm [33]. MWCNTs outperform DWCNTs and SWCNTs in terms of signal-to-noise ratio, however, their sensitivity is influenced by the biological matrix. Their distinct shape makes them attractive for the treatment of cancer since it enables focused therapy and effective drug delivery. MWCNTs containing chemotherapeutic agents were employed in a trial on pancreatic cancer patients as a targeted therapeutic approach. The MWCNTs effective distribution, biocompatibility, and capacity to imitate the extracellular matrix boosted therapeutic efficacy, allowing for better tumor penetration and drug uptake [36].

2.6 *Graphene*

Graphene, a single carbon sheet with sp2-orbits, is used in biocompatible nanocomposites, drug delivery systems, tissue engineering, biomedical imaging, and biosensors due to its strong heat conductivity, optical transparency, and mechanical strength. Its enhanced solubility and compatibility in tissue engineering and drug delivery systems further enhance its applications [37].

2.6.1 Graphene Nanoribbons

Graphene nanoribbons (GNRs) are finite, planar, and almost one-dimensional structures formed by slicing graphene sheets into tiles with a higher aspect ratio, transitioning from semimetals to semiconductors [37]. With increased thermal and electrical conductivity, GNRs provide a copper substitute for integrated circuits. Chemical synthesis, exfoliation, and electron-beam lithography can all be used to produce it [38]. GNRs with oxide functionalization offer potential medical uses in drug delivery, cancer treatment, and DNA delivery. They are useful for biomedicine and diagnostics because they have better mammalian cell uptake than silicon-based transistors. They are appropriate for targeted drug delivery systems because of their large surface area and adjustable optical properties [39].

2.6.2 Graphene Quantum Dots

Graphene is sliced into pieces ranging from 2 to 20 nm to create graphene quantum dots (GQDs), which are less toxic and more environmentally friendly than conventional quantum dots [37]. GDQs, or zero-dimensional nanomaterials, are helpful in bioimaging because they are tiny, photostable, non-toxic, and water-solubilized. They are appropriate for both in vivo and in vitro tumor imaging as well as cellular dynamics because they have robust photoluminescence, photostability, and enhanced renal clearance. Since GQDs are fluorescent, they can be used to create fluorescent probes that can track dynamic changes brought on by oxidative stress and allow for the real-time diagnosis of disease [40].

3 Physicochemical Effects on Carbon Nanomaterial's Ability to Deliver Drugs

Raman spectroscopy, electron microscopy, and atomic force microscopy are crucial characterization techniques for understanding the structural integrity and quality of carbon nanomaterials. These techniques assess physicochemical properties like enzymatic degradation, biocompatibility, biological interaction, biocorona formation, and toxicity, enabling a comprehensive evaluation of the safety and efficacy of drug delivery systems and biomedical implants. This understanding not only allows for the optimization of performance but also aids in identifying any potential risks or limitations that may arise.

3.1 Enzymatic Degradation

Enzymatic degradation of CNs is crucial for immune-competent cells and medical applications. SWCNTs undergo biodegradation through natural enzymatic catalysis, increasing in horseradish peroxidase and low H_2O_2 concentrations for 12 weeks. TEM, Raman, and UV-NIR spectroscopy reveal the disappearance of D- and G-bands and carboxyl group interactions between SWCNTs and oxidizing agents. This understanding opens up possibilities for designing targeted drug delivery systems using SWCNTs. [41]. MWCNT degradation occurs layer-by-layer, and nitrogen can alter the rate through functionalization. The degree of carboxylation, impacted by the duration of oxidative acid treatment, affects the overall rate. HRPs react with oxygen in defective sites, causing CNTs to degrade. MWCNTs exhibit more excellent resistance to HRP degradation, resulting in a longer degradation time than SWCNTs.

Interestingly, myeloperoxidase (MPO) found in neutrophil-extruded extracellular traps (NETs) can influence SWCNT degradation by catalyzing the production of hypochlorous acid, which accelerates their degradation. The presence of functional groups on MWCNTs can decelerate their degradation by HRPs compared to SWCNTs [42]. SWCNT degradation requires NaBr and H_2O_2 addition, similar to human immune systems. CDs have unique properties and kinetics through enzyme oxidation, undergoing peroxide-catalyzed degradation in the presence of a lipase enzyme. These mechanisms generate reactive oxygen species, similar to human immune system oxidative processes. Due to their unique properties, CDs undergo peroxide-catalyzed degradation in the presence of a lipase enzyme, exhibiting distinct degradation kinetics [43].

3.2 Biocompatibility and Biological Interactions

CNs, including carbon nanotubes, graphene, fullerenes, and nanodiamonds, have gained attention in drug delivery due to their physicochemical properties. However, their potential toxicity due to their non-compatibility with biological systems limits their wider acceptance. Surface modification has been introduced to improve compatibility and reduce enzymatic degradation. Surface functionalization reduces cytotoxicity and improves cellular uptake, leading to the exploration of biocompatible coatings or encapsulation techniques to minimize toxicity concerns and maximize drug delivery benefits [44]. Understanding the biological interactions of different sub-classes of CNs with the innate immune system is crucial for drug delivery applications. These interactions can significantly impact the inflammatory response, cytokine production, and complement activation, ultimately influencing the biocompatibility and safety of CNs in various drug delivery systems. Understanding these interactions is essential for drug delivery applications [45]. Functionalized carbon nanomaterials (F-CNs) induce inflammation and generate pro-inflammatory cytokines in monocyte-derived macrophages, making understanding their interaction crucial for drug delivery applications. This understanding of drugs' influence on the immune system is essential for designing effective therapeutic strategies and ensuring safe drug delivery [46]. CDs, with similar dissociation constants to human serum albumin (HSA) and γ-globulins, show higher cell viability and tolerance than polyethyleneimine. They offer a promising alternative to drug delivery systems due to their ability to bind with target proteins without affecting HSA and γ-globulin levels, and their small size and surface modification capabilities enable effective encapsulation and targeted drug delivery [47]. It has been discovered that graphene nanoparticles disrupt macrophages' metabolic processes, causing damage to the mitochondria and raising reactive oxygen species (ROS) levels. Although the precise mode of entry is yet unknown, it could involve adhesion to the exterior cell membrane or passive diffusion across lipid bilayers. Smaller CNs, up to 400 nm in size, are ingested by diffusion, whereas larger CNs are ingested by endocytosis. Increased biofunctionality of CNs can result in increased delivery efficiency by increasing retention periods and decreasing clearance [26] (Table 1).

3.3 *Bio-corona*

Bio-corona is the interaction of biomolecules like lipids and proteins with nanomaterials. Within living systems, this interaction may have toxicological and biological consequences [73]. During the investigation, the interactions between SWCNTs and four important blood proteins—bovine fibrinogen, immunoglobulin, transferrin, and bovine serum albumin—were examined. The results showed that albumin was efficiently adsorbed on the surface of SWCNTs and was essential for RAW264.7 macrophage cellular absorption [74]. A study has found that adding serum proteins to SWCNTs reduces their toxicity to human umbilical vein endothelial and leukemia cells. The presence of serum proteins affects the interaction between SWCNTs and cells, resulting in different binding capabilities. The interaction between SWCNTs and aromatic residues is crucial for tissue binding. The study also found that when exposed to 10% fetal bovine serum, GO exhibits greater protein loading capacity and changes in cytotoxicity, suggesting that covering GO with bovine serum albumin can reduce its cytotoxicity [45]. When inhalation of CNs like SWCNTs occurs, it can lead to bio-corona. This happens when surfactant proteins and lipids are absorbed into the body. The surface of SWCNTs has a coating of phospholipids and hydrophobic alkyl chains, which can increase their uptake by cells. This can potentially cause an immune response through cell surface receptors due to changes in orientation

Table 1 Functionalized carbon nanomaterials and their use in drug delivery

Carbon nanomaterial	Drug used	Functionalized with	References
SWCNT	Paclitaxel	Paclitaxel and SWCNTs	Al Garalleh and Algarni [48]
	Isoniazid	Sulphuric acid and nitric acid	Zomorodbakhsh et al. [49]
MWCNTs	Paclitaxel	Ethylenediamine and phenylboronic acid group	Rathod et al. [50]
	Griseofulvin and sulfamethoxazole	Sulphuric acid and nitric acid	Chen and Mitra [51]
	ABT737	PEG-amine	Kim et al. [52]
	Doxorubicin	TAT peptide and chitosan	Dong et al. [53]
Graphene oxide	Temozolomide	Folic acid	Wang et al. [17]
	Doxorubicin	Transferrin and folic acid	Lu et al. [54]
	miRNA-101	Polyethylene glycol and polyL-argin	Assali et al. [55]
	5-Fluorouracil	Poly (ethyleneimine) and tannic acid	Aliabadi et al. [56]
	Methotrexate	1-Ethyl-3 (3-dimethylamino)- propyl) carbodiimide and n-hydroxysuccinimide	Karimi Shervedani et al. [57]
	Paclitaxel	Human serum albumin and monoclonal antibodies	Deng et al. [58]
	Doxorubicin	Maleimide containing dopamine	Oz et al. [59]
Nanodiamonds	Doxorubicin	Dicumyl peroxide and methyl 3- sulfanylpropanoate	Long et al. [60]
	Paclitaxel	Nitric acid, sodium octanoate, sodium laurate, sodium oleate	Yu et al. [61]
	Cetuximab and cisplatin	N-Hydroxy- succinimide and 1-(3-dimethylaminopropyl)-3-ethylcarbodiimide hydrochloride	Li et al. [62]
Carbon dots	β-Galactosidase	Citric acid, β-galactosidase	Sharma et al. [63]
	Epirubicin and temozolomide	Drug with transferrin	Hettiarachchi et al. [64]
	Doxorubicin	Citric acid, hyaluronic acid, poly(ethylene imine)	Li et al. [65]
Mesoporous	Doxorubicin	Poly (epichlorohydrin) and calcium chloride	Asgari et al. [66]

(continued)

Table 1 (continued)

Carbon nanomaterial	Drug used	Functionalized with	References
	Galantamine	Poly (L-lactic acid), poly (lactide-co glycolide)	Nanaki et al. [67]
Fullerene	Hydroxyurea	Pristine C60 and heterofullerene	Wang et al. [68]
	Ibuprofen	Porphyrin-like transition metal-N4 clusters	Alipour et al. [69]
	Chloroquine	Pristine fullerene and chloroquine	Bagheri Novir and Aram [70]
	Doxorubicin	Toluene	Grebinyk et al. [71]
	N-desmethyl tamoxifen	Glycine and N-desmethyl tamoxifen	Misra et al. [72]

and conformation. When proteins stick to CNs, they can hinder enzyme activity like CYP3A4, which metabolizes drugs. However, coating CNs with BSA can reduce this effect [75].

3.4 Toxicity

CNs hold great promise for biomedical applications such as image processing, sensor development, tissue engineering, and drug delivery. Nevertheless, their bio-persistent properties and asbestos-like pathogenicity raise valid concerns regarding their potential toxicity [76]. The presence of metal impurities can cause cell toxicity during synthesis. The length of CNs is a crucial factor that greatly affects their toxicity. It influences how they are internalized by cells, ultimately leading to inflammation and fibrosis [77]. The dimensions of CNTs are a crucial factor in their impact on cell toxicity. Variances in size, structure, and chemical properties of MWCNT and SWCNT can influence their toxicity to cells. Nevertheless, applying functionalization methods has proven effective in mitigating this concern. Surface modification techniques are increasingly being employed to improve biocompatibility. Despite this, there have been reported cases of cellular phagocytosis and cell apoptosis associated with using acid-treated MWCNTs and taurine-functionalized MWCNTs [78]. According to the findings of in vitro testing, it has been observed that the utilization of fullerene can considerably decrease the cytotoxicity in human macrophages without any effect on alveolar macrophages, indicating its potential in treating diseases involving human macrophages without any adverse effects [79]. A recent study conducted to explore cellular death found that exposure to CNs caused cellular damage in RAW264.7 and human alveolar macrophages, including membrane rupture, enlargement, and IL-1β activation. Graphene induced apoptosis and necrosis, while CNTs and nano graphite caused extensive cellular toxicity [80]. Through laboratory toxicity studies conducted on mouse fibroblasts, it was discovered that neutral CDs-PEG did not produce any abnormalities in cell morphology, cell cycle, or intracellular trafficking. These results indicate that neutral CDs-PEG may be suitable for drug delivery. On the other hand, negative-charged pristine CDs caused increased oxidative stress and cell proliferation despite not entering the cell nucleus. Polyethyleneimine-coated CDs were observed to be highly toxic at lower concentrations [81]. The carbonization degree of CDs affects their toxicity to cells, particularly in terms of general and light-induced toxicity. CDs with a negative charge and size below 5 nm have a higher quantum yield, causing up to 30% toxicity in breast cell lines at concentrations of 1.5–2.0 mg/mL [82].

4 Carbon Nanomaterials in Drug Delivery

CNs have incredible potential in nanotherapeutics and diagnostics, as they regulate and manipulate bio-macromolecular constructs and supramolecular assemblies such as drugs, proteins, DNA/RNA, viruses, and cellular lipid bilayers. Utilizing this technology can significantly enhance our understanding and treatment of human diseases, including cancer and neurological disorders. CNs can be customized to target and deliver drugs to diseased cells, reducing side effects and improving treatment effectiveness. Furthermore, their small size enables them to easily penetrate tissues and cross biological barriers, making them an ideal option for drug delivery. The possibilities of nanoparticles present are truly remarkable [83]. Due to their small size and biological activity, CNs such as fullerenes, CNTs, and graphenes are increasingly employed in drug delivery. These nanoparticles improve drug solubility, stability, and targeted distribution to specific tissues or cells, and their findings have been promising. Furthermore, their distinctive physicochemical properties allow for drug-controlled release, reducing potential side effects and bolsters therapeutic effectiveness [84]. Fullerene is a remarkable carbon nanomaterial that boasts the ability to graft active chemical groups in 3D with pinpoint accuracy. This unique feature allows for precise and rational drug design, bolstered by the efficient encapsulation of drug molecules. With Fullerenes, drug molecules are protected from degradation and efficiently delivered to their intended target sites. Additionally, the tunable properties of Fullerenes make them ideal for targeted drug delivery systems. By functionalizing Fullerenes with ligands that specifically recognize and bind to disease-related biomarkers, they become highly promising agents for biomedical applications [85]. CNs' biological activities depend on their core and chemical modifications. Hydrophobic fullerene molecules have complex behavior, resulting in water-soluble fullerenes capable of carrying drugs and genes. Fullerene derivatives have been found to exhibit superior binding to mitochondria and COS-1 cells compared to lipid-based vectors currently available in the market. The use of DNA-functionalized fullerenes results in the formation of a protective layer around the bound DNA, leading to an extended lifetime in endosomes and promoting chromosomal incorporation. Amino fullerenes are particularly suitable for DNA attachment, as they can lose amino groups or amine binding ability, thus facilitating detachment of DNA in the cytoplasm [86]. Fullerene derivatives increase the effectiveness of slow-release drug delivery and show significant anticancer activity in C60-paclitaxel conjugate, expanding their potential in drug and gene delivery through intact skin [87]. Rouse et al. [88] developed a peptide derivative based on fullerene that has the ability to penetrate both flexible and inflexible skin. The search for safe and effective carriers for transporting genes or drugs has led to an increasing interest in direct drug delivery into cells through the cell membrane. CNs are popular for cellular delivery due to their biocompatibility, targeted delivery, and controlled drug release.

4.1 Delivery of Anticancer Drugs

Drug delivery systems, including microspheres, liposomes, polymers, and carbon nanomaterials, are being explored for their potential to reduce administration times and increase patient compliance in anticancer drug administration. For these substances to be used to their full potential and to prevent potential harm, it is essential to understand their characteristics, reactivity, and cell-to-cell interactions [89]. CNs such as fullerenes, CNTs, GOs, and NDs exhibit outstanding thermal conductivity, superior rigidity, and excellent biocompatibility [90]. CNTs are rapidly progressing in cancer treatment, with targeted drug delivery being the most promising approach. Current achievements include anticancer chemotherapeutics like camptothecin, doxorubicin, gemcitabine, methotrexate, taxanes, immunotherapeutics, and nucleic acids [91]. Due to their high aspect ratio, SWCNTs, nanoscale drug carriers can enhance therapy efficiency and minimize side effects by conjugating with antibodies or ligands of cancer cell surface receptors [92]. Carbon nanotubes can be functionalized (f-CNTs) to develop a drug delivery system for anticancer drugs, with polyethylene glycol being the preferred material due to its increased biocompatibility and dispersion. Functionalization of drugs like doxorubicin and paclitaxel shows high treatment efficiency in both in vitro and in vivo studies [93]. Shi et al. [94] developed a nanocomposite referred to as C60-IONP, which they produced by combining fullerene with iron oxide nanoparticles. To make it more effective at targeting tumors in MCF-7 cells and mice models, they modified the nanocomposite by adding polyethylene glycol and folic acid. As a result, cancer cells in specific regions were selectively destroyed. Authors in [95] discovered that graphene and its derivatives, including graphene oxide, reduced graphene oxide, and GO nanocomposites, are ideal for drug and gene delivery and demonstrated excellent anti-tumor efficacy in animal experiments. Shi et al. [96] developed a GO-IONP-Au nanocomposite with significant superparamagnetic and improved near-infrared optical absorbance that is multifunctional magnetic and plasmonic. In comparison to PEGylated GO, it had increased photothermal cancer ablation effects and great stability while exhibiting less in vitro toxicity.

4.2 Delivery of Anti-tubercular Drugs

The study by Saikia et al. [97] explored the potential of CNTs for drug delivery purposes. The research delved into the electronic and structural properties of CNTs, utilizing density functional theory (DFT) calculations to simulate the interaction between pyrazinamide (PZA) drug molecules and f-SWCNT. The study found that it was indeed possible to form a nanotube-drug conjugate that was thermodynamically feasible. Docking studies also revealed a hydrophobic interaction between the ligand and the receptor. Gallo et al. [98] used DFT calculations to thoroughly investigate the impact of binding the anti-tubercular drugs isoniazid to f-CNTs and f-C60s. According to the results, f-CNTs are more water-soluble than f-CNTs, while f-C60s more readily bind isoniazid. This was found to be advantageous from a thermodynamic perspective. Furthermore, the study found that the global hardness decreases when isoniazid substitution increases in functionalized fullerenes, potentially affecting their stability.

4.3 Delivery of Antifungal Drugs

Amphotericin B (AmB) is an effective antifungal drug administered intravenously to treat systemic fungal infections. However, its poor water solubility leads to severe side effects in mammalian cells, including acute infusion reactions and dose-dependent nephrotoxicity [99]. As per the study, the efficacy of AmB can be improved and antibiotic activity can be regulated by binding it to f-CNT to increase water solubility. Targeted delivery of AmB was successfully achieved using f-MWCNTs, which retained the drug's potent antifungal properties even after binding to MWCNT. Remarkably, the AmB-CNT complex could traverse mammalian cells without causing any damage, highlighting this technique's potential in drug delivery [100]. In a study conducted by Benincasa et al. [101], liposomes containing a combination of Amphotericin B (AmB) and sodium deoxycholate (AmBD) were tested alongside two conjugates of AmB and f-CNTs for their antifungal activity against various fungal strains. The findings indicated that the f-CNT-AmB conjugates exhibited equal or superior minimum inhibitory concentration (MIC) values in comparison to AmB and AmBD. Even Candida strains that were resistant to AmBD were susceptible to f-CNT-AmB1. The researchers theorized that these conjugates functioned through a nonlytic mechanism as they exhibited the most permeabilization effect on fungal strains after extended incubation. Importantly, f-CNT-AmB1 did not demonstrate any significant toxic effects on Jurkat cells.

4.4 Delivery of Anti-inflammatory Drugs

Researchers have found that using SWCNT films as carriers can stimulate the drug Dexamethasone (DEX), which is used to suppress the immune system and reduce inflammation. The drug is rapidly taken up by cells and released due to electrostatic repulsions between SWCNTs and DEX at a potential of −0.8 V. However, the passive release of DEX may be reduced with SWCNTs [102]. Researchers are exploring new methods to deliver anti-inflammatory drugs more accurately and effectively. Ketoprofen, a potent medication for inflammatory conditions like arthritis and toothaches, is being developed using electro-sensitive transdermal DDS. This technology uses a polyethylene oxide-pentaerythritol triacrylate matrix and multiwalled carbon nanotubes to enhance the drug's electrical sensitivity. This allows for precise control over drug release, optimizing dosage, and minimizing side effects.

The higher electrical conductivity of CNTs can optimize dosage and minimize side effects, making them a promising option for anti-inflammatory drug delivery [103].

4.5 Delivery of Neurological Drugs

Nanotherapies, especially those targeting the CNS, have significantly improved disease treatment by crossing the blood–brain barrier and overcoming the challenges of brain-specific agents. Their flexibility offers great promise in this area [104]. Carbon nanomaterial is being explored for treating neurodegenerative diseases due to its advantages like dispersibility, large surface area, easy drug or imaging functionalization, and biocompatibility with neural tissue [105]. Nanotechnology advancements enhance Alzheimer's disease treatment by introducing nanoparticles with high affinity for circulating amyloid-β forms, causing a "sink effect" and improving the condition [104]. CNTs, a type of nanomaterial, have shown potential in treating CNS diseases like Alzheimers. When combined with acetylcholine, SWCNTs can effectively deliver ACh molecules to the brain through nerve axons. This results in a safer and more effective therapy. Even low doses of SWCNTs, as low as 300 mg/kg, can safely carry ACh into the lysosomes of neurons and serve as an effective drug delivery method. However, controlling the dosage is important to avoid any negative effects on lysosomes and mitochondria. Lower doses are still safe and highly effective, making them a viable option for drug carriers [106]. CNTs hold great potential for treating glioblastoma tumors that persist even after aggressive therapy. Despite advancements in tumor therapy and cancer research, treating brain tumors remains a challenge. However, the limited ability of anti-tumor drugs to cross the blood–brain barrier (BBB) has paved the way for new approaches using CNT-based therapies. Recent studies have shown that CNT delivery systems can significantly enhance the efficacy of CpG oligodeoxynucleotide immunotherapy, effectively eradicating gliomas and halting further tumor growth [107]. Lee et al. [108] conducted a study on stroke models. Their goal was to determine whether amine-functionalized SWCNTs could improve survival rates for neurodegenerative diseases. The results were promising: administering SWCNTs through intracerebroventricular injection, without any therapeutic molecules, seemed to lead to better motor function recovery in animals. However, the exact reason behind this effect remains unknown. Another study also showed that amino-functionalized MWCNTs effectively delivered siRNA in a stroke mice model, leading to better behavioral improvement. However, it is important to note that administering CNT alone or complexed with scrambled siRNA showed no neuroprotective activity. The research utilized the intracranial glioma model in mice with plate number GL261 to analyze the absorption of MWCNTs and their potential use in treating brain tumors, while also evaluating their toxicity. The findings indicate that after a single intratumoral injection, around 10–20% of cells exhibited CNT internalization within 24 h. Additionally, approximately 75% of MWCNTs were assimilated into tumor-associated macrophages (MPs), and 30% of tumor MPs tested positive for MWCNTs within 24 h. No significant toxicity was observed in the mice, and only minor alterations in tumor cytokine expression were recorded [109].

4.6 Delivery of Biomolecules, Gene Transfection, and as Biosensors

Due to their high-water solubility properties, CNs are utilized for drug and biomolecule delivery, as proteins covalently bind to SWCNTs and MWCNTs [91]. The field of gene therapy shows great promise in the medical community, as it involves delivering functional genes to target cells for therapeutic purposes. Initially developed for treating hereditary diseases, gene therapy has expanded to include acquired diseases like cancer. A recent breakthrough involves attaching DNA to f-MWCNT amino groups, enhancing nanotube dispersion and efficient gene transfection, eliminating the need for viral vectors. This innovative approach is gaining recognition for its potential to treat many diseases [110]. Recent progress in gene delivery has brought about innovative techniques that do not require the use of viruses. These methods include liposomes, cationic polyplexes, microparticles, and carbon nanomaterials. Among these, carbon nanomaterials have emerged as a promising option due to their biocompatibility, ease of functionalization, and ability to break down easily. They effectively deliver genes directly into cells without any harmful side effects or accumulation in the intended region [111]. Scientists are utilizing nanoparticles consisting of Fullerene C60 and hyperbranched polyethylenimine coatings to deliver genes. This non-viral vector system has high amino group content and is capable of condensing DNA or RNA with negative charges. The Fullerene (C60) nanoparticles, ranging from 3 to 80 nm in diameter, show potential for drug delivery. UV–Vis spectroscopy shows a decrease in the characteristic peak of C60, while the agarose gel electrophoresis assay demonstrates good DNA binding capability. SWCNTs are critical carriers of biologically active molecules in gene therapy [110]. Krajcik et al. [112] functionalized SWCNTs with hexamethylenediamine (HMDA) and poly(diyldimethylammonium) chloride (PDDA) to bind to negatively charged siRNA through electrostatic interactions. PDDA-SWNTs have been used as an effective carrier system for siRNA-mediated gene silencing. Non-functionalized CNTs possess great potential as drug carriers and biosensors. However, it's crucial to understand their binding behavior with proteins to realize their potential in biological studies fully. Research has demonstrated that non-functionalized CNTs can bind with hemoglobin, which makes them effective biosensors for quantifying blood H2S. This is especially important since hemoglobin makes up 37% of total blood volume. Moreover, non-functionalized MWCNTs can bind with methemoglobin, which makes them indispensable for blood H2S quantification and as hemoglobin carriers. In addition, non-functionalized CNTs exhibit selective binding to various protein groups. To fully utilize the potential of CNTs in biological studies and advance the development of new applications, it is crucial to comprehend their binding behaviors with proteins. Gaining an understanding of these behaviors can aid in optimizing their usage [113].

5 Challenges and Limitations of Carbon Nanomaterials

Nanotechnology has emerged as a promising field for drug delivery, with CNs showing great potential for the targeted delivery of drug agents to diseased tissues. However, there are several challenges to overcome to realize the potential of CNs in drug delivery fully. Developing and analyzing CNs has significant challenges and limitations that require thoughtful consideration [114]. These challenges range from cellular toxicity and optimized drug loading to controlled drug release and delivery and predicting the precise pharmacokinetic profile. At the nano level, matter behaves differently than it does at the micro and macro levels, presenting a need for extensive research by experts to understand its behavior fully. CNs, having a large surface area relative to their volume, can cause friction to become more critical, leading to unforeseen consequences for the intended actions of nanoparticles. As a result of their small size, nanoparticles may experience delays or speed-ups in their actions, leading to potential accumulation at unintended sites and triggering immune system reactions. Measuring the concentration and distribution of nanoparticles within biological systems is challenging due to their minute dimensions, further complicated by the intricate interaction between nanoparticles and biological molecules. Predicting the long-term effects of nanoparticles on human health is also complicated by these factors [115]. The impact of nanoparticles on the body depends on their size and how they interact with cells. If nanoparticles are small enough to enter the reticuloendothelial system (RES), they may weaken the immune system. Larger nanoparticles can accumulate in organs, causing toxicity and altering body function. The RES may clear nanoparticles quickly, compromising their therapeutic effectiveness. Dendritic cells can also absorb nanoparticles, leading to inflammation and immune system activation. Understanding nanoparticle size, surface properties, and charge is crucial for optimizing their therapeutic potential in the cellular environment [116]. Another challenge is the limited effectiveness of nanodrugs in treating tumors. Nanodrugs' limited effectiveness in treating tumors is due to factors such as compromised tumor vasculature, local cell density, stromal features, interstitial fluid pressure, and complex tumor tissue architecture. These factors can impede the diffusion of nanodrugs into tumor cells, affecting their therapeutic efficacy. Additionally, the complex architecture of tumor tissue, including fibrotic regions and dense extracellular matrix, can create physical barriers that impede nanodrug movement within the tumor mass. Overcoming these challenges demands the development of novel drug delivery systems and CNs' designs that take into account the specific features of nanoscale matter. By overcoming these challenges, researchers will be able to fully explore the potential of CNs in drug delivery as well as additional applications [117].

6 Conclusions

In recent years, CNs have been extensively studied for their diverse range of uses as nanomaterials. Biomedical engineering is one of the many sectors that finds them appealing due to their unique mechanical, optical, electrochemical, and electrical qualities. CNs have drawn attention because of their adaptable size, shape, and surface characteristics. Nevertheless, it has been observed that CNs may be hazardous to biological systems, which poses a problem for their characterization and design. Researchers have come up with successful chemical modification methods to overcome these limitations for numerous bio-applications such as drug delivery, tissue engineering, biomolecule detection, and cancer therapy. CNs have a lot of promise as a delivery system for drugs, including biomolecules, topical drugs, anticancer, neurological, antimycobacterial, and anti-inflammatory treatments. Still, a number of issues need to be resolved, including drug release control, cellular toxicity, ideal drug loading, and precise pharmacokinetic profile prediction. CNs remain an interesting emerging sector in drug delivery despite these limitations, with the potential to control the stability, solubility, and drug-carrying capacity of drugs while also improving their distribution for a wide range of topical and systemic applications.

References

1. Bachheti RK, Bachheti A, Husen A (2023) Metal and Metal-Oxide Based Nanomaterials (Synthesis, Agricultural, Biomedical and Environmental Interventions). Springer Nature Singapore Pte Ltd., 152 Beach Road, #21-01/04 Gateway East, Singapore 189721, Singapore
2. Chawla S, Singh S, Husen A (2023) Smart nanomaterials targeting pathological hypoxia. Springer Nature Singapore Pte Ltd., 152 Beach Road, #21-01/04 Gateway East, Singapore 189721, Singapore https://doi.org/10.1007/978-981-99-1718-1
3. Husen A, Bachheti RK, Bachheti A (2023) Current Trends in Green Nano-Emulsions (Food, Agriculture and Biomedical Sectors). Springer Nature Singapore Pte Ltd., 152 Beach Road, #21-01/04 Gateway East, Singapore 189721, Singapore
4. Husen A, Siddiqi KS (2023) Advances in smart nanomaterials and their applications. Elsevier Inc., 50 Hampshire St., 5th Floor, Cambridge, MA 02139, USA
5. Jin-Chul K, Madhusudhan A, Husen A (2021) Smart nanomaterials in biomedical applications. Springer Nature Switzerland AG, Gewerbestrasse 11, 6330 Cham, Switzerland. https://doi.org/10.1007/978-3-030-84262-8
6. Madhusudhan A, Purohit SD, Prasad R, Husen A (2023) Functional smart nanomaterials and their theranostics approaches. Springer Nature Singapore Pte Ltd., 152 Beach Road, #21-01/04 Gateway East, Singapore 189721, Singapore
7. Mandal AK, Ghorai S, Husen A (2023) Functionalized smart nanomaterials for point-of-care testing. Springer Nature Singapore Pte Ltd., 152 Beach Road, #21–01/04 Gateway East, Singapore 189721, Singapore
8. Pattnaik S, Pathak K (2016) Mesoporous silica molecular sieve based nanocarriers: transpiring drug dissolution research. Curr Pharm Des 23(3):467–480
9. Sharma G, Nim S, Alle M, Husen A, Kim JC (2022) Nanoparticle-mediated delivery of flavonoids for cancer therapy: prevention and treatment. In: Kim JC, Alle M, Husen A (eds) Smart nanomaterials in biomedical applications. nanotechnology in the life sciences. Springer, Cham. https://doi.org/10.1007/978-3-030-84262-8_3

10. Siddiqi KS, Husen A, Sohrab SS, Osman M (2018) Recent status of nanomaterials fabrication and their potential applications in neurological disease management. Nanoscale Res Lett 13(231):1–17
11. Doane TL, Burda C (2012) The unique role of nanoparticles in nanomedicine: imaging, drug delivery and therapy. Chem Soc Rev 41(7):2885–2911
12. Belay T, Worku LA, Bachheti RK, Bachheti A, Husen A (2023) Nanomaterials: introduction, synthesis, characterization, and applications. In: Husen A, Siddiqi KS (eds) Advances in smart nanomaterials and their applications. Elsevier, Amsterdam, Netherlands, pp 1–21
13. Kumar N, Kumbhat S (2016) Carbon-based nanomaterials. Essentials in Nanoscience and nanotechnology. Wiley, New York, USA, pp 189–236
14. Rauti R, Musto M, Bosi S, Prato M, Ballerini L (2019) Properties and behavior of carbon nanomaterials when interfacing neuronal cells: how far have we come? Carbon 143:430–446
15. Tyagi, S, Rajput P, Sinha A, Pramanik A, Chaubey KK, Jayaraman, Chetan Shrivastva et al (2023). Development strategies and prospects of carbon nanotube as heavy metal adsorbent. In: Bachheti RK, Bachheti A, Husen A (eds) Nanomaterials for environmental and agricultural sectors, pp 59–81. Springer Nature Singapore, Singapore
16. Mohajeri M, Behnam B, Sahebkar A (2018) Biomedical applications of carbon nanomaterials: drug and gene delivery potentials. J Cel Physiol 234:298–319
17. Wang SY, Hu HZ, Qing XC, Zhang ZC, Shao ZW (2020) Recent advances of drug delivery nanocarriers in osteosarcoma treatment. J Cancer 11:69–82
18. Shannahan J (2017) The biocorona: a challenge for the biomedical application of nanoparticles. Nanotechnol Rev 6:345–353
19. Yang D, Ionescu MI (2017) Metal oxide–carbon hyhybrid materials for application in supercapacitors. In: Dubal DP, Gomez-Romero P (eds) Singapore metal oxides in supercapacitors. Elsevier, Amsterdam, Netherlands, pp 193–218
20. Liese A, Hilterhaus L (2013) Evaluation of immobilized enzymes for industrial applications. Chem Soc Rev 42:6236–6249
21. Knoblauch R, Geddes CD (2020) Carbon nanodots in photodynamic antimicrobial therapy: a review. Materials 13:4004
22. Cohen EN, Kondiah PPD, Choonara YE, du Toit LC, Pillay V (2020) Carbon dots as nanotherapeutics for biomedical application. Curr Pharm Des 26:2207–2221
23. Tian XT, Yin XB (2019) Carbon dots, unconventional preparation strategies, and applications beyond photoluminescence. Small 15:e1901803
24. Bondon N, Raehm L, Charnay C, Boukherroub R, Durand JO (2020) Nanodiamonds for bioapplications, recent developments. J Mater Chem B 8:10878–10896
25. Tinwala H, Wairkar S (2019) Production, surface modification and biomedical applications of nanodiamonds: a sparkling tool for theranostics. Mater Sci Eng C 97:913–931
26. Patel KD, Singh RK, Kim HW (2019) Carbon-based nanomaterials as an emerging platform for theranostics. Mater Horiz 6:434–469
27. Plonska-Brzezinska ME (2019) Carbon nano-onions: a review of recent progress in synthesis and applications. Chem Nano Mat 5:568–580
28. Bartkowski M, Giordani S (2021) Carbon nano-onions as potential nanocarriers for drug delivery. Dalton Trans 50:2300–2309
29. Moussa F (2018) Fullerene and derivatives for biomedical applications. In: Narayan R (eds) Nanobiomaterials: nanostructured materials for biomedical applications. Elsevier, Amsterdam, Netherlands, pp 113–136
30. Bepete G, Coleman KS (2019) Carbon nanotubes: electronic structure and spectroscopy. In: Andrews DL, Lipson RH, Nann T (eds) Comprehensive nanoscience and nanotechnology. Elsevier, Amsterdam, Netherlands, pp 205–218
31. Beg S, Rahman M, Jain A, Saini S, Hasnain MS, Swain S, Imam S, Kazmi I, Akhter S (2018) Emergence in the functionalized carbon nanotubes as smart nanocarriers for drug delivery applications. In: Grumezescu AM (eds) Fullerenes, graphenes and nanotubes: a pharmaceutical approach. Elsevier, Amsterdam, Netherlands, pp 105–133

32. Raval JP, Joshi P, Chejara DR, Inamuddin AM, Mohammad A (2018) Carbon nanotube for targeted drug delivery. In Applications of nanocomposite Materials in drug delivery. Elsevier, Amsterdam, Netherlands, pp 203–216
33. Bhatt A, Jain A, Gurnany E, Jain R, Modi A, Jain A (2016) Carbon nanotubes: a promising carrier for drug delivery and targeting. In: Holban AM, Grumezescu AM (eds) Nanoarchitectonics for smart delivery and drug targeting. Elsevier, Amsterdam, Netherlands, pp 465–501
34. Shen C, Brozena AH, Wang Y (2011) Double-walled carbon nanotubes: challenges and opportunities. Nanoscale 3:503–518
35. Punbusayakul N, Talapatra S, Ajayan PM, Surareungchai W (2013) Label-free as-grown double wall carbon nanotubes bundles for Salmonella typhimurium immunoassay. Chem Cent J 7:102–108
36. Matta-Domjan B, King A, Totti S, Matta C, Dover G, Martinez P, Zakhidov A, La Ragione R, Macedo H, Jurewicz I, Dalton A, Velliou EG (2018) Biophysical interactions between pancreatic cancer cells and pristine carbon nanotube substrates: potential application for pancreatic cancer tissue engineering. J Biomed Mater Res 106:1637–1644
37. Crevillen AG, Escarpa A, Garcia CD (2019) Carbon-based nanomaterials in analytical chemistry. In: Crevillen AG, Escarpa A, García CD (eds) Carbon-based nanomaterials in analytical chemistry. Royal Society of Chemistry, Cambridge, United Kingdom, pp 1–36
38. Hernandez Y, Pang S, Feng X, Müllen K (2012) Graphene and its synthesis. In: Matyjaszewski K, Moller M (eds) Polymer science: a comprehensive reference, 10 volume set. Elsevier, Amsterdam, Netherlands, pp 415–438
39. Mousavi SM, Soroshnia S, Hashemi SA, Babapoor A, Ghasemi Y, Savardashtaki A, Amani AM (2019) Graphene nano-ribbon based high potential and efficiency for DNA, cancer therapy and drug delivery applications. Drug Metab Rev 51:91–104
40. Li N, Than A, Sun C, Tian J, Chen J, Pu K, Dong X, Chen P (2016) Monitoring dynamic cellular redox homeostasis using fluorescence-switchable graphene quantum dots. ACS Nano 10:11475–11482
41. Zhao Y, Allen BL, Star A (2011) Enzymatic degradation of multiwalled carbon nanotubes. J Phys Chem A 115:9536–9544
42. Farrera C, Bhattacharya K, Lazzaretto B, Andón FT, Hultenby K, Kotchey GP, Star A, Fadeel B (2014) Extracellular entrapment and degradation of single-walled carbon nanotubes. Nanoscale 6:6974–6983
43. Srivastava I, Sar D, Mukherjee P, Schwartz-Duval AS, Huang Z, Jaramillo C, Civantos A, Tripathi I, Allain JP, Bhargava R, Pan D (2019) Enzyme-catalysed biodegradation of carbon dots follows sequential oxidation in a time dependent manner. Nanoscale 11:8226–8236
44. Medepalli K, Alphenaar B, Raj A, Sethu P (2011) Evaluation of the direct and indirect response of blood leukocytes to carbon nanotubes (CNTs). Nanomed 7:983–991
45. Bhattacharya K, Mukherjee SP, Gallud A, Burkert SC, Bistarelli S, Bellucci S, Bottini M, Star A, Fadeel B (2016) Biological interactions of carbon-based nanomaterials: from coronation to degradation. Nanomed 12:333–351
46. Pescatori M, Bedognetti D, Venturelli E, Ménard-Moyon C, Bernardini C, Muresu E, Piana A, Maida G, Manetti R, Sgarrella F, Bianco A, Delogu LG (2013) Functionalized carbon nanotubes as immunomodulator systems. Biomaterials 34:4395–4403
47. He X, Chen P, Zhang J, Luo TY, Wang HJ, Liu YH, Yu XQ (2019) Cationic polymer-derived carbon dots for enhanced gene delivery and cell imaging. Biomater Sci 7:1940–1948
48. Al Garalleh H, Algarni A (2020) Modelling of paclitaxel conjugated with carbon nanotubes as an antitumor agent for cancer therapy. J Biomed Nanotechnol 16:224–234
49. Zomorodbakhsh S, Abbasian Y, Naghinejad M, Sheikhpour M (2020) The effects study of isoniazid conjugated multi-wall carbon nanotubes nanofluid on mycobacterium tuberculosis. Int J Nanomed 15:5901–5909
50. Rathod V, Tripathi R, Joshi P, Jha PK, Bahadur P, Tiwari S (2019) Paclitaxel encapsulation into dual-functionalized multi-walled carbon nanotubes. AAPS Pharm Sci Tech 20:51

51. Chen K, Mitra S (2019) Incorporation of functionalized carbon nanotubes into hydrophobic drug crystals for enhancing aqueous dissolution. Colloids Surfaces B Biointer 173:386–391
52. Kim SW, Kyung Lee Y, Yeon Lee J, Hee Hong J, Khang D (2017) PEGylated anticancer-carbon nanotubes complex targeting mitochondria of lung cancer cells. Nanotechnology 28:465102
53. Dong X, Sun Z, Wang X, Leng X (2017) An innovative MWCNTs/DOX/ TC nanosystem for chemo-photothermal combination therapy of cancer. Nanomed 13:2271–2280
54. Lu M, Wang YK, Zhao J, Lu H, Stenzel MH, Xiao P (2016) PEG grafted-nanodiamonds for the delivery of gemcitabine. Macromol Rapid Commun 37:2023–2029
55. Assali A, Akhavan O, Adeli M, Razzazan S, Dinarvand R, Zanganeh S, Soleimani M, Dinarvand M, Atyabi F (2018) Multifunctional core-shell nanoplatforms (gold@graphene oxide) with mediated NIR thermal therapy to promote miRNA delivery. Nanomed Nanotech Biol Med 14:1891–1903
56. Aliabadi M, Yunessnia lehi A, Shagholani H, Gerayeli A (2018) Planar polymer-graphene oxide nanohybrid as a 5-fluorouacil carrier in pHresponsive controlled release. J Drug Deliv Sci Tech 43:103–106
57. Karimi Shervedani R, Mirhosseini H, Samiei Foroushani M, Torabi M, Rahsepar FR, Norouzi-Barough L (2018) Immobilization of methotrexate anticancer drug onto the graphene surface and interaction with calf thymus DNA and 4T1 cancer cells. Bioelectrochemistry 119:1–9
58. Deng W, Qiu J, Wang S, Yuan Z, Jia Y, Tan H, Lu J, Zheng R (2018) Development of biocompatible and VEGF-targeted paclitaxel nanodrugs on albumin and graphene oxide dual-carrier for photothermal-triggered drug delivery in vitro and in vivo. Int J Nanomed 13:439–453
59. Oz Y, Barras A, Sanyal R, Boukherroub R, Szunerits S, Sanyal A (2017) Functionalization of reduced graphene oxide via thiol-maleimide "click" chemistry: facile fabrication of targeted drug delivery vehicles. ACS Appl Mater Interf 9:34194–34203
60. Long W, Ouyang H, Wan W, Yan W, Zhou C, Huang H, Liu M, Zhang X, Feng Y, Wei Y (2020) Two in one: simultaneous functionalization and DOX loading for fabrication of nanodiamond-based pH responsive drug delivery system. Mater Sci Eng C 108:110413
61. Yu Y, Yang X, Liu M, Nishikawa M, Tei T, Miyako E (2019) Amphipathic nanodiamond supraparticles for anticancer drug loading and delivery. ACS Appl Mater Interf 11:18978–18987
62. Li D, Chen X, Wang H, Liu J, Zheng M, Fu Y, Yu Y, Zhi J (2017) Cetuximabconjugated nanodiamonds drug delivery system for enhanced targeting therapy and 3D Raman imaging. J Biophoton 10:1636–1646
63. Sharma SK, Micic M, Li S, Hoar B, Paudyal S, Zahran EM, Leblanc RM (2019) Conjugation of carbon dots with β-galactosidase enzyme: Surface chemistry and use in biosensing. Molecules 24:3275
64. Hettiarachchi SD, Graham RM, Mintz KJ, Zhou Y, Vanni S, Peng Z, Leblanc RM (2019) Triple conjugated carbon dots as a nano-drug delivery model for glioblastoma brain tumors. Nanoscale 11:6192–6205
65. Li J, Li M, Tian L, Qiu Y, Yu Q, Wang X, Guo R, He Q (2020) Facile strategy by hyaluronic acid functional carbon dot-doxorubicin nanoparticles for CD44 targeted drug delivery and enhanced breast cancer therapy. Int J Pharm 578:119122
66. Asgari S, Pourjavadi A, Hosseini SH, Kadkhodazadeh S (2020) A pH sensitive carrier based-on modified hollow mesoporous carbon nanospheres with calcium-latched gate for drug delivery. Mater Sci Eng 109:110517
67. Nanaki SG, Spyrou K, Bekiari C, Veneti P, Baroud TN, Karouta N, Grivas I, Papadopoulos GC, Gournis D, Bikiaris DN (2020) Hierarchical porous carbon-PLLA and PLGA hybrid nanoparticles for intranasal delivery of galantamine for Alzheimer's disease therapy. Pharmaceutics 12:227
68. Wang P, Yan G, Zhu X, Du Y, Chen D, Zhang J (2021) Heterofullerene MC59 (M = B, Si, Al) as potential carriers for hydroxyurea drug delivery. Nanomaterials 11:115
69. Alipour E, Alimohammady F, Yumashev A, Maseleno A (2020) Fullerene C60 containing porphyrin-like metal center as drug delivery system for ibuprofen drug. J Mol Model 26:7

70. Bagheri Novir S, Aram MR (2020) Quantum mechanical simulation of chloroquine drug interaction with C60 fullerene for treatment of COVID-19. Chem Phys Lett 757:137869
71. Grebinyk A, Prylutska S, Chepurna O, Grebinyk S, Prylutskyy Y, Ritter U, Ohulchanskyy TY, Matyshevska O, Dandekar T, Frohme M (2019) Synergy of chemo- and photodynamic therapies with C60 fullerenedoxorubicin nanocomplex. Nanomaterials 9:1540
72. Misra C, Kumar M, Sharma G, Kumar R, Singh B, Katare OP, Raza K (2017) Glycinated fullerenes for tamoxifen intracellular delivery with improved anticancer activity and pharmacokinetics. Nanomed (Lond) 12:1011–1023
73. Docter D, Westmeier D, Markiewicz M, Stolte S, Knauer SK, Stauber RH (2015) The nanoparticle biomolecule corona: lessons learned—challenge accepted? Chem Soc Rev 44:6094–6121
74. Fadeel B, Feliu N, Vogt C, Abdelmonem AM, Parak WJ (2013) Bridge over troubled waters: understanding the synthetic and biological identities of engineered nanomaterials. Wiley Interdiscip Rev Nanomed Nanobiotechnol 5:111–129
75. Kapralov AA, Feng WH, Amoscato AA, Yanamala N, Balasubramanian K, Winnica DE, Kisin ER, Kotchey GP, Gou P, Sparvero LJ, Ray P, Mallampalli RK, Klein-Seetharaman J, Fadeel B, Star A, Shvedova AA, Kagan VE (2012) Adsorption of surfactant lipids by single-walled carbon nanotubes in mouse lung upon pharyngeal aspiration. ACS Nano 6:4147–4156
76. Kobayashi N, Izumi H, Morimoto Y (2017) Review of toxicity studies of carbon nanotubes. J Occup Health 59:394–407
77. Aldieri E, Fenoglio I, Cesano F, Gazzano E, Gulino G, Scarano D, Attanasio A, Mazzucco G, Ghigo D, Fubini B (2013) The role of iron impurities in the toxic effects exerted by short multiwalled carbon nanotubes (MWCNT) in murine alveolar macrophages. J Toxicol Environ Health Part A 76:1056–1071
78. Wang X, Guo J, Chen T, Nie H, Wang H, Zang J, Cui X, Jia G (2012) Multi-walled carbon nanotubes induce apoptosis via mitochondrial pathway and scavenger receptor. Toxicol Vitro 26:799–806
79. Yuan X, Zhang X, Sun L, Wei Y, Wei X (2019) Cellular toxicity and immunological effects of carbon-based nanomaterials. Part Fibre Toxicol 16:18–27
80. Figarol A, Pourchez J, Boudard D, Forest V, Akono C, Tulliani JM, Lecompte JP, Cottier M, Bernache-Assollant D, Grosseau P (2015) In vitro toxicity of carbon nanotubes, nano-graphite and carbon black, similar impacts of acid functionalization. Toxicol Vitro 30:476–485
81. Havrdová M, Holá K, Skopalík J, Tománková KB, Petr M, Čépe K, Polakova KM, Tuček J, Bourlinos AB, Zbořil R (2016) Toxicity of carbon dots - effect of surface functionalization on the cell viability, reactive oxygen species generation and cell cycle. Carbon 99:238–248
82. Esfandiari N, Bagheri Z, Ehtesabi H, Fatahi Z, Tavana H, Latifi H (2019) Effect of carbonization degree of carbon dots on cytotoxicity and photo-induced toxicity to cells. Heliyon 5:e02940
83. Mudshinge SR, Deore AB, Patil S, Bhalgat CM (2011) Nanoparticles: emerging carriers for drug delivery. Saud Pharmaceut J 19:129–141
84. Bolskar RD (2012) Fullerenes for drug delivery. In: Bhushan B (ed) Encyclopedia of nanotechnology. Springer, Amsterdam, Netherlandspp, pp 898–911
85. Bogdanović G, Djordjević A (2016) Carbon nanomaterials: biologically active fullerene derivatives. Srp Arh Celok Lek144 (3–4):222–231
86. Yan W, Seifermann SM, Pierrat P, Bräse S (2015) Synthesis of highly functionalized C60 fullerene derivatives and their applications in material and life sciences. Org Biomol 13(1):25–54
87. Zhang LW, Monteiro-Riviere NA (2013) Use of confocal microscopy for nanoparticle drug delivery through skin. J Biomed Opt 18(6):061214
88. Rouse JG, Yang J, Ryman-Rasmussen JP, Barron AR, Monteiro-Riviere NA (2007) Effects of mechanical flexion on the penetration of fullerene amino acid-derivatized peptide nanoparticles through skin. Nano Lett 7:155–160
89. Iannazzo D, Piperno A, Pistone A, Grassi G, Galvagno S (2013) Recent advances in carbon nanotubes as delivery systems for anticancer drugs. Curr Med Chem 20:1333–1354

90. Lim DJ, Sim M, Oh L, Lim K, Park H (2014) Carbon-based drug delivery carriers for cancer therapy. Arch Pharm Res 37:43–52
91. Fabbro C, Ali-Boucetta H, Da Ros T, Kostarelos K, Bianco A, Prato M (2012) Targeting carbon nanotubes against cancer. Chem Commun 48:3911–3926
92. Vittorio O, Raffa V, Cuschieri A (2009) Influence of purity and surface oxidation on cytotoxicity of multiwalled carbon nanotubes with human neuroblastoma cells. Nanomed 5:424–431
93. Lay CL, Liu J, Liu Y (2011) Functionalized carbon nanotubes for anticancer drug delivery. Expert Rev Med Devices 8:561–566
94. Shi J, Wang L, Gao J, Liu Y, Zhang J, Ma R, Liu R, Zhang Z (2014) A fullerene-based multi-functional nanoplatform for cancer theranostic applications. Biomaterials 35:5771–5784
95. Yang K, Feng L, Shi X, Liu Z (2013) Nano-graphene in biomedicine: theranostic applications. Chem Soc Rev 42:530–547
96. Shi X, Gong H, Li Y, Wang C, Cheng L, Liu Z (2013) Graphene-based magnetic plasmonic nanocomposite for dual bioimaging and photothermal therapy. Biomaterials 34:4786–4793
97. Saikia N, Rajkhowa S, Deka RC (2013) Density functional and molecular docking studies towards investigating the role of single-wall carbon nanotubes as nanocarrier for loading and delivery of pyrazinamide antitubercular drug onto pnca protein. J Comput Aided Mol Des 27:257–276
98. Gallo M, Favila A, Glossman-Mitnik D (2007) Dft studies of functionalized carbon nanotubes and fullerenes as nanovectors for drug delivery of antitubercular compounds. Chem Phys Lett 447:105–109
99. Pastorin G (2009) Crucial functionalizations of carbon nanotubes for improved drug delivery: a valuable option? Pharm Res 26(4):746–769
100. Pruthi J, Mehra NK, Jain NK (2012) Macrophages targeting of amphotericin B through mannosylated multiwalled carbon nanotubes. J Drug Target 20(7):593–604
101. Benincasa M, Pacor S, Wu W, Prato M, Bianco A, Gennaro R (2011) Antifungal activity of amphotericin b conjugated to carbon nanotubes. ACS Nano 5:199–208
102. Minnikanti S, Pereira MG, Jaraiedi S, Jackson K, Costa-Neto CM, Li Q, Peixoto N (2010) In vivo electrochemical characterization and inflammatory response of multiwalled carbon nanotube-based electrodes in rat hippocampus. J Neural Eng 7:016002
103. Im JS, Bai B, Lee YS (2010) The effect of carbon nanotubes on drug delivery in an electro-sensitive transdermal drug delivery system. Biomaterials 31:1414–1419
104. Brambilla D, Le Droumaguet B, Nicolas J, Hashemi SH, Wu LP, Moghimi SM, Couvreur P, Andrieux K (2011) Nanotechnologies for alzheimer's disease: Diagnosis, therapy, and safety issues. Nanomed 7:521–540
105. Nunes A, Al-Jamal K, Nakajima T, Hariz M, Kostarelos K (2012) Application of carbon nanotubes in neurology: clinical perspectives and toxicological risks. Arch Toxicol 86:1009–1020
106. Yang Z, Zhang Y, Yang Y, Sun L, Han D, Li H, Wang C (2010) Pharmacological and toxicological target organelles and safe use of single-walled carbon nanotubes as drug carriers in treating alzheimer disease. Nanomed Nanotechnol Biol Med 6:427–441
107. Ghosh R, Bag J, Datta A, Pramanick A, Abubakar IH (2023) Functionalized carbon nanotubes—a boon in treating brain diseases. J Appl Pharm Sci 13(07):032–039
108. Lee HJ, Park J, Yoon OJ, Kim HW, Lee do Y, Kim do H, Lee WB, Lee NE, Bonventre JV, Kim, SS (2011) Amine-modified single-walled carbon nanotubes protect neurons from injury in a rat stroke model. Nat Nanotechnol 6:121–125
109. VanHandel M, Alizadeh D, Zhang L, Kateb B, Bronikowski M, Manohara H, Badie B (2009) Selective uptake of multi-walled carbon nanotubes by tumor macrophages in a murine glioma model. J Neuroimmunol 208:3–9
110. Vashist SK, Zheng D, Pastorin G, Al-Rubeaan KA, Luong JH, Sheu F (2011) Delivery of drugs and biomolecules using carbon nanotubes. Carbon 49:4077–4097
111. Ditto AJ, Shah PN, Yun YH (2009) Non-viral gene delivery using nanoparticles. Expert Opin Drug Deliv 6:1149–1160

112. Krajcik R, Jung A, Hirsch A, Neuhuber W, Zolk O (2008) Functionalization of carbon nanotubes enables non-covalent binding and intracellular delivery of small interfering RNA for efficient knock-down of genes. Biochem Biophys Res Commun 369:595–602
113. Wu XC, Zhang WJ, Sammynaiken R, Meng QH, Yang QQ, Zhan E, Liu Q, Yang W, Wang R (2008) Non-functionalized carbon nanotube binding with hemoglobin. Colloids Surf 65:146–149
114. Li Y, Wang T, Wang J, Jiang T, Cheng G, Wang S (2011) Functional and unmodified MWNTs for delivery of the water-insoluble drug carvedilol—a drug-loading mechanism. Appl Surf Sci 257:5663–5670
115. Wong BS, Yoong SL, Jagusiak A, Panczyk T, Ho HK, Ang WH, Pastorin G (2013) Carbon nanotubes for delivery of small molecule drugs. Adv Drug Deliv Rev 65:1964–2015
116. Maiti D, Tong X, Mou X, Yang K (2019) Carbon-based nanomaterials for biomedical applications: a recent study. Front Pharmacol 9:1401
117. Jha R, Singh A, Sharma PK, Fuloria NK (2020) Smart carbon nanotubes for drug delivery system: a comprehensive study. J Drug Deliv Sci Technol 58:101811

Biomedical Applications of 1D and 2D Carbon-Based Nanomaterials

Tahmeena Khan, Saman Raza, and Qazi Inamur Rahman

Abstract Biomedical applications of nanomaterials have gained attention because of their pivotal role in the improvement of human health. As compared to bulk material, nanomaterials have improved diagnostic and therapeutic applications. Carbon-based nanomaterials have remarkable potential for biomedical applications like biosensing, drug delivery, and gene therapy. Carbon-based one-dimensional (1D) and two-dimensional (2D) materials have excellent thermal, mechanical, and optical properties and possess structural diversity. Carbon has several interesting allotropes, like carbon nanotubes (CNTs), graphene oxide (GO), and graphene quantum dots (GDQs). These allotropes have been associated with numerous biological applications like biosensing and drug delivery. This chapter explores the biomedical applications of 1D and 2D carbon-based nanomaterials, including CNTs, graphene, GO, and reduced graphene oxide (RGO). The structural features and their specific applications in the biomedical field have been elaborated, taking examples from the literature. Potential toxicity risks on living organisms and the environment are also discussed, along with their applications. It is pertinent to comprehend their interaction with living systems to use them for biomedical applications.

Keywords Carbon-based nanomaterials · Graphene · Biomedical · Drug delivery · Biosensing

1 Introduction

Nanotechnology is a fast-emerging field of science and technology that uses substances and techniques at the nanoscale, i.e., 10^{-9} m in dimension [1–9]. The very small size imparts unique characteristics owing to a large surface area to volume

T. Khan (✉) · Q. I. Rahman
Department of Chemistry, Integral University, U.P., Lucknow 226026, India
e-mail: tahminakhan30@yahoo.com

S. Raza
Department of Chemistry, Isabella Thoburn College, U.P., Lucknow 226007, India

A. Bachheti et al. (eds.), *Carbon-Based Nanomaterials*, Smart Nanomaterials Technology, https://doi.org/10.1007/978-981-97-0240-4_10

ratio [10–15]. The special characteristics include optical, magnetic, thermal, electrical, chemical, and biological properties as well [16–18]. Nanomaterials can be classified according to their dimensions (Fig. 1) [19–21].

Zero-dimensional: Nanomaterials having all three dimensions, i.e., x, y, and z, within the nanoscale range are called 0-D. They may have amorphous or crystalline structures and exhibit different shapes and forms. Examples: Quantum dots, fullerenes, and nanoparticles.

- One-dimensional: Nanomaterials having two dimensions within the nanoscale range, but one dimension outside this range belong to this category. They may have an amorphous or crystalline structure that can exist individually or be embedded within another material. Examples include nanofibers, nanotubes, nanorods, and nanowires.
- Two-dimensional: Nanomaterials having two dimensions outside the nanoscale range and only one within this range belong to this category and have a flat, layer, or plate-like shape. They may have amorphous or crystalline structures and may be deposited on or mixed into another material. Examples include Coatings, nanosheets, nanofilms, and nanolayers.
- Three-dimensional: Nanomaterials not within the nanoscale in any dimension are of this type. Although all dimensions of such materials are greater than 100 nm, their bulk is composed of several arrangements of nano-size materials like dispersions of nanoparticles (NPs), bundles of nanowires or nanotubes, or multiple nanolayers, in close association with each other Examples include colloids and free nanoparticles with different morphological features.

Another classification of nanomaterials can be done based on composition, and the main types are as follows:

- Carbon nanomaterials: These are allotropes of carbon having different shapes and forms. For example, 0-D quantum dots and fullerenes, which may be spherical or

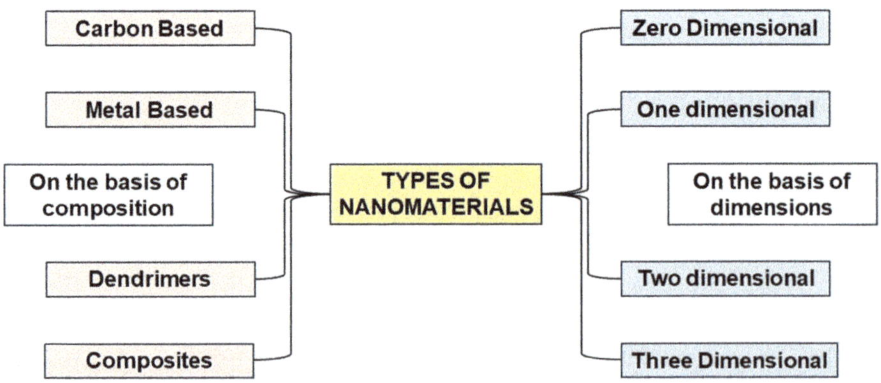

Fig. 1 Different types of nanomaterials

ellipsoidal; 1D carbon nanotubes that have cylindrical structures; 2D graphene, having a flat shape; and 3D nano horns and nanodiamonds.

- Metallic nanomaterials: Nanoparticles of metals like gold, silver, copper, platinum, and metal oxides like TiO_2, ZnO, and quantum dots made of ZnS, PbS, CdSe, and InP belong to this class of nanomaterials. Metallic NPs can have a crystalline or amorphous structure.
- Dendrimers: Their name is derived from the Greek word dendron, which means tree. These nanomaterials are nanosized polymers having a highly ordered structure, with a symmetrical core giving off branched units. The numerous chain ends present on the surface help in performing specific functions.
- Composites: Nanocomposites are formed by mixing certain nanomaterials with other nanomaterials or with bulk-type materials such as organic polymers, biomolecules, and organometallic compounds.

1.1 Preparation of Nanomaterials

Two routes are adopted for the synthesis of NPs: the Top-down and Bottom-up approach (Fig. 2) [22]. The top-down approach involves breaking down large materials into smaller and nanosized particles. Techniques like mechanical cutting, grinding, and milling or laser ablation, thermolysis, pyrolysis, and radiation-induced methods can do this. The limitations of this approach are that it can be expensive due to the specialized equipment required for the preparation of nanoparticles, and the NPs prepared by this method have surface imperfections. The bottom-down approach involves self-assembling small building blocks, i.e., atoms or molecules, to form nanosized particles. This approach uses methods like chemical, electrochemical, and biogenic methods. The NPs formed by this method have an even surface and are more homogenous.

Due to their unique properties, nanomaterials are used for a variety of purposes, such as in the manufacture of electronics, computers, automotive, and fabrics, as well as energy production, construction, packaging, environment protection, and healthcare [23–25]. This chapter explores the biomedical applications of nanoparticles.

2 Biomedical Importance of Nanoparticles

Nanoparticles are now being used extensively in different areas of medicine and healthcare, some of which are discussed below

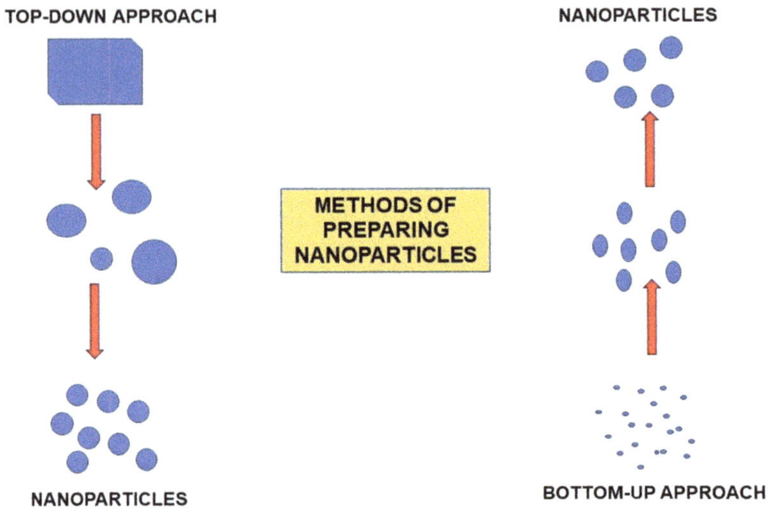

Fig. 2 The two approaches for the preparation of nanoparticles

2.1 *Detection and Diagnosis*

The early and timely detection of disease is usually associated with less severity of symptoms and reduced mortality rate, especially in cases of diseases like cancer which are difficult to treat in advanced stages [26]. Nanomaterials provide faster, easier, and more accurate disease diagnosis and detection than older techniques. Some of the bioimaging techniques used are Magnetic Resonance Imaging (MRI), Computed Tomography (CT), Positron Emission Tomography (PET), Single Photon Emission Computed Tomography (SPECT), and fluorescence imaging. Most techniques use a contrast agent to differentiate between healthy and diseased tissue. NPs are used as contrast agents due to their special properties. They show the better distribution and cellular uptake and can penetrate deeply into the tissues, making tumour detection easier and more accurate. They are also quickly cleared from the body compared to larger dyes [27]. The surface of NPs can also be labelled with ligands that target specific receptors present in tumour cells. For example, labelling the surface of AuNPs with a prostate-specific membrane antigen RNA aptamer increases the CT density during prostate cancer detection [28]. Besides acting as contrast agents, NPs can also be used as carriers of other imaging agents, for example, Her2 antibody-functionalized liposome-encapsulated gold nanoclusters are used to detect human breast cancer cells in blood and tissue by colorimetry [29].

2.2 Drug Delivery

The drugs delivered by conventional methods are not properly absorbed and are excreted early; they are also randomly distributed, which damages normal tissue, causing side effects [30]. So, drugs must be delivered only to the diseased tissue in a controlled manner. This can be achieved by targeted drug delivery which was tried for a long time but had numerous limitations like poor distribution, low efficacy, less selectivity, high cost, and time factor. However, recent research has shown that nanoparticles are very in targeted drug delivery. Targeted therapy is highly useful in the treatment of cancer. Several nanocarriers are being used for this, organic and inorganic, such as liposomes, solid lipid NPs, magnetic NPs, quantum dots, and carbon nanotubes. Recently, the FDA approved doxorubicin loaded on liposomes for the treatment of ovarian cancer, multiple myeloma (brand name Doxil), and metastatic breast cancer (brand name Myocet). Another drug Abraxane has Paclitaxel loaded on albumin-bound nanoparticles and has been approved by FDA for the treatment of various cancers. It is the first protein nanotechnology-based drug that uses albumin, which has a high affinity for binding to hydrophobic molecules making the delivery of water-insoluble drugs to tumour cells possible [31]. Nanocarriers can be used for either active or passive drug targeting. In passive targeting, the nanocarriers increase the drug concentration in the diseased tissue, thus making them more effective. Due to their small size, the NPs can cross cellular barriers; also, they stay longer in the diseased tissue as it has poor lymphatic drainage, increasing the efficacy of the cancer drug they carry [32]. Active targeting involves modifying the nanocarrier's surface with targeting moieties specific for receptors on tumour cells. For example, in a study on lung cancer A549 cells, redox-sensitive NPs of hyaluronic acid–disulfide–vitamin E succinate conjugates were loaded with the drug paclitaxel. Hyaluronic acid, an anionic polysaccharide, helps target tumour cell CD44 receptors. This active targeting enhanced cytotoxicity against A549 mouse xenograft models compared to the drug being used alone [33].

2.3 Therapeutics Agents

Apart from being used as carriers of drugs, NPs show promising therapeutic effects on their own, such as antibacterial, antifungal, antiviral, anti-leishmanial, anticancer, and antioxidant among others.

2.3.1 Anticancer Agents

Cancer treatment has evolved over the years from harmful anticancer drugs associated with severe side effects to recent drugs with fewer adverse effects. The reason for side effects is that the drugs would also target healthy cells, as they were not specific in action. However, more specific drugs are being prepared that target only the diseased tissue. This is where nanomaterials are playing a vital role in recent years. As discussed in the previous section, they are being used for targeted drug delivery and as anticancer agents. Several NPs have shown remarkable anticancer effects, such as metallic NPs of Au, Ag, and Cu [34]. However, the exact mechanism of action of NPs as anticancer agents has yet to be fully understood. It is mainly due to an increased concentration of Reactive Oxygen Species (ROS) in the cancer cell, leading to oxidative stress and apoptosis. For example, AgNPs prepared from cell extracts of the *cyanobacterium Anabaena* doliolum showed anticancer effects due to enhanced levels of ROS [35]. Another example is of AuNPs produced from *Trachyspermum ammi* seed extract, which were shown to inhibit HepG2 cancer cell lines, possibly by ROS-mediated apoptosis [36].

Apart from this, NPs can also show anticancer effects by binding and interacting with proteins present on the surface of and inside the cancer cells. For example, AgNPs produced by biogenic synthesis using longan peel powder exhibited cytotoxicity in vitro when used against human lung cancer H1299 cells and in vivo in the mouse model. This effect is likely due to the binding of NPs leading to inhibiting the transcription factor NF-κB activity and reduced expression of Bcl-2/caspase-3, which plays a crucial role in mitochondrial apoptosis [37]. Demonstrating another mechanistic pathway, CuNPs displayed a strong anticancer effect by causing DNA cleavage and serving as drug delivery agents when used as Transferrin copper nanoclusters [38]. Due to their unique thermal properties, certain NPs can be used in Photothermal therapy, one of the latest methods to treat cancer. It converts light into heat energy using NPs as ablation agents, which kills cancer cells without harming normal cells. For example, AuNPs have exhibited cytotoxicity against HOC313 clone 8 and HSC3 cell lines [39, 40]. NPs are also used to augment the effects of conventional drugs and radiotherapy. For example, using Bi NPs increased radiation therapy's effect by enhancing the cancer cells' radio-sensitivity [41].

2.3.2 Antimicrobials

Bacteria are fast developing resistance to antibiotics for various reasons, and treating such infections has become a challenge for the medical community. NPs also play an important role in this area, and several NPs have been reported to exhibit potent antimicrobial effects. Moreover, NPs are also used as carriers for the effective distribution of known drugs, which augments the drug's efficacy [42]. Their small size permits close interaction with the bacterial cells and causes bactericidal effects. AgNPs have been shown to adhere to the bacterial cell wall, penetrate it, and disrupt cellular pathways and metabolism. They cause DNA damage, increase ROS generation, inhibit protein synthesis, and damage the cell (Fig. 3) [43].

NPs have also been shown to resist efflux from the microbial cell and biofilm formation [44], two mechanisms bacteria use for antibiotic resistance. These reasons make NPs effective against multidrug-resistant bacteria, making them a good

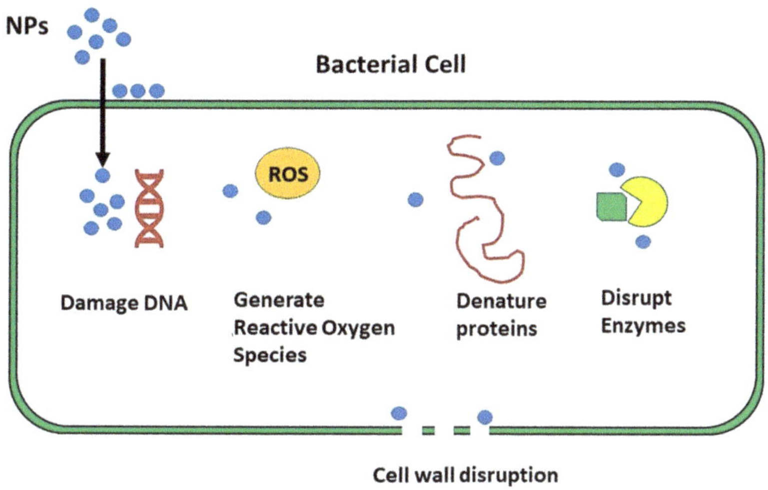

Fig. 3 Mechanism of bactericidal effect of NPs

choice as antimicrobials. For example, pneumonia-causing multidrug-resistant *Pseudomonas aeruginosa* and *Acinetobacter baumanii* were inhibited by AgNPs prepared by biogenic synthesis using extracts of *Sisymbrium irio*, a Saudi Arabian desert plant [45]. Other than AgNPs, ZnO, Au, Fe, and CuNPs and several others have shown good antibacterial effects in various studies, and so, they should be further explored for the development of potent antibiotics against MDR bacteria [15, 46–48].

Fullerenes have several loci that can be functionalized for binding at different biological surfaces and receptors. Studies on functionalized fullerenes show their biocompatibility and low toxicity; many are in clinical trials. They have shown antibacterial as well as antiviral activity in various studies. The virucidal effect of bis(monosuccinimide) derivative of p,p'-bis(2-aminoethyl)diphenyl-C60 against Human Immunodeficiency Virus 1 and 2 is a notable example [49].

Several studies on NPs also show their antifungal effect, possibly by increasing ROS. For example, FeNPs prepared by using seedless pods extract of *Acacia nilotica* showed effective antifungal activity against a species of *Candida*, while nanoxylan obtained from corncob was found to be effective against *Candida albicans, Candida parapsilosis, and Cryptococcus neoformans* [50].

2.3.3 Anti-Leishmanial Agents

Millions of people worldwide are affected by Leishmaniasis, a protozoal disease. Its treatment involves chemotherapy which has many side effects. However, nanomaterials like liposomes, metallic NPs, polymeric NPs, nanotubes, etc., are now being used to treat Leishmaniasis. For example, nano xylan, i.e., xylan-containing AgNPs produced by biogenic synthesis using corncob xylan for reduction and stabilization, successfully decreased *Leishmania amazonensis* promastigote forms. AgNPs produced from *Fusarium oxysporium* also displayed a similar effect against promastigote and amastigote forms of *Leishmania amazonensis* by increased production of ROS leading to apoptosis [51]. Another example is of ZnO NPs produced from the leaf extracts of *Verbena officinalis* and *Verbena tenuisecta.* The NPs exhibited anti-leishmanial activity; the *V. officinalis* ZnO NPs were more effective due to the smaller size and higher phenolic content than those prepared from *V. tenuisecta* [52].

3 Emerging 1D and 2D Nanomaterials and Their Biomedical Applications

Two-dimensional (2D) nanomaterials have a high anisotropy and diverse functionality. The research on the exploration of applications of 2D nanomaterials is still in a nascent stage. However, reports suggest their interaction with biological molecules makes them useful for biomedical applications. 2D nanomaterials are broadly classified as inorganic, organic, or hybrid. Of the important 2D nanomaterials, carbon-based materials, silicates and transition metal dichalcogenides (TMDs), and oxides give improved functionalities, which are attributed to their morphological features and surface-to-charge volume ratios and surface charge [53]. They have enhanced mechanical and optical properties along with variations in morphology and biocompatibility. These many features make them suitable for various applications as drug delivery agents in biosensing and tissue engineering [54]. However, due to their thin morphology, they are thought to be invaluable and need high surface interactions on a small scale. On the contrary, owing to this very feature, they find use in drug delivery systems because of their adsorbing nature. Their surface to volume ratios and high modulus grant improved mechanical strength to 2D-based nanocomposites in biomedical applications as well as their role in gene sequencing and biosensing. Their thin nature also makes them apt for the response of light, which has been utilized in imaging applications and PTT and PDT. The physicochemical interactions of these 2D nanomaterials with tissues still need to be fully understood.

Along with the benefits, the toxic reactions of such materials also need to be fully elucidated. They are thought to cause oxidative stress from free radicals. This harm may be caused by the immune response prompted by them or from their inherent properties. Their elimination from the body is also important as they can cause damage to the liver, kidney, spleen, or lungs. Apoptosis, haemolysis, or thrombosis are other causes of damage. Some 2D nanomaterials also contain metals higher than the trace levels, and in the long term, it may cause harm. The composition can also influence the biocompatibility and adsorption of proteins on their surface, which is also affected by the composition of the nanomaterial. The properties of nanomaterials also depend upon their atomic arrangements. 2D nanomaterials have a few atomic layer thicknesses [55]. Graphene is a typical 2D nanomaterial with a monolayer of carbon atoms having sp^2 hybridization and covalently bonded C atoms arranged in a regular hexagonal manner. Graphene oxide (GO) also has a similar pattern but has a few sp^{3-}bonded C atoms attached to different functional groups above and below the plane. That is why GO is less flat than graphene and polar.

Reduced graphene oxide (RGO) has a structure in between that of graphene and GO and is produced by the reduction of GO. The polarity is reduced upon reduction, and sp^2 hybridization is again attained [56]. 2D clays, LDHs, TMOs, and TMDs are nanomaterials composed of a single crystal unit. Laponite is a 2D nanoclay having three layers and 2 tetrahedral sheets of silica with an internal octahedral layer of Mg^{2+} and Li^+ cations [57]. LDHs are anionic clays having positively charged faces [58] and are structurally similar to brucite. MgO and TiO_2 have octahedral geometry, similar to clays and LDHs, and exist in stacks with interlaying ions that hold the stacks [59]. 2D TMDs have a three-layered structure where the outside layers are bonded covalently to an inner metallic layer.

The 2D nanomaterials are being explored for their biomedical uses because of their ability to interact with cells. They are also utilized in bio-integrated soft electronics for non-invasive in situ monitoring [60]. They are also applied in soft cell culturing for in vivo cell alignment [61], in electrochemical sensing [62], and in therapeutic drug delivery. They are also employed in regenerative medicine and diagnostic imaging [63]. The 2D nanomaterials also show piezoelectric effects and the response to voltage changes (flexoelectric effects), making them suitable for biosensing [64] and bone repair. TMDs have been utilized for bone regeneration, and TMOs like MnO_2 and TiO_2 in targeted drug delivery [65], owing to the restricted release of the adsorbed drug. 2D black phosphorus can also be used for controlled drug delivery because of its absorbance in the near-infrared regions [65]. The nanolayers of 2D materials are bound by van der Waals forces [66], which can sequester biomolecules for a sustained and controlled release. Crystallinity is another feature that is important for biomedical utility. Degradation of LDHs and nanoclays by lysosomes leads to the release of bioactive products which govern cell functioning. Synthetic nanoclays (Laponite) can cause osteogenic differentiation of stem cells [67]. These nanoclays also release stem cell secretomes, stimulate angiogenesis, and regenerate myocardial tissues [68]. LDHs can improve the chondrogenic differentiation of mesenchymal stem cells (MSCs) by controlled release of drugs [60].

4 Carbon-Based Nanomaterials and Their Applications

Carbon-based nanomaterials have gained attention during the last few years [69]. Carbon has several interesting allotropes like carbon nanotubes CNTs, GO, and GDQs. These allotropes have been associated with numerous biological applications like biosensing, drug delivery, etc. [70]. Single-walled carbon nanotubes (SWCNTs) have a cylindrical nanostructure. In contrast, in a multiwalled carbon nanotube (MWCNTs), a few graphite layers are rolled with a spacing of 3.4 Å [71]. Owing to its mechanical and electrical properties, it possesses improved strength, flexibility, and conductivity for different biomolecules and is useful as sensors, medical diagnostic tools, and treatment of different diseases. Graphene is the most studied material because of its interesting inherent properties [72]. Enrichment with different functional groups helps in the selective identification of biological fragments. In addition, the delocalization of π electrons makes it suitable as a drug delivery agent [73]. The moiety also shows activity towards fluorescent dyes and other biological molecules and is used for in vivo imaging and diagnosis. GDQs are zero-dimensional graphene sheets having a lateral dimension of less than 100 nm [74]. It can be obtained from the two-dimensional conversion of the graphene sheets, which endow potential photoluminescent properties. GQDs also possess enhanced biocompatibility and are resistant to photo-bleaching and are also used as biosensors and in cancer therapy [75].

5 1D Carbon-Based Nanomaterials and Their Biomedical Applications

5.1 Carbon Nanotubes (CNTs)

The CNTs are cylindrical with hexagonally arranged carbon atoms. They possess a high aspect ratio varying from a few nanometers to several microns. They are also associated with a large surface area and high curvature, which endow their interaction possible with biomolecules via van der Waals, π–π, and hydrophobic interactions. These features make surface modifications easy to increase the solubility of CNTs in an aqueous medium or govern the covalent bonding for biomedical applications [76]. The biological applications of CNTs can also be enhanced by their optical absorption in the NIR-IR region and fluorescence emission for their use in PTT, photoacoustic imaging, and tissue engineering [77]. Owing to their fibrous nature and mechanical strength, they are also used as supporting materials in tissue engineering and provide electrical conductance to frameworks used for cellular growth and differentiation [78]. CNTs have similar hexagons and have sp^2 hybridization, like graphene. In the formation of CNTs from graphene, bending takes place. Σ-bonds lying in the plane move out of the plane, causing delocalization of π-orbitals outside the nanotube, which induces distortion of the electron cloud and results in a rich π-electron conjugation outside the tubular structure [79]. The CNTs can either be single or multiwalled (SWCNTs or MWCNTs). The SWCNTs have an outer diameter between 0.4 and 2 nm, whereas those of MWCNTs range from 2 to 100 nm [80]. The one-dimensional and tubular morphology provide unique mechanical properties to the CNTs that are higher than steel [81] and have lower density. The CNTs also absorb in the NIR-IR region and are helpful in PTT and photoacoustic imaging.

5.2 CNTs as Biosensors

The morphological and optical properties make CNTs useful as nanoprobes [57]. Their high sensitivity and electron-transfer rate make them suitable for biosensing [82]. They immobilize the biomolecules on their surface and enhance the signal transduction process. CNT-glucose biosensors with the conjugation of glucose oxidase were fabricated. For instance, carbon nanotube non-woven fabrics (CNTFs) were developed to recognize glucose from glucose oxidase-impregnated polyvinyl alcohol solution [83]. Glucose oxidase-coated MWCNTs have been developed to sense glucose electrochemically [84]. CNT-based biosensors have also been developed to detect nitric oxide and epinephrine [85]. Human blood protein sensors from SWCNTs have also been fabricated [86]. The corona phase of the sensors was capable of recognizing fibrinogen. The fluorescent response of SWCNT with a smaller diameter was more prominent. Individual real-time protein efflux was detected from *E. coli* and *P. passersby* using SWCNT [87]. A MWCNT-based device was used to detect arginase-1 [88].

5.3 CNTs in Drug Delivery

CNTs have been utilized as excellent vehicles for the transportation of drugs in living cells because of their non-invasive penetrative nature [89]. Non-covalent interactions make the unrelenting release of the drug possible in acidic conditions at the tumour sites. In contrast, the covalent bonding provides stability to the drug-CNT complex stable and hence makes its release difficult. Chitosan-functionalized CNT with thermo-sensitive polymer, poly-N-Isopropyl acryl amide (NIPAAm), and 1-butyl-3-21 vinyl imidazolium (NIPAAm-) was fabricated for temperature-responsive release of the biomolecules [90]. A hybrid hydrogel was fabricated from sodium alginate (SA), bacterial cellulose, and MWCNTs. CNT has been used as a loading agent for many drugs, including doxorubicin, paclitaxel, docetaxel, and oxaliplatin [91]. Extensive studies have been undertaken to check drug delivery through functionalized CNTs [92]. The cancer cell targeting of doxorubicin (DOX)-loaded folic acid (FA) and estrogen (ES)-anchored PEG-functionalized MWCNTs were checked on MCF-7 tumour-bearing Balb/c mice. After 43 days of dosing, the mice treated with DOX/ES-PEG-MWCNTs were found to have a longer life span.

5.4 CNTs in Cancer Therapy

CNTs are potential candidates for carrying agents of anticancer drugs, genes, and proteins. They also act as efficient photothermal agents because they tend to absorb NIR light. iRGDpolyethyleneimine (PEI) functionalized MWCNT was developed

Table 1 Applications of 1D CNTs and their properties being explored for biomedical applications

S. no	Property	Use	References
1	$\pi-\pi$ interactions	For hydrophobic drug delivery and its release as dependent on pH, RNA delivery	Singh et al. [99]
2	Photothermal effect	Enhanced drug delivery efficacy	Han et al. [100]
3	Supporting material	Coupled with magnetic particles for bioimaging	Liu et al. [101]
4	Topography	Tissue regeneration	Mooney et al. [102]
5	Electrical conductivity	Conductive materials mimics, biosensors	Yu et al. [103]
6	Mechanical properties	Mechanical strength to bone	Chen et al. [104, 105]

and conjugated with candesartan. The functionalized nanostructure was fabricated with plasmid AT (2) [93]. They targeted avb3-integrin and AT1R of tumour endothelium and lung cancer cells. A DOX-loaded MWCNT-magneto fluorescent carbon quantum dot (CDQ) nanocomposite for PTT [94]. The negatively charged nanocomposite resulted in binding with positively charged DOX. The nanocomposite also had a high NIR radiation-absorbing capacity. DOX-loaded TAT-chitosan functionalized MWCNT nanosystem was fabricated for combined chemo and PTT [95]. To facilitate apoptosis, a PEG-coated CNT-ABT737 nano drug was formed to target mitochondria [96]. A gold-nanoparticle-coated CNT ring was developed and found to have improved optical properties and showed amended photoacoustic (PA) signal and photothermal behaviour [97]. The surface plasmon resonance (SPR) absorption in SWNT-Au-PEG-FA nanomaterial exhibited enhanced cytotoxicity [98] (Table 1).

6 2D Carbon-Based Nanomaterials and Their Biomedical Applications

6.1 Graphene

Graphene consists of a single layer of carbon atoms arranged in a hexagonal pattern and forms a honeycomb crystal lattice. Graphene-based nanomaterials have been employed in tissue engineering. A nanocomposite aerogel was synthesized from poly (N-isopropyl acrylamide) (PNIPAM) and graphene [106]. The aerogel was found to have improved electrical and thermo-responsive properties. The cytocompatibility of the hydrogel was also improved by doping with graphene, and it has been found to enhance the proliferation and differentiation of human mesenchymal stem cells (hMSCs) towards osteogenic cells [107]. Graphene coatings have been found to have high differentiation of hMSGs due to their ability to adsorb bioactive molecules [108]. The use of graphene to fabricate 3D porous frameworks for osteogenic differentiation of hMSGs for bone regeneration has been reported [109]. High mechanical strength was the main reason for osteogenic differentiation. The biosensing requires electrical stimulation for functioning. An electrically conductive graphene substrate was engineered to dictate cell viability by facilitating electrical interactions between neural stem cells (NSCs) [93]. A two-fold increase was obtained in the neuron density and excitability on the graphene substrate compared to the control. Another research reported a graphene-based biosensor to detect the attachment of bacteria on tooth enamel [110].

6.2 Graphene Oxide

Graphene oxide (GO) is usually produced from graphite oxide exfoliation. Hydroxyl and phenolic groups are present on its surface. The presence of carbon bound to oxygen differentiates graphene from GO. GO has both sp^2 and sp^3 hybridized carbons in its structure, promoting surface interactions. GO finds use in drug delivery and gene therapy and is associated with antibacterial activity and biosensing. The actual application is more or less linked to its biocompatibility [111]. The oxidation of graphene to GO by Hummer's method reduces the electrical conductivity and mechanical strength making it suitable as a biomedical agent by making it hydrophilic. The process facilitates the material's interactions with biomacromolecules like proteins, extracellular matrix (ECM), and several drugs.

The functionalization of GO also increases cyto and biocompatibility [112]. The force acting between different GO is pH dependent, and this fact can be used to fabricate pH-sensitive nanocomposites. A GO-poly(vinyl alcohol) nanocomposite was formed for sustained drug release. The composite remained in solid form and retained vitamin B12, which was used as the model drug, whereas in alkaline medium, the nanocomposite dissolved and released the captured drug. The release of the drug was facilitated by the ionization of the surface-bound carboxyl groups [113]. A study reported that the doping of GO to methacrylate gelatin enhanced the electrical conductivity of the nanocomposite hydrogel [114]. The electrical conductivity led to a higher proliferation of cardiomyocytes making the material useful for fabricating cardiac patches [115].

In a different study, methacrylate groups were added to the GO surface for covalent crosslinking [116], increasing its mechanical stiffness due to the addition of GO. The fabricated nanocomposite had high cell proliferation. The large surface area and functional groups make GO useful as a nanocarrier for stimuli-responsive nanocomposites [117]. GO-polypyrrole (PPy) polymers have been used for drug delivery. The nanocomposite was found to trap and release twice as much dexamethasone as compared to pure PPy. The incorporation of GO increased the sensitivity of the nanocomposite towards electrical stimulation. Drug loading and release were tunable by altering GO's sonication time. The enhanced drug loading and release kinetics were attributed to the high surface area. The controlled release kinetics and drug loading capacity could be explored for cancer treatment, immunotherapies, and tissue engineering. Hydrogels loaded with GO are also used for gene delivery and treatment of myocardial infarction. Polyethylenimine (PEI) coated GO was found to have enhanced the loading of the VEGF DNA plasmid, thereby showing that the functionalized GO can be used for gene delivery.

6.3 Reduced Graphene Oxide

RGO is usually produced from GO using hydrazine, hydrazine hydrate, or L-ascorbic acid as reducing agents. In the reduction process, many functional groups are removed, leading to some structural defects. RGO has biosensing applications and is also used in PDT and PTT owing to enhanced electrical conductivity compared to GO. RGO has a high surface area and carrier mobility and may attach receptor biomolecules on its surface [118], making them useful as biosensors. Using RGO ultrasensitive label-free field effect transistor (FET), a biosensor has been fabricated, which can sense particular antigens in the prostate even at very low concentrations. Several other FET biosensors using RGO have been developed to detect different biomolecules, including DNA. RGO-based materials are also used as PTT agents because of their cost-effectiveness and cytocompatibility [118]. RGO absorbs approximately sevenfold higher NIR light than GO because of many π–π bonds [119, 120]. This makes RGO even more effective at lower doses as compared to GO. RGO has also been employed in selective cancer cell targeting, suggesting its use in PTT/PDT combined therapy. The chemical structure of graphene, GO, and RGO is given in Fig. 4.

7 Some Specific Biomedical Applications of 2D Carbon-Based Nanomaterials

2D carbon-based nanomaterials, particularly those which are based on graphene, are explored for their biomedical applications. Their large surface area, π–π stacking, and electrostatic interactions facilitate drug loading with solubility issues. Owing to rapid development in nanotechnology, novel materials are being fabricated for diagnosis and therapeutics. Some of the specific applications of 2D carbon-based nanomaterials are discussed below:

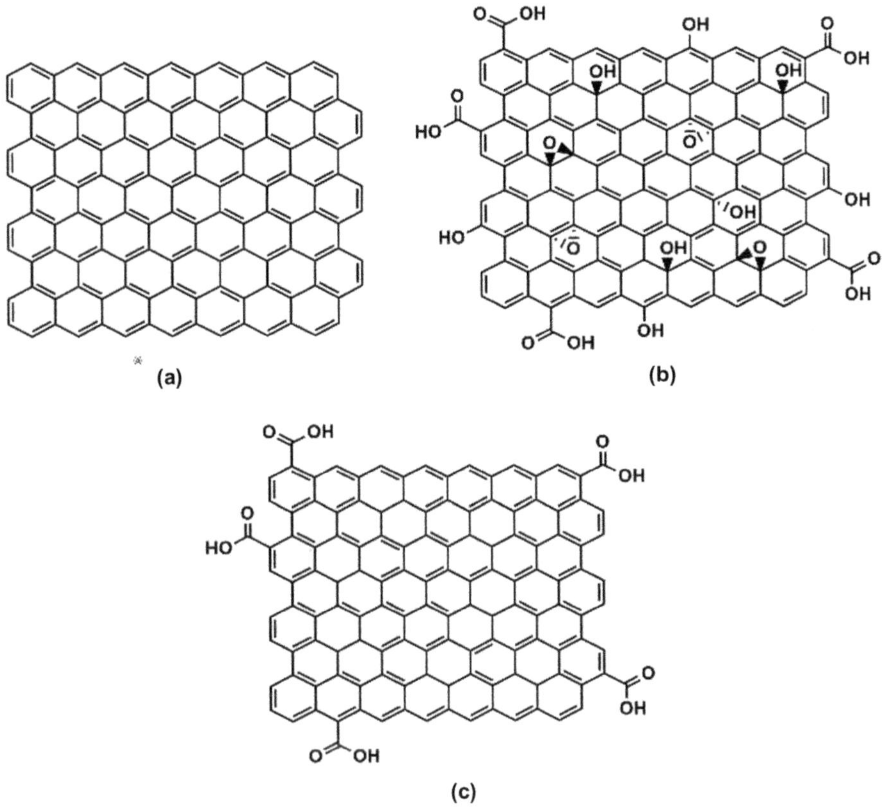

Fig. 4 Structure of **a** graphene, **b** graphene oxide, and **c** reduced graphene oxide

7.1 Drug Delivery

2D carbon-based nanomaterials have been conjugated in drug delivery systems to improve therapeutic applications and reduce toxic effects. Graphene is one of the first materials to be explored in this regard. Single-layered GO or RGO has a very high surface area, making them suitable for drug loading [117]. RO has been used for systematic targeting and effective drug delivery [121]. Physiosorpting through π–π stacking can be useful to load hydrophobic drugs like doxorubicin and docetaxel. Using the nanocarrier approach, graphene has emerged as a promising material showing non-covalent interactions with drug molecules. PEGylated GO loaded with doxorubicin (DOX) caused the real-time release of the drug through π–π interactions. Another study revealed that GO loaded with DOX caused supplementary drug release [122]. A study provided a list of anticancer drugs which could be delivered using graphene nanosheets [123]. Drugs like DOX, camptothecin, paclitaxel, etc., were included in the list. Graphene-based derivatives, including anthracycline antibiotics, quinolone alkaloids, pyrimidine analogs, and polyphenol compounds, were also included in the list. Phototherapeutic nanomaterials have also been studied to target cancer or tumour cells [124]. GO also shows optical absorption in the NIR region and high photothermal conversion. GO hybrid materials have been used in cancer phototherapy [125]. DOX-loaded PEGylated nanographene oxide (NGO-PEG (polyethylene glycol)-DOX) has been used in PTT [126]. DOX-loaded bovine serum albumin-functionalized RGO (DOX-BSA-RGO) nanosheets were used for NIR-induced chemo-photothermal therapy for brain tumour [111]. Hypocretin A (HA), a hydrophobic non-porphyrin photodynamic anticancer drug, was immobilized onto GO through π–π stacking interaction [126]. GO-HA nanomaterial was excited by irradiation and generated singlet oxygen. The nanomaterial testing on HeLa cells revealed high cellular uptake of the drug, and upon light irradiation, the imbued cells died. In a similar study, chlorine 6 (Ce6), a photosensitizer, was loaded on PEG-functionalized GO and delivered to HeLa cells. The nanosheets promoted the delivery of Ce6 molecules upon moderate heating [127]. A nanocomposite was fabricated by combining GO with gold nanoribbon. AuNR-PEG-GO was used for PTT because of the NIR absorption properties of AuNR and GO. The nanocomposite could help introduce additional functional groups to target specific cancer cells [128]. In another study, chitosan (CS)-modified graphene nanogel was developed for sustained drug release. The study searched for a NIR-triggered drug delivery system based on incorporating RGO in a thermo-sensitive nanogel. Upon irradiation with NIR light, a rapid release of DOX was observed from the nanogel [129]. A pH-sensitive drug delivery system was developed by coating nanographene oxide (NGO) with PEG and poly(allyl amine hydrochloride) (PAH); the nanocarrier was loaded with DOX; and the complex showed a pH-responsiveness exhibiting better cellular uptake of the drug in tumour cells and at microenvironmental pH [130]. Some carbon-based nanomaterials used for drug loading are summarized in Table 2.

Table 2 Some carbon-based nanomaterials used for drug loading

S. no.	Nanomaterial and cell line	Drug	References
1	Folic acid-modified GO on MCF-7	Doxorubicin, camptothecin	Zhang et al. [42]
2	Gold nanocluster-decorated RGO on HepG2 cells	Doxorubicin	Wang et al. [131]
3	Poly(N-isopropylacrylamide)-GO on A-5RT3	Camptothecin	Pan et al. [132]
4	Fe_3O_4/RGO, Fe_3O_4/GO on MCF-7	β-Lapachone	Fan et al. [133]
5	Fe_3O_4/graphene nanosheets on HepG2	Fluorouracil	Li et al. [134]
6	Graphene with PEG on U251	Lucanthone	Chowdhury et al. [135]
7	Starch–graphene nanosheets on SW-620	Hydroxycamptothecin	Liu et al. [136]

7.2 Gene Delivery

2D carbon-based nanomaterials are found to interact with biomolecules and identify nucleic acids as well [137]. GO can adsorb nucleobases through π–π interactions and inhibit nucleotides from enzymatic disintegration. A gene delivery vector protects DNA from disintegration. Viral and non-viral vectors are also used for gene delivery. GO complexed with vascular endothelial growth factor-165 (VEGF) proangiogenic gene has been found to act as an efficient delivery agent in myocardial therapy. GO nanosheets are efficient vectors that can be taken up as vectors. Polyethylenimine-GO (PEI-GO) carrier carries the plasmid DNA into HeLa cells with enhanced transfection efficiency [138]. Non-viral gene therapy is also used to treat genetic disorders [139]. GO derivatives have improved siRNA or plasmid DNA (pDNA) penetration in cells protecting the DNA from enzymatic disintegration [140]. The cytotoxicity of cationic polyethyleneimine (PEI) was found to reduce after forming a complex with GO. Preconcentrated PEI/pDNA on adsorbent GO mediated efficient gene delivery and biocompatibility [139]. PEI-GO transferred pDNA to the nucleus. Several carbon vectors like GO/chitosan, GO-PEG, and GO/polyamidoamine (PAMAM) were also found to deliver pDNA and siRNA. Graphene oleate PAMAM showed upgraded green fluorescent protein gene transfecting ability compared to ultrasonic graphene efficiency and GO PAMAM without alterations in oleic. Localized NIR heating of GO-PEG-Ce6 enhanced its efficiency against cancerous cells [127]. Functionalized RGO, when irradiated with NIR, changed the membrane integrity of endosomes and improved intracellular retention of the drug, thereby influencing delivery efficacy [141].

7.3 Biosensing

Biosensing is one of the leading biomedical applications and involves qualitative and quantitative identification of specific analytes. Biosensors can detect biomolecules and chemical analytes [142]. 2D nanomaterials based on carbon have been found to sense these biomolecules due to their electrochemical and optical behaviour and interactions with π–π stacking. Biosensors consist of a biological component (receptor) and an electronic component (transducer). Graphene-based biosensors detected glucose and nicotinamide. They can also detect biomarkers for diagnostic purposes. Biosensors are constructed by employing electrochemical and fluorescence resonance energy transfer (FRET). GO is also employed for biosensors as it has a characteristic G-band in Raman spectra and is water soluble. GO-based biosensors can lower detection limits, have fast response time, and have high sensitivity [104, 105]. A –COOH functionalized GO and polyaniline (PANI) were developed to detect DNA via DVP [143]. A label-free DNA biosensor based on graphene field effect transistors (GEFTs) was developed with single-stranded probe DNA. The biosensor had a broad analytical range with a detection limit of 1 fM for 60-mer DNA oligonucleotides [144]. A single-layer graphene (SLG)-based FET biosensor was developed, and it could detect very low concentrations of DNA (10 fM) [145]. The adsorption and desorption mechanism of single and double-stranded DNA on GO [146]. cDNAs had preferential adsorption on GO, whereas dsDNA had a lower affinity. Another study showed that the adsorption of DNA on GO was length-dependent [147]. A biosensor for identifying Mycobacterium tuberculosis DNA was developed by depositing graphene on an SPR-sensing chip [148]. RGO fabricated with phenyl butyric acid (PBA) was a linker to bind glucose. The radio frequency (RF) sensor device could detect glucose levels between 1 and 4 mM.

7.4 Tissue Engineering

Tissue engineering targets the development of biological substituents which can alter the tissue functioning to maintain its properties and are made of biodegradable materials [149]. Due to its flexibility, elasticity, and adaptability to different surfaces, graphene makes it useful for tissue engineering [150]. GO having different functional groups like hydroxyl, epoxy, and carbonyl can be easily modified and interact with DNA, proteins, and enzymes. RGO and GO nanocomposites are also used in tissue engineering because of the ease of fabrication. GO can also perform specific cellular functions and regulate intercellular interactions [151]. A study demonstrated that GO could cause the differentiation of stem cells. Graphene-based nanomaterials have also been employed in stem cell culture substrate to facilitate cardiomyogenic differentiation process [152]. They have also provided aid in creating a suitable environment for the cell growth for bone regeneration. Another study reported that graphene-based nanomaterials can be used for stem cell-based therapies to treat bone diseases [153]. These scaffolds were also useful for the proliferation of PC-12 cells [154]. GO and fibronectin-RGO-MSC hybrids have improved cardiac function [155].

Some carbon-based nanomaterials used for different therapies are summarized in Table 3.

8 Toxicity Issues of Carbon-Based Nanomaterials

The adverse effects of nanomaterials must be assessed along with their advantages [165]. The nanotoxicology branch evaluates nanomaterials' toxic effects on living beings and the environment [166]. The toxicity is associated with physical and chemical characteristics like particle size, solubility, and surface charge [167]. The nanomaterials owing to their size dimension may pass through cell membranes and cause damage to cells. The high surface area can cause other chemical components to attach to their surface, increasing their reactivity and toxicity. Different effects of carbon-based nanomaterials with mean diameters of 14, 56, and 95 nm were assessed in vitro.

Table 3 Some carbon-based nanomaterials used for different therapies

S. no.	Nanomaterial	Cell	Therapy	Reference
1	NY-ESO-1, CpG-ODNs with MWCNT	Dendritic cells	Immune response	Faria et al. [156]
2	MWCNT	PANC-1	Pancreatic cancer	Mocan et al. [157]
3	CNT		Microbeam radiation therapy	Zhang et al. [158]
4	(PEG-g-PDMA-HA)@RGO	MDAMB-231, A549	Photothermal	Kim et al.
5	(HA)-modified Q- Graphene	A549, MRC-5	Graphene Chemotherapy	Luo et al. [159, 160]
6	Porphyrin immobilized NanoGO	U87MG, HBMEC, ACBRI376	Photothermal	Su et al. [161]
7	MWCNT	HeLa	Photothermal therapy	Sobhani et al. [162]
8	RGO-PEG	U87	Chemo-photothermal photodynamic	Liu et al. [163]
9	GO decorated Ru(II)-PEG complex	A549	Photodynamic-photothermal	Zhang et al. [94]
10	CuS-GO	HeLa	Chemo-photothermal	Han et al. [164]

The findings showed that the low-sized nanomaterials had higher oxidative capacity and caused oxidative stress in alveolar epithelial cells [168]. Their morphology and shape play a key role in toxicity characteristics. Some toxic effects of carbon-based nanomaterials on the human body are shown in Fig. 5.

8.1 Toxicity Issues of Carbon-Based Nanomaterials on Living Organisms

The cytotoxicity of CNMs, graphene, and fullers was assessed against L929 mouse fibroblast cancerous cells by MTT assay [169]. Carbon-based nanomaterials have also shown controversial results [170], possibly due to their adsorptive nature. Studies have also proven that ROS production assays can be affected by some carbon-based nanomaterials. In vivo studies provide more significant information to assess long-term chronic effects [171]. Surface area and chemistry, number of layers, and purity of the nanomaterial have a vital role in deciding the toxicology [172]. The surface

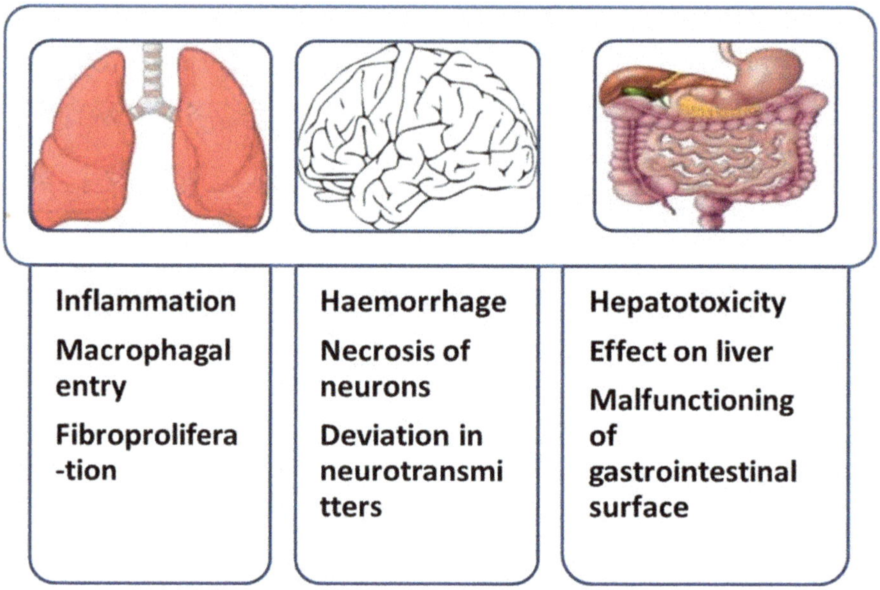

Fig. 5 Some toxic effects of carbon-based nanomaterials on the human body

area of the carbon-based nanomaterial decreases with the increase in the number of layers. The number of layers also decides the bending stiffness. The carbon-based nanomaterials may enter the body through blood circulation by crossing the blood–brain barrier or placental blood–brain barrier due to their small size, charge over the surface, impurities present, corona effects, or physical destruction. Oxidative stress, DNA damage, apoptosis, autophagy, and inflammatory response are some of the mechanisms that have a decisive role in toxicity [173]. The toxicity has also been evaluated against different cell lines. Biocompatible graphene-based nanomaterials have been developed by fabricating microbes and plant extracts [174]. In biological fluids, graphene-based nanomaterials may act as a colloid and form a protein corona. They may interact with the physiological media and result in aggregation and flocculation. The stability of the particle may increase if proteins are adsorbed through a hydrophobic region. On the other hand, adverse reactions may take place with the biodistribution and the interaction with the immune system. The complications caused during the intravenous drug delivery of carbon-based nanomaterials can be overcome by modifications in surface for selective targeting and support biodegradation [172].

8.1.1 Neurotoxicity

Adverse toxic effects on the nervous system cause nephrotoxicity [175]. Few NPs can cross the blood–brain barrier (BBB), which regulates the transport of noxious substances and maintains homeostasis. Functionalized CNTs have been shown to trespass into brain cells [176]. The neurotoxicity of pristine and MWCNTs was assessed in SH-SY5 cells. The findings suggested DNA damage and overexpression of ROS in neurons, causing pro-inflammation with enhanced expression of TNFα, IL-1β, IL-6, and cytokine levels [177]. Studies have also shown that GO can enter into cells via clathrin-channelized endocytosis [178]. It can also cross BBB because of its amphiphilic nature [179].

8.1.2 Cardiotoxicity

Increased haemodynamic flow or thrombosis may cause cardiotoxicity [180]. Graphene-based nanomaterials can cause cytotoxicity in aquatic animals and humans [181]. Pristine GO has been shown to cause cardiac tissue destruction. However, the effect was less pronounced in GO conjugated with Rh2 or arginine (Arg) and lysine (Lys) [182]. Nanodiamonds (NDs) have been found to cause cardiotoxicity in animals [22]. On the other hand, they have been used as biosensors and imaging probes.

8.1.3 Genotoxicity

Alterations in DNA structure may cause genotoxicity [183]. Fibrous carbon-based nanomaterials have been found to cause long-term genotoxicity by interacting with DNA or through inflammatory response or production of ROS. CNTs' surface area and charge density have been shown to play an important role in electrostatic interactions with DNA. Positively charged CNTs may condense DNA [184]. Different studies have shown the geno and non-genotoxic effects of CNTs. Concentration, route of exposure and duration, and dispersing agent used to solubilize the nanomaterial are important factors in determining its activity [185].

Some toxicity effects of carbon-based nanomaterials are given in Table 4.

8.2 Toxicity Issues of Carbon-Based Nanomaterials on the Environment

Carbon-based nanomaterials can cause toxic effects on the environment and biological models. It is pertinent to comprehend their interaction with living systems to use them for biomedical applications. More studies are needed to understand human exposure, as human exposure can lead to disastrous effects on the environment as well as disturb both the abiotic and biotic components [174]. It is also important to know the interaction of these nanomaterials across the membranes to assess the risk hazards upon their release into the environment. The environmental fate and transport of GO have been investigated [194]. The accumulation and deposition of GO in

Table 4 Toxicity effects of carbon-based nanomaterials

S. no.	Nanomaterial	Dose	Cell	Effect	References
1	Pristine graphene; functionalized graphene	75 μg mL^{-1}	RAW264.7	High intracellular uptake	Sasidharan et al. [186]
2	RGO, GONP, RGONP	0.01–100 μg mL^{-1}	MSC	RGO destroyed the cells	Akhavan et al. [119, 120]
3	GO	0–16 μg mL^{-1}	HepG2	Cytotoxicity with damage to the plasma membrane	Matesanz et al. [187]
4	GO, RGO	1–10 μg mL-1	HUVEC	GO exhibited higher toxicity than RGO	Das et al. [188]
5	GO, RGO HUVEC	10 μg mL^{-1}	HUVEC	Increase in intracellular ROS level and mRNA levels	Das et al. [188]
6	GO, RGO	0.0125–12.5 μg cm^{-2}	A549; RAW 264.7	Cellular accumulation of GO	Horva´th et al. [189]
7	GO	100 mg L^{-1}	GLC-82	Changes in miRNA profile	Li et al. [137]
8	Graphene/GO	20 μg mL^{-1}	MDA-MB-231	The activities of ETC complexes were inhibited	Zhou et al. [190]
9	GO; RGO	200 μg mL^{-1}	A549	Uncoated protein caused higher cytotoxicity	Chong et al. [191]
10	GO, sGO	12.5 μg mL^{-1}	PC-12	Inhibition of Aβ peptide monomer	Yang et al. [192]
11	GO, RGO	100 μg mL^{-1}	U87 U118	Decreased cell viability and proliferation	Chwalibog et al. [193]

aquatic environments have also been assessed [135, 195]. The exposure to sunlight substantially impacted its physiological properties and their consequent passage. Another study found that GO could be transformed into RGO by interacting with aquatic entities [196]. The aggregation and resuspension of GO in simulated natural aquatic environments are also reported. The conclusion suggested that the transportation of graphene oxide nanoparticles (GONPs) has a potential impact in natural waters by dipositive ions, natural organic matter (NOM), and hydraulics [197]. The acute effects of GO have been investigated on microbes present in wastewater [198] and showed toxicity between 50 and 300 mg L^{-1}. The findings also suggested that the ROS generated was responsible for the toxicity of GO. The toxic effect factor and fate factor of GO were also assessed in the environment to understand the life cycle impact of GO-based nanomaterials [199]. More studies are needed to know the environmental risks of carbon-based nanomaterials. Practices like exfoliation, CVD growth, and transfer could reduce nanomaterial exposure [200].

9 Future Prospects

The widespread applications of nanotechnology have extended into the field of biomedical sciences. The improvement in this field may take place through (1) disease identification and (2) the invention and implementation of new medicine. These methodologies include biosensing, drug delivery, and nanomedication [201]. The activity of a nanomaterial depends upon its morphology, surface charge, and size ratio, and their interaction with the biomolecules depends on the cell type and the route of administration of the nanomedicine or drug carrier. Future research's main focus will be developing cost-effective methodologies for bulk production, adopting safety measures, and comprehending the targeted biological activity [202]. Safety concerns have emerged as a prime issue and elaborated under nanotoxicology, a field that governs the harmful effects of nanomaterials. Mechanistic insights into nanomaterials' interaction, accumulation, and disintegration within living organisms require thorough understanding. The regulation of the physicochemical properties of nanomaterials is also important, as they have a pivotal role in the toxicological fate of such materials [203]. It has been found that the nanoscale size makes entry into the cell feasible and also facilitates interaction with genetic material. Nanomaterials penetrate the epidermis and cause toxicity. The generation of ROS may also cause the oxidation of biomolecules. Hence modifications in surface morphology can open the possibility of decreasing the toxicity [204]. Controlled dosing can also decrease the negative effect [202]. The scale-up of production can be regulated by modifying the particle size and composition, which has been implemented at the laboratory level in most cases. It is therefore important to implement and innovate techniques for large-scale production, which faces challenges like agglomeration and complex structure designing [205]. To reduce the cost, cooperation in research and availability of data can be of use, and with the advent of technological upgradation, the cost involved in characterization can also be reduced [159, 160].

10 Conclusion

Nanotechnology has found diverse applications in different fields of science and technology; therefore, its exposure will continue and cause beneficial and negative effects on living organisms. Carbon-based nanomaterials are very useful with potential biomedical applications in drug delivery, gene therapy, and tissue engineering because of their mechanical, optical, electrochemical, and electrical properties. They also possess semiconducting properties and π–π stacking, a characteristic of organic substances. Therefore they can interact well with the biomolecules. They have also found applications in PTT and PDT, gene and drug delivery, and antimicrobial agents by incorporating nanohybrids. The targeted therapy has improved therapeutic effects and reduced side effects. As carbon-based nanomaterials are used in different fields for biomedical uses, it is pertinent to make sure that they are safe and require standardized parameters in clinical trials. They have also been associated with toxic side effects. The toxicity is correlated with the composition and morphology of the nanomaterial. The toxicity assessment is very important for implementing regulations required for handling them in healthcare systems. Several chemical modifications have also been suggested and implemented in drug delivery and tissue engineering, as well as in the identification of biomolecules. The future implementations of carbon-based nanomaterials will involve exploring factors like their biocompatibility, solubility, and selectivity. Computational simulation studies can also be taken to assess these nanomaterials' efficacy in clinical trials.

Acknowledgements The authors are thankful to Prof. A.R. Khan, Head, Department of Chemistry, Integral University, Lucknow, and Prof. S. Joshi, Khan, Head, Department of Chemistry, Isabella Thoburn College, Lucknow.

References

1. Bachheti RK, Bachheti A, Husen A (2023) Metal and metal-oxide based nanomaterials (synthesis, agricultural, biomedical and environmental interventions). Springer Nature Singapore Pte Ltd., 152 Beach Road, #21-01/04 Gateway East, Singapore 189721, Singapore
2. Chawla S, Singh S, Husen A (2023) Smart nanomaterials targeting pathological hypoxia. Springer Nature Singapore Pte Ltd., 152 Beach Road, #21-01/04 Gateway East, Singapore 189721, Singapore. https://doi.org/10.1007/978-981-99-1718-1
3. Husen A, Bachheti RK, Bachheti A (2023) Current trends in green nano-emulsions (food, agriculture and biomedical sectors). Springer Nature Singapore Pte Ltd., 152 Beach Road, #21-01/04 Gateway East, Singapore 189721, Singapore
4. Husen A, Siddiqi KS (2023) Advances in smart nanomaterials and their applications. Elsevier Inc., 50 Hampshire St., 5th Floor, Cambridge, MA 02139, USA
5. Jin-Chul K, Madhusudhan A, Husen A (2021) Smart nanomaterials in biomedical applications. Springer Nature Switzerland AG, Gewerbestrasse 11, 6330 Cham, Switzerland. https://doi.org/10.1007/978-3-030-84262-8
6. Madhusudhan A, Purohit SD, Prasad R, Husen A (2023) Functional smart nanomaterials and their theranostics approaches. Springer Nature Singapore Pte Ltd., 152 Beach Road, #21-01/04 Gateway East, Singapore 189721, Singapore

7. Mandal AK, Ghorai S, Husen A (2023) Functionalized smart nanomaterials for point-of-care testing. Springer Nature Singapore Pte Ltd., 152 Beach Road, #21-01/04 Gateway East, Singapore 189721, Singapore
8. Sharma G, Nim S, Alle M, Husen A, Kim JC (2022) Nanoparticle-mediated delivery of flavonoids for cancer therapy: prevention and treatment. In: Kim JC, Alle M, Husen A (eds) Smart nanomaterials in biomedical applications. Nanotechnology in the life sciences. Springer, Cham. https://doi.org/10.1007/978-3-030-84262-8_3
9. Khan T, Rahman QI, Raza S, Zehra S, Ahmad N, Husen A (2023) Nanodimensional materials: an approach toward the biogenic synthesis. In: Husen A, Siddiqi KS (eds) Advances in smart nanomaterials and their applications. Elsevier, pp 523–568
10. Bachheti RK, Fikadu A, Bachheti A, Husen A (2020) Biogenic fabrication of nanomaterials from flower-based chemical compounds, characterization and their various applications: a review. Saudi J Biol Sci. 27(10):2551–2562
11. Bachheti RK, Godebo Y, Bachheti A, Yassin MO, Husen A (2020) Root-based fabrication of metal/metal-oxide nanomaterials and their various applications. In: Husen A, Jawaid M (eds) Nanomaterials for agriculture and forestry applications. Elsevier, pp 135–166. https://doi.org/10.1016/B978-0-12-817852-2.00006-8
12. Bachheti RK, Sharma A, Bachheti A, Husen A, Shanka GM, Pandey DP (2020) Nanomaterials from various forest tree species and their biomedical applications. In: Husen A, Jawaid M (eds) Nanomaterials for agriculture and forestry applications. Elsevier, pp 81–106. https://doi.org/10.1016/B978-0-12-817852-2.00004-4
13. Bachheti A, Bachheti RK, Abate L, Husen A (2022) Current status of Aloe-based nanoparticle fabrication, characterization and their application in some cutting-edge areas. S Afr J Bot 147:1058–1069
14. Bachheti RK, Abate L, Bachheti A, Madhusudhan A, Husen A (2021) Algae-, fungi-, and yeast-mediated biological synthesis of nanoparticles and their various biomedical applications. In: Handbook of greener synthesis of nanomaterials and compounds. Elsevier, Amsterdam, pp 701–734. https://doi.org/10.1016/B978-0-12-821938-6.00022-0
15. Husen A, Rahman QI, Iqbal M, Yassin MO, Bachheti RK (2019) Plant-mediated fabrication of gold nanoparticles and their applications. In: Nanomaterials and plant potential. Springer, Cham, pp 71–110. https://doi.org/10.1007/978-3-030-05569-1_3
16. Baig N, Kammakakam I, Falath W (2021) Nanomaterials: a review of synthesis methods, properties, recent progress, and challenges. Mater Adv 2:1821–2187
17. Jeevanandam J, Barhoum A, Chan YS, Dufresne A, Danquah MK (2018) Review on nanoparticles and nanostructured materials: history, sources, toxicity and regulations. Beilstein J Nanotechnol 9:1050–1074
18. Painuli S, Semwal P, Bacheti A, Bachheti RK, Husen A (2020) Nanomaterials from nonwood forest products and their applications. In: Husen A, Jawaid M (eds) Nanomaterials for agriculture and forestry applications. Elsevier, pp 15–40. https://doi.org/10.1016/B978-0-12-817852-2.00002-0
19. Buzea C, Pacheco II, Robbie K (2007) Nanoparticle classification. Nanomaterials and nanoparticles: sources and toxicity. Biointerphases 26–27
20. Findik F (2021) Nanomaterials and their applications. Period Eng Nat Sci 9:62–75
21. Joudeh N, Linke D (2022) Nanoparticle classification, physicochemical properties, characterization, and applications: a comprehensive review for biologists. J Nanobiotechnol 20:262
22. Wierzbicki M, Sawosz E, Grodzik M, Hotowy A, Prasek M, Jaworski S et al (2013) Carbon nanoparticles downregulate expression of basic fibroblast growth factor in the heart during embryogenesis. Int J Nanomed 8:3427
23. Guo D, Xie G, Luo J (2014) Mechanical properties of nanoparticles: basics and applications. J Phys D Appl Phys 47:13001
24. Holzinger M, Le Goff A, Cosnier S (2014) Nanomaterials for biosensing applications: a review. Front Chem 2:63

25. Todescato F, Fortunati I, Minotto A, Signorini R, Jasieniak J, Bozio R (2016) Engineering of semiconductor nanocrystals for light emitting applications. Materials 9:672
26. Jung DH, Hwang S, Song GW, Ahn CS, Moon DB, Ha TY et al (2016) Survival benefit of early cancer detection through regular endoscopic screening for de novo gastric and colorectal cancers in Korean liver transplant recipients. Transplant Proc 48:145–151
27. Hoshyar N, Gray S, Han H, Bao G (2016) The effect of nanoparticle size on in vivo pharmacokinetics and cellular interaction. Nanomed 11:673–692
28. Alibakhshi A, Abarghooi Kahaki F, Ahangarzadeh S, Yaghoobi H, Yarian F, Arezumand R, Ranjbari J, Mokhtarzadeh A, de la Guardia M (2017) Targeted cancer therapy through antibody fragments-decorated nanomedicines. J Contr Rel 268:323–334
29. Tao Y, Li M, Kim B, Auguste DT (2017) Incorporating gold nanoclusters and target-directed liposomes as a synergistic amplified colorimetric sensor for HER2-positive breast cancer cell detection. Theranostics 7:899–911
30. Builders PF, Arhewoh MI (2016) Pharmaceutical applications of native starch in conventional drug delivery. Starch-Stärke 9–10:864–873
31. Desai S (2016) Nanoparticle albumin-bound paclitaxel (Abraxane®). In: Albumin in medicine. Springer, pp 101–119
32. Haley B, Frenkel E (2008) Nanoparticles for drug delivery in cancer treatment. In: Urologic oncology: seminars and original investigations. Elsevier, pp 57–64
33. Song Y, Cai H, Yin T, Huo M, Ma P, Zhou J, Lai W (2018) Paclitaxel-loaded redox-sensitive nanoparticles based on hyaluronic acid-vitamin E succinate conjugates for improved lung cancer treatment. Int J Nanomed 13:1585–1600
34. Godeto YG, Ayele A, Ahmed IN, Husen A, Bachheti RK (2023) Medicinal plant-based metabolites in nanoparticles synthesis and their cutting-edge applications: an overview. In: Bachheti RK, Bachheti A (eds) Secondary metabolites from medicinal plants. CRC Press, Boca Raton, USA, pp 1–34
35. Singh G, Babele PK, Shahi SK, Sinha RP, Tyagi MB, Kumar AA (2014) Green synthesis of silver nanoparticles using cell extracts of Anabaena doliolum and screening of its antibacterial and antitumor activity. Microbiol Biotechnol 24:1354–1367
36. Perveen K, Husain FM, Qais FA, Khan A, Razak S, Afsar T et al (2021) Microwave assisted rapid green synthesis of gold nanoparticles using seed extract of Trachyspermum Ammi: ROS mediated biofilm inhibition and anticancer activity. Biomolecules 11:197
37. He Y, Du Z, Huang H, Cheng S, Wu W, Zhang K (2016) Effects of green synthesized silver nanoparticles on lung cancer cells in vitro and grown as xenograft tumors in vivo. Int J Nanomed 1879–1887
38. Halevas E, Pantazaki A (2018) Copper nanoparticles as therapeutic anticancer agents. Nanomed Nanotech J 2:119–139
39. Huang X, Jain PK, El-Sayed IH, El-Sayed MA (2006) Determination of the minimum temperature required for selective photothermal destruction of cancer cells with the use of immunotargeted gold nanoparticles. Photochem Photobiol 82:412–417
40. Shi Y, Liu M, Deng F, Zeng G, Wan Q, Zhang X, Wei Y (2017) Recent progress and development on polymeric nanomaterials for photothermal therapy: a brief overview. J Mater Chem B 5:194–206
41. Hossain M, Su M (2012) Nanoparticle location and material-dependent dose enhancement in X-ray radiation therapy. J Phys Chem C 116:23047–23052
42. Zhang L, Pornpattananangku D, Hu CM, Huang CM (2010) Development of nanoparticles for antimicrobial drug delivery. Curr Med Chem 17:585–594
43. El-Batal AI, Mosallam FM, El-Sayyad GS (2008) Synthesis of metallic silver nanoparticles by fluconazole drug and gamma rays to inhibit the growth of multidrug-resistant microbes. J Clust Sci 29:1003–1015
44. Kalishwaralal K, BarathManiKanth S, Pandian SRK, Deepak V, Gurunathan S (2010) Silver nanoparticles impede the biofilm formation by Pseudomonas aeruginosa and Staphylococcus epidermidis. Colloids Surf B Biointerfaces 79:340–344

45. Mickymaray S (2019) One-step synthesis of silver nanoparticles using Saudi Arabian desert seasonal plant sisymbrium Irio and antibacterial activity against multidrug-resistant bacterial strains. Biomolecules 9:662
46. Bachheti RK, Konwarh R, Gupta V, Husen A Joshi A (2019) Green synthesis of iron oxide nanoparticles: cutting edge technology and multifaceted applications. In: Nanomaterials and plant potential. Springer, Cham, pp 239–259. https://doi.org/10.1007/978-3-030-05569-1_9
47. Joshi A, Sharma A, Bachheti RK, Husen A, Mishra VK (2019) Plant-mediated synthesis of copper oxide nanoparticles and their biological applications. In: Nanomaterials and plant potential. Springer, Cham, pp 221–237. https://doi.org/10.1007/978-3-030-05569-1_8
48. Leyu AM, Debebe SE, Bachheti A, Rawat YS, Bachheti RK (2023) Green Synthesis of gold and silver nanoparticles using invasive alien plant *Parthenium hysterophorus* and their antimicrobial and antioxidant activities. Sustainability 15(12):9456
49. Schinazi RF, Sijbesma R, Srdanov G, Hill CL, Wudl F (1993) Synthesis and virucidal activity of a water-soluble, configurationally stable, derivatized C60 fullerene. Antimicrob Agents Chemother 37:1707–1710
50. Da E, Taha A, Afkar E (2018) Green synthesis of iron nanoparticles by *Acacia nilotica* pods extract and its catalytic, adsorption, and antibacterial activities. Appl Sci 8:1922
51. Fanti JR, Tomiotto-Pellissier F, Miranda-Sapla MM, Cataneo AHD, Andrade CGJ, Panis C et al (2028) Biogenic silver nanoparticles inducing Leishmania amazonensis promastigote and amastigote death in vitro. Acta Trop 178:46–54
52. Afridi MS, Hashmi SS, Ali GS, Zia M, Haider Abbasi B (2018) Omparative antileishmanial efficacy of the biosynthesised ZnO NPs from genus Verbena. IET Nanobiotechnol 12:1067–1073
53. Chimene D, Alge DL, Gaharwar AK (2015) Two-dimensional nanomaterials for biomedical applications: emerging trends and future prospects. Adv Mater 7261–7284
54. Gnach A, Lipinski T, Bednarkiewicz A, Rybka J, Capobianco JA (2015) Upconverting nanoparticles: assessing the toxicity. Chem Soc Rev 44:1561–1584
55. Nicolosi V, Chhowalla M, Kanatzidis MG, Strano MS, Coleman JN (2013) Liquid exfoliation of layered materials. . Science 340:1226419
56. Pei S, Cheng HM (2012) The reduction of graphene oxide. Carbon 50:3210–3228
57. Xu M, Liang T, Shi M, Chen H (2013) Graphene-like two-dimensional materials. Chem Rev 113:3766–3798
58. Rives V (2001) Layered double hydroxides: present and future. Nova Publishers
59. Osada M, Sasaki T (2009) Exfoliated oxide nanosheets: a new solution to nanoelectronics. J Mater Chem 19:2503–2511
60. Lee Y, Kim J, Joo H, Raj MS, Ghaffari R, Kim D-H (2017) Wearable sensing systems with mechanically soft assemblies of nanoscale materials. Adv Mater Technol 2:1700053
61. Kang ES, Kim DS, Suhito IR, Lee W, Song I, Kim TH (2018) Two-dimensional material-based bionano platforms to control mesenchymal stem cell differentiation. Biomater Res 22:1–12
62. Nasir MZM, Pumera M (2019) Emerging mono-elemental 2D nanomaterials for electrochemical sensing applications: from borophene to bismuthene. TrAC Trends in Anal Chem 121:115696
63. Meng Z, Stolz RM, Mendecki L, Mirica KA (2019) Electrically-transduced chemical sensors based on two-dimensional nanomaterials. Chem Rev 119:478–598
64. Yang Y, Liu X, Zhu Z, Zhong Y, Bando Y, Golberg D, Yao J, Wang X (2018) The role of geometric sites in 2D materials for energy storage. Joule 2:1075–1094
65. Xu K, Shen X, Chen W, Mu C, Jiang C, Zhao Y, Cai K (2016) Nanosheet-pore topographical titanium substrates: a biophysical regulator of the fate of mesenchymal stem cells. J Mater Chem B 4:1797–1810
66. Zhan H, Guo D, Xie G (2019) Two-dimensional layered materials: from mechanical and coupling properties towards applications in electronics. Nanoscale 11:13181–13212
67. Gaharwar AK, Mihaila SM, Swami A, Patel A, Sant S, Reis RL et al (2013) Bioactive silicate nanoplatelets for osteogenic differentiation of human mesenchymal stem cells. Adv Mate 25:3329–3336

68. Waters R, Pacelli S, Maloney R, Medhi I, Ahmed RP, Paul A (2016) Stem cell secretome-rich nanoclay hydrogel: a dual action therapy for cardiovascular regeneration. Nanoscale 8:7371–7376
69. Mukhopadhyay S, Maiti D, Saha A, Devi PS (2016) Shape transition of TiO_2 nanocube to nanospindle embedded on reduced graphene oxide with enhanced photocatalytic activity. Cryst Growth Des 16:6922–6932
70. Bhattacharya K, Mukherjee SP, Gallud A, Burkert SC, Bistarelli S, Bellucci S et al (2016) Biological interactions of carbon-based nanomaterials: from coronation to degradation. Nanomed 12:333–351
71. Eatemadi A, Daraee H, Karimkhanloo H, Kouhi M, Zarghami N, Akbarzadeh A et al (2014) Carbon nanotubes: properties, synthesis, purification, and medical applications. Nanoscale Res Lett 9:393–405
72. Sun D, Yan X, Lang J, Xue Q (2013) High-performance supercapacitor electrode based on graphene paper via flame-induced reduction of graphene oxide paper. J Power Sources 22:52–58
73. Pattnaik S, Swainb K, Lin Z (2016) Graphene and graphene-based nanocomposites: biomedical applications and biosafety. J Mater Chem B 4:7813–7831
74. Song L, Shi J, Lu J, Lu C (2015) Structure observation of graphene quantum dots by single-layered formation in layered confinement space. Chem Sci 6:4846–4850
75. Chen F, Gao W, Qiu X, Zhang H, Liu L, Luo Y (2018) Graphene quantum dots in biomedical applications: recent advances and future challenges. Front Lab Med 1:192–199
76. Gao Z, Varela JA, Groc L, Lounis B, Cognet L (2016) Toward the suppression of cellular toxicity from single-walled carbon nanotubes. Biomater Sci 4(2):230–244
77. Hong G, Diao S, Antaris AL, Dai H (2015) Carbon nanomaterials for biological imaging and nanomedicine therapy. Chem Rev 115:10816–10906
78. Harrison BS, Atala A (2007) Carbon nanotube applications for tissue engineering. Biomater 28:344–353
79. Zhang D-B, Akatyeva E, Dumitrică T (2011) Bending ultrathin graphene at the margins of continuum mechanics. Phy Rev Lett 106:255503
80. Bekyarova E, Ni Y, Malarkey EB, Montana V, McWilliams JL, Haddon RC et al (2005) Applications of carbon nanotubes in biotechnology and biomedicine. J Biomed Nanotech 1:3–17
81. Han ZJ, Rider AE, Ishaq M, Kumar S, Kondyurin A, Bilek MM et al (2013) Carbon nanostructures for hard tissue engineering. RSC Adv 3:11058–11072
82. Lin Y, Lu F, Tu Y, Ren Z (2004) Glucose biosensors based on carbon nanotube nanoelectrode ensembles. Nano Lett 4:191–195
83. Zhu L, Deng C, Chen P, Dong X, Su YH, Yuan Y et al (2014) Glucose oxidase biosensors based on carbon nanotube non-woven fabrics. Carbon 67:795–796
84. Gaitán CG, Rosas RR, Morallón E, Amorós DC (2017) Effects of the surface chemistry and structure of carbon nanotubes on the coating of glucose oxidase and electrochemical biosensors performance. RSC Adv 7:26867–26878
85. Mphuthi NG, Adekunle AS, Ebenso EE (2016) Electrocatalytic oxidation of epinephrine and norepinephrine at metal oxide doped phthalocyanine/MWCNT composite sensor. Sci Rep 6:2693
86. Bisker G, Dong J, Park HD, Iverson NM, Ahn J, Nelson JT (2016) Protein-targeted corona phase molecular recognition. Nat Commun 7:10241
87. Landry MP, Ando H, Chen AY, Cao J, Kottadiel VI, Chio L et al (2017) Single-molecule detection of protein efflux from microorganisms using fluorescent single-walled carbon nanotube sensor arrays. Nat Nanotechnol 12:368–377
88. Baldo S, Buccheri S, Ballo A, Camarda M, LaMagna A, Castagna ME et al (2016) Carbon nanotube-based sensing devices for human Arginase-1 detection. Sens Biosensing Res 7:168–173
89. Panczyk T, Wolski P, Lajtar L (2016) Coadsorption of doxorubicin and selected dyes on carbon nanotubes. Theoretical investigation of potential application as a pH-controlled drug delivery system. Langmuir 32:4719–4728

90. Kang JH, Kima HS, Shin US (2017) Thermo conductive carbon nanotube-framed membranes for skin heat signal-responsive transdermal drug delivery. Polym Chem 8:3154–3163
91. Lee PC, Lin CY, Peng CL, Shieh MJ (2016) Development of a controlled-release drug delivery system by encapsulating oxaliplatin into SPIO/MWNT nanoparticles for effective colon cancer therapy and magnetic resonance imaging. Biomater Sci 4:1742–1753
92. Liu Z, Fan AC, Rakhra K, Sherlock S, Goodwin A, Chen X et al (2009) Supramolecular stacking of doxorubicin on carbon nanotubes for in vivo cancer therapy. Angew Chem 48:7668–7672
93. Tang M, Song Q, Li N, Jiang Z, Huang R, Cheng G (2013) Enhancement of electrical signaling in neural networks on graphene films. Biomaterials 34:6402–6411
94. Zhang DY, Zheng Y, Tan CP, Sun JH, Zhang W, Ji LN et al (2017) Graphene oxide decorated with Ru(II)–polyethylene glycol complex for lysosome-targeted imaging and photodynamic/photothermal therapy. ACS Appl Mater Interfaces 9:6761–6771
95. Dong X, Sun Z, Wang X, Leng X (2017) An innovative MWCNTs/DOX/TC nanosystem for chemo-photothermal combination therapy of cancer. Nanomed Nanotech Biol Med 13:2271–2280
96. Kim SW, Lee YK, Lee JY, Hong JH, Khang D (2017) PEGylated anticancer-carbon nanotubes complex targeting mitochondria of lung cancer cells. Nanotechnology 28:465102
97. Song J, Wang F, Yang X, Ning B, Harp MG, Culp SH et al (2016) Gold nanoparticle coated carbon nanotube ring with enhanced Raman scattering and photothermal conversion Property for theranostic applications. J Am Chem Soc 138:7005–7015
98. Bao Z, Liu X, Liu Y, Liu H, Zhao K (2016) Near-infrared light-responsive inorganic nanomaterials for photothermal therapy. AJPS 11:349–364
99. Singh S, Mehra NK, Jain NK (2016) Development and characterization of the paclitaxel loaded riboflavin and thiamine conjugated carbon nanotubes for cancer treatment. Pharm Res 33:1769–1781
100. Han S, Kwon T, Um JE, Haam S, Kim WJ (2016) Highly selective photothermal therapy by a phenoxylated-dextran-functionalized smart carbon nanotube platform. Adv Healthcare Mater 5:1147–1156
101. Liu X, Marangon I, Melinte G, Wilhelm C, Menard-Moyon C, Pichon BP et al (2014) Design of covalently functionalized carbon nanotubes filled with metal oxide nanoparticles for imaging, therapy, and magnetic manipulation. ACS Nano 8:11290–11304
102. Mooney E, Mackle JN, Blond DJ-P, O'Cearbhaill E, Shaw G, Blau WJ et al (2012) The electrical stimulation of carbon nanotubes to provide a cardiomimetic cue to MSCs. Biomater 33:6132–6139
103. Yu W, Jiang X, Cai M, Zhao W, Ye D, Zhou Y et al (2014) A novel electrospun nerve conduit enhanced by carbon nanotubes for peripheral nerve regeneration. Nanotech 25:165102
104. Chen C-S, Soni S, Le C, Biasca M, Farr E, Chen EY et al (2012) Human stem cell neuronal differentiation on silk-carbon nanotube composite. Nanoscale Res Lett 7:1
105. Chen D, Feng H, Li J (2012) Graphene oxide: preparation, functionalization, and electrochemical applications. Chem Rev 112:6027–6053
106. Qiu L, Liu D, Wang Y, Cheng C, Zhou K, Ding J, Truong VT, Li D (2014) Mechanically robust, electrically conductive and stimuli-responsive binary network hydrogels enabled by superelastic graphene aerogels. Adv Mater 26:3333–3337
107. Nayak TR, Andersen H, Makam VS, Khaw C, Bae S, Xu X et al (2011) Graphene for controlled and accelerated osteogenic differentiation of human mesenchymal stem cells. ACS Nano 28:4670–4678
108. Lee WC, Lim CHY, Shi H, Tang LA, Wang Y, Lim CT, Loh KP (2011) Origin of enhanced stem cell growth and differentiation on graphene and graphene oxide. ACS Nano 5:7334–7341
109. Crowder SW, Prasai D, Rath R, Balikov DA, Bae H, Bolotin KI, Sung H-J (2013) Three-dimensional graphene foams promote osteogenic differentiation of human mesenchymal stem cells. Nanoscale 5:4171–4176
110. Mannoor MS, Tao H, Clayton JD, Sengupta A, Kaplan DL, Naik RR et al (2012) Graphene-based wireless bacteria detection on tooth enamel. Nat Comm 3:763

111. Yang K, Feng L, Hong H, Cai W, Liu Z (2013) Preparation and functionalization of graphene nanocomposites for biomedical applications. Nat Protoc 8:2392–2403
112. Reshma SC, Mohanan PV (2014) Graphene: a multifaceted nanomaterial for cutting edge biomedical application. Int J Med Nano Res 1(003)
113. Goenka S, Sant V, Sant S (2014) Graphene-based nanomaterials for drug delivery and tissue engineering. J Control Rel 173:75–88
114. Shin SR, Jung SM, Zalabany M, Kim K, Zorlutuna P, Kim SB, Nikkhah M, Khabiry M, Azize M, Kong J, Wan KT (2013) Carbon-nanotube-embedded hydrogel sheets for engineering cardiac constructs and bioactuators. ACS Nano 7:2369–2380
115. Shin SR, Aghaei-Ghareh-Bolagh B, Dang TT, Topkaya SN, Gao X, Yang SY, Jung SM, Oh JH, Dokmeci MR, Tang X, Khademhosseini A (2013) Cell-laden microengineered and mechanically tunable hybrid hydrogels of gelatin and graphene oxide. Adv Mater 25:6385–6391
116. Cha C, Shin SR, Gao X, Annabi N, Dokmeci MR, Tang X, Khademhosseini A (2014) Controlling mechanical properties of cell-laden hydrogels by covalent incorporation of graphene oxide. Small 10:514–523
117. Liu J, Cui L, Losic D (2013) Graphene and graphene oxide as new nanocarriers for drug delivery applications. Acta Biomater 9:9243–9257
118. Cai B, Wang S, Huang L, Ning Y, Zhang Z, Zhang GJ (2014) Ultrasensitive label-free detection of PNA–DNA hybridization by reduced graphene oxide field-effect transistor biosensor. ACS Nano 8:2632–2638
119. Akhavan O, Ghaderi E, Aghayee S, Fereydooni Y, Talebi A (2012) The use of a glucose-reduced graphene oxide suspension for photothermal cancer therapy. J Mater Chem 22:13773–13781
120. Akhavan O, Ghaderi E, Akhavan A (2012) Size-dependent genotoxicity of graphene nanoplatelets in human stem cells. Biomaterials 33:8017–8025
121. Nayak EbPK, Recent advances in graphene research (InTech; 2016)
122. Wu H, Lu W, Shao J-J, Zhang C, Wu M-B, Li B-H, Yang Q-H (2014) PH-dependent size, surface chemistry and electrochemical properties of graphene oxide. Carbon 67:795
123. Shim G, Kim M-G, Park JY, Oh Y-K (2016) Graphene-based nanosheets for delivery of chemotherapeutics and biological drugs. Adv Drug Deliver Rev 105:205–227
124. Chitgupi U, Qin Y, Lovell JF (2017) Targeted nanomaterials for phototherapy. . Nanotheranostics 1:38–58
125. Some S, Gwon AR, Hwang E, Bahn G-H, Yoon Y et al (2014) Cancer therapy using ultrahigh hydrophobic drug-loaded graphene derivatives. Sci Rep 4:6314
126. Zhang W, Guo Z, Huang D, Liu Z, Guo X, Zhong H (2011) Synergistic effect of chemo-photothermal therapy using pegylated graphene oxide. Biomater 32:8555–8561
127. Tian B, Wang C, Zhang S, Feng L, Liu Z (2011) Photothermally enhanced photodynamic therapy delivered by nano-graphene oxide. ACS Nano 5:7000–7009
128. Dembereldorj U, Choi SY, Ganbold E-O, Song NW, Kim D, Choo J, Lee SY, Kim S, Joo S-W (2013) Gold nanorod assembled pegylated graphene-oxide nanocomposites for photothermal cancer therapy. Photochem Photobiol 90:659–666
129. Wang C, Mallela J, Garapati US, Ravi S, Chinnasamy Girard VY et al (2013) A chitosan-modified graphene nanogel for noninvasive controlled drug release. Nanomed Nanotechnol Biol Med 9:903–911
130. Feng L, Li K, Shi X, Gao M, Liu J, Liu Z (2014) Smart pH-responsive nanocarriers based on nano-graphene oxide for combined chemo- and photothermal therapy overcoming drug resistance. Adv Healthc Mater 3:1261–1271
131. Wang C, Li J, Amatore C, Chen Y, Jiang H (2011) Wang X-M (2011) Gold nanoclusters and graphene nanocomposites for drug delivery and imaging of cancer cells. Angew Chem Int Edit 50:11644–11648
132. Pan Y, Bao H, Sahoo NG, Wu T, Li L (2011) Water-soluble poly(n-isopropyl acrylamide)-graphene sheets synthesized via click chemistry for drug delivery. Adv Funct Mater 21:2754–2763

133. Fan X, Jiao G, Zhao Jin WP, Li X (2013) Magnetic Fe_3O_4–graphene composites as targeted drug nanocarriers for pH-activated release. Nanoscale 5:1143–1152
134. Li J, Lyv Z, Li Y, Liu H, Wang J, Zhan W et al (2015) A theranostic prodrug delivery system based on Pt(IV) conjugated nano-graphene oxide with a synergistic effect to enhance the therapeutic efficacy of Pt drug. Biomaterials 5:12–21
135. Chowdhury I, Hou W-C, Goodwin D, Henderson M, Zepp RG, Bouchard D (2015) Sunlight affects aggregation and deposition of graphene oxide in the aquatic environment. Water Res 78:37–46
136. Liu K, Wang Y, Li H, Y. Duan Y (2015) A facile one-pot synthesis of starch functionalized graphene as nano-carrier for pH sensitive and starch-mediated drug delivery. Colloid Surf B 128:86–93
137. Li Y, Wu Q, Zhao Y, Bai Y, Chen P, Xia T, Wang D (2014) Response of microRNAs to in vitro treatment with graphene oxide. ACS Nano 8:2100–2110
138. Feng L, Yang X, Shi X, Tan X, Peng R, Wang J, Liu Z (2013) Polyethylene glycol and polyethylenimine dual-functionalized nano-graphene oxide for photothermally enhanced gene delivery. Small 9:1989–1997
139. Niidome T, Huang L (2002) Gene therapy progress and prospects: nonviral vectors. Gene Ther 9:1647–1652
140. Bao H, Pan Y, Ping Y, Sahoo NG, Wu T, Li L, Li J, Gan LH (2011) Chitosan-functionalized graphene oxide as a nanocarrier for drug and gene delivery. Small 7:1569–1578
141. Kim H, Lee D, Kim J, Kim T-I, Kim WJ (2013) Photothermally triggered cytosolic drug delivery via endosome disruption using a functionalized reduced graphene oxide. ACS Nano 7:6735–6746
142. Nanda SS, Papaefthymiou GC, Yi DK (2015) Functionalization of graphene oxide and its biomedical applications. Crit Rev Solid State Mater Sci 40:291–315
143. Cheng D, Yang L, Li X, Zhou J, Chen Q, Yan S et al (2017) An electrochemical DNA sensing platform using carboxyl functionalized graphene as the electrode-modified material. J Electrochem Soc 164:H345–H351
144. Ping J, Vishnubhotla JR, Vrudhula A, Johnson ATC (2016) Scalable production of high-sensitivity, label-free DNA biosensors based on backgated graphene field effect transistors. ACS Nano 10:8700–8704
145. Zheng XT, Ananthanarayanan A, Luo KQ, Chen P (2015) Glowing graphene quantum dots and carbon dots: properties, syntheses, and biological applications. Small 11:1620–1636
146. Park JS, Goo NI, Kim DE (2014) Mechanism of DNA adsorption and desorption on graphene oxide. Langmuir 30:12587–12595
147. Huang Z, Liu J (2018) Length-dependent diblock DNA with poly-cytosine (Poly-C) as high-affinity anchors on graphene oxide. Langmuir 34:1171–1177
148. Prabowoa BA, Aloma A, Secarioa MK, Masimd FCP, Lai HC, Hatanakad K et al (2016) Graphene-based portable SPR Sensor for the detection of mycobacterium tuberculosis DNA strain. Procedia Eng 168:541–545
149. Song P, Zhang X, Sun M, Cui X, Lin Y (2012) Graphene oxide modified TiO_2 nanotube arrays: enhanced visible light photoelectrochemical properties. Nanoscale 4:1800–1804
150. Wang L, Lu C, Zhang B, Zhao B, Wu F, Guan S (2014) Fabrication and characterization of flexible silk fibroin films reinforced with graphene oxide for biomedical applications. RSC Adv 4:40312–40320
151. Li D, Liu T, Yu X, Wu D, Su Z (2017) Fabrication of graphenebiomacromolecule hybrid materials for tissue engineering application. Polym Chem 8(30):4309–4321
152. Park SY, Park J, Sim SH, Sung MG, Kim KS, Hong BH, Hong S (2011) Enhanced differentiation of human neural stem cells into neurons on graphene. Adv Mater 23:H263–H267
153. Zhou M, Zhai Y, Dong S (2009) Electrochemical sensing and biosensing platform based on chemically reduced graphene oxide. Anal Chem 81:5603–5613
154. Nasim G, Hamidreza G, Mahshid K, Mohammadhossein F (2017) A facile one-step strategy for development of a double network fibrous scaffold for nerve tissue engineering. Biofabrication 9:025008

155. Park S, An J, Jung I, Piner RD, An SJ, Li X, Velamakanni A, Ruoff RS (2009) Colloidal suspensions of highly reduced graphene oxide in a wide variety of organic solvents. Nano Lett 9:1593–1597
156. Faria PCB, Santos LI, Coelho JP, Ribeiro HB, Pimenta MA, Ladeira LO et al (2014) Oxidized multiwalled carbon nanotubes as antigen delivery system to promote superior CD8C T cell response and protection against cancer. Nano Lett 14:5458–5470
157. Mocan T, Matea CT, Cojocaru I, Ilie I, Tabaran FA, Zaharie F et al (2014) Photothermal treatment of human pancreatic cancer using PEGylated multiwalled carbon nanotubes induces apoptosis by triggering mitochondrial membrane depolarization mechanism. J Cancer 5:679–688
158. Zhang L, Yuan H, Inscoe C, Chtcheprov P, Hadsell M, Lee Y et al (2014) Nanotube X-ray for cancer therapy: a compact microbeam radiation therapy system for brain tumor treatment. Expert Rev Anticancer Ther 14:1411–1418
159. Luo J, Wu M, Gopukumar D, Zhao YJ (2016) Big data application in biomedical research and health care: a literature review. Biomed Inform Insights 8:S31559
160. Luo Y, Cai X, Li H, Lin Y, Du D (2016) Hyaluronic acid-modified multifunctional Q-graphene for targeted killing of drug-resistant lung cancer cells. ACS Appl Mater Interfaces 8:4048–4055
161. Su S, Wang J, Vargas E, Wei J, Zaguilaìn RM, Sennoune SR et al (2016) Porphyrin immobilized nanographene oxide for enhanced and targeted photothermal therapy of brain cancer. ACS Biomater Sci Eng 2:1357–1366
162. Sobhani Z, Behnam MA, Emami F, Dehghanian A, Jamhiri I (2017) Photothermal therapy of melanoma tumor using multiwalled carbon nanotubes. Int J Nanomed 12:4509–4517
163. Liu J, Liu K, Feng L, Liu Z, Xu L (2017) Comparison of nanomedicine-based chemotherapy, photodynamic therapy, and photothermal therapy using reduced graphene oxide for the model system. Biomater Sci 5:331–340
164. Han L, Hao YN, Wei X, Chen XW, Shu Y, Wang JH (2017) Hollow copper sulfide nanosphere–doxorubicin/graphene oxide core–shell nanocomposite for photothermal-chemotherapy. ACS Biomater Sci Eng 3:3230–3235
165. Hola K, Zhang Y, Wang Y, Giannelis EP, Zboril R, Rogach AL (2014) Carbon dots—emerging light emitters for bioimaging, cancer therapy and optoelectronics. Nano Today 9:590–603
166. Ai J, Biazar E, Jafarpour M, Montazeri M, Majdi A, Aminifard S, et al (2011) Nanotoxicology and nanoparticle safety in biomedical designs. Int J Nanomed 6:1117–1127
167. Walters C, Pool E, Somerset V (2016) Nanotoxicology: a review. In: Toxicology-new aspects to this scientific conundrum, vol 10, pp 5772–6475
168. Koike E, Kobayashi T (2006) Chemical and biological oxidative effects of carbon black nanoparticles. Chemosphere 65:946–951
169. Kumar V, Sharma N, Maitra S (2017) In vitro and in vivo toxicity assessment of nanoparticles. Int Nano Lett 7:243–256
170. Madani SY, Mandel A, Seifalian AM (2013) A concise review of carbon nanotube's toxicology. Nano Rev 4:21521
171. Dhawan A, Sharma V (2010) Toxicity assessment of nanomaterials: methods and challenges. Anal Bioanal Chem 398:589–605
172. Sanchez VC, Jachak AH, Hurt RH, Kane AB (2012) Biological interactions of graphene-family nanomaterials: an interdisciplinary review. Chem Res Toxicol 25:15–34
173. Ou L, Song B, Liang H, Liu J, Feng X, Deng B et al (2016) Toxicity of graphene-family nanoparticles: a general review of the origins and mechanisms. Particle and fibre Toxicol 13:1–24
174. Syama S, Mohanan PV (2016) Safety and biocompatibility of graphene: a new generation nanomaterial for biomedical application. Int J Biol Macromol 86:546–555
175. Costa LG, Pellacani C, Guizzetti M (2017) Vitro and alternative approaches to developmental neurotoxicity. In: Reproductive and developmental toxicology, 2nd edn., Elsevier, pp 241–253
176. Bussy C, Al-Jamal KT, Boczkowski J, Lanone S, Prato M, Bianco A et al (2015) Microglia determine brain region-specific neurotoxic responses to chemically functionalized carbon nanotubes. ACS Nano 9:7815–7830

177. Visalli G, Curro M, Iannazzo D, Pistone A, Ciarello MP, Acri G et al (2017) In vitro assessment of neurotoxicity and neuroinflammation of homemade MWCNTs. Environ Toxicol Pharmacol 56:121–128
178. Jaeschke H, Gores GJ, Cederbaum AI, Hinson JA, Pessayre D, Lemasters JJ (2002) Mechanisms of hepatotoxicity. Toxicol Sci 65:166–176
179. Hu X, Wei Z, Mu L (2017) Graphene oxide nanosheets at trace concentrations elicit neurotoxicity in the offspring of zebrafish. Carbon 117:182–191
180. Brana I, Tabernero J (2010) Cardiotoxicity. Annals Oncol Off J Europ Soc Med Oncol 21:vii173–v179
181. Manjunatha B, Park SH, Kim K, Kundapur RR, Lee SJ (2018) Pristine graphene induces cardiovascular defects in zebrafish (Danio rerio) embryogenesis. Environ Pollut 243:246–254
182. Zare-Zardini H, Taheri-Kafrani A, Ordooei M, Amiri A, Karimi-Zarchi M (2018) Evaluation of toxicity of functionalized graphene oxide with ginsenoside Rh2, lysine and arginine on blood cancer cells (K562), red blood cells, blood coagulation and cardiovascular tissue: in vitro and in vivo studies. J Taiwan Inst Chem Eng 93:70–78
183. Pellevoisin C, Bouez C, Cotovio J (2018) Cosmetic industry requirements regarding skin models for cosmetic testing. Elsevier, Skin tissue models, pp 3–37
184. Lindberg HK, Falck GC-M, Suhonen S, Vippola M, Vanhala E, Catalan J et al (2009) Genotoxicity of nanomaterials: DNA damage and micronuclei induced by carbon nanotubes and graphite nanofibres in human bronchial epithelial cells in vitro. Toxicol Lett 186:166–173
185. Demir E, Marcos R (2018) Toxic and genotoxic effects of graphene and multiwalled carbon nanotubes. J Toxicol Environ Health, Part A 1–16
186. Sasidharan A, Panchakarla LS, Sadanandan AR, Ashokan A, Chandran P et al (2012) Hemocompatibility and macrophage response of pristine and functionalized graphene. Small 8:1251–1263
187. Matesanz M-C, Vila M, Feito M-J, Linares J, Gonçalves G, Vallet-Regi M, Marques P, Portolés M-T (2013) The effects of graphene oxide nanosheets localized on f-actin filaments on cell-cycle alterations. Biomaterials 34:1562–1569
188. Das S, Singh S, Singh V, Joung D, Dowding JM et al (2013) Oxygenated functional group density on graphene oxide: its effect on cell toxicity. Part Syst Charact 30:148–157
189. Horva'th L, Magrez A, Burghard M, Kern K, Forro' L, Schwaller B (2013) Evaluation of the toxicity of graphene derivatives on cells of the lung luminal surface. Carbon 64:45–60
190. Zhou H, Zhang B, Zheng J, Yu M, Zhou T et al (2014) The inhibition of migration and invasion of cancer cells by graphene via the impairment of mitochondrial respiration. Biomaterials 35:1597–1607
191. Chong Y, Ge C, Yang Z, Garate JA, Gu Z, Weber JK, Liu J, Zhou R (2015) Reduced cytotoxicity of graphene nanosheets mediated by blood-protein coating. ACS Nano 9:5713–5724
192. Yang Z, Ge C, Liu J, Chong Y, Gu Z, Jimenez-Cruz CA, Chai Z, Zhou R (2015) Destruction of amyloid fibrils by graphene through penetration and extraction of peptides. Nanoscale 7:18725–18737
193. Chwalibog A, Jaworski S, Sawosz E, Kutwin M, Wierzbicki M et al (2015) In vitro and in vivo effects of graphene oxide and reduced graphene oxide on glioblastoma. Int J Nanomed 10:1585–1596
194. Chowdhury I, Duch MC, Mansukhani ND, Hersam MC, Bouchard D (2013) Colloidal properties and stability of graphene oxide nanomaterials in the aquatic environment. Environ Sci Technol 47:6288–6296
195. Chowdhury SM, Sutherland C, Sanchez Z, Chaudhary P, Kumar S et al (2015) Graphene nanoribbons as a drug delivery agent for lucanthone mediated therapy of glioblastoma multiforme. Nanomed Nanotechnol Biol Med 1:109–118
196. Zhao J, Wang Z, White JC, Xing B (2014) Graphene in the aquatic environment: adsorption, dispersion, toxicity and transformation. Environ Sci Technol 48:9995–10009
197. Hua Z, Tang Z, Bai X, Zhang J, Yu L, Cheng H (2015) Aggregation and resuspension of graphene oxide in simulated natural surface aquatic environments. Environ Pollut 205:161–169

198. Ahmed F, Rodrigues DF (2013) Investigation of acute effects of graphene oxide on wastewater microbial community: a case study. J Hazard Mater 33–39:256–257
199. Deng Y, Li J, Qiu M, Yang F, Zhang Yuan JC (2016) Deriving characterization factors on freshwater ecotoxicity of graphene oxide nanomaterial for life cycle impact assessment. Int J Life Cycle Ass 22:222–236
200. Lee Y-H, Lai Y-H (2016) Synthesis, characterization, and biological evaluation of anti-her2 indocyanine green-encapsulated pegcoated plga nanoparticles for targeted phototherapy of breast cancer cells. PLoS ONE 11:e0168192
201. Elahi M, Kamali M, Baghersad MHJT (2018) Recent biomedical applications of gold nanoparticles: a review. Talanta 184:537–556
202. Amin YM, Hawas AM, El-Batal A, Elsayed SHHE (2015) Evaluation of acute and subchronic toxicity of silver nanoparticles in normal and irradiated animals. Br J Pharmacol Toxicol 6:22–38
203. Ajdary M, Moosavi M, Rahmati M, Falahati M, Mahboubi M, Mandegary A et al (2018) Health concerns of various nanoparticles: a review of their in vitro and in vivo toxicity. Nanomaterials 8:634
204. Choi AO, Cho SJ, Desbarats J, Lovrić J, Maysinger DJJ (2007) Quantum dotinduced cell death involves Fas upregulation and lipid peroxidation in human neuroblastoma cells. J Nanobiotech 5:1
205. Colombo S, Beck-Broichsitter M, Botker JP, Malmsten M, Rantanen J, Bohr AJA (2018) Transforming nanomedicine manufacturing toward quality by design and microfluidics. Adv Drug Deliv Rev 128:115–131
206. For ME, Gopinath SCB (2017) Feasibility of graphene in biomedical applications. Biomed Pharmacother 94:354–361
207. Hopley EL, Salmasi S, Kalaskar DM, Seifalian AM (2014) Carbon nanotubes leading the way forward in new generation 3D tissue engineering. Biotech Adv 32:1000–1014
208. Kuila T, Bose S, Mishra AK, Khanra P, Kim NH, Lee JH (2012) Chemical functionalization of graphene and its applications. Prog Mater Sci 57(7):1061–1105
209. Lee SS, Choi GE, Lee HJ, Kim Y, Choy JH, Jeong B (2017) Layered double hydroxide and polypeptide thermogel nanocomposite system for chondrogenic differentiation of stem cells. ACS Appl Mater Interfaces 9:42668–42675
210. Li J, Tan S, Kooger R, Zhang C, Zhang Y (2014) Micrornas as novel biological targets for detection and regulation. Chem Soc Rev 43:506–517
211. Tîlmaciu CM, Morris MC (2015) Carbon nanotube biosensors. Front Chem 3:59
212. Wang Y, Xia Y (2004) Bottom-up and top-down approaches to the synthesis of monodispersed spherical colloids of low melting-point metals. Nano Lett 4:2047–2050
213. Weaver CL, LaRosa JM, Luo X, Cui XT (2014) Electrically controlled drug delivery from graphene oxide nanocomposite films. ACS Nano 8:1834–1843
214. Yang Y, Asiri AM, Tang Z, Du D, Lin Y (2013) Graphene-based materials for biomedical applications. Mater Today 16:365–373
215. Zhang L, Xia J, Zhao Q, Liu L, Zhang Z (2010) Functional graphene oxide as a nanocarrier for controlled loading and targeted delivery of mixed anticancer drugs. Small 6:537–544
216. Zou Y, Liu Y, Yang Z, Zhang D, Lu Y, Zheng M et al (2018) Effective and targeted human orthotopic glioblastoma xenograft therapy via a multifunctional biomimetic nanomedicine. Adv Mater 30:e1803717

A Recent Update of Graphene-Based Nanomaterials for Biomedical Applications: Focusing on Drug Delivery and Tissue Engineering

Sevgi Kemeç Aslan and Kaan Hürkan

Abstract In recent times, there has been a surge in the use of nanomaterials within biomedical. These materials have gained significant interest for their potential in several applications, including the development of artificial organs and tissues, drug delivery systems, and gene therapy. This increasing interest may be attributed to the unique properties of nanomaterials, namely their decreased dimensions and expanded surface area. Graphene and its derivatives have garnered significant attention across various areas, with a particular emphasis on biomedical applications. This preference stems from the exceptional structural, mechanical, optical, thermal, and electrical properties of graphene and its substantial surface area. These characteristics facilitate efficient and effective drug-loading processes. The use of graphene and its derivative materials in biomedical settings has significant importance in several domains, including tissue engineering, biosensor applications, and gene transfer, alongside its established role in controlled drug release applications. In this section, drug delivery and tissue engineering issues are specifically addressed. Graphene and its derivatives have exceptional characteristics as drug delivery carriers, rendering them very promising for applications involving materials with superior loading efficiency, targeting capabilities, imaging capabilities, sensing abilities, and stimulus-sensitive release qualities. These materials, which also aid in the restoration of the extracellular matrix or provide structural reinforcement, enhance the adhesion, proliferation, survival, and derivatives of stem cells, hence playing an important role in tissue engineering. The pre-use modification of graphene and its derivative-based materials used in biomedical implementations is crucial for mitigating concerns with "toxicity" and "biosafety," in addition to the many advantages they provide.

S. Kemeç Aslan (✉)
Department of Chemical Engineering, Faculty of Engineering, and Natural Sciences, Bursa Technical University, 16310 Bursa, Türkiye
e-mail: sevgi.kemec@btu.edu.tr

K. Hürkan
Department of Agricultural Biotechnology, Faculty of Agriculture, Iğdır University, 76000 Iğdır, Türkiye

A. Bachheti et al. (eds.), *Carbon-Based Nanomaterials*, Smart Nanomaterials Technology, https://doi.org/10.1007/978-981-97-0240-4_11

Keywords Graphene · Graphene-based nanomaterials (GbNs) · Biomedical application · Drug delivery · Tissue engineering

1 Introduction

The utilization of nanomaterials in biomedical applications, including gene therapy and the creation of artificial organs and tissues to replace damaged or defective organs and tissues, has recently gained popularity as a result of advancements in biotechnological applications [2, 8, 9, 15–17, 19–21, 28]. Graphene is widely used in biomedical applications because of its mechanical and chemical features such as huge surface area, great thermal conductivity, high electron movability at room temperature, extremely high light transmittance, single-atomic thickness, and chemical resistance (Fig. 1) [11, 25, 29].

Graphene is the single-atomic plane of graphite. Graphene is defined as a single-atom thin cage consisting of six covalently bonded, two-dimensional (2D), honeycomb-shaped carbon atoms connected to each other by sp^2 bonds. Graphene-based nanomaterials (GbNs) include graphene oxide (GO), reduced graphene oxide (rGO), and graphene quantum dots (GQD). In addition, graphene-based nanomaterials include a variety of structures, including graphene nanosheets, graphene

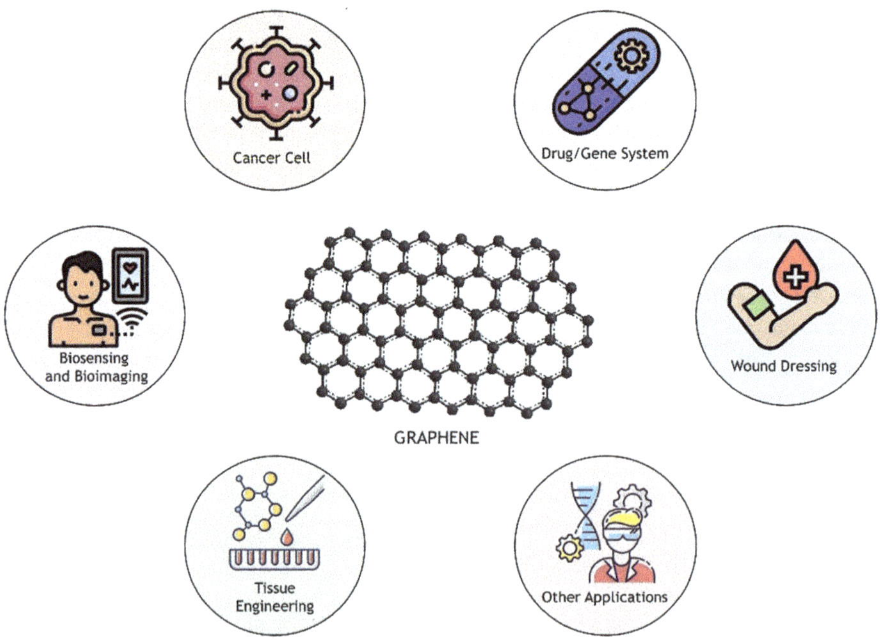

Fig.1 A schematic illustration of the utilization of graphene in biomedical implementation

nanoribbons, single-layer graphene, and multi-layer graphene, which display a wide range of morphologies. The investigation of graphene oxide and reduced graphene oxide preceded that of graphene [23, 26].

Graphene-based nanomaterials possess exceptional attributes that set them apart from other nanomaterials, making them very notable throughout several domains of application. The expansive surface area of nanoscale graphene and its derivatives has been widely acknowledged for its potential to provide favorable mechanical, electrical, chemical, and magnetic properties. In recent times, the use of graphene has expanded beyond its conventional applications, owing to an enhanced comprehension of its diverse chemical characteristics. This has led to its adoption in both biological contexts and the improvement of high-performance energy storage and generation systems [27, 37, 48].

Examples of application areas of graphene include graphene-based composite materials, sensors, field effect transistors, alternative energy devices, wearable electronic textiles, and multifunctional fabrics. This new generation of nanomaterial also has great potential in medical applications thanks to its extraordinary properties [1, 7, 33, 40]. Particularly, its biocompatibility and biological and chemical properties have enabled graphene to be used as a biomaterial. Researchers' hunt for compounds with varied biological properties that can be employed as scaffolds has expanded. Graphene and its derivatives, like graphene oxide, have been proven to be the most effective nanomaterials in biomedical research applications in recent years. It has features like biocompatibility, hardness, flexibility, and thermal properties. Because of its unique features, this 2D nanomaterial has been shown to have a vast range of applications in biomedicine and electronics. Because of its physicochemical qualities, huge surface area, and biocompatibility, graphene was considered suitable for application in stem cell-based tissue engineering [13, 35, 46]. Conductive nano textiles are also employed in electromagnetic shielding and nano optoelectronics due to their qualities. In materials research, graphene is sometimes referred to as a "supermaterial" or "miracle material" due to its remarkable features [24].

Besides the impressive properties of graphene, there are many methods to produce graphene. Various methods are applied for direct graphene synthesis, as well as graphene oxide synthesis and reduction.

2 Graphene and Graphene Oxide Synthesis

The process of producing graphene and graphene oxide (GO) involves the material's fragmentation into nano-sized particles using external mechanical or chemical methods. A number of techniques have been improved for the synthesis of graphene, a single sheet of graphite (Fig. 2). The most famous methods for the separation of graphite layers, known as exfoliation, include micromechanical techniques, chemical vapor deposition, and thermal/chemical reduction of GO [5].

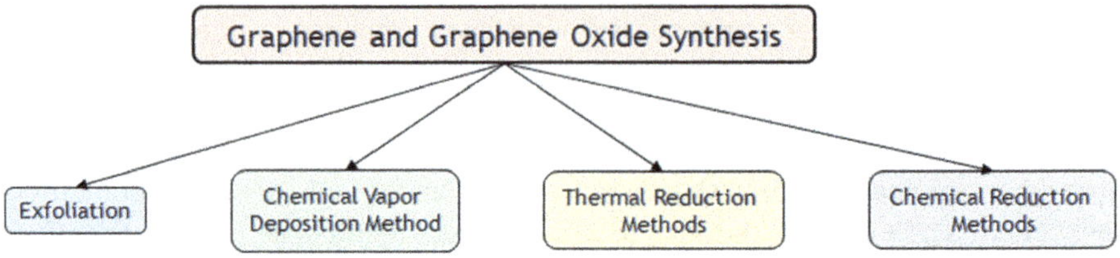

Fig.2 Graphene and graphene oxide synthesis

2.1 Exfoliation

The exfoliation (Micromechanical peeling) process is straightforward and requires no extra equipment to create immaculate graphene. The only thing you will need is a sticky tape. Graphite chips are sandwiched between sticky tape, and the surface is gradually peeled away. The graphite is continuously peeled, causing the split thin flakes to stick to the surface with an almost atomically pristine flat surface [10, 24].

2.2 Chemical Vapor Deposition Method

This method can obtain single-layer and two-layer graphene sheets, and the graphene produced is intact. Graphene produced by this process is less costly per unit area and is limited by the size of the substrate [3, 36].

2.3 Thermal Reduction Methods

Reduction is a very important step in the preparation of metallic catalysts. When the reduction process is not carried out correctly, the catalyst may sinter or the optimum reduction state cannot be reached [22].

2.4 Chemical Reduction Methods

Chemical reagents are reduced by chemical interactions with GO. The elimination of oxygen-containing groups produces aggregation when graphene oxide is reduced to graphene. Many methods and reducing agents based on these methods are used to reduce graphene oxide by chemical methods. Hydrazines, borohydrides, aluminum hydrides, and sulfur-containing reducing agents are the most commonly used ones. Other reducing agents, however, include ascorbic acid, glucose, alkaline solutions, pyrrole, and hydroxylamine. The Modified Hummers method is the best method for obtaining graphene from graphite. This technique may produce around 80% of single-layer reduced graphene oxide sheets [41, 43, 45].

3 The Graphene/Graphene Derivatives-Based Materials in Biomedical Applications

Graphene-based nanomaterials (GbNs) are biocompatible, have a high drug charge capability with a huge surface area, and have essential mechanical strength and chemical stability in the body for biomedical applications. These properties make these nanoparticles appropriate for medication delivery applications [4, 47]. Furthermore, because of their large surface area, electrochemical, and optical characteristics, these nanomaterials are employed in biosensor applications, which play a significant role in the early identification of illnesses [14, 29]. Additionally, since their functional groups are simple to change, they are excellent candidates for a diverse variety of applications in biology. Due to their wide range of applications, GbNs are anticipated to change a number of sectors, including optics, electronics, thermodynamics, and mechanics. Tissue engineering has arisen as an interdisciplinary discipline focused on the restoration of injured tissues and organs. The addition of nanoparticles to the scaffold material allows for the advancement of this area, and graphene-based nanomaterials may also be employed for this purpose [30, 47]. In bioimaging for illness diagnosis, graphene-based nanomaterials provide more sensitive detection by boosting contrast and improving the picture using features such as optics, thermal conductivity, and electrical conductivity [47].

The significance of toxicity and biosafety concerns has grown in relation to the initial implementation of graphene oxide (GO) in biological contexts. Graphene materials are exclusively comprised of carbon. The examination of the conduct of carbon derivatives, namely rGO and GO, inside biological systems, together with assessing their removal kinetics from the human body, has significant significance. The reason behind the material loss in GbNs may be attributed to the several chemical phases involved in their manufacturing process. These stages include functionalization steps such as metal doping, addition of functional groups, and oxidation. This suggests that certain graphene derivatives under consideration for bioapplications include carbon, metals, and contaminants. According to the structural characteristics of GbNs, it is observed that graphene possesses hydrophobic properties. Consequently, in order to utilize graphene as a biomedical material, it becomes necessary to modify its functional groups. This modification may exhibit both covalent and non-covalent functionality. The introduction of non-covalent functionalization has significantly improved several properties, including dispersibility, biocompatibility, reactivity, binding capacity, and sensing capabilities. To facilitate the evolution of a reliable colloidal solution of graphene oxide (GO) for prospective medicinal uses, hydrogen bonds are established among polar functional groups and water molecules on the surface of GO. The predominant factors that render GO more favorable than other carbon-based compounds are its stability in aqueous solutions and colloidal suspensions [14, 34, 39].

Tissue engineering applications, where graphene and materials made from it are used in drug delivery and which are important in biomedical applications, are discussed in detail.

3.1 Drug Delivery

The graphene-based structure can support a high number of carriers due to its huge surface area and sp^2 hybridization (Fig. 3). Graphene-based nanomaterials are suitable for binding to DNA, antibodies, proteins, and genes due to their properties such as their ability to cross cell membranes, planar a large-surface-area structure offering many binding sites for drug targeting, high loading capacity, chemical and mechanical stabilities, and superior-excellent biocompatibility and conductivity. However, hydroxylamine, glucose, phenylhydrazine, ascorbic acid, hydroquinone, and alkaline solutions are also reducing agents. They have also made them useful carriers for various biomolecules, including small drug molecules. Drugs can be loaded noncovalently onto GbN by interactions such as hydrogen bonding, $\pi - \pi$, or hydrophobic interactions [4, 47].

Graphene-based nanomaterials exhibit responsiveness to several stimuli, including changes in temperature, pH levels, magnetic fields, electric fields, acoustic waves, and radiation. Two methods of medication delivery are possible using graphene-based nanostructures. Temperature, ultrasound, and magnetic/electric/light fields are examples of the first category, whereas pH, enzyme concentration, and redox are examples of the second [4, 38]. The aforementioned techniques serve to augment the bioavailability of the medicine, expand its capacity to traverse physical barriers, reduce the dosage of the drug, and mitigate the adverse effects associated with the administration of pharmaceuticals [44]. Graphene-based nanomaterials are increasingly being used in drug delivery applications, which bodes well for biomedicine and therapeutic applications.

The graphene oxide surface (GO) exhibits functional groups like hydroxyl, carboxyl, and epoxy, enabling targeting or therapeutic molecules to be attached. This characteristic highlights the potential suitability of GO as a material for many biological applications. Photothermal therapeutic agents derived from graphene have been used in studies to enhance cytotoxicity in drug delivery systems.

Camptothecin (CPT), a medication that cannot readily dissolve in water, is delivered using Janus-type graphenes. The functionalization of graphene sheets involves the introduction of hydrophilic catechol-containing pyrrolidine rings, resulting in the hydrophobicity of the untreated side. In water-dispersed amphiphilic graphene sheets, bilayer superstructures self-organize with hydrophilic outside spokes and hydrophobic interior cavities. It can be used in cancer drug delivery systems [6]. Methotrexate (MTX), an anticancer medication, was studied by electrochemically immobilizing the drug transport effectiveness of graphene nanoparticles, which

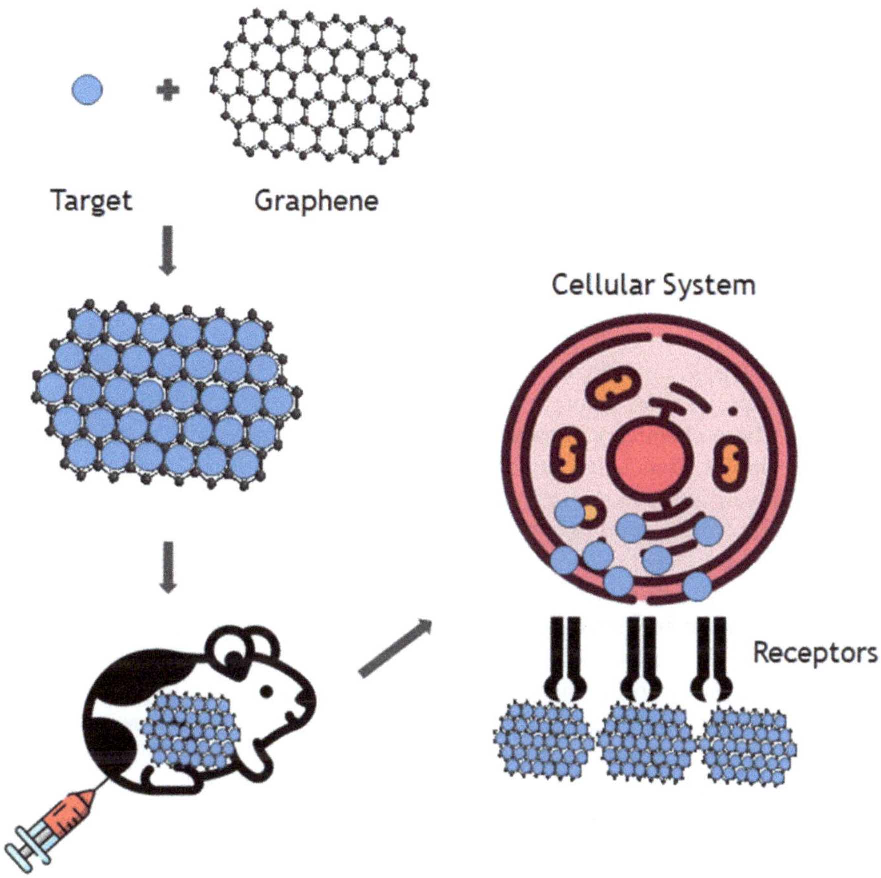

Fig.3 An example demonstration of graphene-mediated target/drug delivery

exhibit successful carrier qualities and cellular contact, against 4T1 cancer cells in vitro [18].

Sun and his colleagues immobilized and targeted the CD20+ antigen on the graphene oxide surface of cancer cells by forming a polyethylene glycone bond with it. Doxorubicin (DOX), an anticancer drug, establishes a strong connection with the GO surface, as shown in some studies in the literature. Its release is more common in tumors or acidic environments than in healthy tissues. In order to load and unload vitamin B12, Bai and colleagues created a pH-sensitive GO/polyvinyl alcohol hydrogel, which they favored as a test substance. PVA and GO sheets created composite hydrogels. The GO sheets showed characteristics of a two-dimensional macromolecular structure. The gel-sol transition of GO/PVA hydrogels was seen to be induced by changes in pH. In their study, Zhao et al. devised a unique reductive trigger switch that demonstrates compatibility with the intracellular milieu of tumor tissues. This switch was included in a carrier system known as cross-linked GO-PEG (cysteine polymethacrylic acid cross-linked nano GO polyethylene glycol) [12, 39].

Graphene nanoparticles are excellent drug delivery carrier materials. To name a few, high loading efficiency, targeting, imaging, sensing, stimulus-responsive release, and other structure-related special properties. Multifunctional drug delivery systems are the future of graphene-based nano-drug delivery systems.

3.2 Tissue Engineering

Tissue engineering, an extensively specialized field with significant consequences for medical applications, incorporates the use of graphene and derivatives as crucial materials. These chemicals often serve to restore the extracellular matrix or offer structural support. Stem cells, possessing the potential for self-renewal and deriving into several cell lineages, serve as the principal foundation for tissue engineering endeavors. In order to optimize the use of stem cells, it is imperative to take into account precise controls encompassing biochemical and physical microenvironmental signals. Because of its unique characteristics, graphene holds promise as a perfect candidate for enhancing stem cell adhesion, proliferation, survival, and differentiation. Recent research has investigated the capacity of graphene oxide (GO) as a catalyst for promoting stem cell derivatives. The stem cells generated from human adipose cultured on gelatin hydroxyapatite (HA)-coated graphene oxide (GO) showed better cell adhesion, multiplication, and differentiation, in addition to considerably greater levels of osteogenic marker proteins. Similarly, the MC3T3-E1 cell line was grown utilizing a nanocarrier consisting of GO, poly(lactic-co-glycolic acid) (PLGA), and hydroxyapatite (HA) generated from mouse cranial parietal osteoblast progenitor cells. This culture method improved cellular adhesion and mineralization, increased osteogenic potential, alkaline phosphatase activity, and expression of osteogenesis-related genes. Additionally, it was shown that mouse mesenchymal stem cells exhibited the presence of osteogenic markers during two weeks of cultivation on graphene oxide sheets in a medium conducive to osteogenesis. The findings presented in this study indicate that the application of GO (graphene oxide) may have a stimulatory effect on the process of osteogenic differentiation in stem cells [13, 30, 32].

The optoelectronic properties of developing nanomaterials have attracted significant interest due to their exceptional characteristics, which facilitate advancements in tissue engineering for the development of artificial tendons, heart valves, and other types of tissues. Upon entering the human body, these chemicals have been seen to engage with many tissues, particularly emphasizing the immune system [42]. Moreover, owing to their exceptional conductivity, nanomaterials based on graphene exhibit significant promise in the realm of nerve regeneration. Recent research findings have demonstrated the efficacy of integrating graphene-based nanomaterials with mechanically rigid and flexible polycaprolactone (PCL) materials, polydopamine (PDA) as a cell migration mediator, and arginine-glycine-aspartate (RGD) in enhancing cell adhesion. Furthermore, this combination has shown promise in developing novel porous graphene nerve channels. Recent research has shown that the implementation of these scaffolds improves the attachment of Schwann cells, promotes the regeneration of damaged peripheral nerves in living organisms, and boosts the electrical conduction properties of cells, such as voltage propagation and local currents.

4 Conclusion

The potential of nanomaterials in the biological area has shown great promise in recent times. Graphene and its derivatives possess unique structural features, including a two-dimensional structure, biocompatibility, and exceptional drug-loading capacity. These attributes make them very adaptable for many biomedical applications. Furthermore, these materials exhibit remarkable characteristics in terms of their structural, optical, thermal, mechanical, and electrical properties. The use of graphene and its derivative materials in the field of biomedicine has great significance across several areas, including tissue engineering, biosensor applications, gene transfer, and its recognized function in controlled drug release applications. This section focuses on the particular aspects related to medication delivery and tissue engineering. Graphene and its derivatives have remarkable properties as carriers for drug delivery, making them very promising for many applications that need materials with enhanced loading efficiency, targeting capabilities, imaging capabilities, sensing skills, and stimulus-responsive release properties. These materials can assist in the repair of the extracellular matrix and provide structural reinforcement.

Additionally, they have the capacity to improve the adhesion, multiplication, survival, and differentiation of stem cells. As a result, they play a pivotal role pertaining to tissue engineering. The pre-application alteration of graphene and its derivative-based materials used in biomedical contexts is of utmost importance in addressing concerns about "toxicity" and "biosafety," alongside their many benefits.

References

1. Abbasi H, Antunes M, Velasco JI (2019) Recent advances in carbon-based polymer nanocomposites for electromagnetic interference shielding. Prog Mater Sci 103:319–373
2. Bachheti RK, Bachheti A, Husen A (2023) Metal and metal-oxide based nanomaterials (Synthesis, agricultural, biomedical and environmental ınterventions). Springer Nature Singapore Pte Ltd., 152 Beach Road, #21-01/04 Gateway East, Singapore 189721, Singapore
3. Bahri M, Gebre SH, Elaguech MA, Dajan FT, Sendeku MG, Tlili C, Wang D (2023) Recent advances in chemical vapour deposition techniques for graphene-based nanoarchitectures: from synthesis to contemporary applications. Coord Chem Rev 475:214910
4. Bai RG, Husseini GA (2019) Graphene-based drug delivery systems. Elsevier Inc.
5. Bedeloğlu A, Taş M (2016) Graphene and ıts production methods. Afyon Kocatepe Univ J Sci Eng 16:544–554

6. Bekiari V, Karakassides A, Georgitsopoulou S, Kouloumpis A, Gournis D, Georgakilas V (2018) Self-assembly of one-side-functionalized graphene nanosheets in bilayered superstructures for drug delivery. J Mater Sci 53:11167–11175
7. Bhanuprakash L, Parasuram S, Varghese S (2019) Experimental investigation on graphene oxides coated carbon fibre/epoxy hybrid composites: mechanical and electrical properties. Compos Sci Technol 179:134–144
8. Chawla S, Singh S, Husen A (2023) Smart nanomaterials targeting pathological hypoxia. Springer Nature Singapore Pte Ltd., 152 Beach Road, #21-01/04 Gateway East, Singapore 189721, Singapore. https://doi.org/10.1007/978-981-99-1718-1
9. Das S, Mitra S, Khurana SMP, Debnath N (2013) Nanomaterials for biomedical applications. Front Life Sci 7:90–98
10. Dasari BL, Nouri JM, Brabazon D, Naher S (2017) Graphene and derivatives—synthesis techniques, properties and their energy applications. Energy 140:766–778
11. Dasari Shareena TP, McShan D, Dasmahapatra AK, Tchounwou PB (2018) A review on graphene-based nanomaterials in biomedical applications and risks in environment and health. Nanomicro Lett 10:1–34
12. Depan D, Shah J, Misra RDK (2011) Controlled release of drug from folate-decorated and graphene mediated drug delivery system: synthesis, loading efficiency, and drug release response. Mater Sci Eng, C 31:1305–1312
13. Foo ME, Gopinath SCB (2017) Feasibility of graphene in biomedical applications. Biomed Pharmacother 94:354–361
14. Georgakilas V, Tiwari JN, Kemp KC, Perman JA, Bourlinos AB, Kim KS, Zboril R (2016) Noncovalent functionalization of graphene and graphene oxide for energy materials, biosensing, catalytic, and biomedical applications. Chem Rev 116:5464–5519
15. Husen A, Bachheti RK, Bachheti A (2023) Current trends in green nano-emulsions (Food, agriculture and biomedical sectors). Springer Nature Singapore Pte Ltd., 152 Beach Road, #21-01/04 Gateway East, Singapore 189721, Singapore
16. Husen A, Siddiqi KS (2023) Advances in smart nanomaterials and their applications. Elsevier Inc., 50 Hampshire St., 5th Floor, Cambridge, MA 02139, USA
17. Jin-Chul K, Madhusudhan A, Husen A (2021) Smart nanomaterials in biomedical applications. Springer Nature Switzerland AG, Gewerbestrasse 11, 6330 Cham, Switzerland. https://doi.org/10.1007/978-3-030-84262-8
18. Karimi Shervedani R, Mirhosseini H, Samiei Foroushani M, Torabi M, Rahsepar FR, Norouzi-Barough L (2018) Immobilization of methotrexate anticancer drug onto the graphene surface and interaction with calf thymus DNA and 4T1 cancer cells. Bioelectrochemistry 119:1–9
19. Mabrouk M, Das DB, Salem ZA, Beherei HH (2021) Nanomaterials for biomedical applications: production, characterisations, recent trends and difficulties. Molecules 26:1–27
20. Madhusudhan A, Purohit SD, Prasad R, Husen A (2023) Functional smart nanomaterials and their theranostics approaches. Springer Nature Singapore Pte Ltd., 152 Beach Road, #21-01/04 Gateway East, Singapore 189721, Singapore
21. Mandal AK, Ghorai S, Husen A (2023) Functionalized smart nanomaterials for point-of-care testing. Springer Nature Singapore Pte Ltd., 152 Beach Road, #21-01/04 Gateway East, Singapore 189721, Singapore
22. Mcallister MJ, Li J, Adamson DH, Schniepp HC, Abdala AA, Liu J, Herrera-alonso OM, Milius DL, Car R, Prud RK, Aksay IA (2007) Expansion of graphite. Society 19:4396–4404
23. Perreault F, Fonseca De Faria A, Elimelech M (2015) Environmental applications of graphene-based nanomaterials. Chem Soc Rev 44:5861–5896
24. Phiri J, Gane P, Maloney TC (2017) General overview of graphene: production, properties and application in polymer composites. Mater Sci Eng B Solid State Mater Adv Technol 215:9–28
25. Priya Swetha PD, Manisha H, Sudhakaraprasad K (2018) Graphene and graphene-based materials in biomedical science. Part Part Syst Charact 35:1–29
26. Pumera M (2010) Graphene-based nanomaterials and their electrochemistry. Chem Soc Rev 39:4146–4157

27. Schäffel F (2012) The atomic structure of graphene and ıts few-layer counterparts. Graphene: fundamentals and emergent applications 5–59
28. Sharma G, Nim S, Alle M, Husen A, Kim JC (2022) Nanoparticle-mediated delivery of flavonoids for cancer therapy: prevention and treatment. In: Kim JC, Alle M, Husen A (eds) Smart nanomaterials in biomedical applications. Nanotechnology in the Life Sciences. Springer, Cham. https://doi.org/10.1007/978-3-030-84262-8_3
29. Shen H, Zhang L, Liu M, Zhang Z (2012) Biomedical applications of graphene. Theranostics 2:283–294
30. Shin SR, Li YC, Jang HL, Khoshakhlagh P, Akbari M, Nasajpour A, Zhang YS, Tamayol A, Khademhosseini A (2016) Graphene-based materials for tissue engineering. Adv Drug Deliv Rev 105:255–274
31. Siddiqi KS, Husen A, Sohrab SS, Osman M (2018) Recent status of nanomaterials fabrication and their potential applications in neurological disease management. Nanoscale Res Lett 13(231):1–17
32. Solanki A, Shah S, Yin PT, Lee KB (2013) Nanotopography-mediated reverse uptake for siRNA delivery into neural stem cells to enhance neuronal differentiation. Sci Rep 3
33. Song WL, Gong C, Li H, Cheng XD, Chen M, Yuan X, Chen H, Yang Y, Fang D (2017) Graphene-based sandwich structures for frequency selectable electromagnetic shielding. ACS Appl Mater Interfaces 9:36119–36129
34. Sontakke AD, Tiwari S, Purkait MK (2023) A comprehensive review on graphene oxide-based nanocarriers: synthesis, functionalization and biomedical applications. FlatChem 38:100484
35. Wu SY, An SSA, Hulme J (2015) Current applications of graphene oxide in nanomedicine. Int J Nanomedicine 10:9–24
36. Xu S, Zhang L, Wang B, Ruoff RS (2021) Chemical vapor deposition of graphene on thin-metal films. Cell Rep Phys Sci 2:100372
37. Yang H, Zheng D, Tang W, Bao X, Cui H (2023) Application of graphene and its derivatives in cementitious materials: an overview. J Build Eng 65:107054
38. Yang K, Feng L, Liu Z (2016) Stimuli responsive drug delivery systems based on nano-graphene for cancer therapy. Adv Drug Deliv Rev 105:228–241
39. Yang X, Zhang X, Liu Z, Ma Y, Huang Y, Chen Y (2008) High-efficiency loading and controlled release of doxorubicin hydrochloride on graphene oxide. J Phys Chem C 112:17554–17558
40. Yapici MK, Alkhidir TE (2017) Intelligent medical garments with graphene-functionalized smart-cloth ECG sensors. Sensors (Switzerland) 17:1–12
41. Yazıcı M, Tiyek İ, Ersoy MS, Alma MH, Dönmez U, Yıldırım B, Salan T, Karataş Ş, Uruş S, Karteri İ, Yıldız K (2016) Synthesis of graphene oxide (GO) by modified hummer's methods and its characterization. Gazi Üniversitesi Fen Bilimleri Dergisi Part C: Tasarım ve Teknoloji 4:41–48
42. Yin PT, Shah S, Chhowalla M, Lee KB (2015) Design, synthesis, and characterization of graphene-nanoparticle hybrid materials for bioapplications. Chem Rev 115:2483–2531
43. Yu H, Zhang B, Bulin C, Li R, Xing R (2016) High-efficient synthesis of graphene oxide based on improved hummers method. Sci Rep 6:1–7
44. Yu X, Cheng H, Zhang M, Zhao Y, Qu L, Shi G (2017) Graphene-based smart materials. Nat Rev Mater 2
45. Zaaba NI, Foo KL, Hashim U, Tan SJ, Liu WW, Voon CH (2017) Synthesis of graphene oxide using modified hummers method: solvent influence. Procedia Eng 184:469–477
46. Zhang H, Grüner G, Zhao Y (2013) Recent advancements of graphene in biomedicine. J Mater Chem B 1:2542–2567
47. Zhao H, Ding R, Zhao X, Li Y, Qu L, Pei H, Yildirimer L, Wu Z, Zhang W (2017) Graphene-based nanomaterials for drug and/or gene delivery, bioimaging, and tissue engineering. Drug Discov Today 22:1302–1317
48. Zhen Z, Zhu H (2017) Structure and properties of graphene. Elsevier Inc.

Carbon Nanomaterials in Drug and Gene Delivery Potential: Focus on Fungal Infections

Piyush Kumar, Ritu Singh, Gurpreet Kaur, and Sonam Chawla

Abstract Fungi are eukaryotic microorganisms, unicellular or multicellular, and several species are responsible for dormant/mild/life-threatening clinical implications in humans, animals and plants. The current chapter identifies the global burden of fungal infections and prospects for the utility of carbon nanomaterials in improving treatment regimes for the most prevalent fungal infections in humans. Major hurdles in clinically addressing fungal infections are drug resistance, metabolic and gene level homology of the fungi with humans leading to off-target effects, and lastly poor pharmacokinetics of the known anti-fungals. Carbon nanomaterials have emerged as promising candidates for drug and gene delivery due to their unique physicochemical properties, including high surface area, tunable surface chemistry, and biocompatibility. We have described in detail the recent literature wherein carbon nanomaterials–carbon nanotubes, carbon nanodots, graphene and graphene oxide, and fullerene are elaborated in terms of their properties suitable for drug and gene delivery, functionalization and efficacy in addressing fungal infections. The applications of carbon nanomaterials in addressing fungal infections via improved drug pharmacokinetics, controlled/sustained release of drug and improved efficacy due to synergistic effect with anti-fungal drugs has been elaborated. Lastly we have discussed the challenges of using carbon nanomaterials in anti-fungal regimes in terms of optimization required for drug loading and release kinetics, safety considerations and potential toxicity, and regulatory hurdles for translating carbon nanomaterial based interventions.

Keyword ■■■

P. Kumar · R. Singh · S. Chawla (✉)
Department of Biotechnology, Jaypee Institute of Information Technology, Sector 62, Noida, India
e-mail: falsecontact.schawla@gmail.com

G. Kaur (✉)
Department of Biotechnology, Chandigarh College of Technology (CCT), Chandigarh Group of Colleges (CGC), Landran, Mohali, Punjab, India
e-mail: falsegurpreet.dipas@gmail.com

A. Bachheti et al. (eds.), *Carbon-Based Nanomaterials*, Smart Nanomaterials Technology, https://doi.org/10.1007/978-981-97-0240-4_12

1 Introduction

1.1 Background on Fungal Infections and Their Impact on Human Health

Fungi are a type of eukaryotic microorganisms that exist in various forms, including yeast, mold, or dimorphic forms. While yeasts are unicellular and reproduce through budding, fungi are multicellular organisms and consist of elongated filaments called hyphae that grow through apical extension. Fungi can be readily encountered in various natural habitats, such as aquatic environments, terrestrial soils, and the atmosphere. They tend to thrive in warm and humid climates, and can easily multiply in these conditions [30]. While the majority of fungi are not harmful, certain species can lead to human infections, varying in severity from mild superficial conditions to potentially life-threatening ones. In addition, certain fungal species act as pathogens for plants, causing significant damage to agriculture. This primer seeks to provide a comprehensive understanding of what distinguishes fungal pathogens from other pathogens, as well as the difficulties associated with fighting against fungal infections [97].

Fungal pathogens are widely acknowledged to have a significant impact on both plant and animal life. In fact, a recent study has provided comprehensive information on the alarming and exceptional effect of these pathogens on the extinction of species, disruption of ecosystems, and food security [28]. Fungal infections typically begin in the lungs due to the breathing in polluted air, or contact through the skin or nails, and mucous membranes such as the vagina, mouth, and sinuses, through physical contact, such as direct exposure or by means of wounds. In rare cases, fungal infections may also occur in the gastrointestinal tract due to the ingestion of contaminated food or water [35]. Although superficial fungal infections, which are generally easy to treat, affect most individuals at some point in their lives, millions of people worldwide are at risk of progressing life-threatening invasive infections that pose significant challenges in terms of diagnosis and cure. Superficial fungal infections are commonly caused by different strains of *Candida* species., which rank as the second most frequent causative agents of fungal infections globally [50]. On the other hand, the occurrence of invasive fungal infections is comparatively lower than that of superficial infections, but the latter poses more significant concerns due to the elevated mortality rates that are associated with them. Invasive fungal infections, caused by many species of fungi, result in approximately 1.5 million fatalities annually. Surprisingly, the top 10 invasive fungal diseases are responsible for causing a higher number of deaths compared to tuberculosis or malaria. The overwhelming majority (over 90%) of documented fatalities resulting from fungal infections can be attributed to a selected group of species of four genera: *Aspergillus, Cryptococcus, Pneumocystis and Candida* [9, 14].

As mentioned, fungal pathogens significantly affect the health of plants and animals, their effect on human health is not as widely acknowledged, and fatalities caused by these infections are frequently disregarded. For instance, the World Health

Organization does not have a program that specifically focuses on fungal infections, and most public health agencies, apart from the U.S. Centers for Disease Control and Prevention (CDC), undertake minimal or no surveillance of mycological diseases [14, 27]. The high mortality rate associated with these invasive fungal infections is of particular concern, with rates often exceeding 50%, even with the availability of several antifungal drugs. It is true that funding allocated to medical mycology research is significantly inadequate in comparison with funding for other infectious diseases, nevertheless, this could also be due to a lower number of funding applications in this field. For instance, out of the total amount spent in the past five years on research for immunology and infectious diseases by organizations such as the Wellcome Trust (a charitable organization of the U.K.), the U.K. Medical Research Council, and the U.S. National Institutes of Health (NIH), only a small percentage, ranging from 1.4% to 2.5%, was specifically allocated to medical mycology [14].

Each year, fungi that cause diseases are the primary suspect for approximately 13 million infections and 1.5 million fatalities worldwide, mainly observed in individuals with compromised immune systems. The infections can deteriorate rapidly in such patients, leading to significant illness and mortality. However, despite these alarming trends, fungal infections have historically been overlooked in public health discussions, and funding for research on these pathogens remains significantly lower than that for other diseases with comparable mortality rates. Current deficiencies in broad clinical recognition and the absence of standardized protocols for the diagnosis and management of fungal infections persist. Furthermore, due to the vague symptoms of severe fungal disease, diagnoses are often delayed, leading to chronic underdiagnosis of these infections and a wide range of prognosis for affected patients [13, 46].

Dermatophytes are the primary cause of fungal infections, leading to commonly known conditions like athlete's foot (which affects approximately one in five adults), scalp ringworm (prevalent among kids of age between 3 yr–5 yr and estimated to impact 200 million peoples globally), and nail infections (affecting around ~10% of the global population, with an increased incidence of approximately ~50% among individuals aged 70 years and older) [111]. The occurrence of each specific infection is also influenced by socioeconomic conditions, geographical region, and cultural practices. Mucosal infections affecting the oral and genital tracts are also prevalent, with vulvovaginal candidiasis (or thrush) being particularly common. As a matter of fact, it is estimated that approximately 50 to 75% of women during their childbearing age will experience at least one occurrence of vulvovaginitis, with approximately 5 to 8% of women (equivalent to around 75 million) experiencing at least four episodes per year [112]. In regions of the world with limited access to healthcare, HIV/AIDS is responsible for approximately 10 million instances of oral thrush and approximately 2 million occurrences of esophageal fungal infections each year [14]. Oral infections are also prevalent in infants [112] and denture wearers [62], individuals using inhaled steroids for asthma [29], leukemia and transplant patients, as well as individuals who have received radiotherapy for cancers in the head and neck region [61].

Healthy individuals have effective immune mechanisms that prevent fungal infections, and the rise in invasive fungal diseases over the past few decades is primarily attributed to increases in immunosuppressive conditions like HIV/AIDS, as well as contemporary immunosuppressive treatments and invasive medical procedures [56, 107].

AIDS is the primary risk factor for cryptococcosis, while people who have undergone solid organ transplantation and received immunosuppressive medications are also at an increased risk of acquiring the infection [56, 107]. According to recent estimates by the CDC, the prevalence of cryptococcal meningitis is alarmingly high, with approximately one million cases reported annually worldwide. Tragically, this devastating disease claims the lives of over 620,000 individuals in sub-Saharan Africa alone. Death rates vary across regions, with AIDS patients experiencing rates ranging from 15 to 20% in the US, while in Latin America and sub-Saharan Africa, the rates soar to an alarming 55 to 70%, even with the availability of treatment [80]. An overview of the human fungal infections and their clinical implications is depicted in Fig. 1.

Both eukaryotic mammalian cells and fungal cells exhibit notable similarities in their cellular structures. They both possess a phospholipid bilayer plasma membrane, as well as share common organelles like the nucleus, ribosomes, Golgi apparatus, mitochondria, and endoplasmic reticulum. Nevertheless, fungal cells exhibit unique structural and functional characteristics due to their possession of specific cytoplasmic organelles and biosynthetic pathways [107]. Notably, fungal cells are encased in a double-layered wall structure called the fungal envelope, comprising an inner plasma membrane and an outer cell wall [56]. Unlike mammalian cells, fungal cells have an outer cell wall primarily consisting of glycoproteins and polysaccharides, specifically glucan and chitin. The fungal envelope serves to protect fungal cells

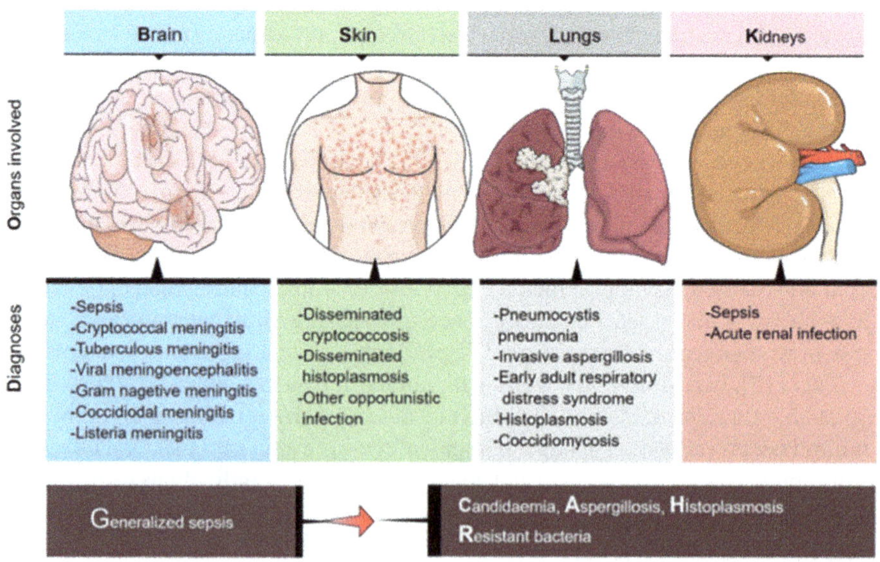

Fig. 1 Clinical implications of most prevalent fungal infections [9]

from environmental stresses and facilitates various signaling pathways [56]. While the cell membranes of both mammalian and fungal cells exist, they differ in terms of their lipid compositions. Although mammalian cells are characterized by membranes rich in cholesterol, fungal cells predominantly contain ergosterol. Ergosterol plays a role similar to cholesterol in membranes of mammalian cells, serving as a critical component for membrane function and biogenesis [107]. Targeting ergosterol provides a fungi-specific approach for designing antifungals that distinguish fungal cells from mammalian cells. In spite of these variations, fungi exhibit metabolic similarities to mammalian cells, although they do have a few specific targets associated with pathogens. For more comprehensive information on structures and classifications of fungal cells, detailed descriptions can be found in books such as those authored by Webster et al. [107], [58].

1.2 Need for Effective Drug and Gene Delivery Systems for Fungal Infections

Fungal infections can be challenging to treat, especially when the infection has developed resistance to conventional antifungal drugs. This is increasingly concerning due to the rising prevalence of drug-resistant fungal infections. Scientists are investigating novel drug delivery methods to more accurately and efficiently treat fungal infections [30, 54]). The economic burden of these infections is also considerable, with estimated healthcare costs of up to $7 billion annually in the United States alone [107].

The current treatments for fungal infections include antifungal drugs like amphotericin B, fluconazole, and itraconazole [75]. However, these drugs have several limitations, including poor solubility, limited bioavailability [72], and the development of drug resistance [22]. In addition, they may cause significant side effects, such as liver toxicity and nephrotoxicity [103]. Furthermore, some fungal species, such as *Aspergillus* and *Candida*, are recognized for their resistance to multiple antifungal agents, making treatment of these infections even more challenging [14].

Liposomes, which are spherical structures made of lipids, offer a promising strategy for treating fungal infections. These structures can encapsulate drugs or genetic material, protecting them from degradation. Furthermore, they can be engineered to target particular cells or tissues. In a recent study, liposomes loaded with an antifungal drug known as amphotericin B were shown to be successful in treatment of various fungal infections, including those caused by drug-resistant strains (Chen et al. 2020).

Hydrogels are being studied as a potential approach for drug delivery systems to treat fungal infections. Hydrogels are polymer networks that can absorb water and release drugs or genetic material over time. They can also be modified to selectively target particular tissues or cells. A recent study demonstrated that a hydrogel containing fluconazole, an antifungal medication, was successful in treating a skin infection caused by *Candida albicans* fungus (Sukumar et al. 2021).

Although the outcomes are encouraging, there are still numerous obstacles that require attention. These challenges include the potential toxicity of other nanomaterials [73], regulatory hurdles [77], long-term stability and degradation issues [52], and optimization of drug loading and release kinetics [64]. Given these limitations, the development of more effective drug and gene delivery systems is urgently needed for the treatment of fungal infections.

1.3 Overview of Carbon Nanomaterials and Their Unique Properties

Carbon, the sixth element on the periodic table, possesses an average atomic mass of 12 amu. The distinct allotropes of carbon arise from variations in the arrangement of carbon atoms [70]. Within the realm of medical science and emerging technologies, carbon nanostructures play a crucial role. These materials, such as zero-dimensional fullerenes, one-dimensional carbon nanotubes (CNTs), and two-dimensional graphene, share a common characteristic: carbon atoms exhibiting sp^2 hybridization arranged in a hexagonal pattern [11, 67].

Carbon nanomaterials have emerged as excellent contenders for drug and gene delivery because of their remarkable physicochemical properties, including high surface area, tunable surface chemistry, and biocompatibility [87]. Carbon nanomaterials also offer several advantages compared to traditional drug delivery systems, such as improved drug solubility and stability, targeted drug delivery, and reduced toxicity [78]. Carbon nanomaterials have brought about a new era in biomedicine with applications ranging from biosensors and bioimaging contrast agents to regenerative medicine, tissue engineering, drug delivery, and nonviral gene transfer. Recent research has highlighted the potential of CNMs in these fields [67].

The use of carbon-based nanomaterials is increasing in popularity in the fields of science and technology. The carbon family encompasses a range of unique properties that have been extensively utilized in various medical applications, such as biosensing and drug delivery. Various carbon allotropes have emerged, including well-known allotropic phases such as amorphous and lesser-known amorphous phases like fullerenes, graphene, carbon nanotubes (CNTs), and carbon nanofibers (CNFs) [1].

Despite the positive contributions of CNMs, their potential applications have not yet progressed to the clinical trial stage. Major concerns include their low aqueous solubility, limited biodegradability, and potential biosafety issues. It is important to note that, in general, the toxicity of CNMs is low when administered via intravenous injection, and their biosafety is influenced by their specific features. Due to their small size and extensive surface area [85], nanomaterials exhibit greater reactivity and catalytic activity than their larger counterparts. However, their application in

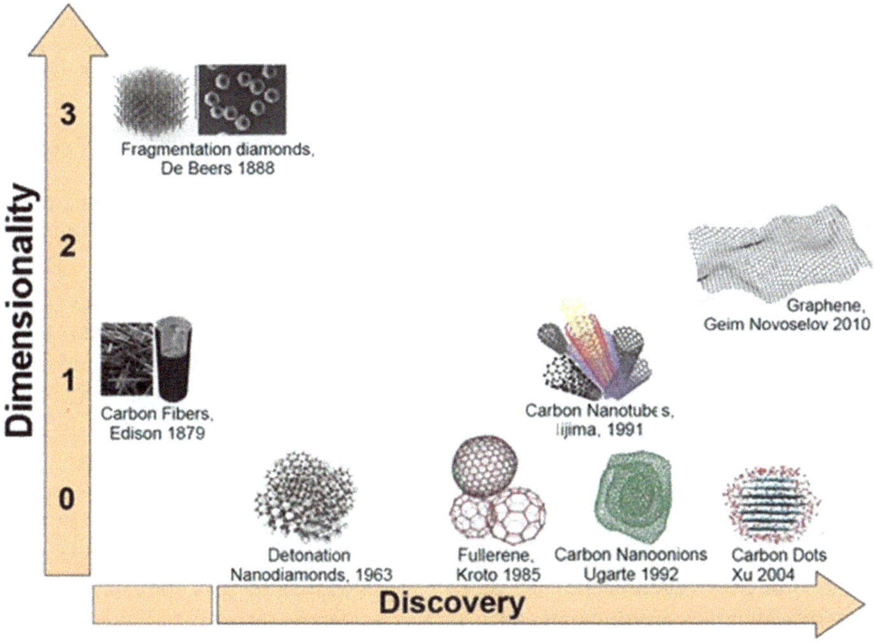

Fig. 2 A timeline of discovery of carbon nanomaterials and advancements in their dimensionality is illustrated by Speranza in his seminal review (Speranza 2021). (Creative Commons License)

nanomedicine may face a limitation in terms of their introduction into the human body without being absorbed via the digestive system, especially after intravenous or interstitial injections [67]. A perspective on the timeline of discovery and the advancement in the dimensionality of carbon nanomaterials is depicted in the Fig. 2 (Speranza 2021).

1.4 Rationale for Exploring Carbon Nanomaterials in Drug and Gene Delivery for Fungal Infections

Carbon nanomaterials, including carbon nanotubes, fullerenes, and graphene have high surface area, high loading capacity, and capability to penetrate the cell barriers [122]. These properties make them attractive for drug delivery applications, especially in the treatment of fungal infections.

Fungal infections are a growing global health concern, with limited treatment options available. Traditional antifungal drugs are often associated with severe side effects, drug resistance, and limited efficacy [107]. Therefore, there is a need for novel therapeutic strategies for the treatment of fungal infections.

Extensive research has been conducted on carbon nanomaterials as potential drug carriers for antifungal agents [65]. These materials can undergo functionalization with various drug compounds to improve their effectiveness and reduce side effects [16]. Additionally, carbon nanomaterials can target fungal cells selectively because of their small size and surface properties [85]. This allows for more effective drug delivery and improved therapeutic outcomes.

In recent years, many studies have documented the utilization of carbon nanomaterials for the delivery of antifungal drugs, including amphotericin B and fluconazole, with promising results [108]. For instance, research by Das et al. reported the successful encapsulation of amphotericin B within polyethylene glycol-coated carbon nanotubes, resulting in enhanced drug solubility, stability, and antifungal activity against *Aspergillus fumigatus* (Das et al. 2015). Similarly, graphene oxide nanosheets have been shown to improve the efficacy of fluconazole against *Candida albicans* by promoting the drug's targeted delivery at the infection site (Liu et al. 2014). For instance, carbon nanotubes functionalized with polyethyleneimine have been shown to effectively deliver small interfering RNA molecules targeting essential fungal genes, resulting in significant inhibition of fungal growth [10].

Carbon nanomaterials have been explored not just for drug delivery purposes but also for their potential in gene delivery applications to treat fungal infections. Gene therapy holds a promising approach to specifically target fungal pathogenesis by addressing key genes and pathways involved in the infection process. Carbon nanomaterials can be used as carriers for gene delivery, allowing for efficient delivery of therapeutic genes to target fungal cells [88].

Overall, the distinctive physicochemical characteristics of carbon nanomaterials make them attractive candidates for drug and gene delivery in fungal infections treatment [88, 122]. Further research in this area may lead to the development of more effective and targeted therapies for fungal infections.

2 Types of Carbon Nanomaterials for Drug and Gene Delivery

2.1 Carbon Nanotubes (CNTs)

In 1991, Sumio Iijima, a Japanese scientist, made the groundbreaking discovery of carbon nanotubes (CNTs). Carbon nanotubes (CNTs) are carbon allotropes with a one-dimensional structure at the nanoscale. They possess an extraordinary length-to-diameter ratio, exceeding one million. CNTs are created by the rolling of a graphene sheet into a seamless cylindrical shape, typically with a diameter on the scale of a nanometer (nm) [86]. Their unique tubular structure has opened up new avenues for their utilization in diverse fields, including mechanical electronics, biomedical sciences, structural engineering, thermal management, optical technologies, and numerous other areas of medicine and engineering. This distinctive characteristic of CNTs enables novel approaches and makes them valuable in diverse applications across various disciplines [86, 122].

Carbon nanotubes (CNTs) can be categorised into three distinct types viz., single-walled CNTs (SWCNTs), double-walled CNTs (DWCNTs), and multi-walled CNTs (MWCNTs). SWCNTs are composed of mono-layer sheets of graphite seamlessly rolled into a cylindrical tube. On the other hand, MWCNTs exhibit an arrangement where SWCNTs are concentrically positioned inside one another, resembling the concentric rings found in a tree trunk [49, 86, 95, 122]. CNTs can be manufactured through three primary methods, namely arc discharge, chemical vapour deposition (CVD) and laser ablation [15, 86, 122].

2.1.1 Properties and Characteristics Relevant to Drug and Gene Delivery

CNTs exhibit a wide range of promising properties that contribute to their potential applications including their application as carriers in drug and gene delivery. These properties include an enhanced area of surface, spike-shaped structure which aids in easily penetrating the cell wall, exceptional mechanical strength, dynamic engagement with cargo, high capacity for drug loading, remarkable optical and electrical characteristics, excellent stability, acceptance by the body, and the site-specific release of therapeutic agents. Additionally, CNTs possess rich electronic properties, are extremely light-weight, chemically stable, and exhibit excellent thermal stability. However, It is significant to highlight that CNTs also have negative attributes, notably their lack of biodegradability and potential toxicity [49, 70, 86, 95, 122].

The exceptional properties of CNTs facilitate their efficient uptake by various cell types. Their spike-shaped structure, for instance, aids CNTs to readily penetrate cell membranes, making them highly suitable for applications involving drug and gene delivery. Consequently, the uptake capabilities of CNTs make them well-suited for various biomedical applications, especially in the fields of drug delivery and gene therapy [122]. Numerous mechanisms have been recommended to explain the internalisation of CNTs, including penetration through the cellular barrier and passive or active uptake through endocytosis mechanisms. These mechanisms are categorised as independent energy pathways (direct penetration) and dependent energy pathways (endocytosis) [70, 86, 122]. CNTs can passively enter cells by diffusing across the lipid bilayer of the cellular membrane, utilising their high aspect ratio and spike-shaped structure to resolve impediments. Endocytosis, on the other hand, involves several pathways such as cell engulfment (phagocytosis), cell drinking (pinocytosis) (particularly micro-pinocytosis), clathrin-mediated endocytosis or Receptor-mediated cellular internalization, caveolae-dependent internalization, and non-clathrin/non-caveolin-mediated routes. These pathways contribute to the internalisation of CNTs by cells [70, 122].

2.1.2 Functionalization and Modification Approaches

Raw CNTs possess highly hydrophobic surfaces, rendering them insoluble in aqueous solutions. To overcome this challenge, functionalization of CNTs is employed. Functionalization is a chemical synthesis process that facilitates the incorporation of desired functional groups onto the surfaces of CNTs, resulting in the production of functionalized carbon nanotubes (f-CNT). This process enhances the solubility of CNTs and enables their application in various fields [49].

For functionalization and modification approaches extensive research has been conducted on carbon nanotubes (CNTs) for their potential use in drug and gene delivery. The objective of functionalizing CNTs is not only to enhance their physical properties, such as solubility and dispersibility but also to enhance their bio-performance. Poor dispersibility and accumulation of CNTs can increase their cytotoxicity inside the body. Therefore, surface functionalization is a promising approach to minimise the CNT-mediated cytotoxicity by enhancing their cellular uptake. Moreover, the quality of cellular uptake is directly influenced by CNT functionalization [122].

CNT functionalization can be broadly Categorized as either covalent or non-covalent. [49, 66, 70, 122]. In covalent functionalization, a chemical moiety is affixed to the surface of the nanoparticle through sharing at least one pair of electrons [70]. There are three main methods used for covalent functionalization: direct reaction of molecules or polymer chains with pristine CNT surfaces, functionalization of premodified CNTs, and oxidation of CNTs followed by functionalization [49, 122]. In contrast, non-covalent functionalization involves encasing CNTs with amphiphilic surfactants or polymers, forming micelle-like structures on the CNT surface. Another common non-covalent functionalization approach is achieved through p-p bonding, where pyrene molecules or single strands of DNA are stacked onto the CNT surface. Non-covalent functionalization is a straightforward process that Can be performed using gentle techniques such as sonication and room temperature, without affecting the structural integrity of CNTs and preserving their optical and electrical properties [49, 66].

Furthermore, apart from the covalent and non-covalent nature of functionalization, the specific functional groups used can also have an impact on the cellular uptake process. Surface functionalization improves the dispersion of CNTs and enables the conjugation of various nucleic acids on their surface [122].

A recent study by Thakur et al. introduced Microwave-assisted functionalization. Microwave chemistry provides an eco-friendly strategy to functionalize multi-walled carbon nanotubes (MWCNTs) by reducing reaction time and minimising solvent usage. Compared to traditional methods, microwave irradiation yields improved results in terms of functional group incorporation. Functional groups, including carboxylic acid and epoxy groups, can be incorporated onto the carbon atoms of MWCNTs using microwave irradiation. These functionalized MWCNTs can then be utilised for conjugating vitamins, proteins, and polymers. Microwave-assisted functionalization of multi-walled carbon nanotubes (MWCNTs) includes the attachment of biomolecules such as lysozymes, arginine, dopamine, amino uracil, and riboflavin. This process involves a combination of noncovalent adsorption and covalent attachment. This versatile functionalization enables MWCNTs to be applied in various fields such as drug delivery, biosensor development, and organocatalysis, highlighting their practical utility [101].

2.1.3 Examples of CNT-Based Drug and Gene Delivery Systems for Fungal Infections

Antifungal agents can be categorised into polyenes (e.g., natamycin, amphotericin B, and nystatin), triazoles (e.g., itraconazole, posaconazole, voriconazole, fluconazole, and ravuconazole), azoles/imidazoles (e.g., ketoconazole, miconazole, and econazole), and echinocandins (e.g., caspofungin, micafungin, and anidulafungin) [5]. While for the treatment of persistent fungal infections, Amphotericin B (AmB) is a highly effective and powerful antifungal agent, its high toxicity towards mammalian cells necessitates the use of a drug delivery vehicle to enhance water solubility and prevent aggregation [89]. Carbon nanotubes (CNTs) possess the unique ability to traverse biological barriers and enter cells via a nanoneedle-like mechanism, facilitating the transportation and delivery of cargo molecules or therapeutics to target tissues [49]. Researchers, such as [115], have utilised CNTs in the delivery of AmB, covalently linking the drug to CNTs and a luminescent stain for investigating cellular absorption and adverse effects in human Jurkat lymphoma T-cells and various fungal cells. This CNT-AmB system demonstrated reduced adverse effects on mammalian cells while exhibiting enhanced antifungal activity compared to AmB alone. Furthermore, carbon nanotubes possess the unique ability to penetrate living cells without causing harm or cell death, making them suitable for drug incorporation [108].

2.2 Graphene and Graphene Oxide (GO)

The journey of an extraordinary material began in the year preceding 2004, when Geim and Novoselov conducted a simple yet remarkable experiment. They applied Scotch tape onto a solid graphite surface, leading to the discovery of a material that would later earn them the Nobel Prize in Physics in 2010. Graphene is an incredibly thin layer of graphite, consisting of a solitary atom thickness and forming a two-dimensional molecular sheet with a hexagonal lattice structure. It has a thickness of approximately 1 nm and lateral dimensions that can range from a few nanometers to several microns. Graphene exhibits remarkable properties, exhibiting strength that exceeds diamond and conductivity that outperforms copper, and astonishingly thinner than paper. Its structure is composed of sp^2 carbon networking, making it a highly versatile material [23, 26, 51, 58, 70].

The graphene family of nanostructures encompasses various forms such as few-layer graphene (FLG), graphene nanosheets (GNS), Graphene Oxide (GO), reduced graphene oxide (rGO),, graphene foams, graphene quantum dots (GQDs), and ultra-thin graphite. These structures vary in terms of the number of walls, diameter, length, and surface characteristics they possess. The structure of graphene consists of a single layer of carbon atoms that are sp^2 bonded, while few-layer graphene consists of 2 to 5 well-defined stacked layers. GNS is a monolayer sheet involving a perpendicular dimension below 100 nm and is not integrated into any carbon material. GQDs, on the contrary, are graphene nanosheets possessing a narrow lateral dimension, usually below 10 nm, crucial for photoluminescence investigations. Graphene Oxide (GO) is a derivative of graphene that has been modified with a significant oxygen presence, typically exhibiting a C/O ratio below 3.0. By reducing the oxygen content through processes such as chemical, microwave, photo-chemical, or microbial techniques, GO can be converted into rGO [70].

Graphene serves as a fundamental unit for diverse graphitic materials, exhibiting different geometries based on its arrangement. It can adopt spherical structures known as fullerenes in zero-dimensional form, form one-dimensional carbon nanotubes (CNTs) by rolling, create a single-layer two-dimensional structure known as graphene, and arrange in a three-dimensional stacked formation like graphite. These variations in wall number, diameter, length, and surface chemistry allow for the versatility of graphene, similar to how fullerenes and CNTs possess their own distinct characteristics [26, 58].

A variety of synthetic techniques are utilised in the manufacturing of graphene and graphene oxide (GO). These methods can be categorised into two main groups: bottom-up approaches and top-down approaches. Bottom-up techniques encompass epitaxial growth, chemical vapour deposition (CVD), and Hummer's method. Epitaxial growth involves the deposition of carbon atoms onto a substrate to form graphene, while CVD synthesises graphene by decomposing hydrocarbon gases on a metal catalyst. Hummer's method, on the other hand, employs oxidation and exfoliation steps to produce GO through a chemical process.

In contrast, top-down approaches involve the exfoliation of bulk graphite to obtain graphene and GO. Mechanical exfoliation also known as "Scotch-tape" technique, graphene was first synthesized using this technique, it utilises adhesive tape to mechanically separate graphene layers from graphite. Chemical exfoliation involves the use of chemical agents to detach graphene layers from graphite. Electrochemical exfoliation utilises electrochemical processes to exfoliate graphite into graphene layers. Arc discharge is another top-down method in which high-intensity electric arc vaporises graphite, resulting in the production of graphene. These diverse techniques offer different avenues for synthesising graphene and GO, thereby providing versatility in the production processes [26, 36].

2.2.1 Properties and Characteristics Relevant to Drug and Gene Delivery

Graphene and graphene oxide (GO) exhibit a range of properties and characteristics that make them highly relevant for drug and gene delivery applications. These materials possess an extensive area of surface, exceptional mechanical strength, and remarkable optical properties such as photoluminescence. Graphene, with its remarkable mechanical strength, exceptional electrical conductivity, impressive thermal properties, and notable magnetic characteristics, has garnered attention as a potential candidate for various engineering fields. Additionally, its acceptable biocompatibility, large surface area, and easy biofunctionalization have attracted researchers in the biomedical engineering realm, particularly in drug/gene delivery and tissue engineering [42, 70, 71].

Graphene and GO possess distinct attributes, namely their planar structure, extensive surface area, enduring chemical and mechanical stability, remarkable conductivity, and excellent biocompatibility, which establish them as highly prospective biomaterials for biomedical applications. Their planar structure enables the immobilisation of various substances such as drugs, metals, biomolecules, fluorescent probes, and cells, making them ideal for cutting-edge drug delivery systems and diverse therapeutic applications. Graphene's physical attributes, include exceptional tensile strength, excellent electrical and thermal conductivity, swift charge transport, and a significant surface-to-volume ratio, have made it highly desirable in nanoelectronics, renewable energy systems, and medical science [36, 58].

To address challenges related to graphene's instability factor, process manageability, storage feasibility, and production expenses, researchers have explored the creation of graphene derivatives that mimic its properties. One such derivative is graphene oxide (GO), which possesses favourable attributes like excellent fluidic processability, strong water affinity, customizable surface properties, and a significant ratio of surface area to volume. GO's oxidative functionalization opens up possibilities for specific applications In the realm of biomedicine, notable advancements can be seen in drug delivery, tissue engineering, wound healing, and medical devices.

Graphene's intrinsic flexibility, excellent optical transparency, exceptional properties such as elevated electrical and thermal conductivity, speedy charge carrier mobility, a considerable specific surface area, and a biocompatible nature make it an appealing choice for nanomedicine and biomedical applications. Extensive research has been conducted to evaluate its appropriateness as a drug delivery platform in anti-cancer/gene delivery, biosensing, bioimaging, antibacterial applications, cell culture, and tissue engineering. Compared to carbon nanotubes (CNTs), graphene offers advantages such as lower cost, ease of fabrication and modification, higher surface area, and absence of toxic metal particles, making it a competitor in drug delivery applications with lower toxicity and superior biocompatibility [36, 51].

GO, as a precursor for graphene, exhibits attractive characteristics such as electronic properties, sp2 $\pi - \pi$ interaction, and the ability to quench fluorescence. It boasts features like its convenient synthesis process, superior water dispersibility, robust colloidal stability, versatile surface functionalization and compatibility with biological environments. making it highly conducive for bioapplications. GO's sp2 $\pi - \pi$ interaction with hydrophobic drugs facilitates drug loading, enabling its use as a drug delivery carrier. Additionally, GO finds utility in gene delivery and biological sensing due to the fluorescence quenching properties of labeled DNA [23, 51].

Graphene's impressive properties, stemming from its unique chemical structure and geometry, encompass a substantial specific surface area, a notable Young's modulus, and remarkable breaking strength, excellent electrical and thermal conductivity, biocompatibility, and an ultrahigh surface-to-volume ratio. These qualities render graphene suitable for a myriad of applications, including nanocomposites, energy research, catalysis, biomaterials, and nanomedicines. Its planar structure allows for the loading of various exogenous molecules, such as therapeutic drugs, metals, and fluorescent probes, enabling the detection of intracellular components [26].

In conclusion, the properties and characteristics of graphene and graphene oxide make them highly promising materials for drug and gene delivery applications. Their high surface area, mechanical strength, optical properties, and biocompatibility contribute to their suitability in various biomedical fields. As the field of graphene research continues to evolve, further advancements in the field of pharmaceutical delivery systems, biosensors, and tissue engineering scaffolds are expected, harnessing the unique properties of these materials to enhance healthcare and biomedical technologies [23, 42, 51, 58, 70].

2.2.2 Functionalization and Modification Approaches

Functionalization of graphene and graphene oxide (GO) is crucial to overcome their inherent limitations such as limited dispersion and is highly resistant to chemical reactions. Two main approaches for functionalization are covalent and non-covalent methods. Covalent functionalization involves reactions at single atoms, radical reactions, and cycloaddition reactions. Polymers like PEG, chitosan, polymethylmethacrylate, and polythiophene are commonly used for covalent attachment, enhancing solubility, physical robustness, thermal conductivity, ability to conduct electricity, ability to deliver drugs, and biosensor construction. Silane groups, folic acid, sulfonic acid, porphyrins, and aspartic acid are among the diverse chemical attachments explored [26, 58, 120].

Non-covalent modification of graphene and GO includes π-π stacking, Van der Waals forces, hydrogen and ionic bonds, and coordination bonds. Extensive research has been conducted on non-covalent attachment using a range of materials such as drugs, photosensitizers, fluorescent probes, nucleic acids, pyrene derivatives, and polymers. The goal is to improve solubility in polar solvents and modify electronic properties. In one study, graphene oxide was functionalized with dextran and chitosan using a layer-by-layer method, resulting in a nanocomposite for drug delivery and photothermal therapy. Doxorubicin (DOX) was successfully loaded into GO nanosheets via electrostatic interactions and π-π stacking. Another study utilised reduced-GO Polyethylene glycol (PEG) and branched polyethylenimine (BPEI) were covalently attached to the surface, providing functionalization for cytosolic drug delivery [26, 58].

Graphene has also been functionalized with nucleic acids for bio-sensing applications. Multilayer graphene was functionalized with DNA to detect the p53 protein, while graphene quantum dots (GQDs) functionalized with DNA were capable of detecting toxic Ag ions. Non-covalent bonding using poly(sodium 4-styrenesulfonate) (PSS) prevented agglomeration of GO and reduced graphene in aqueous media. Graphene and polymers were combined to create composites with thermo-responsive properties through non-covalent interaction. Antibody-modified graphene field effect transistors were used for cancer prognosis by non-covalently binding a pyrene derivative [26, 74].

Functionalized graphene-based materials find diverse applications in biomedicine, including tissue engineering, drug delivery, biosensors, and photothermal therapy. The presence of oxygen groups on graphene oxide allows for easy functionalization, altering physicochemical properties. Graphene-based materials exhibit higher surface areas and structural defects, enabling interactions with polymers and enhancing mechanical properties. In tissue engineering, GO-reinforced scaffolds promote cell growth and proliferation. Graphene-based nanomaterials have been used for controlled drug delivery and pH-responsive release systems. Additionally, their high electron transport mobility and conductivity make them suitable for biosensors, often with polymer coatings to enhance stability and biocompatibility[74].

In a specific study, a polyamidoamine dendrimer functionalized with graphene oxide was used for co-delivery of doxorubicin (DOX) and MMP-9 shRNA in breast cancer therapy. The graphene oxide-based delivery system showed pH-controlled drug release and significant reduction of MMP-9 protein in cancer cells. It has demonstrated superior biocompatibility when compared to conventional gene vectors, making it a promising strategy for tumour therapy [37].

Overall, functionalization and modification approaches enhance the properties and expand the potential applications of graphene and graphene oxide in drug and gene delivery systems, offering opportunities for innovative biomedical solutions [26, 37, 58, 74, 120].

2.2.3 Examples of Graphene-Based Drug and Gene Delivery Systems for Fungal Infections

Graphene-based nanomaterials (GBNs) have garnered significant interest within the domain of antifungal research due to their potential in combating fungal infections, which pose risks to human health and cause substantial economic losses. Various types of GBNs, including graphene, graphene oxide (GO), and reduced graphene oxide (rGO), have been synthesized and extensively evaluated for their antifungal properties [4, 38].

Research has indicated that rGO exhibits broad-spectrum antifungal activity against pathogenic fungi. For instance, a study conducted by Sawangphruk et al. investigated the antifungal efficacy of rGO against three distinct species of fungi namely *A. niger*, *A. oryzae*, and *F. oxysporum*. The results demonstrated the effectiveness of rGO, with significant inhibitory concentrations (IC50 values) of 50, 100, and 100 μg/mL for *F. oxysporum, A. niger,* and *A. oryzae,* respectively. The interaction between rGO nanosheets and fungal cell walls, along with the formation of reactive oxygen species (ROS) by rGO, were proposed as the probable mechanisms underlying its antifungal activity.

On the other hand, GO exerts its antifungal effects primarily through its sharp edges, which induce stress on the plasma membrane of pathogenic cells. Chen et al. investigated the interaction mechanism of GO with bacterial and fungal pathogens, including *P. syringae, X. campestris pv. Undulosa, F. graminearum, and F. oxysporum.* The study revealed that GO inhibited approximately 90% of bacteria and suppressed 80% of macroconidia germination, accompanied by cell swelling and lysing. The antimicrobial properties of GO against bacterial and fungal pathogens were attributed to the aggregation of GO nanosheets, which perturbed cell membranes, reduced membrane potential, and caused electrolyte leakage. Moreover, the photothermal properties of GO in the near-infrared (NIR) region demonstrated its potential as an effective photothermal material for antifungal treatment, particularly in wound healing applications [4].

GO has also exhibited efficacy as an antifungal agent against specific fungal species such as *S. cerevisiae, C. utilis,* and *P. chrysosporium.* Zhu et al. evaluated the cytotoxicity of GO on S. cerevisiae and observed dose-dependent effects, including decreased mitochondrial transmembrane potential accompanied by heightened ROS generation, and changes in the expression of apoptosis-related genes. Additionally, GO was found to stimulate the growth of *P. chrysosporium* at low concentrations while inducing disruption of fiber structure at higher concentrations [4, 38].

Comparing various nanomaterials, including single-walled carbon nanotubes (SWCNTs), multi-walled carbon nanotubes (MWCNTs), and activated carbon (AC), SWCNTs demonstrated the strongest antifungal activity against *F. graminearum* and *F. poae.* GO and rGO also exhibited notable antifungal effects, while fullerene (C60) and AC showed no significant activity. These nanomaterials exert their antifungal effects by deposition on spore surfaces, preventing water uptake, and inducing plasmolysis [4].

To enhance the antimicrobial properties of GO, researchers have explored the synergistic effects of combining GO with other nanoparticles, particularly silver nanoparticles (AgNPs). The combination of carbon nanoscrolls with AgNPs has shown prolonged antifungal activity against species such as *C. albicans* and *C. tropicalis*, while GO-AgNPs demonstrated clear inhibition zones. Another study focused on the antifungal activity of GO-AgNPs nanocomposites against the phytopathogen *F. graminearum,* both in vitro and in vivo. The results indicated that the nanocomposite displayed potent fungicidal activity, effectively inhibiting mycelial growth, spore germination, and DON (deoxynivalenol) production. The combination of GO with antifungal drugs, such as Mancozeb, Cyproconazole, and Difenoconazole, has also shown synergistic effects, enhancing the fungicidal activity against pathogenic fungi [38].

Furthermore, GBNs have been investigated for their potential application in plant disease management. Fungal diseases significantly impact crop yields, and the development of sustainable and effective control strategies is crucial for agricultural productivity. Studies have demonstrated the antifungal activity of GBNs, particularly GO and rGO, against plant pathogens. For instance, GO-based nanocomposites exhibited inhibitory effects against *F. graminearum*, which causes Fusarium head blight (FHB) in wheat and other small grains. The combination of GO with the fungicides Carbendazim and Triadimenol resulted in enhanced antifungal activity against *F. graminearum*. The nanocomposites reduced the viability of fungal spores, impaired mycelial growth, and decreased DON production, highlighting their potential in agricultural applications [2].

In summary, graphene-based nanomaterials, including graphene, graphene oxide, and reduced graphene oxide, possess significant potential as antifungal agents. These nanomaterials exhibit broad-spectrum antifungal activity by disrupting fungal cell walls, inducing ROS generation, and perturbing cell membranes. The combination of graphene oxide with other nanoparticles, such as silver nanoparticles, enhances their antimicrobial efficacy. Moreover, GBNs have shown promise in plant disease management, effectively inhibiting the growth and toxin production of plant pathogens. Continued research and development in this field hold great potential for the utilisation of graphene-based nanomaterials in combating fungal infections and improving agricultural practices [2, 4, 38].

2.3 *Carbon Dots and Nanodots*

While isolating single-wall carbon nanotubes (SWCNTs) from carbon soot using the arc discharge method, [117] serendipitously discovered an innovative category of luminescent carbon nanoparticles known as carbon dots (CDs) [117].

Within the realm of carbon nanoparticles, carbon dots (CDs) stand out as zero-dimensional structures with diverse functional groups on their surfaces, making them highly hydrophilic and suitable for surface modifications. These versatile fluorescent nanomaterials have found applications in optical sensing, bioimaging, drug delivery, photocatalysis, and electrocatalysis. Their tunable photoluminescence, efficient multiphoton upconversion, and unique optical properties have attracted significant scientific interest. CDs are synthesised through a simple and low-cost process and exhibit excellent biocompatibility [45] [12].

C-dots, also known as carbon quantum dots (CQDs), are small spherical nanocarriers with high biological activity, water solubility, and the ability to form conjugates with other substances. Their utilization spans across an extensive range of applications, including fluorescent labelling, cell imaging, targeted drug delivery, and detection of heavy metal ions [12]. Unlike metallic quantum dots, C-dots are biocompatible and safe for living tissues. They offer advantages such as photoluminescence and small size, making them valuable in research fields [43].

CDs possess remarkable physicochemical and optical properties, making them ideal for cancer therapy, bioimaging, and other applications. They have adjustable optical properties, low cytotoxicity, and a simple synthesis method. CDs show promise in sensing, catalysis, deep tissue imaging, and drug and gene delivery. Their ability to pass the blood–brain barrier and target brain cancer glioma tissue makes them valuable for diagnosing and treating neurological disorders [8, 100].

Overcoming initial challenges, breakthroughs have been achieved in the synthesis of high-quality CDs with features like multicolor emission, narrow bandwidth, and resistance to photobleaching. These advancements have sparked interest in their use as printing inks and functional nanocomposites [59].

There are numerous types of carbon dots (CDs) utilized in drug and gene delivery systems for fungal infections. These include graphene quantum dots (GQDs), carbon quantum dots (CQDs), carbon nanodots (CNDs), and carbonised polymer dots (CPDs). GQDs consist of single or few layers of graphene sheets and possess an improved crystal lattice compared to other CDs. CQDs possess graphite structures with multiple layers linked by surface functional groups. On the other hand, CNDs are carbon nanoparticles that exhibit high carbonization but lack evident crystallinity. CPDs have a unique structure with a carbon core and polymer shell, offering higher stability and better compatibility. The optical properties of CDs vary depending on their type, with GQDs and CQDs influenced by size and surface/edge structures, while CPDs rely on molecular states and crosslink enhanced emission. The diversity of CDs is driven by the availability of different carbon sources and synthetic strategies used for their preparation [12, 59, 81].

When it comes to the synthesis of carbon dots (CDs), two main approaches are employed: top-down and bottom-up methods. Top-down methods implement techniques such as laser ablation, arc-discharge, hydrothermal treatment, and electrochemical oxidation to break down bulk carbon precursors into nanosized particles. On the contrary, bottom-up methods rely on hydrothermal synthesis, microwave pyrolysis, and solution chemistry to produce graphene quantum dots and amorphous carbon nanodots by dehydrating, polymerizing, and carbonizing small carbon molecules [63].

CD synthesis can also involve the introduction of additional atoms like nitrogen, boron, and sulphur through suitable doping techniques. The discovery of CDs in 2004 led to the evolution of cost-effective and eco-friendly methods for their fabrication. Top-down methods offer better efficiency and surface properties but require complex procedures, expensive chemicals, and specialised instruments. In contrast, bottom-up approaches provide accurate manipulation of the structure and distribution of size, as well as convenient surface passivation. Green synthetic methods using biomolecules, biological species, and plant materials as precursors have emerged as simple and eco-friendly approaches for large-scale CD production [81].

The selection of appropriate methods is crucial to ensure the production of high-quality fluorescent CDs for diverse applications. Factors such as quantum yield, surface chemistry, and purification requirements should be considered. Overall, the synthesis of carbon dots involves top-down and bottom-up approaches, each with its own advantages and challenges [6, 8, 12, 43, 68, 104].

2.3.1 Properties and Characteristics Relevant to Drug and Gene Delivery

Carbon dots (CDs) possess exceptional attributes that render them well-suited for biomedical applications. In terms of electrochemical properties, CD nanoparticles outperform other carbon-based nanomaterials in terms of charge transferability, electroconductivity, surface area, and exhibit reduced toxicity. Doping CDs with heteroatoms further improves their electronic attributes and electrocatalytic activity. CDs also contribute to stability enrichment, acting as active centers during electrochemical reactions. CDs display absorption in the UV region and fluorescence properties, including up-conversion and down-conversion fluorescence. Their emission properties can be controlled by adjusting physicochemical parameters. CDs demonstrate chemical stability, long emission lifetimes, and resistance to photobleaching. They also possess room-temperature phosphorescence and can generate chemiluminescence and electrochemiluminescence. Furthermore, CDs offer a broad spectrum of optical properties suitable for various bio-applications [33, 63], [3]; [68], [12, 43, 59].

To enhance the quantum yield (QY) of CDs, different techniques, including metal doping, inorganic salt doping, and surface passivation, have been utilized due to their influence on the fluorescence mechanism involving factors like the quantum size effect, surface defects, emissive traps, and radiative recombination of excitons. Nitrogen doping has shown promising results in increasing the QY of CDs. The fluorescence properties of CDs are also influenced by the size of the particles, pH of the medium, and surface functional groups. Smaller CDs with a large surface-to-volume ratio exhibit higher QY due to quantum confinement. The emission wavelength of CDs is also influenced by their size, with smaller CDs emitting UV light and larger CDs emitting near-infrared (NIR) light. The pH of the medium and surface functional groups can modulate the fluorescence properties of CDs [33], [3].

Biocompatibility and toxicity are important considerations for the use of CDs in biological applications. In vitro cytotoxicity studies have shown that surface-passivated CDs exhibit low toxicity. Functionalization with non-toxic molecules enhances the biocompatibility of CDs. However, additional research is needed to assess and evaluate their in vivo biocompatibility extensively. The biodistribution, clearance, and tumor uptake of CDs can be influenced by the route of administration [33, 81].

The adsorption characteristics of carbon quantum dots (CQDs) and graphene quantum dots (GQDs) are influenced by several factors, including the size of the dots. CQDs demonstrate significant optical absorption in the ultraviolet (UV) range, whereas GQDs display absorption shoulders within the 270–390 nm range. CQDs and GQDs both exhibit tunable photoluminescence properties, with GQDs generally exhibiting higher quantum yield (QY) due to their layered structure and improved crystallinity. The photoluminescence of CQDs and GQDs is affected by factors such as aromatic structures, emissive traps, quantum confinement effect (QCE), surface defects, and oxygen-containing groups. Additionally, parameters such as size, excitation wavelength, solvent polarity, pH, surface oxidation, surface functionalization, and heteroatom doping influence the photoluminescent behavior of CQDs and GQDs.

The size of GQDs affects their shape, absorption characteristics, and photoluminescence properties. The pH of the solution also influences the photoluminescence intensity. Solution concentration and reaction temperature during synthesis can affect the photoluminescence properties of GQDs [3]; [81].

CDs exhibit significant photoluminescence properties, which make them promising for biomedical applications. The fluorescence emission is influenced by various factors such as their size, morphology, abundance, chemical composition, and internal arrangement. In terms of biocompatibility, CDs offer advantages over semiconductor quantum dots (QDs) by being non-toxic, environmentally friendly, and highly soluble in water. Extensive research has shown that CDs are generally non-cytotoxic and exhibit low accumulation in organs. In vivo studies have revealed no significant toxicity, clinical symptoms, or weight loss. The rapid clearance of CDs from the body via various injection routes makes them suitable for clinical applications. The surface features of CDs, particularly when combined with bio-compatible conjugates, contribute to their stability and safe excretion from the body [68].

Carbon dots demonstrate a distinctive fluorescence characteristic ascribed to surface irregularities and the impact of quantum size phenomena. Energy centers formed by these surface defects contribute to numerous discharges, while the size quantization effect, influenced by the particle size, ensures remarkable stability and resistance to photobleaching. Moreover, C-dots display fluorescence emission dependent on the excitation, enabling customizable emission properties. Certain C-dots even exhibit up-conversion fluorescence, emitting wavelengths shorter than the excitation wavelength. CDs demonstrate significant absorption in the ultraviolet (UV) range (280–360 nm), which can be ascribed to the electronic transitions of $C = O/C = N$ or $C = C$ bonds. The manipulation of surface modifications, functional groups, π-conjugated regions, and the carbon core's oxygen/nitrogen content allows for the precise regulation of CDs' absorption and emission characteristics. To enhance the photoluminescence efficiency and quantum yield, one can employ techniques such as surface passivation, modification, and introducing nitrogen and sulfur through element doping [12].

In conclusion, Carbon dots exhibit distinct characteristics that set them apart, including enhanced electrochemical properties, optical absorption, fluorescence properties, and photoluminescence. CDs exhibit high charge transferability, electroconductivity, and lower toxicity compared to other carbon-based nanomaterials. The doping of CDs with heteroatoms further improves their electronic attributes and electrocatalytic activity. The fluorescence properties of carbon dots (CDs) are influenced by various factors, including the quantum size effect, surface defects, the presence of emissive traps, and the process of radiative recombination of excitons. Biocompatibility studies have shown that surface-passivated CDs exhibit low toxicity, but further research is needed to evaluate their in vivo biocompatibility extensively. The adsorption and photoluminescence properties of CDs are influenced by several parameters including size, wavelength of excitation, pH level, solvent characteristics, surface oxidation state, surface functionalization, and the presence of heteroatom doping. CDs offer advantages such as adjustable luminescence, enhanced quantum efficiency, non-toxicity, biocompatibility, and cost-effectiveness. Understanding and controlling these properties are crucial for harnessing the potential of CDs in various applications [33, 63], [3] [68], [12, 43, 59].

2.3.2 Functionalization and Modification Approaches

Carbon dots (CDs) have garnered considerable interest in various bio-applications because of their distinct characteristics. Several modification approaches have been explored to optimise and improve the performance of CDs. Doping is one such technique that involves introducing dopant atoms such as nitrogen or sulphur into the carbon matrix. This tuning of the emission wavelength enhances the photoluminescence and alters the electronic structure of CDs. Surface functionalization is another approach that involves attaching chemical moieties or organic compounds onto the superficial layer of CDs. Amino-modifications are commonly used to improve stability, alter surface chemistry, and optimize optical properties. Functionalized CDs can be further tailored by conjugating them with targeting moieties for selective binding in biological systems [3].

Polymer capping is a technique where CDs are coated with a polymer layer, eliminating surface defects and enhancing photoluminescence. Composite formation involves incorporating CDs into other materials, synergistically enhancing the properties of both components. Core–shell structures are created by surrounding CDs with another material, providing additional functionalities such as stability, controlled release, or biocompatibility.

Surface passivation of CDs has been shown to enhance photoluminescence. Citric acid-derived CDs with more carboxyl groups exhibit higher quantum yield. Passivation with ethyleneimine, polyethylene glycol (PEG), or bovine serum albumin (BSA) improves stability and fluorescence. Doping with nitrogen, sulphur, and boron improves optical properties and allows for applications such as fluorescent MRI probes and detection of specific ions [43].

In the realm of gene delivery, PEI-passivated CDs have been developed as highly efficient dual-functional nano gene vectors. PEI is pivotal to surface passivation and DNA condensation, leading to heightened efficiency in gene delivery with diminished cytotoxic effects compared to pristine PEI. The optimization of CD-PEI vectors relies on finding the appropriate pyrolysis time for a balance between CD formation and preservation of PEI functionality [57].

Theranostic nanoagents based on CDs have shown promise in targeting and treating lung cancer. These nano agents absorb and emit specific wavelengths of light, enabling diagnosis and therapy. The application of folate as a targeting moiety enhances the specificity of treatment by targeting overexpressed folate receptors on cancer cells. The theranostic nanoagents demonstrate proficient complexity in their interaction with siRNAs and liberate them under conditions of reduced environment, resulting in suppressed gene expression and reduced cell viability [116].

In summary, various modification strategies have been explored to enhance the characteristics of carbon dots. Doping, modifying the surface, incorporating polymers, creating composites, and developing core–shell structures are essential strategies for customizing the optical, electronic, stability, and targeting properties of CDs. These modifications enable their effective utility in a wide range of applications within the field of biotechnology, including bioimaging, biosensing, drug delivery, and gene therapy [3]; [43], [57, 116].

2.3.3 Examples of Carbon Dot-Based Drug and Gene Delivery Systems for Fungal Infections

In recent times, the escalating prevalence of yeast infections and the emergence of drug resistance have raised concerns in the medical community. To address these challenges, carbon dots have garnered consideration for their potential antimicrobial and antifungal properties. Various studies have investigated the application of C-dots as effective agents against fungal pathogens. For instance, glucose-derived C-dots doped with nitrogen have demonstrated strong antifungal effects against *Candida albicans, Ammophilus fumigatus,* and *Penicillium citrinum.* Similarly, C-dots doped with sulfur have shown inhibitory effects on the development of *Fusarium solani.* Researchers have also utilized metabolites from baker's yeast and doped C-dots with nitrogen and iodine as peroxidase mimics for disinfection against *C. albicans.* Moreover, multifunctional C-dots have been synthesized, exhibiting anticancer and antifungal properties. These C-dots possess lower cytotoxicity, durable fluorescent properties and remarkable photothermal conversion efficiency, making them suitable for imaging-based photothermal treatments. Importantly, C-dots demonstrate intrinsic antifungal properties even at lower concentrations, and recent studies have demonstrated the biodegradability of vitamin C-derived C-dots while maintaining their bactericidal and fungicidal properties. The degradation of C-dots into harmless byproducts suggests their capability as antifungal agents [69, 79]. A notable theranostic application of carbon dots in antifungal development was demonstrated by Li and co-workers, wherein they conjugated red-emissive carbon dots (CD) with positively charged guanidine groups and antifungal amphotericin B. The modified carbon dots accumulated specifically in the planktonic and candida biofilms and exhibited significantly reduced invasiveness, as observed in photomicrographs as well as reduced enzymatic and cytokine markers of infection [55] (Fig. 3).

In addition to their antimicrobial properties, C-dots have shown promise in gene delivery applications. Their capacity to engage in electrostatic interactions with functionalized C-dots carrying a positive charge and nucleic acids possessing a negative charge makes them suitable gene carriers. Positively charged C-dots synthesized using porphyria polysaccharide and ethylenediamine precursors have been employed for promoting the differentiation of adult stem cells into neurons through non-viral

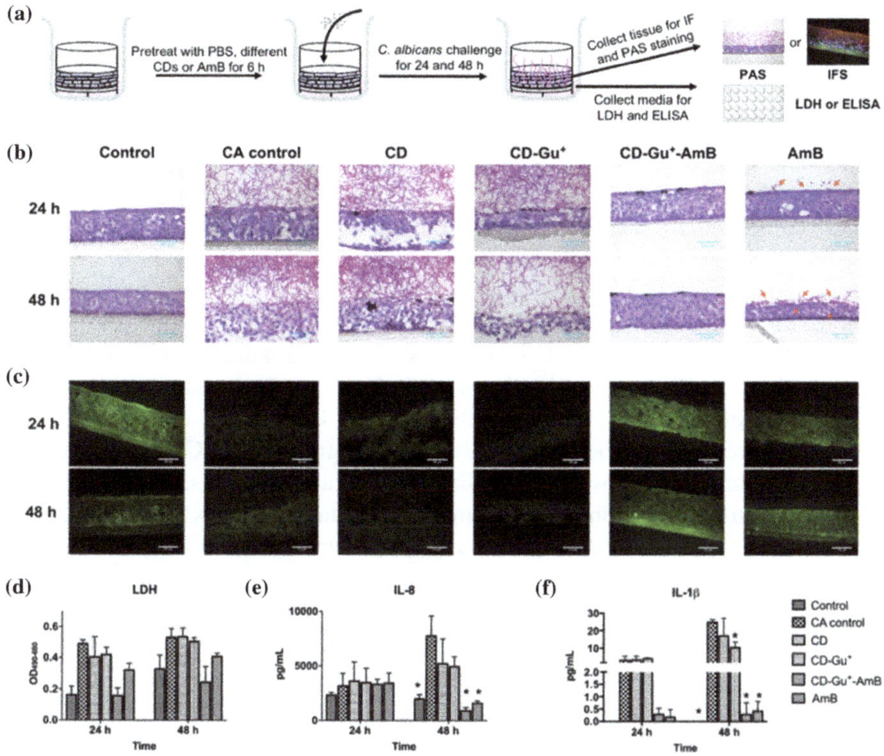

Fig. 3 (**A**) Schematic diagram of the experimental design. After pretreating Reconstituted Human Oral Epithelia (RHOE) with PBS, different Carbon Dots (CD), and Amphotericin B for 6 h, RHOE were subsequently challenged with *C. albicans* SC5314 (4 × 104 CFU/mL) for 24 and 48 h. At each time point, the tissue and culture media were harvested for the following biological analysis. The cryosection of the collected tissue samples was performed with (**B**) periodic acid-Schiff (PAS) and (**C**) immunofluorescent (IF) staining (green representing E-cadherin) (scale bar is 50 μm). The red arrows indicate the yeast, hyphae, or pseudohyphae located on/in RHOE. The amount of released lactate dehydrogenase (LDH) in the culture media at 24 and 48 h was assessed by measuring the absorbance of reduced tetrazolium salt (INT) via a coupled enzymatic reaction at 490 and 680 nm (**D**). The levels of IL-8 (E) and IL-1β (F) in the media were measured with ELISA. The asterisk indicates the significant differences as compared to the CA control groups, *p < 0.05 [55]

gene delivery. These gene delivery systems hold potential for therapeutic applications [69].

2.4 Fullerenes

Fullerenes, specifically C60, are nanoscale carbon structures and the third known carbon allotrope, following graphite and diamond. They possess unique photoelectrochemical features and exhibit diverse shapes such as hollow spheres, ellipsoids, and tubes. The most common and symmetrical form is C60, also known as buckyball or footballene, which consists of 12 pentagons and 20 hexagons. Fullerenes have high structural and chemical stability due to their exceptional symmetry. However, their limited solubility in organic solvents and inability to dissolve in water limit their use in biological systems, necessitating modifications [1, 67, 70, 106], [24, 101].

In 1985, researchers Robert F. Curl, Jr., Richard E. Smalley, and Sir Harold W. Kroto made a significant scientific breakthrough at Rice University by uncovering the existence of fullerene, with a specific focus on C60. It was a groundbreaking achievement, resulting in the Nobel Prize in Chemistry in 1996. C60's symmetrical structure, resembling an icosahedron with pentagons and hexagons, made it a symbol of symmetry in chemistry and a key member of the nanomaterials family. Its unique physical and chemical properties found applications in material science, electronics, and nanotechnology. However, challenges arise in biomedical applications due to its hydrophobic nature and tendency to form toxic aggregates when in contact with water. Various methods, including chemical modification, incorporation into supramolecular structures, solvent exchange, and stirring in water, have been devised to enhance the ability to dissolve in water. Macromolecules such as calixarene, polyvinylpyrrolidone, lecithin, cyclodextrins, phosphatidylcholine liposomes, and surfactants have been used to prepare stable water dispersions [24, 106].

Fullerenes possess exceptional optical and chemical properties, making them suitable for photothermal therapy and other applications. They exhibit high symmetry, strong UV absorption, and moderate visible light absorption due to their conjugated π-electron system. The unique properties of fullerenes offer opportunities for various chemical transformations and exploration of their biological and pharmaceutical applications [101].

Initially, the synthesis of fullerenes involved using laser-induced carbon vaporization in an oxygen-free environment. However, advancements in the field led to the production of fullerene C60 through induction heating of graphite as a carbonization source and microwave irradiation of polyaromatic hydrocarbons [67].

In conclusion, fullerenes, especially C60, represent a significant advancement in carbon nanostructures. Their symmetrical and stable nature, along with their unique properties, have led to diverse applications in different fields. However, their limited solubility in water has prompted the development of modification techniques to enhance their water solubility and biomedical usability. Ongoing research aims to explore the capability of functionalized fullerenes for various medical utilities, particularly in nanomedicine [1, 67, 70, 106], [24, 101].

2.4.1 Properties and Characteristics Relevant to Drug and Gene Delivery

Fullerenes, with their unique properties and structural characteristics, have garnered significant attention as potential candidates for effective drug delivery systems and biomedical applications. The comparable dimensions of carbon nanostructures to biological structures make them well-suited for this purpose, and their easy synthesis enables their integration with various therapeutic agents such as chemotherapeutics, antitumor drugs, and antibodies, further enhancing their potential in the field [1].

Fullerenes, specifically the C60 molecule, exhibit a distinctive geometry known as a truncated icosahedron, resembling a soccer ball with pentagonal and hexagonal faces. The crystallization of C60 forms a highly symmetrical structure with a fivefold axis, which was previously unknown in crystallography. The diameter of the C60 molecule is approximately 0.710 nm, with outer and inner π electron cloud diameters of 0.34 nm and 0.35 nm, respectively. The high hydrophobicity, cohesive force between molecules, photoactivity, electron accepting and releasing capabilities, and reactivity of fullerenes allow for structural modifications through various chemical transformations [24].

Fullerenes have demonstrated intriguing properties for biological applications, particularly in photodynamic therapy and antioxidant treatment. They exhibit extended lifetimes of triplet excited states and demonstrate high efficacy in transferring energy to oxygen molecules, leading to the generation of reactive oxygen species (ROS) under light illumination. This ability to generate ROS offers a potential route for photodynamic ablation of tumour cells. Additionally, fullerenes act as radical scavengers or antioxidants by engaging with dynamic, unstable molecules within the body and effectively eliminating their presence. The presence of p-p carbon bonds in fullerenes facilitate chemical conjugations with active macromolecules, peptides, and chemical compounds. Recent efforts have focused on developing water-soluble fullerenes through conjugation with biopolymers, designing tumour-specific fullerenes to improve photodynamic anticancer treatment, and synthesising biocompatible fullerenes for antioxidant therapy [119].

In summary, fullerenes possess unique properties and characteristics that make them promising candidates for drug and gene delivery systems. Their comparable dimensions to biological structures, ease of synthesis, and ability to be linked with therapeutic agents contribute to their potential for efficient drug delivery. Furthermore, the geometric structure of fullerenes, along with their photo-responsive properties, radical-capturing abilities, and chemical conjugation capabilities, provide opportunities for various biomedical applications [1, 24, 119].

2.4.2 Functionalization and Modification Approaches

Functionalization and modification of fullerenes have been explored to enhance their solubility and biocompatibility for various biomedical applications. Fullerenes are naturally hydrophobic and insoluble in water, limiting their use. To address this, functionalization with amino, carboxyl, or hydroxyl groups has been shown to increase water solubility and biocompatibility [67]. Synthetic derivatives of fullerenes have also been developed to overcome their hydrophobic nature and low solubility in polar solvents like water [1]. Chemical derivatization allows for the attachment of functional groups, with exohedral addition being a versatile technique [101]. Strategies such as complexation with cyclodextrins, sulfocalix, polyvinylpyrrolidone, or polyethylene glycol, as well as the preparation of micelles and liposomes, have been employed to enhance aqueous solubility [101]. Introducing multipolar groups like amino acids, carboxylic acids, and hydroxyl groups has also been explored [101].

Water-soluble fullerene derivatives have shown promising biological activity and potential for gene therapy [101]. Tetraamino fullerene synthesis has been achieved through a simple route, and these derivatives have demonstrated the ability to attach to DNA effectively and condense different sizes of DNA fragments [101]. The internalization of fullerene/DNA complexes occurs through endocytosis, and lysosomal degradation of the complexes can be delayed using chloroquine [101]. Amino fullerenes synthesized with piperazine derivatives have shown DNA binding ability and protection of DNA in the cytoplasm [101]. Tetrapiperidino fullerene epoxide (TPFE) has exhibited high transfection efficiency and reduced cytotoxicity, making it a preferred gene delivery carrier [101].

Fullerene C60 derivatives prepared through Hirsch-Bingel chemistry have been researched as possible scaffolds for gene or drug delivery [101]. Positive charge density on the surface of these derivatives enhances plasmid DNA transfection, and the flexibility of linkers between the fullerene core and amino groups plays a role in their efficiency [101]. Strategies to prevent aggregation of nanoplexes include using phosphate buffers or NaCl [101]. The hydrophobic nature of carbon nanostructures has facilitated their preparation and cellular uptake [101].

Fullerene derivatives, including C60, have shown potential as therapeutic and diagnostic agents, with applications in gene therapy and magnetic resonance imaging [101]. These advancements provide opportunities for the utilisation of fullerene derivatives in gene and drug delivery systems, besides in diagnostic imaging [99].

2.4.3 Examples of Fullerene-Based Drug and Gene Delivery Systems for Fungal Infections

Fullerenes, particularly functionalized fullerenes such as C60, have emerged as promising nanocarriers for drug and gene delivery. They possess optimal dimensions and water-repellent surfaces, making them suitable for conjugating anticancer drugs. Conjugation of drugs to fullerenes has shown optimal release characteristics and substantial antitumor effects. Fullerenols, which are highly soluble in polar solvents, have been explored for delivering anticancer drugs and overcoming drug resistance. Moreover, biocompatible and water-soluble fullerenes coated on nanoparticles exhibit pH-sensitive drug release and cellular uptake detection capabilities. In gene delivery, fullerene derivatives efficiently deliver nucleic acids into target cells, protecting them from degradation and facilitating gene expression. These fullerene-based systems hold promise for transfection and therapeutic applications without acute toxicity [67].

Fullerene derivatives have demonstrated potential in gene delivery. Tetra(piperazino)fullerene epoxide (TPFE) has exhibited higher transfection rates than Lipofectin, a lipid-based system. TPFE is non-toxic and effectively delivers genes, such as the mouse insulin 2 gene, reducing glucose levels in mice. TPFE has also been utilized for siRNA delivery to treat sepsis, forming complexes that release siRNAs into lung cells. An additional fullerene incorporating amine groups derivative, C60-Dex-NH2, forms micelle-like aggregates to protect siRNAs from degradation. Upon being subjected to visible light, C60-Dex-NH2 generates reactive oxygen species, enhancing gene silencing efficiency. Evidence from experiments conducted in controlled environments and living organisms has proven significant gene silencing effects using C60-Dex-NH2 [70].

Fullerenes have shown promise not only as drug carriers but also as antibacterials. In drug delivery, they enhance the bioavailability of peptides and hydrophobic drugs and reduce toxicity when conjugated with anticancer drugs such as doxorubicin and cisplatin. Fullerenes have also demonstrated potential in DNA delivery by improving transfection efficiency and reducing cytotoxicity. As antibacterials, fullerenes exhibit antimicrobial activity against various bacterial strains, including Gram-positive organisms. They disrupt cell walls, inhibit growth, and offer protection against bacterial infections. Different functionalized fullerenes have specific antimicrobial properties, with some showing efficacy against Mycobacterium avium and Mycobacterium tuberculosis. Fullerene derivatives with positive charges have proven more efficient in association with bacterial membranes, and their bactericidal effects are associated with the production of reactive oxygen species. However, further investigation is necessary to comprehensively understand the underlying mechanisms and explore potential applications of fullerenes in the field of drug delivery and antibacterial treatments [101].

Water-soluble fullerenes have been established as structures for drug and gene delivery, demonstrating the ability to pass through cell membranes and serve as intracellular delivery agents. Polyhydroxylated fullerenes, known as fullerenols, have been used to deliver doxorubicin and protect against oxidative damage in rat models. Fullerenol administration has shown protective effects on the liver and heart, mitigating doxorubicin-induced damage. Additionally, fullerenols have been found to reduce neutrophil count and cardiovascular damage resulting from chronic doxorubicin toxicity. Covalently crosslinked doxorubicin-PEG-C60 conjugates exhibit time-dependent cytotoxicity against breast cancer, melanoma, and lung carcinoma cells, inducing apoptosis through G2-M phase block [53].

Fullerenes have attracted profound concern for the expansion of safe and targeted mechanisms for drug and gene delivery. Despite the challenges posed by cellular membranes as barriers to uptake, the small size of fullerenes (<5 nm) has made them intriguing candidates. Fullerenes can be modified by decorating hydrophilic entities around their cage structure, rendering them soluble in aqueous solutions. These water-soluble fullerenes can readily cross cellular membranes without causing damage. Further modifications allow fullerenes to serve as versatile nanocarriers for targeting, treating, and preventing diseases with high biocompatibility [21].

3 Applications and Results in Fungal Infection Therapy

3.1 Antifungal Drug Delivery Using Carbon Nanomaterials

3.1.1 Enhancement of Drug Solubility, Stability, and Bioavailability

Efficient treatment outcomes in the field of fungal infection therapy depend greatly on improving drug solubility, stability, and bioavailability. Carbon nanomaterials (CNMs) have shown great potential to overcome these challenges [31].

The limited solubility often encountered in antifungal drug formulations poses a significant challenge, resulting in compromised therapeutic effectiveness. However, CNMs possess distinctive attributes, including a high surface area and versatile surface chemistry, that have the potential to substantially enhance drug solubility [87]. The encapsulation of antifungal drugs within CNMs offers a promising strategy to enhance their solubility, leading to improved dispersion and distribution throughout the body [86]. The enhanced solubility enables higher drug concentrations to be achieved at the affected area, resulting in enhanced antifungal activity [84].

CNMs significantly contribute to the improvement of antifungal drug stability, which is essential for ensuring their effectiveness during the entire treatment duration. These nanomaterials provide a protective environment that effectively shields the encapsulated drugs from degradation, thereby improving their stability. This protective effect offered by CNMs is particularly advantageous in extending the shelf life of antifungal formulations and maintaining the drugs' activity until they are administered to patients. The preservation of antifungal drug potency is greatly facilitated by CNMs, which effectively prevent drug degradation. This preservation is pivotal in maintaining the effectiveness of the agents and ultimately leads to enhanced therapeutic outcomes.

CNTs (carbon nanotubes) have emerged as promising candidates for drug delivery of antifungal agents. One notable example is the encapsulation of the antifungal drug Amphotericin B within CNTs, which has demonstrated excellent preservation of its high antifungal activity. Studies have demonstrated that conjugating Amphotericin B with CNTs enhances its potency against Candida compared to the drug alone [89].

Bioavailability pertains to the portion of the administered medication that successfully enters the bloodstream and its intended destination. CNMs can be engineered to enhance drug bioavailability by improving their pharmacokinetic properties. The unique surface characteristics of CNMs allow for functionalization or modification, enabling targeted drug delivery to specific fungal pathogens or infected tissues. This targeted approach minimizes systemic side effects, while maximizing drug concentrations at the site of action, resulting in enhanced therapeutic efficacy against fungal infections [34].

Current limitations in ocular therapy can potentially be overcome through the use of nanotechnology-based approaches, particularly those involving biodegradable polymers. These strategies have garnered significant interest as they offer promising solutions for enhancing the effectiveness of ocular treatments. Extensive investigations have been carried out on different types of nanoparticles for ocular drug delivery, including polymeric nanoparticles, liposomes, and nanotransferosomes. Among these, polymer-based nanoscale particles have demonstrated significant capability in providing controlled release of drugs, leading to improved bioavailability at the targeted ocular sites. Reported studies have indicated that ocular drug delivery utilizing particulate systems demonstrates the potential for entrapping particles in the eye's mucin, resulting in prolonged residence and reduced drainage. Furthermore, the utilization of drug-incorporated nanoparticles has been demonstrated to enhance bioavailability through the promotion of prolonged drug delivery and improved drug assimilation [5].

3.1.2 Controlled and Sustained Drug Release

Fungal infections present a substantial worldwide health challenge, and the successful administration of antifungal medications plays a vital role in achieving favorable treatment results. Conventional drug delivery methods frequently encounter obstacles such as inadequate drug solubility, restricted stability, and insufficient bioavailability. In recent years, carbon based nanomaterials have gained recognition in terms of a promising option for antifungal drugs delivery. This is attributed to their unique properties and their ability to enable controlled and sustained release of drugs [41].

Enhancing drug solubility and stability is a significant advantage of utilizing carbon nanomaterials for antifungal drug delivery [31]. Carbon nanomaterials, like carbon nanotubes, graphene, and fullerenes, possess high surface area and versatile surface chemistry [122], which make them capable of enhancing the dissolution characteristics of hydrophobic antifungal medications. The incorporation of drugs into carbon nanomaterials enhances their solubility, leading to improved dispersion and distribution throughout the body. Additionally, this encapsulation offers protection against degradation, thereby increasing the stability of the drugs during storage and administration [89].

The utilization of carbon nanomaterials provides the benefit of controlled and sustained release of drugs, which is crucial for maximizing therapeutic effectiveness. By modifying and functionalizing the surface of these nanomaterials, the release kinetics of the encapsulated drugs can be precisely adjusted. Researchers have investigated different strategies, including pH-responsive and stimuli-responsive systems, to achieve controlled drug release specifically at the site of fungal infection. Such controlled release mechanisms enable the sustained presence of drugs at the target site, ensuring prolonged antifungal activity and reducing the need for frequent drug administration [20, 42].

Carbon nanomaterials possess distinct characteristics that enable precise drug delivery to the specific site of a fungal infection [115]. The small size and surface properties of carbon nanomaterials facilitate efficient penetration of fungal cells, As a result, augmenting the therapeutic potency of the medication [15]. Incorporating ligands with specific affinity for fungal cell membrane receptors through the functionalization of carbon nanomaterials allows for active targeting. This active targeting approach enhances the specificity and selectivity of drug delivery, minimizing off-target effects and maximizing the drug's level of concentration specifically at the infected area [42].

To achieve controlled drug release, one effective approach involves the utilization of stimuli-responsive systems. Carbon nanomaterials can be functionalized with components that respond to specific environmental cues at the site of infection. Various stimuli, including pH, temperature, and enzymatic activity, can trigger the responsive elements incorporated into carbon nanomaterials, resulting in controlled release of the encapsulated drug. By integrating these responsive components into carbon nanomaterials, researchers can achieve targeted drug delivery and enhance the therapeutic effectiveness in treating fungal infections [20]. For instance, pH-responsive systems can exploit the naturally acidic environment commonly associated with fungal infections. Upon encountering the acidic pH, nanomaterials undergo structural or surface property modifications, resulting in the release of the encapsulated drug. This pH-triggered release mechanism ensures targeted drug delivery to the site of infection, enhancing its effectiveness while minimizing off-target effects [42, 93].

An alternative approach involves surface modification of carbon nanomaterials using polymers to regulate the liberation of the encapsulated medication. These polymers can create a protective barrier around the nanomaterials, leading to a gradual drug release and prolonged presence in the body. Through careful selection of polymer composition and adjustment of its properties, researchers have the ability to attain meticulous command over the rate and duration of drug release, leading to a continuous and prolonged release of the medication. This sustained release strategy is instrumental in maintaining optimal therapeutic drug concentrations within the body. By avoiding the need for frequent dosing, it ensures a continuous antifungal effect and enhances treatment outcomes [58].

The distinctive physicochemical characteristics of carbon nanomaterials offer potential solutions to overcome the drawbacks associated with traditional drug delivery systems in treating fungal infections. However, additional research is necessary to optimize the development, synthesis, and characterization of drug delivery systems based on carbon nanomaterials. Furthermore, comprehensive evaluation of their safety and effectiveness through preclinical and clinical studies is crucial. Ongoing advancements in this area hold significant potential for improving antifungal therapy and effectively combating fungal infections [65].

3.1.3 Synergistic Effects of Combination Therapies

Combination therapy, which involves administering multiple drugs with different mechanisms of action, is a more effective approach in disease treatment compared to single-agent therapies. Graphene and graphene oxide (GO) have unique characteristics that customize them for developing multifunctional drug delivery platforms. These platforms can deliver multiple drugs, target specific cells, and track nanocarriers within cells. They also have possible uses in cancer biomarker sensing and biomedical imaging. Current investigations revealed the benefits of co-delivering multiple drugs using graphene-based carriers, leading to improved targeting efficacy and enhanced therapeutic effectiveness. The concept of combined therapy holds promise for cancer treatment, where diverse approaches can improve efficacy and minimize side effects. Further research exploring the combination of graphene drug carriers with other therapies may lead to advanced nanocarriers for cancer treatment [58].

In recent times, scientists have extensively explored the application of Graphene Oxide (GO) nanocomposites for their antifungal activity. The combination of Graphene oxide with other nanoparticles has shown a synergistic effect, leading to the formulation of more efficient antimicrobial products. Among these nanocomposites, Ag nanocomposites have garnered significant attention.

To enhance the fungicidal properties of carbon nanoscrolls (CNSs), researchers filled them with nanoparticles of silver (AgNPs) and compared their effectiveness with the GO-AgNPs nanocomposite. The results showed that the CNSs-AgNPs exhibited prolonged antifungal activity against *Candida albicans* and *Candida tropicalis* when compared to the GO-AgNPs nanocomposites. Interestingly, no inhibition zone was observed for GO alone, while clear inhibition zones were observed for the GO-AgNPs samples.

The findings highlight the potential of Ag nanocomposites and the enhanced fungicidal activity of CNSs when loaded with silver nanoparticles. The absence of inhibition zones for GO alone suggests the crucial role of silver nanoparticles in conferring antimicrobial properties. This research demonstrates the significance of utilizing GO nanocomposites, particularly in combination with AgNPs, for the production of more effective approaches to combat fungal infections. As the incubation time increased, a noticeable trend was observed where a decrease in the size of the inhibition zone was observed, accompanied by an increase in the viable fungal colony count. Interestingly, when comparing the inhibition zones of *Candida albicans* and *Candida tropicalis* under the identical GO-AgNPs concentration and incubation period, it was observed that the inhibition zone of *C. albicans* was considerably smaller than that of *C. tropicalis*. This suggests that the antifungal activity of GO-AgNPs may have varying effectiveness against different fungal species, even when exposed to identical conditions.

In a study conducted by Chen et al. they examined the antifungal properties of a nanocomposite consisting of GO-AgNPs against the phytopathogen Fusarium graminearum. The results revealed that the GO-AgNPs nanocomposite exhibited significantly greater efficacy, with three to seven times higher potency compared to pure AgNPs and GO individually. The observed antifungal mechanism was attributed to the remarkable synergistic effect of GO-AgNPs, which involved inflicting physical damage to the fungal cells and generating reactive oxygen species through chemical means. This combined action contributed to the heightened effectiveness of the nanocomposite in inhibiting the growth and survival of *F. graminearum*, thereby demonstrating its potential as a promising antifungal agent.

Karimi et al., conducted a study to explore the potential of G-TiO_2 nanocomposite as a novel approach for fabricating cotton fabric with antibacterial and antifungal properties, without any associated toxicity. The findings of the study demonstrated that the cotton fabric coated with G-TiO_2 nanocomposite exhibited remarkable bactericidal and fungicidal activity against bacteria such as E. coli and S. aureus, as well as fungi like *C. albicans*. These results highlight the effectiveness of G-TiO_2 nanocomposite as a viable solution for developing cotton fabrics with enhanced antimicrobial properties for various applications.

In the context of enhancing bacterial decomposition and improving antimicrobial properties, the addition of G to TiO_2 nanoparticles was investigated to promote increased interactions of nanoparticles and microorganisms. Interestingly, fabrics treated solely with GO did not exhibit any antimicrobial activity. The Graphene oxide-borneol (GOB) composite, obtained through grafting borneol onto graphene oxide, exhibited a remarkable antifungal effect on *M. racemosus*. However, when compared to the GO nanosheet, the GOB composite did not demonstrate significant antifungal activity. Notably, the GOB composite exhibited a sustained antifungal effect over an extended period, as evidenced by the inhibited germination of fallen spores even after 5 days [4].

3.2 Gene Delivery for Antifungal Therapy

3.2.1 Delivery of Antifungal Genes or RNA Interference Molecules

Extensive research has been undertaken regarding treatment of diseases through gene therapy, which involves introducing nucleic acids to modify, regulate, or repair specific genetic targets responsible for disease manifestations. Therapeutic nucleic acids such as pDNA, siRNA, and miRNA have been utilized to either enhance or inhibit the expression of proteins, leading to therapeutic effects. However, gene delivery remains a critical and challenging aspect of gene therapy.

Gene delivery systems can be categorised as two main categories: viral vectors and non-viral vectors. Viral vectors, which encompass retroviruses, lentiviruses, herpes simplex viruses, RNA viruses, and adenoviruses, have demonstrated high transfection efficiencies. However, their clinical utilization is restricted due to concerns regarding potential disease-causing effects, limited genetic payload capacity, systemic instability, inability to target specific cell populations, and high production costs. In contrast, non-viral delivery vectors offer several advantages, including a higher level of safety, controllable chemical properties, decreased toxicity, and the capability to deliver genes without size limitations. Nevertheless, non-viral methods have faced challenges in clinical settings, mainly attributed to limitations such as the inability to cross the nuclear membrane efficiently, lower transfection efficiencies, and inadequate transgene expression.

Various nanovectors, including peptides, liposomes, polymers, dendrimers, micro and nanoparticles, besides carbon nanotubes (CNTs), have been investigated as potential gene therapy vehicles. However, delivering nucleic acids therapeutically in vivo poses challenges, including instability against endogenous enzymes and poor cellular membrane penetration [47].

The optimal delivery of nucleic acids to their target tissues has gained significant attention, with a focus on utilizing nanocarriers. The objective is to transport foreign genetic material into the nucleus of specific cells, enabling the production of desired protein products. An optimal carrier should accurately deliver a predetermined quantity of genetic material to particular cell types, ensuring the correct intensity and duration of transgene expression to rectify the anomaly without inducing immune responses or adverse effects. This allows for the expression of the gene product without toxicity.

Carbon nanotubes (CNTs) have demonstrated promising capabilities in vector delivery because of their elongated shape and capacity to traverse plasma membranes. Additionally, their nanoneedle properties facilitate diffusion into the cytosol and provide protection against enzyme degradation, safeguarding the delivered gene. Taking advantage of these attributes, CNTs of varying radii have been encapsulated with polyallylamine, resulting in positively charged nanocarriers capable of conjugating with GFP plasmid. This approach exhibited potent delivery to 3T3 cells. Furthermore, Ohta and colleagues [76] conducted a study where they devised a nanocarrier system utilizing single-walled CNTs. These CNTs were made with polycationic and amphiphilic peptides that were modified using PEG [39].

Green carbon dots (CDs) show great potential as carriers for delivering biomolecules and foreign genes into host cells because of their desirable characteristics like less toxicity to cells and biodegradability. Additionally, the use of fluorescent CDs enables the tracking and visualization of the gene delivery pathway.

However, the utilization of green CDs as gene carriers remains largely unexplored. Liu et al. (46) employed graft copolymer CDs synthesized from lignin and DMAEA to deliver pRL-CMV plasmid in various cell types such as Cos -7, MDA -MB -231, and HeLa cells, showing efficiency dependent on DMAEA chain length and cell type. Das et al. (158) used CDs derived from chitosan, PEI-coated PEG, and amino-functionalized silica (AFS) to deliver dsRNA in *Aedes aegypti* larvae, achieving gene knockdown. Ethylenediamine-passivated CDs derived from Porphyra polysaccharides (PP) demonstrated efficient delivery of genes responsible for neuronal differentiation in ectodermal mesenchymal stem cells. The potential of green CDs in gene therapy remains underexplored, despite their non-toxic nature and the opportunity they present for curing diseases and facilitating tissue repair or regeneration [100].

RNA Interference

RNA interference (RNAi) refers to a mechanism of gene suppression facilitated by small interfering RNAs (siRNAs) and microRNAs (miRNAs), which act by binding to and degrading specific mRNA molecules, leading to the inhibition of gene expression. Effective delivery of siRNAs into cells and tissues has been a subject of extensive research in recent years. One major advantage of siRNA therapeutics is their ability to be designed for highly specific inhibition of target genes, without the need for complex protein purification or cellular expression systems. Clinical trials for siRNA-based drugs have shown promising results, although challenges such as activation of the innate immune system and potential off-target effects need to be addressed. Strategies such as chemical modifications of siRNA guide strands have been explored to minimize off-target effects. Another critical challenge in siRNA therapy is the efficient delivery of siRNAs into cells, as their large size and hydrophilic nature hinder passive diffusion across cell membranes [47].

3.2.2 Overcoming Antifungal Drug Resistance Using Gene Therapy Approaches

Fungi have the ability to adapt and develop resistance to fungicides and antifungals through various mechanisms, for instance, issues like insufficient target binding, heightened stress response, excessive efflux pump expression, and the development of biofilms. This resistance hinders the effectiveness of antifungal treatments. Examples include Botrytis cinerea, which exhibits rapid resistance development due to modifications in its target sites and efficient drug efflux transportation, and *Aspergillus fumigatus*, which exhibits intrinsic resistance to certain antifungals due to gene mutations. *Cryptococcus neoformans*, on the other hand, possesses intrinsic resistance through the calcineurin pathway. Resistance can also arise through hybridization between different fungal species, leading to increased fitness and drug resistance in the offspring. Cross-resistance, where a fungus becomes resistant to multiple antifungal agents, is another significant concern. In addition, the emergence of new fungal species, such as *Candida auris*, further complicates antifungal treatment strategies due to their high tolerance to existing drugs. These various mechanisms of antifungal resistance emphasize the need for integrated and innovative approaches to combat fungal diseases in both agricultural and clinical settings [114].

The emergence of antimicrobial drug resistance is a widespread and inevitable phenomenon in the microbial world. While antifungal drug resistance has not progressed at the same pace as antibacterial resistance, the economic impact of fungal infections remains unacceptably high. The limited availability of effective antifungal agents emphasizes the need to address antifungal resistance as a priority in improving treatment approaches for fungal infections. Inappropriate use of antifungal drugs is a major contributor to the development of resistance, as fungi are forced to adapt under selective pressure. Several studies have demonstrated the link between improper antifungal drug exposure and the development of resistance., highlighting the importance of employing appropriate and effective antifungal therapies to minimize the promotion of drug resistance. Additionally, the dosage and spectrum of action of antimicrobial agents can have significant consequences for the human microbiome, as an imbalance in microbial diversity can lead to detrimental effects and the development of tolerance in less-susceptible fungi. Taking these factors into consideration, it is crucial to explore potential strategies to overcome antifungal resistance [96].

The field of gene therapy has made significant progress, with more than 33 gene-therapy drugs/products receiving clinical approval from the FDA and an excess of 3500 ongoing clinical trials ongoing worldwide [92]. Nevertheless, the utilization of non-viral vectors for gene delivery poses several hurdles, encompassing complications tied to blood circulation, coating for phagocytosis, mononuclear phagocyte system internalization, tissue protection mechanisms, entrapment within endosomes, efficacy of gene transfer, selectivity, duration of gene expression, and overall safety. Carbon-based nanocarriers of nucleic acid face the risk of degradation or capture before reaching the targeted cell. Surface functionalization has emerged as a strategy to overcome these barriers. Additionally, the size of nanomaterials influences the internalization of cells, while the process itself presents an additional obstacle. Once nanovectors enter cells, they may degrade within the enzyme-rich and the low pH conditions within intracellular endosomes and lysosomes. To address these challenges, it is necessary to delve into the manufacturing process, surface functionalization, size, and cellular uptake of each carbon-based nanostructure in more detail [70].

4 Challenges and Future Perspectives

4.1 Safety Considerations and Potential Toxicity of Carbon Nanomaterials

Carbon nanomaterials, particularly carbon nanotubes (CNTs), possess distinctive characteristics that render them appealing for diverse applications [88]. However, along with their remarkable characteristics, concerns regarding their possible harmful effects on human health and the environment have arisen [86]. Extensive research has been conducted to assess the safety considerations and potential risks associated with these materials.

The toxicity of CNTs is shaped by multiple elements, including their size, functionalization, impurities, and solubilizing agents. These factors can alter the protein structures, increase antigenicity, and generate autoimmune effects. Oxidative stress, membrane injury, and genotoxicity are recognized as major mechanisms underlying CNT toxicity [88]. The presence of metal catalyst residues, which are used in CNT synthesis, and the generation of reactive oxygen species (ROS) is associated with the occurrence of oxidative damage and genotoxicity. Furthermore, post-fabrication treatments and the external attributes of CNTs, such as their shape, length, and aggregation, can induce immune responses and lead to tissue deposition [88].

Several adverse effects have been associated with CNT exposure. Studies have demonstrated that CNTs can cause DNA damage, elicit inflammatory responses, induce oxidative stress, and contribute to fibrosis and granuloma formation. They have also shown adverse effects on the cardiovascular and reproductive systems. Furthermore, there is evidence suggesting that CNTs have the potential to be carcinogenic and alter gene expression profiles. Despite these concerns, CNTs are still utilized in biomedical applications, and researchers are actively exploring various functionalization strategies to mitigate their harmful effects [88].

It is important to note that the toxicity of CNTs is not universal and can vary depending on their specific properties. Factors such as physical dimensions, dose, length-to-diameter ratio, purity, and surface chemistry can significantly influence the cytotoxicity of CNTs. Therefore, it is crucial to avoid generalizations and consider the unique characteristics of each type of CNT when assessing their toxicity profile [122].

In addition to the potential risks to human health, carbon nanomaterials (CNMs) also raise concerns regarding their impact on the environment. Studies have shown that CNMs, including CNTs, graphene, and fullerene, exhibit cytotoxicity, genotoxicity, and oxidative stress in various cellular systems [83]. Inhalation of CNTs can lead to lung damage, inflammation, and changes in lung function. Moreover, CNMs can negatively affect the development and survival of aquatic organisms and interfere with microbial processes. Understanding the toxic mechanisms of CNMs and implementing measures to limit their release into the environment are crucial for minimizing their risks [83].

Reactive oxygen species-induced multi-walled carbon nanotubes (ROS-MWCNTs) have been a subject of interest in research. These nanotubes have been found to induce cell apoptosis through the generation of ROS, disruption of lysosomal membrane integrity, permeability of the mitochondria's membrane, and activation of apoptotic pathways. The production of ROS precedes other cellular responses, and studies have demonstrated the correlation between the production of reactive oxygen species (ROS) and adverse cellular effects. Oxidative stress represents a prevalent mechanism underlying cytotoxic effects of carbon nanotubes. In response to MWCNT exposure, rapid ROS production occurs, leading to dose-dependent cytotoxicity associated with oxidative stress [121].

Overall, the safety considerations and potential toxicity of carbon nanomaterials, particularly carbon nanotubes, are important areas of research. Understanding the specific properties and characteristics of these materials is crucial for evaluating their potential risks and developing strategies to mitigate their harmful effects. Continued investigation and comprehensive assessments are necessary to ensure the ethical and accountable utilization of these nanomaterials in various applications.

Strategies to overcome the toxicity of carbon nanotubes (CNTs) involve modulating their physicochemical characteristics, particularly their surface charge. Altering the surface charge can reduce the toxic effects of CNTs when interacting with biological systems. Researchers have developed statistical models to establish the relationship between surface charge and toxicity using zebrafish models as an in vivo evaluation system. By closely monitoring zebrafish bioactivity after exposure to systematically modified CNTs, they identified surface charge as the most significant predictor of zebrafish mortality. This property-risk relationship provides a basis for designing novel CNTs with lower toxicity. Additionally, it is important to thoroughly explore the materials' structural characterization, uptake by cells, elimination from the body, and associated mechanisms. Further research is necessary to understand the toxicity mechanisms and develop comprehensive test systems for the risk assessment of newly developed CNTs [86].

4.2 Regulatory Aspects and Hurdles in Translating Carbon Nanomaterial-Based Therapies to the Clinic

The clinical translation of carbon nanomaterials (CNMs) in nanomedicine presents various obstacles that must be overcome for their successful implementation. One crucial aspect is the standardisation of efficacy, toxicity, safety, and formulation, which is currently lacking [53]. New clinical translational methods and protocols are necessary to assess the safety, effectiveness, and superiority of CNM-based nanodrugs compared to conventional medicines [53]. However, recruitment difficulties and administrative hurdles often hinder the progress of potential nanomedicines to advanced clinical phases [53].

To achieve clinical translation of carbon nanomaterials, in vivo studies play a critical role. These studies evaluate factors such as in vivo dynamics, molecular alterations, oxidative burden, and degradability of CNMs [53]. It is essential to optimize the synthesis methods to develop CNMs with desired physicochemical properties [53]. CNMs face challenges related to poor solubility, in vivo drug breakdown, and lack of tissue selectivity, which limit their success in clinical trials [53]. Nonetheless, CNMs have shown promising biomedical effects, including anticancer, antibacterial, and antiviral activities [53]; [90].

The use of toxic chemicals in CNM preparation poses risks to biological systems, emphasising the need for alternative and efficient synthesis techniques [53]. Commercially available CNMs often contain impurities that affect their properties, making risk assessment more complex [53]. The biological effects of CNMs are also influenced by surface functionalization, which can modify their solubility and interaction with cells [53]. Proper guidelines for safe handling should be established to address concerns regarding occupational exposure to CNMs, primarily through inhalation [53].

In order to successfully translate CNMs into clinical applications, it is crucial to address these challenges and optimize CNM properties and synthesis techniques. This optimization process will contribute to the safe implementation of CNMs in various biomedical applications [53]. Regulation and risk assessment of nanomaterial-based products are essential to ensure their safety [53]. Several organisations, including the FDA, EPA, NIOSH, OSHA, and CPSC, have assessed the environmental, health, and safety implications of nanomaterials and implemented regulations accordingly [53]. The FDA, in particular, has the authority to regulate a variety of products, including drugs, medical devices, and biologic products [53].

When considering the clinical translation of nanoparticles (NPs), certain factors must be carefully evaluated. The necessity of using NPs to address specific clinical problems should be assessed, as they may introduce additional complexity and potential toxicity [101]. Intellectual property constraints can also hinder the efficient clinical translation of NPs.

Carbon nanotubes and C60 buckyballs have attracted significant attention among different types of NPs. Carbon nanotubes, in particular, possess unique properties with exceptional length-to-diameter ratios, but they face challenges related to aggregation, non-specific binding to plasma proteins, and potential toxicity [53]. C60 buckyballs have shown promise in drug delivery and imaging applications, but their toxicity and long-term effects still need to be carefully investigated [53].

The clearance of NPs from the body is a critical consideration for their clinical translation. Small NPs can be excreted through urine, while larger NPs tend to accumulate in organs of the reticuloendothelial system (RES) [18]. Factors such as size, shape, surface charge, and surface modifications can affect the clearance of NPs from the blood and their uptake into the RES [18].

The clinical translation of NPs also requires regulatory guidelines to ensure their safety and efficacy. The FDA has established guidelines, including the requirement that metal-containing NPs should not be retained in the body for prolonged periods [18]. Ultrasmall quantum dots (QDs) that are rapidly cleared by renal filtration and eliminated in urine have been developed to meet this criterion [18]. Organic surface coatings have also been explored to minimise nonspecific uptake and retention of NPs after intravenous injection [18]. NPs that effectively address clinical problems, avoid prolonged retention in the body, and are eliminated through renal clearance are more likely to be clinically translated [18].

In conclusion, the clinical translation of carbon nanomaterials and nanoparticles faces several challenges and requires addressing issues such as standardisation, toxicity, formulation, and safety [53] [90], [101]; [18]. Optimising CNM properties and synthesis techniques, conducting in vivo studies, and developing appropriate guidelines for safe handling are crucial steps for successful clinical translation [53]. Regulatory agencies play a vital role in ensuring the safety and efficacy of nanomaterial-based products, and guidelines have been implemented to mitigate the potential risks associated with nanotechnology [53]. Understanding the biocompatibility, pharmacokinetics, and biophysical features of NPs is essential for their clinical translation [18]. The match between the biophysical features of NPs and the targeted disease under treatment is a key consideration [18]. By addressing these challenges and following regulatory guidelines, the clinical translation of nanomaterial-based therapies can be facilitated [53]; [18].

4.3 Long-Term Stability and Degradation Issues

Carbon nanomaterials can degrade through chemical and biodegradation processes. Chemical degradation involves photodegradation and reactions with reagents, while biodegradation includes enzymatic, cell, and bacterial degradation. Chemical degradation requires strong oxidising agents but may result in secondary pollution. Enzymatic degradation is widely studied and involves the use of specific enzymes to break down nanomaterials. Cells and bacteria can also degrade these materials. Graphitized carbon materials require environmentally friendly degradation methods. Microbial bacteria and enzymes have been shown to promote oxidation reactions. Surface modification, doping, and "design degradation" approaches enhance biodegradation. Gaining a comprehensive understanding of effective and environmentally-friendly degradation techniques for carbon nanomaterials holds significant significance due to their environmental impact [83].

[98] conducted a research study demonstrating that selective attachment of specific molecules onto multi-walled carbon nanotubes (MWCNTs) through covalent interactions, such as coumarin derivatives and catechols, accelerates their degradation compared to simply oxidized CNTs. The presence of hydroxyl groups in these functional molecules is pivotal in enhancing the catalytic action of horseradish peroxidase (HRP). Additionally, the researchers showed that the enzyme xanthine oxidase, when combined with hydrogen peroxide, can efficiently degrade purine-modified multi-walled carbon nanotubes (MWCNTs).This work emphasizes the importance of surface functionalization to modulate the enzymatic biodegradability of CNTs for potential biomedical applications [98].

4.4 Optimization of Drug Loading and Release Kinetics

In recent studies, researchers have explored the optimization of drug loading and release kinetics using carbon nanomaterials as carriers. The research conducted by Farahani et al. in [25] focused on MWNTs-PEG as a controlled release carrier for the drug doxorubicin (DOX). They discovered that the adsorption process was greatly influenced by the pH level, achieving a high adsorption efficiency of 96% under optimal conditions. A pseudo-second-order kinetic model was found to govern the adsorption mechanism, indicating chemisorption involvement. The adsorption isotherms conformed to the Freundlich model, suggesting a heterogeneous surface. DOX release from MWCNTs-PEG exhibited pH dependence, resulting in a more rapid release under acidic conditions.

Additional research by Ghaedi et al. [32] explored the sorption of Triamterene using MWCNT and SWCNT as eco-friendly adsorbents. Both adsorbents demonstrated high adsorption capacity, efficiently removing Triamterene within a short period. The experimental data exhibited the closest agreement with the Langmuir isotherm model. The researchers developed predictive models for Triamterene removal efficiency, with the artificial neural network (ANN) model outperforming the multiple linear regression (MLR) model.

Vasheghani Farahani and Javadi, in their 2016 study, prepared and characterized M-CS-MWNTs as an adsorbent for controlled release of the drug methotrexate (MTX). The pH was identified as the primary factor influencing the adsorption process. The experimental data exhibited conformity with the pseudo-second-order model for kinetics, while the adsorption isotherms were well characterized by the Freundlich model. The spontaneous and exothermic nature of MTX adsorption, along with the dominance of electrostatic interactions, was noted.

Chudoba et al., in a research conducted in 2020, modified the surface of multi-walled carbon nanotubes (MWCNTs) to improve their efficiency as drug carriers. Surface modification significantly augmented the adsorption capability of MWCNTs and molecular dynamics simulations indicated a poor tendency of DOX to adsorb on pristine hydrophobic CNTs. Kinetic studies on oxidized MWCNTs highlighted the influence of chemisorption and intra-particle diffusion in the adsorption process. MD simulations revealed DOX aggregation, occurring through adsorption on nanotube walls and self-aggregation of free DOX molecules.

These studies collectively contribute to the understanding of optimizing drug loading and release kinetics using carbon nanomaterials, with insights into factors affecting adsorption, kinetics models, and the role of surface modification [19, 25, 32, 105].

4.5 *Advancements in Personalized and Targeted Therapy Using Carbon Nanomaterials*

Advancements in personalized and targeted therapy using The utilization of carbon nanomaterials has demonstrated remarkable promise in enhancing the efficiency of drug delivery and therapeutic outcomes. Extensive research has been conducted on carbon nanotubes (CNTs) for their potential in targeted delivery of anticancer drugs, such as doxorubicin (DOX). Functionalization of CNTs with targeting agents, including folic acid and peptides, enhances their uptake into tumor cells and increases cytotoxicity. In vivo experiments have demonstrated the effectiveness of targeted DOX-CNT formulations in reducing tumor volume and improving survival rates [15].

Graphene and graphene oxide (GO) have also been explored as nanocarriers for targeted drug delivery. Conjugation of graphene with ligands like folic acid and transferring has shown improved tumor targeting and increased cytotoxicity. These nanocarriers have also demonstrated potential in gene delivery and responsive drug release systems, allowing for on-demand drug release triggered by internal or external stimuli [58].

Graphene-based nanomaterials (GFNs) have remarkable properties that enable their accumulation in specific diseased regions, improving the effectiveness of therapeutic agents. Active targeting strategies involving the modification of GFNs with specific ligands have shown promise in site-specific drug delivery. Receptors like epidermal growth factor receptor (EGFR), CD44, and folate receptor (Fr) have been targeted using antibodies, peptides, or natural molecules, enhancing cellular uptake and tumor targeting. Mitochondria-targeting nanocarriers and Fr-mediated targeted therapy have also been explored, showing potential in cancer treatment and inflammation therapy [44].

CNTs offer advantages in targeted drug delivery, which includes their notable drug loading capacity and efficient cellular uptake. Functionalization techniques enable customization for biomedical applications, and surface coating allows for site-specific drug recognition. CNTs have shown successful drug delivery, including pH-responsive systems and targeted therapy approaches. They have demonstrated potential in brain targeted drug delivery, gene therapy, and sustainable release systems [86].

Another carbon nanomaterial, carbon dots (CDs), has shown promise in targeted drug delivery and imaging. Conjugation of CDs with ligands like folic acid and RGD peptides enables specific drug targeting. Amino acid-functionalized CDs and folate-functionalized CDs have shown potential in protein sensing and cancer cell targeting. Triple-conjugated CDs incorporating transferrin and anti-cancer drugs have shown enhanced cytotoxicity against brain tumor cells. These findings highlight the potential of CDs in targeted drug delivery and therapeutic applications [6].

In summary, carbon nanomaterials, including CNTs, graphene, GFNs, and CDs, offer significant advancements in personalized and targeted therapy. They enable enhanced drug delivery to specific sites, improved therapeutic outcomes, and potential applications in gene therapy. These nanomaterials offer a bright outlook for future advancements of targeted drug delivery and precision medicine.

References

1. Abu Owida H, Turab NM, Al-Nabulsi J (2023) Carbon nanomaterials advancements for biomedical applications. Bullet Electr Eng Inf 12(2):891–901. https://doi.org/10.11591/eei.v12i2.4310
2. Ahmad V, Ansari MO (2022) Antimicrobial activity of graphene-based nanocomposites: synthesis, characterization, and their applications for human welfare. Nanomaterials (Basel, Switzerland) 12(22):4002. https://doi.org/10.3390/nano12224002
3. Alaghmandfard A, Sedighi O, Tabatabaei Rezaei N, Abedini AA, Malek Khachatourian A, Toprak MS, Seifalian A (2021) Recent advances in the modification of carbon-based quantum dots for biomedical applications. Mater Sci Eng C Mater Biol Appl 120(111756):111756. https://doi.org/10.1016/j.msec.2020.111756
4. Aliamradni V, Sadat Abolmaali S, Borandeh S (2019) Antifungal and antibacterial properties of graphene-based nanomaterials: a mini-review. J Nanostruct 9(3):402–413

5. Almehmady AM, El-Say KM, Mubarak MA, Alghamdi HA, Somali NA, Sirwi A, Algarni R, Ahmed TA (2022) Enhancing the antifungal activity and ophthalmic transport of fluconazole from PEGylated polycaprolactone loaded nanoparticles. Polymers 15(1). https://doi.org/10.3390/polym15010209
6. Amer Ridha A, Pakravan P, Hemati Azandaryani A, Zhaleh H (2020) Carbon dots; the smallest photoresponsive structure of carbon in advanced drug targeting. J Drug Deliv Sci Technol 55(101408):101408. https://doi.org/10.1016/j.jddst.2019.101408
7. Anzar N, Hasan R, Tyagi M, Yadav N, Narang J (2020) Carbon nanotube-a review on synthesis, properties and plethora of applications in the field of biomedical science. Sens Int 1(100003):100003. https://doi.org/10.1016/j.sintl.2020.100003
8. Ashrafizadeh M, Mohammadinejad R, Kailasa SK, Ahmadi Z, Afshar EG, Pardakhty A (2020) Carbon dots as versatile nanoarchitectures for the treatment of neurological disorders and their theranostic applications: a review. Adv Coll Interface Sci 278(102123):102123. https://doi.org/10.1016/j.cis.2020.102123
9. Bajpai VK, Khan I, Shukla S, Kumar P, Rather IA, Park Y-H, Huh YS, Han Y-K (2019) Invasive fungal infections and their epidemiology: measures in the clinical scenario. Biotechnol Bioprocess Eng BBE 24(3):436–444. https://doi.org/10.1007/s12257-018-0477-0
10. Behnam B, Shier WT, Nia AH, Abnous K, Ramezani M (2013) Non-covalent functionalization of single-walled carbon nanotubes with modified polyethyleneimines for efficient gene delivery. Int J Pharm 454(1):204–215. https://doi.org/10.1016/j.ijpharm.2013.06.057
11. Belay T, Worku LA, Bachheti RK, Bachheti A, Husen A (2023) Nanomaterials: introduction, synthesis, characterization, and applications. In: Husen A, Siddiqi KS (eds) Advances in smart nanomaterials and their applications. Elsevier, Amsterdam, Netherlands, pp 1–21
12. Biswal MR, Bhatia S (2021) Carbon dot nanoparticles: exploring the potential use for gene delivery in ophthalmic diseases. Nanomaterials (Basel, Switzerland) 11(4):935. https://doi.org/10.3390/nano11040935
13. Bongomin F, Gago S, Oladele R, Denning D (2017) Global and multi-national prevalence of fungal diseases—estimate precision. J Fungi (Basel, Switzerland) 3(4):57. https://doi.org/10.3390/jof3040057
14. Brown GD, Denning DW, Gow NAR, Levitz SM, Netea MG, White TC (2012) Hidden killers: human fungal infections. Sci Transl Med 4(165):165rv13–165rv13. https://doi.org/10.1126/scitranslmed.3004404
15. Chadar R, Afzal O, Alqahtani SM, Kesharwani P (2021. Carbon nanotubes as an emerging nanocarrier for the delivery of doxorubicin for improved chemotherapy. Coll Surf B Biointerfaces 208(112044):112044. https://doi.org/10.1016/j.colsurfb.2021.112044
16. Chandrasekaran R, Krishnan M, Bupesh G, Chacko S, Gawade O, Hasan S, George E, Vijayakumar TS, Sundaram M, Sagadevan S (2023) Prospective features of functional 2D nanomaterial graphene oxide in the wound healing process. J Drug Deliv Sci Technol 82(104352):104352. https://doi.org/10.1016/j.jddst.2023.104352
17. Chen B, Liu M, Zhang L, Huang J, Yao J, Zhang Z (2011) Polyethylenimine-functionalized graphene oxide as an efficient gene delivery vector. J Mater Chem 21(21):7736. https://doi.org/10.1039/c1jm10341e
18. Choi HS, Frangioni JV (2010) Nanoparticles for biomedical imaging: fundamentals of clinical translation. Mol Imaging 9(6):291–310. https://doi.org/10.2310/7290.2010.00031
19. Chudoba D, Łudzik K, Jażdżewska M, Wołoszczuk S (2020) Kinetic and equilibrium studies of doxorubicin adsorption onto carbon nanotubes. Int J Mol Sci 21(21):8230. https://doi.org/10.3390/ijms21218230
20. Das SS, Bharadwaj P, Bilal M, Barani M, Rahdar A, Taboada P, Bungau S, Kyzas GZ (2020) Stimuli-responsive polymeric nanocarriers for drug delivery, imaging, and theragnosis. Polymers 12(6):1397. https://doi.org/10.3390/polym12061397
21. Dellinger A, Zhou Z, Connor J, Madhankumar AB, Pamujula S, Sayes CM, Kepley CL (2013) Application of fullerenes in nanomedicine: an update. Nanomedicine (Lond) 8(7):1191–1208. https://doi.org/10.2217/nnm.13.99

22. Denning DW, Bromley MJ (2015) Infectious disease. How to bolster the antifungal pipeline. Science (New York, N.Y.) 347(6229):1414–1416. https://doi.org/10.1126/science.aaa6097
23. Di Santo R, Digiacomo L, Palchetti S, Palmieri V, Perini G, Pozzi D, Papi M, Caracciolo G (2019) Microfluidic manufacturing of surface-functionalized graphene oxide nanoflakes for gene delivery. Nanoscale 11(6):2733–2741. https://doi.org/10.1039/c8nr09245a
24. Djordjević A, Bogdanović G, Dobrić S (2006) Fullerenes in biomedicine. J B.U.ON. Off J Balk Union Oncol 11(4):391–404
25. Farahani BV, Behbahani GR, Javadi N (2015) Functionalized multi walled carbon nanotubes as a carrier for doxorubicin: drug adsorption study and statistical optimization of drug loading by factorial design methodology. J Braz Chem Soc. https://doi.org/10.5935/0103-5053.20150318
26. Farjadian F, Abbaspour S, Sadatlu MAA, Mirkiani S, Ghasemi A, Hoseini-Ghahfarokhi M, Mozaffari N, Karimi M, Hamblin MR (2020) Recent developments in graphene and graphene oxide: properties, synthesis, and modifications: a review. Chem Sel 5(33):10200–10219. https://doi.org/10.1002/slct.202002501
27. Fisher MC, Gow NAR, Gurr SJ (2016) Tackling emerging fungal threats to animal health, food security and ecosystem resilience. Philos Trans Royal Soc Lond Ser B Biol Sci 371(1709). https://doi.org/10.1098/rstb.2016.0332
28. Fisher MC, Henk DA, Briggs CJ, Brownstein JS, Madoff LC, McCraw SL, Gurr SJ (2012) Emerging fungal threats to animal, plant and ecosystem health. Nature 484(7393):186–194. https://doi.org/10.1038/nature10947
29. Fukushima C, Matsuse H, Tomari S, Obase Y, Miyazaki Y, Shimoda T, Kohno S (2003) Oral candidiasis associated with inhaled corticosteroid use: comparison of fluticasone and beclomethasone. Annal Allergy Asthma Immunol Off Publ Am Coll Allergy Asthma Immunol 90(6):646–651. https://doi.org/10.1016/S1081-1206(10)61870-4
30. Garvey M, Rowan NJ (2023) Pathogenic drug resistant fungi: a review of mitigation strategies. Int J Mol Sci 24(2). https://doi.org/10.3390/ijms24021584
31. Gaur M, Misra C, Yadav AB, Swaroop S, Maolmhuaidh FÓ, Bechelany M, Barhoum A (2021). Biomedical applications of carbon nanomaterials: fullerenes, quantum dots, nanotubes, nanofibers, and graphene. Materials 14(20). https://doi.org/10.3390/ma14205978
32. Ghaedi AM, Ghaedi M, Pouranfard AR, Ansari A, Avazzadeh Z, Vafaei A, Tyagi I, Agarwal S, Gupta VK (2016) Adsorption of Triamterene on multi-walled and single-walled carbon nanotubes: artificial neural network modeling and genetic algorithm optimization. J Mol Liq 216:654–665. https://doi.org/10.1016/j.molliq.2016.01.068
33. Ghosal K, Ghosh A (2019) Carbon dots: the next generation platform for biomedical applications. Mater Sci Eng C Mater Biol Appl 96:887–903. https://doi.org/10.1016/j.msec.2018.11.060
34. Ghosh S, Thongmee S, Mostafavi E (2023) Chemically modified carbon nanotubes in drug delivery. In: Chemically modified carbon nanotubes for commercial applications (pp 331–345). Wiley. https://doi.org/10.1002/9783527838790.ch14
35. Gisi U (2022) Crossover between the control of fungal pathogens in medicine and the wider environment, and the threat of antifungal resistance. Plant Pathol 71(1):131–149. https://doi.org/10.1111/ppa.13429
36. Grant JJ, Pillai SC, Hehir S, McAfee M, Breen A (2021) Biomedical applications of electrospun graphene oxide. ACS Biomater Sci Eng 7(4):1278–1301. https://doi.org/10.1021/acsbiomaterials.0c01663
37. Gu Y, Guo Y, Wang C, Xu J, Wu J, Kirk TB, Ma D, Xue W (2017) A polyamidoamne dendrimer functionalized graphene oxide for DOX and MMP-9 shRNA plasmid co-delivery. Mater Sci Eng C Mater Biol Appl 70(Pt 1):572–585. https://doi.org/10.1016/j.msec.2016.09.035
38. Gungordu Er S, Edirisinghe M, Tabish TA (2023) Graphene-based nanocomposites as antibacterial, antiviral and antifungal agents (adv. Healthcare mater. 6/2023). Adv Healthc Mater 12(6):2370026. https://doi.org/10.1002/adhm.202370026
39. Hamimed S, Jabberi M, Chatti A (2022) Nanotechnology in drug and gene delivery. Naunyn-Schmiedeberg's Arch Pharmacol 395(7):769–787. https://doi.org/10.1007/s00210-022-02245-z

40. Hasanzadeh A, Radmanesh F, Kiani J, Bayandori M, Fatahi Y, Aref AR, Karimi M (2019) Photoluminescent functionalized carbon dots for CRISPR delivery: synthesis, optimization and cellular investigation. Nanotechnology 30(13):135101. https://doi.org/10.1088/1361-6528/aafbf9
41. Holmannova D, Borsky P, Svadlakova T, Borska L, Fiala Z (2022) Carbon nanoparticles and their biomedical applications. Appl Sci (Basel, Switzerland) 12(15):7865. https://doi.org/10.3390/app12157865
42. Hoseini-Ghahfarokhi M, Mirkiani S, Mozaffari N, Abdolahi Sadatlu MA, Ghasemi A, Abbaspour S, Akbarian M, Farjadian F, Karimi M (2020) Applications of graphene and graphene oxide in smart drug/gene delivery: is the world still flat? Int J Nanomed 15:9469–9496. https://doi.org/10.2147/IJN.S265876
43. Jaleel JA, Pramod K (2018) Artful and multifaceted applications of carbon dot in biomedicine. J Control Release Off J Control Release Soc 269:302–321. https://doi.org/10.1016/j.jconrel.2017.11.027
44. Jiang C, Zhao H, Xiao H, Wang Y, Liu L, Chen H, Shen C, Zhu H, Liu Q (2021) Recent advances in graphene-family nanomaterials for effective drug delivery and phototherapy. Expert Opin Drug Deliv 18(1):119–138. https://doi.org/10.1080/17425247.2020.1798400
45. Kailasa SK, Bhamore JR, Koduru JR, Park TJ (2019) Carbon dots as carriers for the development of controlled drug and gene delivery systems. In Biomedical applications of nanoparticles. Elsevier, pp 295–317
46. Kainz K, Bauer MA, Madeo F, Carmona-Gutierrez D (2020) Fungal infections in humans: the silent crisis. Microbial Cell (Graz, Austria) 7(6):143–145. https://doi.org/10.15698/mic2020.06.718
47. Karimi M, Solati N, Ghasemi A, Estiar MA, Hashemkhani M, Kiani P, Mohamed E, Saeidi A, Taheri M, Avci P, Aref AR, Amiri M, Baniasadi F, Hamblin MR (2015) Carbon nanotubes part II: a remarkable carrier for drug and gene delivery. Expert Opin Drug Deliv 12(7):1089–1105. https://doi.org/10.1517/17425247.2015.1004309
48. Karki N, Tiwari H, Tewari C, Rana A, Pandey N, Basak S, Sahoo NG (2020) Functionalized graphene oxide as a vehicle for targeted drug delivery and bioimaging applications. J Mater Chem B Mater Biol Med 8(36):8116–8148. https://doi.org/10.1039/d0tb01149e
49. Kaur J, Gill GS, Jeet K (2019) Applications of carbon nanotubes in drug delivery. In Characterization and biology of nanomaterials for drug delivery. Elsevier, pp 113–135
50. Kelly BP (2012) Superficial fungal infections. Pediatr Rev 33(4):e22-37. https://doi.org/10.1542/pir.33-4-e22
51. Kim H, Namgung R, Singha K, Oh I-K, Kim WJ (2011) Graphene oxide-polyethylenimine nanoconstruct as a gene delivery vector and bioimaging tool. Bioconjug Chem 22(12):2558–2567. https://doi.org/10.1021/bc200397j
52. Kumar JR, Muralidharan S, Parasuraman S (2014) Antifungal agents: new approach for novel delivery systems. J Pharm Sci Res 6(5):229
53. Lalwani G, Sitharaman B (2013) Multifunctional fullerene- and metallofullerene-based nanobiomaterials. Nano LIFE 03(03):1342003. https://doi.org/10.1142/s1793984413420038
54. Lee Y, Puumala E, Robbins N, Cowen LE (2021) Antifungal drug resistance: molecular mechanisms in Candida albicans and beyond. Chem Rev 121(6):3390–3411. https://doi.org/10.1021/acs.chemrev.0c00199
55. Li X, Huang R, Tang FK, Li WC, Wong SSW, Leung KC, Jin L (2019) Red-emissive guanylated polyene-functionalized carbon dots arm oral epithelia against invasive fungal infections. ACS Appl Mater Interfaces 18;11(50):46591–46603. https://doi.org/10.1021/acsami.9b18003
56. Lionakis MS, Drummond RA, Hohl TM (2023) Immune responses to human fungal pathogens and therapeutic prospects. Nat Rev Immunol. https://doi.org/10.1038/s41577-022-00826-w
57. Liu C, Zhang P, Zhai X, Tian F, Li W, Yang J, Liu Y, Wang H, Wang W, Liu W (2012) Nano-carrier for gene delivery and bioimaging based on carbon dots with PEI-passivation enhanced fluorescence. Biomaterials 33(13):3604–3613. https://doi.org/10.1016/j.biomaterials.2012.01.052

58. Liu J, Cui L, Losic D (2013) Graphene and graphene oxide as new nanocarriers for drug delivery applications. Acta Biomater 9(12):9243–9257. https://doi.org/10.1016/j.actbio.2013.08.016
59. Liu J, Li R, Yang B (2020) Carbon dots: a new type of carbon-based nanomaterial with wide applications. ACS Cent Sci 6(12):2179–2195. https://doi.org/10.1021/acscentsci.0c01306
60. Liu P (2013) Modification strategies for carbon nanotubes as a drug delivery system. Ind Eng Chem Res 52(38):13517–13527. https://doi.org/10.1021/ie402360f
61. Lu S-Y (2021) Oral candidosis: pathophysiology and best practice for diagnosis, classification, and successful management. J Fungi (Basel, Switzerland) 7(7):555. https://doi.org/10.3390/jof7070555
62. Manikandan S, Vinesh E, Selvi DT, Kannan RK, Jayakumar A, Dinakaran J (2022). Prevalence of Candida among denture wearers and nondenture wearers. J Pharm Bioallied Sci 14(Suppl 1):S702–S705https://doi.org/10.4103/jpbs.jpbs_781_21
63. Mansuriya BD, Altintas Z (2021) Carbon dots: classification, properties, synthesis, characterization, and applications in health care-an updated review (2018–2021). Nanomaterials (Basel, Switzerland) 11(10):2525. https://doi.org/10.3390/nano11102525
64. Marangon M, De Souza Carvalho-Wodarz C, Duarte MC, Lehr CM (2018) Nanotechnology-based drug delivery systems for treatment of fungal diseases: a review. J Pharm Sci 107(9):2337–2345
65. Mathur M, Devi VK (2017) Potential of novel drug delivery systems in the management of topical candidiasis. J Drug Target 25(8):685–703. https://doi.org/10.1080/1061186X.2017.1331352
66. Merum S, Veluru JB, Seeram R (2017) Functionalized carbon nanotubes in bio-world: applications, limitations and future directions. Mater Sci Eng B Solid-State Mater Adv Technol 223:43–63. https://doi.org/10.1016/j.mseb.2017.06.002
67. Mohajeri M, Behnam B, Sahebkar A (2018) Biomedical applications of carbon nanomaterials: drug and gene delivery potentials: MOHAJERI et al. J Cell Physiol 234(1):298–319.https://doi.org/10.1002/jcp.26899
68. Mohammadinejad R, Dadashzadeh A, Moghassemi S, Ashrafizadeh M, Dehshahri A, Pardakhty A, Sassan H, Sohrevardi S-M, Mandegary A (2019) Shedding light on gene therapy: carbon dots for the minimally invasive image-guided delivery of plasmids and noncoding RNAs-a review. J Adv Res 18:81–93. https://doi.org/10.1016/j.jare.2019.01.004
69. Mohammed J, Desu PK, Namratha JR, Rao GK (2023) Applications of carbon dots (CDs) in drug delivery. Adv Pharmacol Pharm 11(1):36–45. https://doi.org/10.13189/app.2023.110104
70. Mostafavi E, Zare H (2022) Carbon-based nanomaterials in gene therapy. OpenNano 7(100062):100062. https://doi.org/10.1016/j.onano.2022.100062
71. Mousavi SM, Hashemi SA, Ghasemi Y, Amani AM, Babapoor A, Arjmand O (2019) Applications of graphene oxide in case of nanomedicines and nanocarriers for biomolecules: review study. Drug Metab Rev 51(1):12–41. https://doi.org/10.1080/03602532.2018.1522328
72. Nami S, Aghebati-Maleki A, Aghebati-Maleki L (2021) Current applications and prospects of nanoparticles for antifungal drug delivery. EXCLI J 20:562–584. https://doi.org/10.17179/excli2020-3068
73. Nel A, Xia T, Mädler L, Li N (2006) Toxic potential of materials at the nanolevel. Science (New York, N.Y.), 311(5761):622–627. https://doi.org/10.1126/science.1114397
74. Nurunnabi M, Parvez K, Nafiujjaman M, Revuri V, Khan HA, Feng X, Lee Y-K (2015) Bioapplication of graphene oxide derivatives: drug/gene delivery, imaging, polymeric modification, toxicology, therapeutics and challenges. RSC Adv 5(52):42141–42161. https://doi.org/10.1039/c5ra04756k
75. Official American A (n.d.) Thoracic society statement: treatment of fungal infections in adult pulmonary and critical care patients
76. Ohta T, Hashida Y, Yamashita F, Hashida M (2016) Development of novel drug and gene delivery carriers composed of single-walled carbon nanotubes and designed peptides with PEGylation. J Pharm Sci 105(9):2815–2824. https://doi.org/10.1016/j.xphs.2016.03.031

77. Onugwu AL, Nwagwu CS, Onugwu OS, Echezona AC, Agbo CP, Ihim SA, Emeh P, Nnamani PO, Attama AA, Khutoryanskiy VV (2023) Nanotechnology based drug delivery systems for the treatment of anterior segment eye diseases. J Control Release Off J Control Release Soc 354:465–488. https://doi.org/10.1016/j.jconrel.2023.01.018
78. Pandey J, Dubey R, Kate A, Prasad B, Sinha A, Mishra MS (2022) Nanomedicines: a focus on nanomaterials as drug delivery system with current trends and future advancement. Drug Res 72(7):355–366. https://doi.org/10.1055/a-1824-4619
79. Parhama S, Parham S-S, Nur H (2023) Chapter 11 carbon dots in antibacterial, antiviral, antifungal, and antiparasitic agents. In Carbon dots in biology. De Gruyter, pp 241–258
80. Park BJ, Wannemuehler KA, Marston BJ, Govender N, Pappas PG, Chiller TM (2009) Estimation of the current global burden of cryptococcal meningitis among persons living with HIV/AIDS. AIDS (London, England) 23(4):525–530. https://doi.org/10.1097/QAD.0b013e328322ffac
81. Pathak R, Punetha VD, Bhatt S, Punetha M (2023) Multifunctional role of carbon dot-based polymer nanocomposites in biomedical applications: a review. J Mater Sci 58(15):6419–6443. https://doi.org/10.1007/s10853-023-08408-4
82. Patil R, Bhosale R (2016) Nanostructured materials for antifungal applications: possibilities and challenges. In Nano-antimicrobials. Elsevier, pp 357–383
83. Peng Z, Liu X, Zhang W, Zeng Z, Liu Z, Zhang C, Liu Y, Shao B, Liang Q, Tang W, Yuan X (2020) Advances in the application, toxicity and degradation of carbon nanomaterials in environment: a review. Environ Int 134(105298):105298. https://doi.org/10.1016/j.envint.2019.105298
84. Prajapati SK, Malaiya A, Kesharwani P, Soni D, Jain A (2022) Biomedical applications and toxicities of carbon nanotubes. Drug Chem Toxicol 45(1):435–450. https://doi.org/10.1080/01480545.2019.1709492
85. Qindeel M, Barani M, Rahdar A, Arshad R, Cucchiarini M (2021) Nanomaterials for the diagnosis and treatment of urinary tract infections. Nanomaterials (Basel, Switzerland) 11(2):546. https://doi.org/10.3390/nano11020546
86. Rahamathulla M, Bhosale RR, Osmani RAM, Mahima KC, Johnson AP, Hani U, Ghazwani M, Begum MY, Alshehri S, Ghoneim MM, Shakeel F, Gangadharappa HV (2021) Carbon nanotubes: current perspectives on diverse applications in targeted drug delivery and therapies. Materials 14(21):6707. https://doi.org/10.3390/ma14216707
87. Rajakumar G, Zhang X-H, Gomathi T, Wang S-F, Azam Ansari M, Mydhili G, Nirmala G, Alzohairy MA, Chung I-M (2020) Current use of carbon-based materials for biomedical applications—a prospective and review. Processes (Basel, Switzerland) 8(3):355. https://doi.org/10.3390/pr8030355
88. Raphey VR, Henna TK, Nivitha KP, Mufeedha P, Sabu C, Pramod K (2019) Advanced biomedical applications of carbon nanotube. Mater Sci Eng C Mater Biol Appl 100:616–630. https://doi.org/10.1016/j.msec.2019.03.043
89. Rosen Y, Elman NM (2009) Carbon nanotubes in drug delivery: focus on infectious diseases. Expert Opin Drug Deliv 6(5):517–530. https://doi.org/10.1517/17425240902865579
90. Saleem J, Wang L, Chen C (2018) Carbon-based nanomaterials for cancer therapy via targeting tumor microenvironment. Adv Healthcare Mater 7(20):e1800525. https://doi.org/10.1002/adhm.201800525
91. Scheinberg DA, Grimm J, Heller DA, Stater EP, Bradbury M, McDevitt MR (2017) Advances in the clinical translation of nanotechnology. Curr Opin Biotechnol 46:66–73. https://doi.org/10.1016/j.copbio.2017.01.002
92. Shahryari A, Burtscher I, Nazari Z, Lickert H (2021) Engineering gene therapy: advances and barriers (adv. Therap. 9/2021). Adv Ther 4(9):2170023. https://doi.org/10.1002/adtp.202170023
93. Sharma S, Javed MN, Pottoo FH, Rabbani SA, Barkat MA, Harshita Sarafroz M, Amir M (2019) Bioresponse inspired nanomaterials for targeted drug and gene delivery. Pharm Nanotechnol 7(3):220–233.https://doi.org/10.2174/2211738507666190429103814

94. Simon J, Flahaut E, Golzio M (2019) Overview of carbon nanotubes for biomedical applications. Materials 12(4):624. https://doi.org/10.3390/ma12040624
95. Singh A, Hua Hsu M, Gupta N, Khanra P, Kumar P, Prakash Verma V, Kapoor M (2020) Derivatized carbon nanotubes for gene therapy in mammalian and plant cells. ChemPlusChem 85(3):466–475. https://doi.org/10.1002/cplu.201900678
96. Srinivasan A, Lopez-Ribot JL, Ramasubramanian AK (2014) Overcoming antifungal resistance. Drug Discov Today Technol 11:65–71. https://doi.org/10.1016/j.ddtec.2014.02.005
97. Sun S, Hoy MJ, Heitman J (2020) Fungal pathogens. Curr Biol CB 30(19):R1163–R1169. https://doi.org/10.1016/j.cub.2020.07.032
98. Sureshbabu AR, Kurapati R, Russier J, Ménard-Moyon C, Bartolini I, Meneghetti M, Kostarelos K, Bianco A (2015) Degradation-by-design: surface modification with functional substrates that enhance the enzymatic degradation of carbon nanotubes. Biomaterials 72:20–28. https://doi.org/10.1016/j.biomaterials.2015.08.046
99. Taghavi S, Abnous K, Taghdisi SM, Ramezani M, Alibolandi M (2020) Hybrid carbon-based materials for gene delivery in cancer therapy. J Control Release Off J Control Release Soc 318:158–175. https://doi.org/10.1016/j.jconrel.2019.12.030
100. Tejwan N, Saha SK, Das J (2020) Multifaceted applications of green carbon dots synthesized from renewable sources. Adv Coll Interface Sci 275(102046):102046. https://doi.org/10.1016/j.cis.2019.102046
101. Thakral S, Thakral NK (2013) Potential medical applications of fullerenes: an overview. In Bio-nanotechnology. Blackwell Publishing Ltd, pp 424–441
102. Thakur CK, Karthikeyan C, Abou-Dahech MS, Altabakha MMAM, Al Shahwan MJS, Ashby CR Jr, Tiwari AK, Babu RJ, Moorthy NSHN (2023) Microwave-assisted functionalization of multi-walled carbon nanotubes for biosensor and drug delivery applications. Pharmaceutics 15(2). https://doi.org/10.3390/pharmaceutics15020335
103. Tragiannidis A, Gkampeta A, Vousvouki M, Vasileiou E, Groll AH (2021) Antifungal agents and the kidney: pharmacokinetics, clinical nephrotoxicity, and interactions. Expert Opin Drug Saf 20(9):1061–1074. https://doi.org/10.1080/14740338.2021.1922667
104. Tuerhong M, Xu Y, Yin X-B (2017) Review on carbon dots and their applications. Fenxi Huaxue 45(1):139–150. https://doi.org/10.1016/s1872-2040(16)60990-8
105. Vasheghani Farahani B, Javadi N (2016) Kinetic and thermodynamic studies of methotrexate adsorption on chitosan-modified magnetic multi-walled carbon nanotubes. Monatshefte Fur Chemie 147(12):2051–2060. https://doi.org/10.1007/s00706-016-1753-3
106. Verner RF, Benvegnu C (eds) (2012) Handbook on fullerene: synthesis, properties & applications. Nova Science
107. Vitiello A, Ferrara F, Boccellino M, Ponzo A, Cimmino C, Comberiati E, Zovi A, Clemente S, Sabbatucci M (2023) Antifungal drug resistance: an emergent health threat. Biomedicines 11(4):1063. https://doi.org/10.3390/biomedicines11041063
108. Waghule T, Sankar S, Rapalli VK, Gorantla S, Dubey SK, Chellappan DK, Dua K, Singhvi G (2020) Emerging role of nanocarriers based topical delivery of anti-fungal agents in combating growing fungal infections. Dermatol Ther 33(6):e13905. https://doi.org/10.1111/dth.13905
109. Wang H, Boghossian AA (2023) Covalent conjugation of proteins onto fluorescent single-walled carbon nanotubes for biological and medical applications. Mater Adv 4(3):823–834. https://doi.org/10.1039/d2ma00714b
110. Wang X, Zhou Z, Chen F (2017) Surface modification of carbon nanotubes with an enhanced antifungal activity for the control of plant fungal pathogen. Materials 10(12). https://doi.org/10.3390/ma10121375
111. White TC, Findley K, Dawson TL Jr, Scheynius A, Boekhout T, Cuomo CA, Xu J, Saunders CW (2014) Fungi on the skin: dermatophytes and Malassezia. Cold Spring Harb Perspect Med 4(8):a019802–a019802. https://doi.org/10.1101/cshperspect.a019802
112. Willems HME, Ahmed SS, Liu J, Xu Z, Peters BM (2020) Vulvovaginal candidiasis: a current understanding and burning questions. J Fungi (Basel, Switzerland) 6(1):27. https://doi.org/10.3390/jof6010027

113. Wong BS, Yoong SL, Jagusiak A, Panczyk T, Ho HK, Ang WH, Pastorin G (2013) Carbon nanotubes for delivery of small molecule drugs. Adv Drug Deliv Rev 65(15):1964–2015. https://doi.org/10.1016/j.addr.2013.08.005
114. Woods M, McAlister JA, Geddes-McAlister J (2023) A One Health approach to overcoming fungal disease and antifungal resistance. WIREs Mechanisms of Disease. https://doi.org/10.1002/wsbm.1610
115. Wu W, Wieckowski S, Pastorin G, Benincasa M, Klumpp C, Briand J-P, Gennaro R, Prato M, Bianco A (2005) Targeted delivery of amphotericin B to cells by using functionalized carbon nanotubes. Angewandte Chemie (International Ed. in English) 44(39):6358–6362. https://doi.org/10.1002/anie.200501613
116. Wu Y-F, Wu H-C, Kuan C-H, Lin C-J, Wang L-W, Chang C-W, Wang T-W (2016) Multi-functionalized carbon dots as theranostic nanoagent for gene delivery in lung cancer therapy. Sci Rep 6(1):21170. https://doi.org/10.1038/srep21170
117. Xu X, Ray R, Gu Y, Ploehn HJ, Gearheart L, Raker K, Scrivens WA (2004) Electrophoretic analysis and purification of fluorescent single-walled carbon nanotube fragments. J Am Chem Soc 126(40):12736–12737. https://doi.org/10.1021/ja040082h
118. Yazdani S, Mozaffarian M, Pazuki G, Hadidi N, Gallego I, Puras G, Pedraz JL (2022) Design of double functionalized carbon nanotube for amphotericin B and genetic material delivery. In Research square. https://doi.org/10.21203/rs.3.rs-1958531/v1
119. Youn YS, Kwag DS, Lee ES (2017) Multifunctional nano-sized fullerenes for advanced tumor therapy. J Pharm Investig 47(1):1–10. https://doi.org/10.1007/s40005-016-0282-8
120. Yousefi M, Dadashpour M, Hejazi M, Hasanzadeh M, Behnam B, de la Guardia M, Shadjou N, Mokhtarzadeh A (2017) Anti-bacterial activity of graphene oxide as a new weapon nanomaterial to combat multidrug-resistance bacteria. Mater Sci Eng C Mater Biol Appl 74:568–581. https://doi.org/10.1016/j.msec.2016.12.125
121. Yuan X, Zhang X, Sun L, Wei Y, Wei X (2019) Cellular toxicity and immunological effects of carbon-based nanomaterials. Part Fibre Toxicol 16(1):18. https://doi.org/10.1186/s12989-019-0299-z
122. Zare H, Ahmadi S, Ghasemi A, Ghanbari M, Rabiee N, Bagherzadeh M, Karimi M, Webster TJ, Hamblin MR, Mostafavi E (2021) Carbon nanotubes: smart drug/gene delivery carriers. Int J Nanomed 16:1681–1706. https://doi.org/10.2147/IJN.S299448
123. Zhang M, Zhao X, Fang Z, Niu Y, Lou J, Wu Y, Zou S, Xia S, Sun M, Du F (2017) Fabrication of HA/PEI-functionalized carbon dots for tumor targeting, intracellular imaging and gene delivery. RSC Adv 7(6):3369–3375. https://doi.org/10.1039/c6ra26048a
124. Zygouri P, Athinodorou AM, Spyrou K, Simos YV, Subrati M, Asimakopoulos G, Vasilopoulos KC, Vezyraki P, Peschos D, Tsamis K, Gournis DP (2023) Oxidized-multiwalled carbon nanotubes as non-toxic nanocarriers for hydroxytyrosol delivery in cells. Nanomaterials (Basel, Switzerland) 13(4). https://doi.org/10.3390/nano13040714

Functionalized Carbon Nanotubes Biomedical Applications and Toxicological Implications

Urvashi Goyal, Vineeta Singh, Kundan Kumar Chaubey, Shalini Sharma, Gaurav Bhardwaj, and Krishan Raj Singh

Abstract Carbon nanotubes (CNTs) are one of the most studied carbon-based materials. Due to their distinct physicochemical characteristics, CNTs can have a variety of biomedical applications. However, their toxicity is a significant issue with CNTs' utilization in biomedical applications. Hence, many research groups are still focusing on this problem with carbon nanotubes. Functionalization of CNT improves the application of CNTs in the pharmaceutical area. Due to their distinctive qualities, including their high surface area, extraordinary mechanical strength, and great biocompatibility when suitably modified, functionalized CNTs have drawn a lot of attention in biomedical research. Here, in this chapter, we present a brief overview of carbon nanotubes and the key synthesis techniques. Further, the chapter also describes how functionalized CNTs are used in biomedicine and the variables and mechanisms impacting their toxicity. Due to their unique combination of mechanical, optical, and electrical properties, carbon nanotubes have attracted attention for use in a wide range of biomedical applications, including drug delivery, gene therapy, biosensors, and tissue engineering.

Keywords Carbon nanotubes · Single-walled CNTs · Multi-walled CNTs · Chemical vapour deposition · Functionalization of CNTs · CNTs · 3D bio-printing

---

U. Goyal
Department of Microbiology, Sanjay Gandhi Postgraduate Institute of Medical Sciences, Lucknow 226014, India

V. Singh
Department of Biotechnology, Institute of Engineering and Technology, Dr. A.P.J. Abdul Kalam Technical University, Lucknow 226021, India

K. K. Chaubey (✉) · G. Bhardwaj · K. R. Singh
School of Basic and Applied Sciences, Sanskriti University, Mathura, Uttar Pradesh 281401, India
e-mail: kundan2006chaubey@gmail.com

S. Sharma
Department of Veterinary Physiology and Biochemistry, Lala Lajpat Rai University of Veterinary and Animal Sciences, Hisar, Haryana 125004, India

A. Bachheti et al. (eds.), *Carbon-Based Nanomaterials*, Smart Nanomaterials Technology, https://doi.org/10.1007/978-981-97-0240-4_13

1 Introduction

Carbon-based nanostructures called carbon nanotubes (CNTs) have unique properties, like semiconducting to superconducting power and metallic electron transport characteristics [40]. They also have remarkable mechanical, chemical, electrical, and structural qualities. High elastic modulus and tensile strength are well-known properties of CNTs [5, 16]. CNTs are particularly helpful in creating novel goods on commercial scales because of their physicochemical characteristics, and significant research is being done in this field [36]. In modern times, nanotechnology offers tremendous prospects. In developing novel products, carbon-based nanomaterials are getting popular in related industries. They can be utilized, for instance, in tissue engineering, quantum dots, radiation oncology devices, biosensors, nanofluidic systems, and also show their ability in drug discovery and drug delivery [45]. Besides this, CNTs have applications in the biomedical field like medical nanorobots, nanosensors in devices for detecting biomolecules, and biomedical implants [4, 34].

Biomedical research can be revolutionized by CNTs [45]. According to several research, CNTs may be used in the biomedical sector [4]. To improve illness diagnosis and therapy, CNTs can be employed to develop novel materials with special features. The biological characteristics, behavior, and performance of CNTs must be thoroughly investigated. CNTs should follow clearly established environmental, biological, and safety profiles when produced on a big scale. CNTs' chemical characteristics can change, lacking purity, crystallinity, and well-defined structures. Depending on the synthesis, functionalization, or purification process employed to create them, they can vary significantly in shape, purity, and structure [17, 28].

2 Structure of Carbon Nanotubes

Massive collections of carbon nanoparticles having hollow fiber-like structure along with high aspect ratios, and dimensions typically less than 100 nm make up carbon nanotubes. High electrical and thermal conductivities, high gas adsorption, and high tensile strength must all be maintained. CNTs could be refined or manufactured. Long carbon atom chains are the building blocks of CNTs, which are joined with other individual carbon atoms through sp^2- hybridization. In general, the sp^2 hybridized carbon atom, like in diamond, is significantly stronger than the sp^3 hybridized one. The quantity of graphite layers on the surfaces of carbon nanotubes determines their classification and the special properties they possess. Graphene, with its monolayer makes up single-walled Carbon Nano Tubes (SWCNTs).

In contrast, its double layer makes up double-walled CNTs (DWCNTs), and numerous layers of graphite make up multi-walled CNTs (MWCNTs), all wrapped into a tubular structure. CNT widths can range from a few nanometers (nm) to tens nm, yet their lengths can range from less than 1 μm (μm) to a few millimeters (mm).

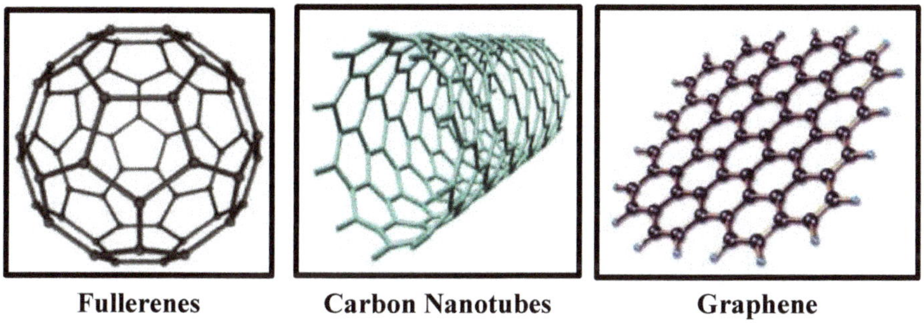

Fig. 1 Some of the basic CNTs structures

Based on their morphologies, CNTs can be divided into nanotorus, nanobuds, and nanohorns [4] (Fig. 1).

3 Synthesis

For the production of nanotubes, a variety of physical, chemical, and electrochemical methods have been employed. Some of the major techniques used to create CNTs are listed in Fig. 2. Several precursors (acetylene, benzene, carbon monoxide, ethylene, xylene) are used in developing CNTs (Fig. 3). In order to prevent negative environmental effects, safer synthesis and functionalization processes can reduce the use of poisonous and dangerous ingredients and equipment (Table 1).

3.1 Arc Discharge

Generally, high-purity graphite electrodes are employed in the arc discharge procedure. Typically, the anode is composed of either pure graphite or a combination of other metals. The synthesis reaction is conducted under low pressure (30–130 torr or 500 torr) and controlled reaction environment developed by inert and reactant gas. Lowering the inter electrodes distance maintains a current flow between 50 and 150 A (Szabo et al. 2010). An arc discharge between the target and counter electrodes prepares the sample. In this process, some of the most popular electrodes utilized in the reactor are graphite plates, C-coated Cu grids, and Ni-coated glass slides. Final CNTs are collected at the cathode. No additional form of preparation or pretreatment is necessary for this procedure (Gacem et al. 2022). Some examples are listed in Table 1, showing the successful use of this method.

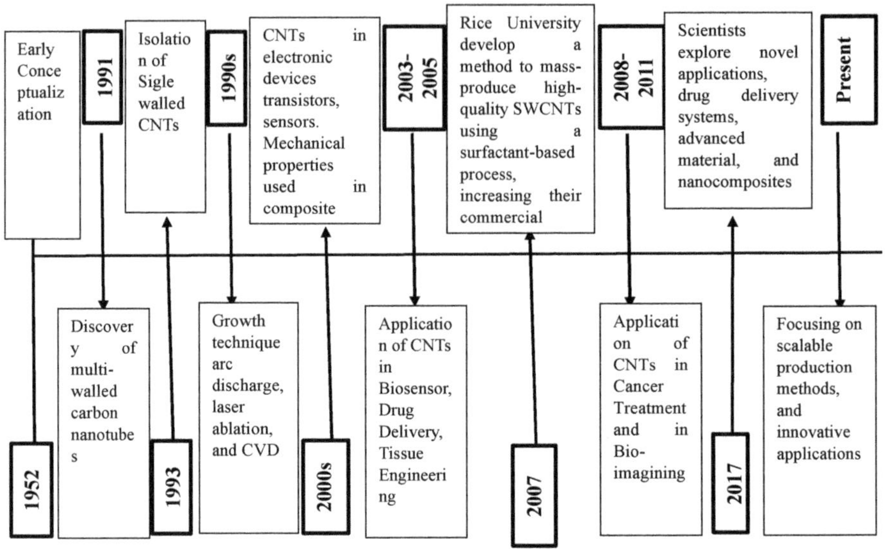

Fig. 2 Timelines for major milestones in the development of carbon nanotubes

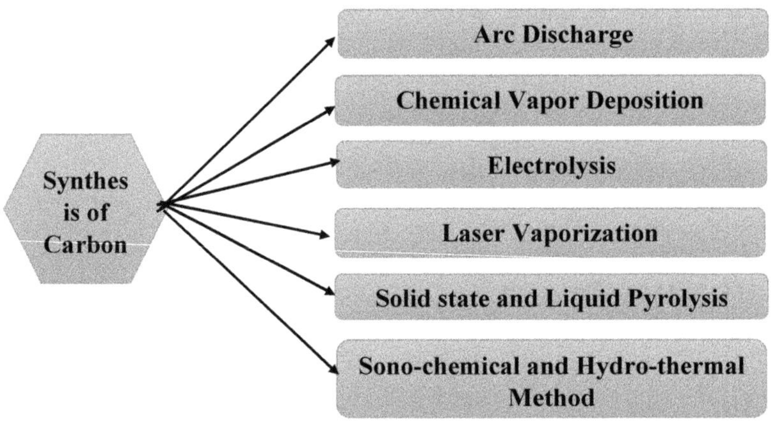

Fig. 3 Processes that are utilised to make CNTs

3.2 Chemical Vapour Deposition (CVD)

This process relies on the catalytic degradation of various suitable hydrocarbon or CO (carbon monoxide) feedstock where transition metals act as catalysts. The experiment is conducted in a horizontal or vertical flow furnace with atmospheric pressure (Szabo et al. 2010). Guo et al. created the pulsed laser-ablation method for making SWCNTs at Rice University. He used the oven laser-vaporization apparatus for the production of CNT. The target is placed in a reaction furnace at a high temperature (1200 °C), and

Table 1 Major techniques used in the development of CNTs

	Principle	Method	Common applications	References
Arc discharge	The reactor was designed with metallic electrode like tungsten electrode	Plasma source and carbon source are Argon and benzene, respectively alternate current (AC) arc discharge plasma	Production of carbon nanostructures	Zhao et al. (2011)
	Two electrodes of rectangular type C-Ni-Y was used in the experiment	A bipolar pulsed current circuit was designed, and the constant current and constant pulse duration were used	Synthesis of SWCNT	Maria and Mieno (2015)
Chemical vapor deposition (CVD)	Chemical reactions in the vapour phase were used in the CVD method to deposit thin films or coatings onto a substrate	In CVD, the required materials are delivered as precursor gases into a heated chamber where they react to create a solid thin layer on the substrate surface	In the semiconductor sector, CVD is widely employed for the deposition of thin films required for the creation of integrated circuits. Advanced materials, optical components, and coatings are also produced using it	Sabzi et al. [35]
Electrolysis	An electrical current is used to stimulate a chemical reaction that is not spontaneously occurring in the process known as electrolysis. It entails either the deposition of compounds from solutions or the breakdown of a chemical into its component parts	Electrolysis typically involves two electrodes immersed in an electrolyte solution. When an electric current is applied, chemical reactions take place at the electrodes, resulting in the separation of elements or the formation of new compounds	Electrolysis is used in various applications, including electroplating, water splitting to produce hydrogen and oxygen, and the purification of metals	Kumar and Himabindu [19]

(continued)

Table 1 (continued)

	Principle	Method	Common applications	References
Laser ablation	Laser ablation is a process in which a high-energy laser beam is focused on a material, leading to the removal or vaporization of material from the surface	A laser beam with sufficient energy is directed at the target material, causing it to rapidly heat up and either vaporize or be physically removed from the surface	Laser ablation is used in various fields, including material processing, medical procedures (e.g., laser surgery), and the analysis of solid samples in analytical chemistry techniques like laser-induced breakdown spectroscopy (LIBS)	Lawrence et al. [22]
Solid vapour deposition	A process called solid vapor deposition (SVD) is used to sublimate solid source materials in order to produce thin films or coatings onto a substrate	In SVD, a solid source material is heated to a temperature at which it sublimes (changes directly from a solid to a vapor) and then creates a thin film by condensing on the substrate surface	SVD is employed in the production of thin-film semiconductors, optical coatings, and protective layers for various applications. It is particularly useful for materials that are difficult to deposit using other methods	Baptista et al. [3]
Sonochemical method	It creates cavitation bubbles using high-frequency sound waves to produce extremely harsh conditions for chemical reactions	Ultrasonic generator produces sound waves, bubbles form and collapse, leading to chemical reactions	Nanoparticle synthesis, water treatment, drug delivery, material modification	Price et al. [31]

a laser beam (532 nm) is passed over the target for vaporization. The argon gas from the high-temperature zone swept the soot created by the laser-assisted vaporization. It settled onto a copper collector retained downstream, just outside the furnace. After this development, various advancement has been made in the technique as per the requirements. Some of them are summarized in the table.

3.3 Electrolysis

The primary step in this process is electrowinning or electroextraction of alkali metals (Li, Na, K) or alkaline-earth metals (Mg, Ca) from their ore. Generally, metal ore acts as an anode, and as a result of electrolysis, chloride salts of metals are deposited on a graphite cathode to produce carbon tubes (Szabo et al. 2010). For each electrolyte composition, a specific optimum temperature is required for synthesizing nanotubes, and the purity of CNTs was found dramatically dropping on either side of this temperature. After the completion of the electrolysis process, the ionic salt is in distilled water to extract the carbonaceous material, which is finally separated by filtration. Zhou et al. [48] synthesized the CNT with the help of the electrochemical deposition method and used Fe/Ni alloy-coated nanoparticles as an anode and cathode.

3.4 Laser Ablation

This method is based on Physical Vapour Deposition technique where a graphite target is placed in the middle of a quartz chamber filled with argon gas. A laser source is used to vaporize the target at 1200 °C. Guo et al. (1995) created the pulsed laser-ablation method for making single-wall carbon nanotubes. Smalley et al. (1996) devised an effective laser ablation technique for the large scale production of SWNTs in 1996. In comparison to those made using the arc procedure, laser-produced nanotubes are purer (up to 90%), and their structure is comparatively better graphitized.

3.5 Sonochemical Method

Sonochemical synthesis is a technique that uses intense ultrasonic radiation (20 kHz–10 MHz) to induce chemical reactions in molecules. Prince et al. (2016), utilizing substantially lower quantities of diluted nitric and sulfuric acids than previously reported, used multi-walled carbon nanotubes (CNTs) in an experiment and treated them with 20 kHz ultrasonic waves. The observations revealed an ideal set of sonication settings (in this case, 30 min at 12 W cm^2) to avoid nanotube aggregate formation and to enable effective dispersion in ethanol or chitosan.

In these processes, the carbon precursor is recombined into CNTs using either thermal energy or an electrical breakdown of gas (arc discharge) [15] originally described an approach to getting CNTs called arc discharge. The main drawback of this approach is that the catalyst, which is frequently a metal like Fe, Ni, or Co, is present as an impurity and creates yet another carbon product. Due to their propensity to evaporate following laser treatment, the laser ablation process delivers a better yield of CNTs and a smaller proportion of metallic contaminants. Furthermore, it is possible to control the CNTs' diameter better. Due to the expensive equipment, it is not an economically viable method [30]. Chemical vapor deposition (CVD), which is less expensive, uses lower temperatures and pressures, and produces the best yield of CNTs compared to other synthesis techniques, is the most widely used approach. Although they have a relatively high number of structural flaws, these moderate parameters have the best control over the dimension of the CNTs that are synthesized (Gupta et al. 2019). In addition to dimensionality, purity is the most crucial aspect of the biological application of CNTs. As a result, the CVD approach is most frequently utilized to create CNTs for use in biological investigations [14] (Nomanbhay et al. 2013).

4 Materials

Generally, inorganic carbon materials are used to synthesize carbon nanomaterials, but one of the major drawbacks of this resource is its nonrenewable nature. Besides this, their manufacturing process, especially conventional methods, frequently results in the development of hazardous or toxic waste/chemicals, potentially negatively influencing the environment. Thus, for the easier, safer, less expensive, and more sustainable manufacturing of these nanostructures there is a need to find some alternative renewable carbon resources and parallelly to develop new innovative strategies and technologies for these environmental benefits (Mustafavi et al. 2022).

Recently, researchers started using various plant extracts as natural precursors. Plant extracts help create porous, activated carbons with lots of nucleation sites that are environmentally acceptable and can be used to create CNT-based nanostructures (Table 3). Precursors of cellulosic materials, including rice husk, algae, coconut shell, palm kernel shell, lotus stem, leaves, and date seeds, are frequently used in biomass production. Biopolymers like glucose, chitin, chitosan, gelatin, and starch have also been used as renewable carbon sources (Malini et al. 2023).

In comparison to transition metals, the porous surfaces of the activated carbons showed a greater surface area. As a result, CNT nucleation sites grew larger, and activated carbon could degrade hydrocarbons at a lower temperature. Reports suggest a case where transition metals are used as catalysts in a process that takes place between B700 and 1200 °C. However, by using plant extract, this temperature condition has been reduced up to 575 °C (Makgabutlane et al. 2021).

5 Functionalization Techniques of CNTs

The basic forms in which the CNTs are synthesized have some intrinsic restrictions like insolubility, and poorly dispersible i.e., aggregation and toxicity, which limit its applications in diverse fields. Therefore, it is imperative to modify their surfaces to improve their natural features in order to achieve improved solubility, greater biocompatibility, deeper penetration, and lower cytotoxicity in biological systems (Tang et al. 2021). CNT functionalization is the process of modifying CNTs by attaching a medicinal molecule or ligand to their sides or walls. This process helps CNTs improve biocompatibility, increase encapsulation propensity, and multimodal drug delivery and imaging (Singh et al. 2016). This process can be done by two means i.e., covalent and non-covalent functionalization.

6 Functionalization of CNTs

6.1 Noncovalent Functionalization

Various small to large drug molecules can adsorb non-covalently onto the surfaces of CNTs using this technique. Physical adsorption plays a critical role in this process. CNT structure serves as a reservoir to adsorb various drug molecules in a host–guest interaction manner. It involves Van der Waals contacts, stacking interactions, and hydrophobic interactions between the chains of the adsorbed molecules and the surfaces of CNTs, and it only slightly harms the surface of CNTs (Tang et al. 2021). Antibodies, aptamers, polymers, and surfactants are some examples successfully used in the noncovalent functionalization of CNTs. The hydrophobic forces (interactions) are the primary driving forces for the attachment of lipophilic pharmaceuticals onto CNTs if the adsorbed molecule is one of the lipophilic drug moieties. The possibility of surface modification of CNTs without changing their electrical structure makes noncovalent functionalization of the materials very appealing. A simple method of noncovalent functionalization is to expose the vapours of functionalization species onto the CNTs so that they noncovalently link to the nanotube surface (Fig. 4).

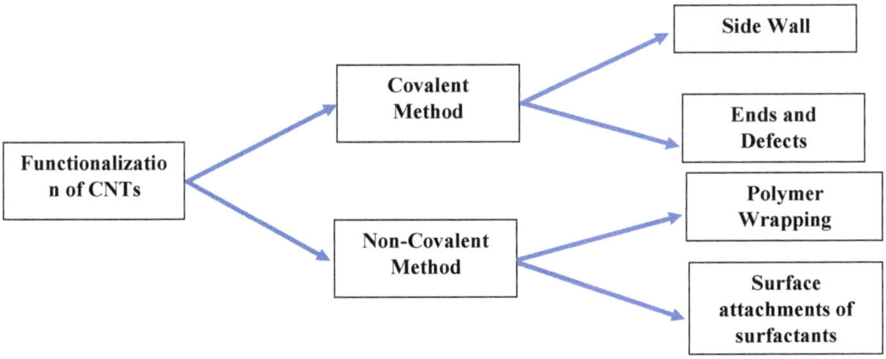

Fig. 4 Functionalization methods of CNTs

6.2 *Covalent Functionalization*

The method of attaching chemical compounds to the tubular structure of carbon nanotubes (CNTs) by forging covalent connections that share at least one pair of electrons with the CNT. A change from sp2 to sp3 hybridization and the accompanying loss of the p-conjugation system on the graphene layer occur in a direct covalent sidewall functionalization. This process can be produced by interaction with specific chemicals that exhibit strong chemical reactivity. Defect functionalization of CNTs is another technique. Strong acids that cause oxidative damage to the structure of nanotube are used to oxidise the CNTs, leaving holes that are functionalized with oxygenated functional groups. It is also possible to make CNTs soluble in water by adding hydrophilic groups to their external walls and tips. Some of the strategies used in the covalent functionalization are shown in Figs. 5 and 6.

7 Advancement in CNTs—3D Bioprinting

A biomaterial is accurately extruded in layers using 3D bioprinting technology, providing the user with better control over the structure and microarchitecture of a printed scaffold. As a result, it makes it possible to produce something that looks very much like original tissue. As was previously noted, every aspect of cellular behavior, like metabolism, proliferation, and migration, is influenced by the local environment within the tissue. As a result, this is an important factor for both mechanical support and correct cellular function [13] (Axpe et al. 1976).

The biomaterial suitable for the survival of living cells is generally used to develop hydrogel matrix used in bioprinting, and this matrix with live cells is known as bioink. A proper bioink should possess several critical qualities, such as printability, stability of the printed structure, appropriate mechanical strength, bio and cytocompatibility, and degradability. Due to the requirement of bio- and cytocompatible, the majority of currently utilised bio inks are based on natural polymers such alginate, agarose, gelatin, fibrin, chitosan, hyaluronate, and collagen; however, synthetic polymers like poloxamer (pluronic) or polyethylene glycol (PEG) are also used. These materials, except agarose and alginate, naturally resemble the ECM. They have acceptable printability properties since they are relatively soft hydrogels but lack the mechanical stability to withstand the same continuous loads as a cartilage or bone substitute. As a result, two techniques are used to create a composite scaffold: crosslinking and augmentation with stronger materials (Axpe et al. 1976).

Physical and chemical crosslinking are the two forms. Hydrogen bonds, ionic or hydrophobic contacts, and hydrophobic interactions support physical crosslinking. Some hydrogel requires treatment after printing, alginate-based printed object is kept in calcium salt solution (such as $CaCl_2$) for physical cross-linking. Because covalent connections are formed due to chemical crosslinking, as-prepared scaffolds

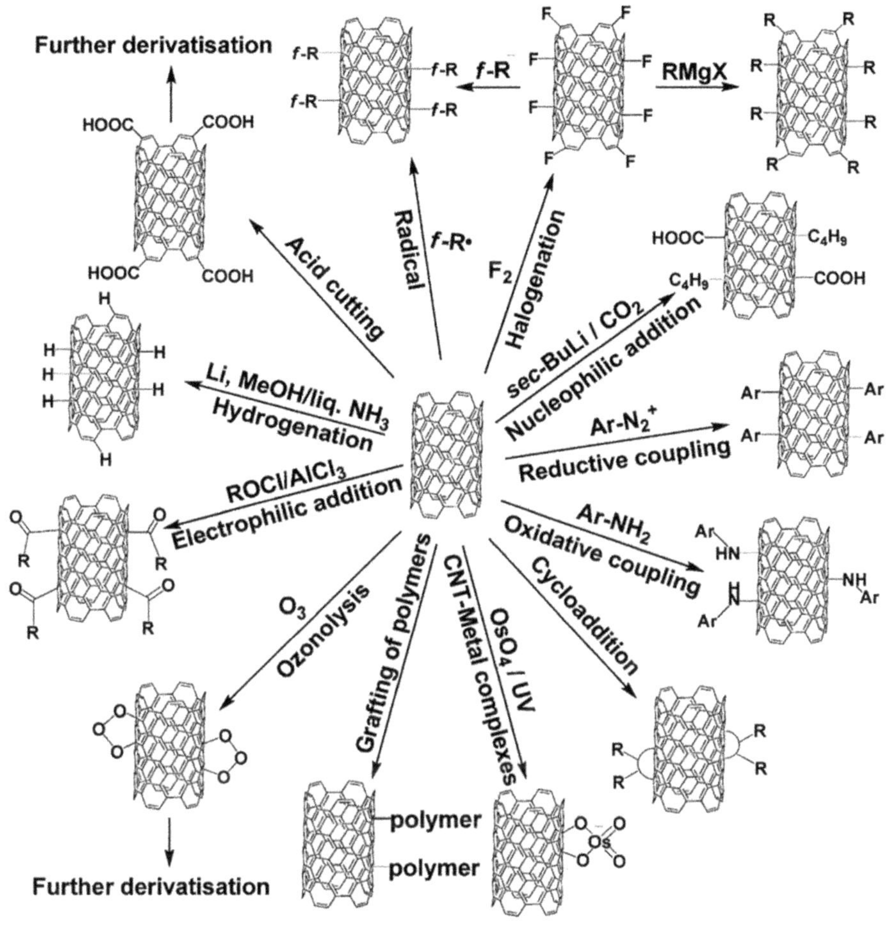

Fig. 5 Surface functionalization of carbon nanotubes [42]

have a higher mechanical strength (Klotz et al. 2016). Different chemical modifications of polymers are employed for this form of crosslinking, such as cellulose, gelatin, methacrylated or diacrylated PEG moieties, and others that go through radical polymerization after being exposed to UV light (Poldervaart et al. 2017) [23, 39, 44].

8 Biomedical Application of Carbon Nanotubes

Carbon nanotubes (CNTs) have shown promise in several biomedical applications because of their peculiar properties (Table 2). With references to the pertinent studies, the following are some salient points about the biological applications of CNTs:

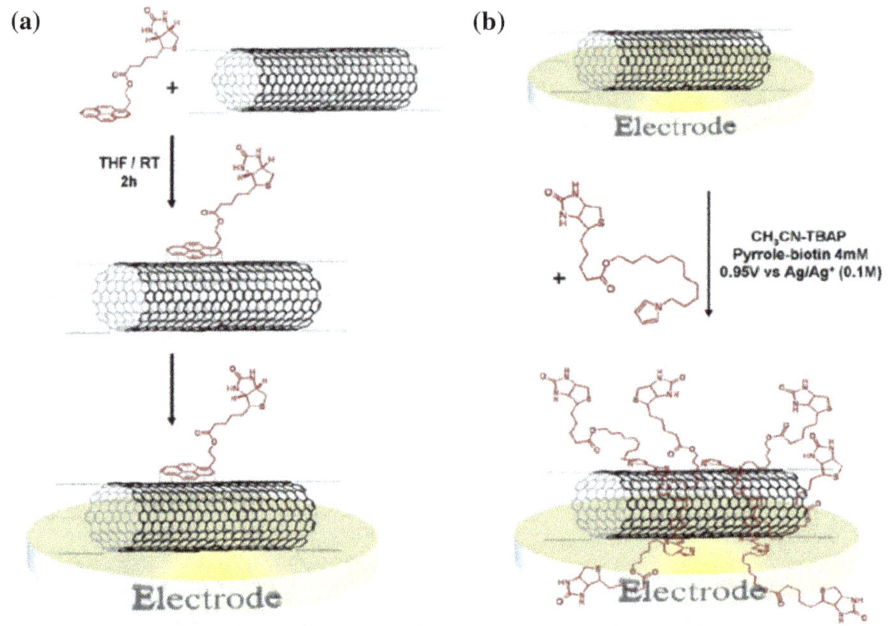

Fig. 6 Biotin functionalization on nanotubes using **a** π–π stacking interaction between pyrene-biotin derivatives and the nanotube side walls; **b** Electropolymerization of the pyrrole-biotin monomer on the SWCNT [47]

1. **CNT Synthesis and Types:** CNTs can be developed with diverse methods, such as electric-arc discharge, laser ablation, and catalytic CVD. They come in a variety of shapes, including single-wall CNTs (SWCNTs) and multi-walled CNTs (MWCNTs), each with their special characteristics [27].
2. **Biocompatibility:** CNTs have a fair amount of biocompatibility, which makes them appropriate for various biomedical applications [26].
3. **Diagnostic Equipment:**
   - CNTs are utilized as sensing components in biosensors to find infections and biomarkers. They improve sensitivity and speed up electron transmission, increasing detection accuracy [32].
   - Agents for imaging: CNTs can serve as contrast agents in imaging processes, including photoluminescent imaging and photoacoustic imaging, improving tissue visualisation [46, 41].
4. **Tissue engineering:** CNTs can boost neural signal transmission and encourage cellular proliferation. They are employed to develop 3D designs that enable cell proliferation and tissue regeneration in tissue engineering applications [9, 24, 25].
5. **Targeted Therapies:**

Table 2 Some plant-based renewable precursors/catalysts for the synthesis of CNTs

Techniques	CNT-based nanostructures	Plant-based resources	Refs.
Pyrolysis method	SWCNTs	Olive oil (*Olea europaea*)	Abdel Hamid et al. [1]
	MWCNTs	Rice straw (raw rice straw & neutral pulp)	Awasthi et al. [2]
Spray pyrolysis method	SWCNTs	Eucalyptus oil	Ghosh et al. [11]
	MWCNTs	Turpentine oil (pine tree)	Awasthi et al. [2]
	MWCNTs	*Azadirachta indica* (neem oil)	Kumar et al. [21]
CVD	MWCNTs	Coconut oil (*Cocos nucifera* Linn.)	Paul and Samdarshi [29]
	MWCNTs	Fresh bamboo culms	Zhu et al. [49]
	MWCNTs	*Cinnamomum camphora* (camphor)	Kumar and Ando [18]
	MWCNTs	*Cynodon dactylon*, Rosa, *Azadirachta indica, Juglans regia*	
Spray pyrolysis-assisted CVD	Branched nitrogen (N)-doped CNTs	Sesame oil (*Sesamum indicum*)	Kumar et al. [20]
Thermal catalytic CVD	Vertically aligned CNTs; SWCNTs	Natural palm oil	Suriani et al. [38]
–	CNTs-Cu/ZnO nanocomposites	*Brassica juncea* L	Qu et al. [33]

Table 3 General properties of Carbon Nano Tubes (CNTs)

Properties	Description
Family	CNT belongs to the third allotropic form of carbon
Types	Single-walled (SW) CNT, double-walled (DW) CNT and multi-walled (MW) CNT
Categories	Based on its atomic structure, it can be categorized as sidewall and end cap
Aspect Ratio	This ration for CNT is higher than 3
Solubility	CNT are found soluble in polar organic solvents like ethanol (C_2H_5OH), isopropyl alcohol ($(CH_3)_2CHOH$), etc. and insoluble in inorganic solvents
Synthesis Method	Electric arc discharge, chemical vapor deposition (CVD) and Laser vaporization/ablation are the most common methods used of synthesis
Purification	Oxidation, centrifugation, chromatography, and ultrasonication are commonly used for the purification
Modification	Covalent method and non-covalent method
Flexibility	CNT are flexible structure and can be easily bent without showing any type of damage
Conductivity	CNT structure provides high electrical and thermal conductivity

- **Drug Delivery:** CNTs are utilised as drug delivery vehicles. Drugs like doxorubicin and hydroxy camptothecin can be transported by them to specific target locations, improving treatment efficiency while reducing side effects (Liu et al. 2007) [43]
- **Gene Delivery:** CNTs are used in gene therapy due to their ability to transfer nucleic acids (DNA plasmids and short interfering RNA). into cells, such as They present a potentially effective strategy for gene silencing therapy (Bartholomeusz et al. 2009) [37].

Anticancer Therapies: Photothermal therapy (PTT) can be utilized to treat cancer using CNTs. They generate heat by absorbing near-infrared (NIR) light, which can be utilized to destroy tumours or improve the delivery of anticancer medications (Wang et al. 2014) [8].

The challenges in CNT biomedical applications include attaining good CNT dispersion in solvents (particularly in water), assuring high purity to reduce potentially harmful effects, and dealing with the issue of increased viscosity when CNTs are distributed in fluids [12].

Safety and Toxicity: Research is still being done to understand better the long-term impacts and potential toxicity of CNTs in biological applications [12].

Surface Functionalization: Whether covalent or non-covalent, surface functionalization of CNTs is essential for changing their characteristics for specific biomedical applications [7, 47].

Future study: To ensure that CNTs are used safely and effectively in biomedical applications, more study is required to examine the destiny and potential degradation of CNTs in biological contexts [50].

While noting the difficulties and current research in this sector, these considerations illustrate the various and potential functions that carbon nanotubes can play in the field of biomedicine, from diagnostics to targeted therapeutics and tissue engineering.

9 Limitations of CNT

Pristine carbon nanotubes' (CNTs') limits in their biomedical applications due to poor solubility and the significance of functionalizing CNTs to make them more soluble and useful in various applications. Here, there are more details and citations about the restrictions on pure CNTs:

1. Low Solubility: Pristine CNTs are notorious for being poorly soluble in organic and aqueous solvents, restricting their use in biomedical applications [6].
2. Biocompatibility: Due to their insolubility and certain toxicity issues, pristine CNTs can only have a limited biocompatibility. For its application in biomedicine, modified CNTs must maintain their biocompatibility.

Table 4 Advantages and disadvantages of carbon nanotubes

Advantages	Disadvantages
Larger surface area	Insolubility
Higher thermal conductivity	Toxicological assessments
Higher mechanical strength	Complex standardization
Higher stability and functionality	Adverse effects
Mass production	Dose-dependent toxicity
Crossing biological barriers	Immunogenicity
Large surface area	Tissue tolerance/accumulation

3. Aggregation: Unmodified CNTs tend to group in solutions, which can cause uneven dispersion and make it difficult for them to be incorporated into biological systems.
4. Challenges in Functionalization: CNT functionalization can be a difficult procedure, and selecting the right functionalization technique is crucial to preserving their structural integrity and desired attributes [10].
5. Biological Interactions: Pristine CNTs may interact in unforeseen ways with biological molecules and systems, which could negatively affect biomedical applications.
6. Limited Versatility: Without the necessary functionalization, pristine CNTs would not have the qualities needed for some biomedical uses, such as medication delivery or tissue engineering (Table 4).

10 Toxicity

One of the most extensively studied forms of allotropic carbon is carbon nanotubes (CNTs). They have diverse physicochemical features that make them suitable for usage in various biomedical applications. The small size of CNT and the large surface area per unit volume are major factors for its toxicity [36]. According to reports, CNTs are cancer-causing and can result in lung tumors. In animal tests, it has been found that exposure to CNTs through inhalation can cause malignant mesothelioma (Kobayashi et al. 2017). Therefore, although CNTs are used in many different applications, their toxicity remains a significant issue that needs further study. According to some investigations, CNTs generate oxidative stress as they travel through the lipid bilayer membrane of cells, which might result in inflammation and cytotoxicity. One of the most likely reasons CNTs are hazardous to cells is because they produce free radicals, which lead to oxidative stress (Mohanta et al. 2019). Potential controls for these mechanisms include a variety of physicochemical components (metal impurities, length, size, solubilizing agents, CNTs functionalization, and agglomeration) that may result in oxidative stress and dangerous signaling pathways [36]. We must thoroughly investigate these issues in order to achieve progress in every area if we would like to utilize the potential of CNT fully.

11 Conclusion

In conclusion, carbon nanotubes (CNTs) are a special material with unique characteristics like strength and electrical and thermal conductivity. Reports suggest the application of CNT in cell growth and recolonization in the nerve cells. Once the CNT has been functionalized with organic, metallic, or biologicals to enhance its biocompatibility, it opens the direction of its uses in biomedical fields such as biosensors, targeted medication delivery, etc. However, the toxicity and pharmacokinetics of CNTs must be studied more thoroughly because they still have certain harmful properties. The application of bioprinting in developing 3D porous structure of hydrogel further widens the application area of CNTs.

References

1. Abdel Hamid Z, Abdul Azim A, Abdel Mouez F, Abdel Rehim SS (2017) Challenges on synthesis of carbon nanotubes from environmentally friendly green oil using pyrolysis technique. J Anal Appl Pyrolysis 126:218–229
2. Awasthi K, Kumar R, Tiwari RS, Srivastava ON (2010) Large scale synthesis of bundles of aligned carbon nanotubes using a natural precursor: turpentine oil. J Exp Nanosci 5:498–508
3. Baptista A, Silva F, Porteiro J, Míguez J, Pinto G (2018) Sputtering physical vapour deposition (pvd) coatings: a critical review on process improvement and market trend demands. Coatings 8(11):402
4. Beg S, Rizwan M, Sheikh AM, Hasnain MS, Anwer K, Kohli K (2011) Advancement in carbon nanotubes: basics, biomedical applications and toxicity. J Pharm Pharmacol 63:141–163
5. Belay T, Worku LA, Bachheti RK, Bachheti A, Husen A (2023) Nanomaterials: introduction, synthesis, characterization, and applications. In: Husen A, Siddiqi KS (eds) Advances in smart nanomaterials and their applications. Elsevier, Amsterdam, Netherlands, pp 1–21
6. Bianco A (2005) All in the graphene family—a recommended nomenclature for two-dimensional carbon materials. Carbon 65:1–6
7. Bortolamiol T, Lukanov P, Galibert A-M, Soula B, Lonchambon P, Datas L, Flahaut E (2014) Double-walled carbon nanotubes: quantitative purification assessment, balance between purification and degradation and solution filling as evidence of opening. Carbon 78:79–90
8. Burke A, Ding X, Singh R, Kraft RA, Levi-Polyachenko N, Rylander MN, Fisher J (2009) Long-term survival following a single treatment of kidney tumors with multiwalled carbon nanotubes and near-infrared radiation. Proc Natl Acad Sci USA 106:12897–12902
9. Correa-Duarte MA, Wagner N, Rojas-Chapana J, Morsczeck C, Thie M, Giersig M (2004) Fabrication and biocompatibility of carbon nanotube-based 3D networks as scaffolds for cell seeding and growth. Nano Lett 4:2233–2236
10. Georgakilas V, Kordatos K, Prato M, Guldi DM, Holzinger M, Hirsch A (2002) Functionalization of carbon nanotubes. J Am Chem Soc 124(5):760–761
11. Ghosh P, Afre RA, Soga T, Jimbo T (2007) A simple method of producing single-walled carbon nanotubes from a natural precursor: eucalyptus oil. Mater Lett 61:3768–3770

12. Grady BP (2006) The use of solution viscosity to characterize single-walled carbon nanotube dispersions. Macromol Chem Phys 207:2167–2169
13. Gungor-Ozkerim PS, Inci I, Zhang YS, Khademhosseini A, Dokmeci MR (2018) Bioinks for 3D bioprinting: an overview. Biomaterials Science 6:915–946
14. Ibrahim KS (2013) Carbon nanotubes-properties and applications: a review. Carbon Lett 14:131–144
15. Iijima S (1991) Helical microtubules of graphitic carbon. Nature 354:56–58
16. Jafar A, Alshatti Y, Ahmad A (2016) Carbon nanotube toxicity: the smallest biggest debate in medical care. Cogent Medicine 3:1217970
17. Karousis N, Papi RM, Siskos A, Vakalopoulou P, Glezakos P, Sarigiannis Y, Stavropoulos G, Kyriakidis DA, Tagmatarchis N (2009) Peptidomimetic-functionalized carbon nanotubes with antitrypsin activity. Carbon 47:3550
18. Kumar M, Ando Y (2007) Carbon nanotubes from camphor: an environment-friendly nanotechnology. J Phys Conf Ser 61:129
19. Kumar SS, Himabindu V (2019) Hydrogen production by PEM water electrolysis–a review. Mater Sci Energy Technol 2(3):442–454
20. Kumar R, Singh RK, Tiwari RS (2016) Growth analysis and high-yield synthesis of aligned-stacked branched nitrogen-doped carbon nanotubes using sesame oil as a natural botanical hydrocarbon precursor. Mater Des 94:166–175
21. Kumar R, Tiwari RS, Srivastava ON (2011) Scalable synthesis of aligned carbon nanotubes bundles using green natural precursor: neem oil. Nanoscale Res Lett 6:92
22. Lawrence J, Pou J, Low DKY, Toyserkani E (2010) Laser ablation. In: Advances in laser materials processing. Woodhead publishing series in welding and other joining technologies. Woodhead Publishing, pp 575–628
23. Li J, Chen M, Fan X-Q, Zhou H (2016) Recent advances in bioprinting techniques: approaches, applications, and future prospects. J Transl Med 14:271
24. Lovat V, Pantarotto D, Lagostena L, Cacciari B, Grandolfo M, Righi M, Ballerini L et al (2005) Carbon nanotube substrates boost neuronal electrical signaling. Nano Lett 5:1107–1110
25. Mazzatenta A, Giugliano M, Campidelli S, Gambazzi L, Businaro L, Markram H, Ballerini L (2007) Interfacing neurons with carbon nanotubes: electrical signal transfer and synaptic stimulation in cultured brain circuits. J Neurosci 27:6931–6936
26. Mohajeri M, Behnam B, Sahebkar A (2018) Biomedical applications of carbon nanomaterials: drug and gene delivery potentials. J Cell Physiol 234:298–319
27. Monthioux M, Serp P, Caussat B, Flahaut E, Razafinimanana M, Valensi F, Laurent C, Peigney A, Mesguich D, Weibel A et al (2017) Carbon nanotubes. In: Bhushan B (ed) Springer handbook of nanotechnology. Springer Handbooks, Berlin Germany, pp 193–247
28. Pagona G, Tagmatarchis N (2006) Carbon nanotubes: materials for medicinal chemistry and biotechnological applications. Curr Med Chem 13:1789
29. Paul S, Samdarshi SK (2011) A green precursor for carbon nanotube synthesis. New Carbon Mater 26:85–88
30. Prasek J, Drbohlavova J, Chomoucka J, Hubalek J, Jašek O, Adam V, Kizek R (2011) Methods for carbon nanotubes synthesis—review. J Mater Chem 21:15872–15884
31. Price GJ, Nawaz M, Yasin T, Bibi S (2018) Sonochemical modification of carbon nanotubes for enhanced nanocomposite performance. Ultrason Sonochem 40:123–130
32. Punbusayakul N, Talapatra S, Ajayan PM, Surareungchai W (2013) Label-free as-grown double wall carbon nanotubes bundles for salmonella typhimurium immunoassay. Chem Cent J 7:102
33. Qu J, Luo C, Cong Q, Yuan X (2014) Recycling of the hyperaccumulator Brassica juncea L.: synthesis of carbon nanotube-Cu/ZnO nanocomposites. Mater Cycles Waste Manag 16:162–166
34. Roldo M, Fatouros DG (2013) Biomedical applications of carbon nanotubes. Annu Rep Prog Chem Sect C: Phys Chem 109:10–35
35. Sabzi M, Mousavi Anijdan SH, Shamsodin M, Farzam M, Hojjati-Najafabadi A, Feng P, Park N, Lee U (2023) A review on sustainable manufacturing of ceramic-based thin films by chemical vapor deposition (CVD): reactions kinetics and the deposition mechanisms. Coatings 13:188

36. Saleemi MA, Hosseini Fouladi M, Yong PVC, Chinna K, Palanisamy NK, Wong EH (2020) Toxicity of carbon nanotubes: molecular mechanisms, signaling cascades, and remedies in biomedical applications. Chem Res Toxicol 34(1):24–46
37. Sanz V, Tilmacîu C, Soula B, Flahaut E, Coley HM, Silva SRP, McFadden J (2011) Chloroquine-enhanced gene delivery mediated by carbon nanotubes. Carbon 49:5348–5358
38. Suriani AB, Azira AA, Nik SF, Nor RM, Rusop M (2009) Synthesis of vertically aligned carbon nanotubes using natural palm oil as carbon precursor. Mater Lett 63:2704–2706
39. Szymański T, Mieloch AA, Richter M, Trzeciak T, Florek E, Rybka JD, Giersig M (2020) Utilization of carbon nanotubes in manufacturing of 3D cartilage and bone scaffolds. Materials 13(18):4039
40. Wang SD, Chang MH, Cheng JJ, Chang HK, Lan KMD (2005) Unusual morphologies of carbon nanoparticles obtained by arc discharge in deionized water. Carbon 43(6):1322–1325
41. Welsher K, Liu Z, Sherlock SP, Robinson JT, Chen Z, Daranciang D, Dai H (2009) A route to brightly fluorescent carbon nanotubes for near-infrared imaging in mice. Nat Nanotechnol 4:773–780
42. Wu HC, Chang X, Liu L, Zhao F, Zhao Y (2010) Chemistry of carbon nanotubes in biomedical applications. J Mater Chem 20(6):1036–1052
43. Wu W, Li R, Bian X, Zhu Z, Ding D, Li X, Hu Y (2009) Covalently combining carbon nanotubes with anticancer agent: preparation and antitumor activity. ACS Nano 3:2740–2750
44. Wu D, Yu Y, Tan J, Huang L, Luo B, Lu L, Zhou C (2018) 3D bioprinting of gellan gum and poly(ethylene glycol) diacrylate based hydrogels to produce human-scale constructs with high-fidelity. Mater Design 160:486–495
45. Xue Y (2017) Carbon nanotubes for biomedical applications. In: Industrial and applied carbon nanotubes, vol 4, pp 323–346
46. De La Zerda A, Zavaleta C, Keren S, Vaithilingam S, Bodapati S, Liu Z, Oralkan O (2008) Carbon nanotubes as photoacoustic molecular imaging agents in living mice. Nat Nanotechnol 3:557–562
47. Zhou Y, Fang Y, Ramasamy RP (2019) Non-covalent functionalization of carbon nanotubes for electrochemical biosensor development. Sensors 19:392
48. Zhou Z, Zhao Z, Qiu J (2006) Synthesis of branched carbon nanotubes from coal. Carbon 44(7):1321–1324
49. Zhu J, Jia J, Kwong FL, Leung Ng DH, Tjong SC (2012) Synthesis of multiwalled carbon nanotubes from bamboo charcoal and the roles of minerals on their growth. Biomass Bioenergy 36:12–19
50. Zhu Y, Sun Q, Liu Y, Ma T, Su L, Liu S, Liang F (2018) Decorating gold nanostars with multiwalled carbon nanotubes for photothermal therapy. R Soc Open Sci 5:180159

Role of Carbon Nanomaterials in Air Pollution Remediation

Addisu Tamir Wassie, Rakesh Kumar Bachheti, Archana (Joshi) Bachheti, and Azamal Husen

Abstract Globally, air pollution is a huge problem that impacts both the natural environment and the well-being of humans. The most recent advances in nanotechnology can help with potential remedies for the problems associated with air pollution and significantly improve the selectivity and efficacy of air cleansing systems, resulting in cost-effectiveness and more long-term performances. With the use of nanotechnology, air pollution could be reduced through a variety of methods, including using nanocatalysts, nanoadsorbents, nanosensors, and so on. The nanoparticles can adsorb or absorb several airborne pollutants. By employing the arrangement and size of the materials at the nanoscale form, nanotechnology is an innovative branch of science that may solve a wide range of environmental problems. For the reason of their benign nature, large surface area, ease of biodegradation, and mainly relevant environmental monitoring properties, carbon nanomaterials are distinctive. However, the use of carbon nanomaterials and nanotechnology is confronted with a number of difficulties, including production costs, toxicity, environmental hazards, and public

A. T. Wassie · R. K. Bachheti (✉)
Department of Industrial Chemistry, College of Natural and Applied Sciences, Addis Ababa Science and Technology University, Addis Ababa, Ethiopia
e-mail: rkbachheti@gmail.com

A. T. Wassie
Department of Chemistry, College of Natural Sciences, Wollo University, Dessie, Ethiopia

A. Bachheti
Department of Environment Science, Graphic Era University, Dehradun, Uttarakhand 248002, India

R. K. Bachheti
Department of Allied Sciences, Graphic Era Hill University, Society Area, Clement Town, Dehradun, Uttarakhand 248002, India

A. Husen
Department of Biotechnology, Smt. S. S. Patel Nootan Science and Commerce College, Sankalchand Patel University, Visnagar, Gujarat 384315, India

Department of Biotechnology, Graphic Era (Deemed to Be University), Dehradun, Uttarakhand 248002, India

Wolaita Sodo University, Wolaita, Ethiopia

A. Bachheti et al. (eds.), *Carbon-Based Nanomaterials*, Smart Nanomaterials Technology, https://doi.org/10.1007/978-981-97-0240-4_14

acceptance. So, CNMs have advantages and disadvantages discussed in this chapter. The main emphasis of this chapter is the usage of CNMs in applications for air pollution remediation. Mainly, we discussed different types of carbon nanomaterial structures, properties, and removal mechanisms of different air pollutants.

Keywords Carbon nanomaterials · Air pollution · Remediation · Adsorption · Mechanisms · Challenge

1 Introduction

When contaminants that are bad for the environment and human health are released into the air from different sources, it is considered air pollution. In other words, suspended gases, liquids, and solid particles produce air pollution. These various particles and gases are produced by factories and industries, which are the key springs of organic compounds, hydrocarbons, carbon monoxide, and chemicals. Agricultural activities emitting the greatest dangerous gases like ammonia and fertilizers, pesticides, and insecticide usage can release atmospheric contaminant destructive chemicals. Automobiles and trucks also emit the main causes of greenhouse gases and affect individual health. Mining activities, wildfires, dust, mold spores, volcanoes, and combustion of fossil fuels, which can emit a huge amount of SO_2 and CO released by the partial burning of fossil fuels, also result in air pollution [26–29]. The term "aerosols" refers to compact and liquefied particles dispersed in the atmosphere, which may cause air pollution [48, 71]. Humans are susceptible to air pollution-related illnesses, allergies, and even death. Diseases that are caused by air pollution can have short or long-term effects. Exposure to air pollution in the long term can cause health effects, including heart disease or cardiovascular damage, lung cancer, harm to the liver, blood, and spleen, damage to the nervous system, brain, and kidneys, and respiratory diseases such as asthma and emphysema finally it causes death. Some scientists also suspect air pollutants cause birth defects [43, 46, 49]. Short-term exposure to air pollution also causes health effects such as headache, coughing, dizziness, wheezing/difficulty breathing, fatigue, and irritation to the eyes, nose, and throat [17]. Air pollution not only harms humans but also can harm other living creatures, such as plants and animals, and can harm the natural surroundings, such as changing climate, habitat degradation, and O_3 depletion (caused by the release of hydrochlorofluorocarbons, halons, and chlorofluorocarbons in the atmosphere which is the main reason of weakening of the ozone layer). They hampered the plant growth and developmental process [15, 16, 30, 31, 62, 63]. Air pollution can also damage the built environment. For instance, when hazardous substances like SO_2 and oxides of nitrogen escape into the environment during the burning of fossil fuels and react with the water droplets, they become acidic and fall as acid rain, which can harm plants and animals as well as people [17, 71].

To reduce all the above-listed air pollution problems, effective air pollution remediation strategies must be considered. Some common and effective air pollution remediation strategies include conserving energy at home, work, or everywhere, using public transportation, bike, or walking whenever possible. Moreover, eliminating contaminants, including chemical ingredients that are arbitrated to have a bad environmental impact. Consequently, lowering harmful exhaust from industrial processes, reducing exhaust from automobiles and engines with new, stricter exhaust rules and cleaner-burning fuel, and tackling indoor air pollution with cooperative programs are all necessary emission reduction measures for effective air pollution remediation strategies [49]. Geothermal, solar, wind, and other clean energy sources are also used to reduce air pollution to a greater extent. Utilizing these resources is something that many nations, including India, have done to improve the environment [38]. The additional steps also taken to reduce air pollution are listed as follows:

a. By less frequently utilizing fire and items associated with fire.
b. Industrial releases are one of the key providers of air pollution. Thus, it is possible to minimize its impacts by regulating or treating the pollutants at the source. For example, if a specific raw material interacts in a manner that produces contaminants, that initial raw material can be substituted with one that does not.
c. Fuel substitution is another way to reduce air pollution. Vehicles fueled by compressed natural gas have replaced those powered by petrol and diesel in several nations worldwide. The majority of vehicles that do not entirely function with low-emission engines do so.
d. Even though the World has a wide variety of practices aimed at improving air excellence, most are either unnoticed or inappropriately applied because many vehicles still need to operate on the road without having undergone vehicle emissions testing.
e. Modifying and maintaining existing equipment to decrease the discharge of contaminants is another technique for decreasing air pollution companies carry. In some cases, it is impossible to control pollution at its source. The pollutants can then be controlled with process control devices.
f. Air pollution can be efficiently managed via the dilution of air pollutants.
g. Typically, the best and last option for reducing the effects of air pollution is to plant trees. Plants and trees considerably minimize the number of air pollutants. Planting trees in areas with high pollution levels will be ideal and highly effective.

One of the most advantageous in air pollution monitoring is nanotechnology, in which various nanomaterials are used as nanosensors, nano adsorbents, nanocatalysts, and so on. The nanomaterials can absorb many air pollutants. Carbon-based nanomaterials, encompassing various allotropes such as nanotubes, graphene, and C60 fullerenes, are particularly well-liked because of their wide applications in electronics, medication delivery, sensors, energy storage, and other sectors. Contemporary industrial growth and agricultural practices increase the possibility of serious environmental degradation inside civilization. The discussion of preventing the deterioration of the environment is timely. In this context, carbon-based nanoparticles can eliminate gaseous air contaminants. Carbon nanoparticles' well-known properties, such as their enormous, precise floor area and mesoporous nature, support their use in air pollutant monitoring systems [20, 67]. Therefore, the main focus of this study is the use of different CNMs as nanocatalysts, nanosensors, and nano adsorbents for air pollution remediation purposes.

Additionally, many elements influencing the effectiveness of the carbon nanomaterials are examined. The application of widely studied popular carbon nanoparticles is reviewed. Significant research efforts have been made to improve the use of carbon nanomaterials in a diversity of environmental remediation uses. To the greatest extent of our understanding, there still needs to be more information available on using carbon nanoparticles for applications that aim to remediate air pollution in a comprehensive form.

2 Common Types of Carbon Nanomaterials

According to nanotechnology, nanoscale adsorbents, sometimes called nano adsorbents, might significantly alleviate most current issues, including poor air quality. The nanoparticles may absorb numerous airborne contaminants. Various nano adsorbents have been developed and categorized to detect and adsorb contaminants such as bacteria, volatile organic compounds (VOCs), and metallic oxides [67]. Hereinafter, the frequently used pivotal carbon nanomaterials due to their distinctive electronic and mechanical belongings for eliminating air pollutants are discussed.

2.1 Carbon Nanotube (CNT)

CNTs are a type of carbon nanomaterials with long, thin carbon cylinders that were discovered by Sumio Iijima, in 1991. They are massive macromolecules that stand out for their distinctive shape, size, and exceptional physical characteristics. CNTs can be compared to a graphite sheet rolled into a cylinder. A graphite sheet is folded into a tube to form single-walled CNTs, as shown in Fig. 1. CNTs can be categorized as chiral, zigzag, or armchair (Fig. 1) based on the orientation of their hexagons. Concentric graphene tubes are wrapped into many layers to form multi-walled CNTs (MWCNT) [42]. These interesting structures have attracted much attention in recent years, and a lot of study has been done to comprehend them. The physical characteristics of CNTs are still being uncovered and debated right now. The numerous types of nanotubes, which are classified by their length, diameter, twist, or chirality, have different thermal, electrical, and structural characteristics. CNTs can have numerous cylindrical walls, or cylinders on outside other cylinders, to add novelty to the situation in addition to having a single cylindrical wall [68].

The application of CNTs as a nano adsorbent for various air pollutants has been the subject of numerous investigations. The better physical characteristics of carbon-based nanostructures, such as a large surface area, high reactivity, small size, good biocompatibility, a larger pore diameter, good surface function, and a large pore

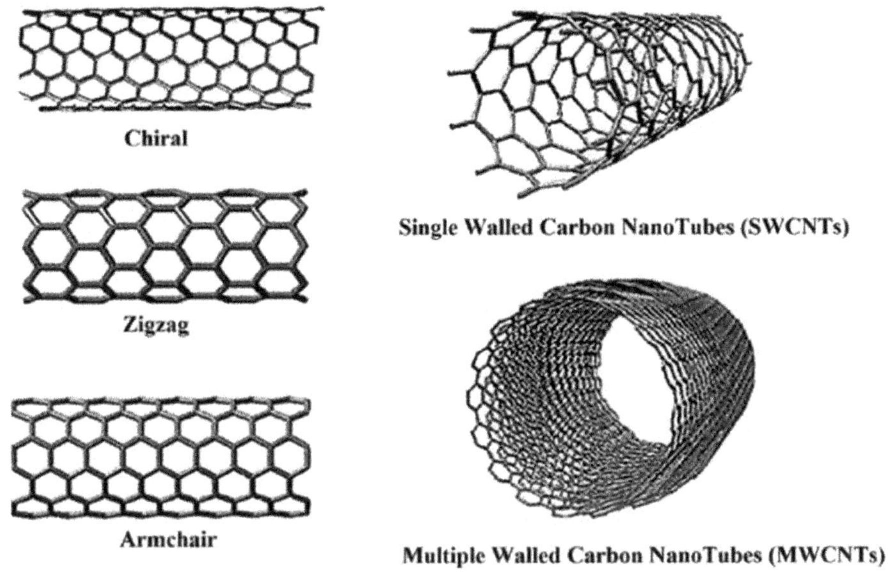

Fig. 1 Different shapes of carbon nanotubes [42]

volume, make them suitable for usage in the industry as nano adsorbents with higher capability, affinities, and selective operation [20]. CNTs are utilized to regulate the release of carbon dioxide and methane from automobile and industrial chimney releases. CNTs can also capture greenhouse gases produced by power plants and coal mines. CNTs, due to their improved characteristics, such as their attraction for selective pollutant adsorption, can be employed as nano adsorbents for the departure of various air pollutants. The surface characteristics managed aforestated contaminants by increasing the reactive sites that could bind with specific airborne pollutants [24]. Additionally, adding functional compounds to the nanoadsorbents for expanding the active sites could alter the adsorptive properties of these nanomaterials [41].

According to Gulati et al. [23] study, there are various ways to clean the air, but employing CNTs as adsorbents is still one of the best and most dependable options. This is because CNTs have an excellent optimal adsorption ability, which makes them ideal for removing contaminants from the air. Also, the various forms of CNTs as single-walled CNTs (SWCNTs), MWCNTs, and others, can be functionalized and organized for the elimination of hazardous gases like ozone, oxides of sulfur and nitrogen, and VOCs, among others [23]. There have been several investigations on surface modification methods, including metal/metal oxide impregnation and functional groups grafting during acid treatment. Covalent or non-covalent functionalization techniques can often be used to achieve surface changes on CNTs, as shown in Fig. 2. The surface modification of CNTs for covalent bond functionalizing depends on a chemical reaction between the conjugation of hydrophilic molecules and the carbon atoms on the surface of CNTs. Contrarily, the non-covalent functionalizing

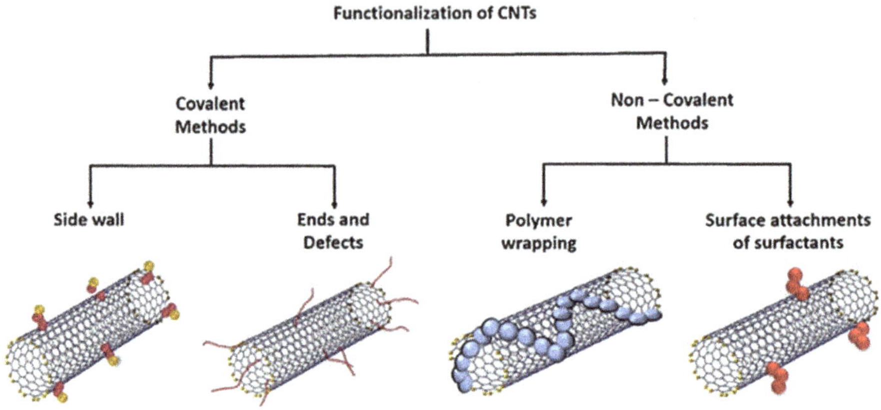

Fig. 2 Functionalizing techniques of CNT [35]

of CNTs entails polymer wrapping and surfactant surface attachment. In addition, the physical adsorption of the surfactant molecules on the hydrophobic portion of CNMs is a necessary step in the process of surfactant functionalization [35].

2.2 Graphene

For us to live in the future, there is a strong desire to safeguard the environment in every manner. Therefore, nowadays, various substances like graphene are being researched for their potential use in environmental air pollution remediation applications. Graphene is the finest form of Carbon and maintains its original structure, allowing for responses from all directions to each atom. Graphene's edge atoms have unusually high chemical reactivity [36]. It is one of the carbon-based nanomaterials and an effective air pollution remediation candidate substance. Carbon atoms in sp^2 hybridization and bonding to three neighboring carbon atoms form the honeycomb-shaped crystal structure of graphene (Fig. 3) [75]. Graphene has 2600 m^2/g specific surface area as reported by Bayantong et al. [9]. Graphene serves as the fundamental precursor for all other dimensionalities of graphitic materials. It can be heaped into three-dimensional graphite, rolled into one-dimensional nanotubes, or wrapped into zero-dimensional fullerenes. Consequently, graphene is just one layer of graphite and has a high inherent property related to carbon nanotubes. Particularly, graphene is thought to be the most powerful, thinnest, single-layer material with low-dimensional physics, optical transparency, better mechanical flexibility, and low coefficient of thermal expansion, as well as most thermal and electricity-conductive substances in existence [51, 83]. Consequently, they have potential uses in filtration, photovoltaics, biological engineering, electronics, and energy storage [9, 32].

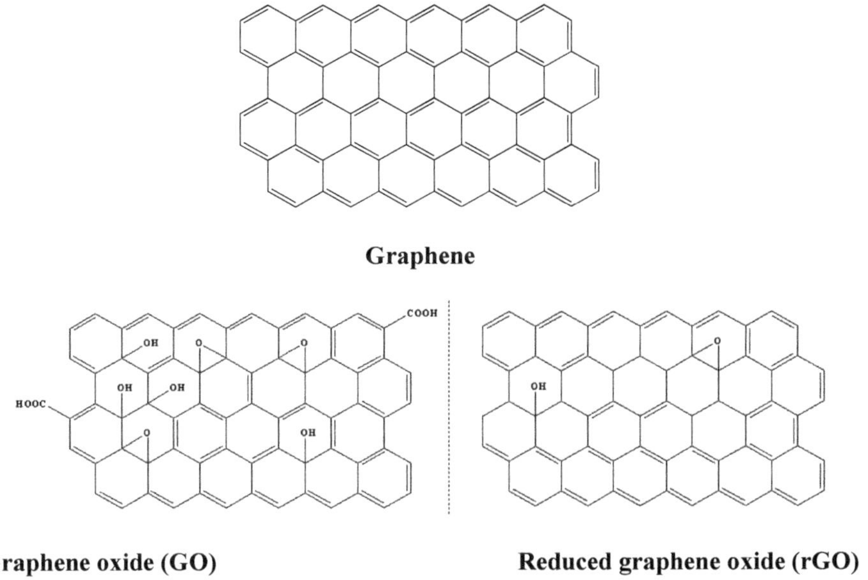

Fig. 3 Structure of Graphene, GO, and rGO [21]

Petal et al. [58] study shows that environmental contamination is one of the biggest intimidations in the world. So graphene and its derivatives lead to various environmental remediation like air pollution sensing and detection. Numerous sensors and effective adsorbents built of graphene can be used to find and get rid of environmental toxins. Graphene has a variety of derivatives. For instance, graphene oxide (GO), reduced graphene oxide (rGO) (Fig. 3), graphene nanoribbons, and so on. The existence of these outspreads the application of graphene [58]. Graphene oxide is one of the promising CNMs for cleaning air pollutants and serves as a hydrogen storage medium in photocatalyst water splitting [36]. Additionally, graphene oxide has outstanding adsorption for cations and basic compounds due to its strong acidity and the presence of many functional groups. Graphene exhibits high chemical adsorption and a hydrophobic surface due to strong π-π interaction. Different nanocomposites can be created by modifying GO or graphene with metal oxides or organic materials to improve adsorption capacity and separation effectiveness. A viable strategy to increase adsorption capacity will be to activate graphene into porous carbonaceous material. Because of these factors, graphene is a preferred material for use in air pollution cleanup applications [3].

2.3 Fullerenes

Carbon compounds are present everywhere. They are in our body, the pencil we use to write with, the table we read from, and the book we read. Fullerene is an allotropic form of Carbon. It is a magical material. Due to its radical scavenging characteristics, fullerene is used in cosmetics as an additive. As seen in Fig. 4, it is composed of fused rings of 5 to 6 carbon atoms joined by single and double bonds to form a closed or partially closed mesh [66]. It was discovered by a scientist named Buckminster, which is why the name Buckminsterfullerene is another name for fullerene. It has a shape similar to a football's. The entire complex resembles the ball that the world enjoys playing in a sport that is widely recognized and celebrated. Fullerenes, particularly the C_{60} sphere, are very symmetrical. Fullerenes are made of pentagonal (or rarely heptagonal) rings, which prevent the sheet from being planar, much like graphite, which is made of a sheet of connected hexagonal rings. They are called buckyballs or buckytubes depending on their shape [56]. Nanotubes are referred to as cylindrical fullerenes. Depending on how many carbon atoms are contained in the molecule, it can exist in main varieties of fullerenes, which are C_{60}, C_{70}, C_{80}, C_{90}, C_{70}, C_{120}, C_{185}, C_{540}, C_{720}, etc. [70]. Fullerene C_{60} is prevalent and does not share an edge of two pentagons (Fig. 4). The average C–C bond distance of a C_{60} molecule is 1.44 A^0. The changes in temperature cause changes in the behavior and structure of fullerene. For instance, the transformation of fullerene into the C_{70} is due to increasing temperature, and at different pressures, the structure of the fullerene also changes. Fullerene has 7.61 eV ionization enthalpy and has an electron affinity between 2.60 and 2.80 eVs. C_{60} looks like an electrophile (it can accept three or more electrons) in chemical reactions and can perform as an oxidizing agent. When doped with alkaline earth or alkali metals, fullerenes can show superconductivity properties. Furthermore, they have ferromagnetism characteristics and are soluble in organic solvents due to the presence of abound carbon molecules [37, 44].

Different shapes and surface patterns have a remarkable influence on the catalytic functions of fullerene, which in turn affects catalyst performance and durability. Metal oxides (MOs) adsorbed on the surfaces of C_{60} can show gas-sensing capabilities in their surface structures. On certain surfaces, greenhouse gases can be collected and held in storage. The adsorption of gases on the MO-C_{60}-fullerene complex is superior to the adsorption of gases on fullerene alone in terms of effectiveness. Metal oxides (MOs) such as NiO, ZnO, and Cu2O occupy various locations when exposed

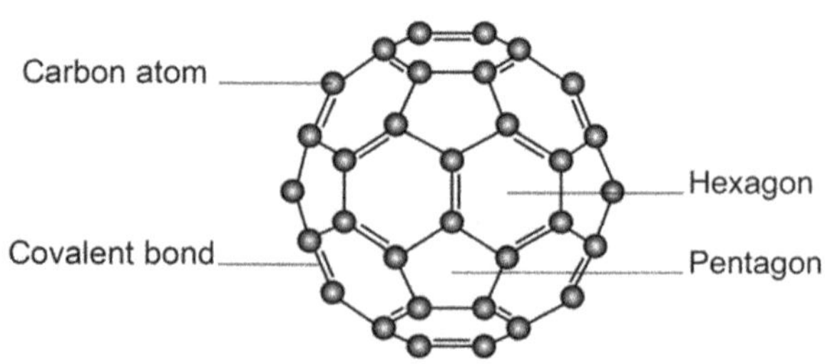

Fig. 4 . Structure of fullerene C_{60} [66]

to C60-fullerene at various positions between rings. The creation of quasi-rings gives the adsorbed molecules more stability and contributes to a cleaner and greener environment [1].

2.4 *Carbon Nanofibers (CNFs)*

Carbon nanofibers (Fig. 5), also known as stacked-cup carbon nanotubes, are solid carbon fibers that differ from CNTs in that they do not contain a hollow chamber. Unlike continuous and Carbon fibers with high micrometers in diameter, CNFs are noncontinuous filaments based on sp^2 and linear form. The relatively high surface area (1877 $m^2\ g^{-1}$), flexibility, high stiffness, homogeneous mesoporous nature, and distribution of CNFs are their distinguishing features from nanostructured materials yet recorded [65].

According to Perez-Rodriguez et al. [60] study, CNFs are nanometer-sized filaments. Their diameters range between 3 and 100 nm and their length varies from 5 to 100 μm. They are formed by stacking graphene sheets in a specific orientation. The core of the filament is rod-like, discontinuously hollow, with a structure resembling bamboo, allowing for the development of tubes. Additionally, the angle between the graphene layers and the fiber growth axis can vary from 0 degrees, which are parallel nanofibers, to 90 degrees, which are platelet nanofibers, with the most common values falling between those two producing fishbone or herringbone nanofibers [60].

Due to their special small-size qualities, carbon nanofibers have better mechanical and chemical belongings than frequently used carbon fibers. As a result, there is a lot of concern about carbon nanofibers in many areas, including catalytic activities, different pollutant adsorption, structural strengthening, nanodevice preparations, etc. [25]. Energy storage, electrochemical catalysis, and the production of sensors are just a few of the many uses for CNFs. CNFs can be added to a variety of matrices,

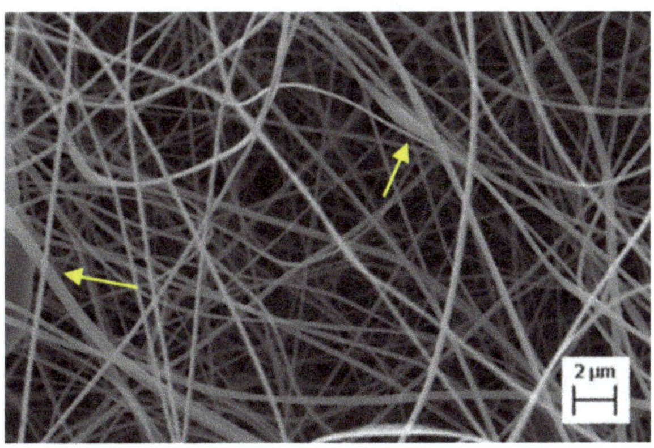

Fig. 5 The carbon nanofiber's structure is obtained from the electrospinning process [2]

such as thermoplastics, thermosets, elastomers, ceramics, and metals, to give them outstanding mechanical characteristics, high electrical conductivity, and high thermal conductivity [8].

3 Mechanisms of Air Pollution Remediation

3.1 Adsorption Activity of Carbon Nanomaterials (CNMs)

Removing air pollutants from mobile sources and other sources is a crucial worldwide concern. At this time, nanotechnology is on an upward trend in its diverse environmental remediation applications. Nanotechnologies that can be used in air pollution remediation are CNMs. CNMs are the best substance for removing any air pollutants in pristine and modified form. This is because of its outstanding physical and chemical characteristics, which permit it to have superior accomplishment in remediating inorganic and organic air contaminants. CNM technologies, including disinfection, hybrid catalysis, adsorption, membrane filtration, monitoring, and sensing, can be applied to remediate air contaminants [12].

Adsorption methods of air cleaning, including those for gas, liquid, and organic and inorganic pollutants adsorption (Fig. 6), have been widely employed in environmental remediation. Adsorption is a process in which the adsorbates in a liquid or gaseous bulk attach to an adsorbent's surface but do not diffuse within it [12]. Adsorption depends on the shape of CNTs. The shape of CNTs is classified as SWCNTs and MWCNTs. SWCNTs differ, having fewer outer and inner diameters than MWCNTs [11]. The bundle geometry and tube diameter of MWCNTs, among other morphologies, have an impact on how atoms or molecules adsorb on them. CNTs having a minimum size, a greater defined surface area, and a special layered structure will have a greater potential for performing the adsorption [72]. Therefore, MWCNTs are more effective than SWCNTs at adsorbing the adsorbate. However, since each pollutant has a unique ability for adsorption because of differences in the bonding networks between its molecules, various contaminants have various adsorption capabilities. The bonding interactions among pollutants' molecules, such as hydrogen bonding interaction, electrostatic interaction, or/and hydrophobic interaction, are the mechanisms by which organic chemicals are adsorbed to CNTs. In addition, the numerous molecule–molecule bonding interactions can influence the mobility characteristics of nanotubes [72].

Additionally, there is heavy metal adsorption between the functional groups and metal ions of CNTs by physical adsorption, electrostatic attraction, surface precipitating, and complexation. In comparison to a single CNT, a bundle of CNTs affords substantially greater adsorption of heavy metals on the outer layer or inside of the NT. External and internal sites are the two different types of adsorption sites found in CNT bundles [6]. While SWCNT's adsorption performance and transmission of charges

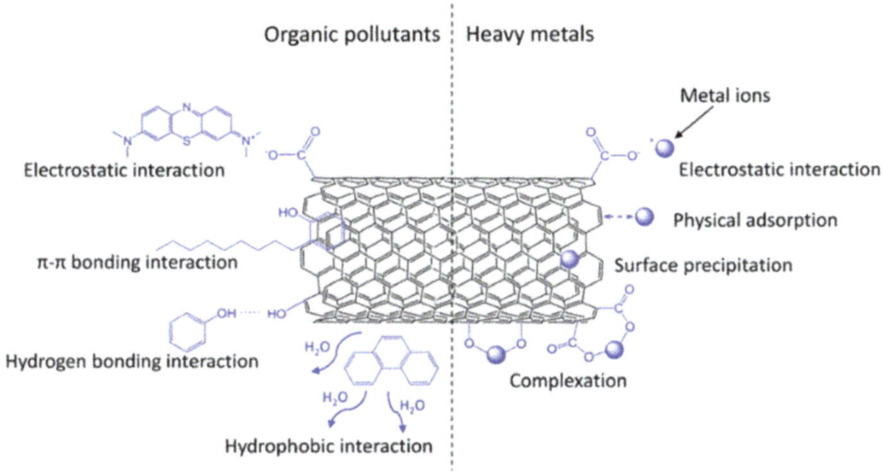

Fig. 6 Adsorption mechanisms of organic contaminants and heavy metals by CNTs [72]

are insufficient and insignificant, CNT in bundles has a strong adsorption performance in interstitial and groove sites, suggesting that the inside site is a little more energetic than the outside site. It follows that increasing the number of nanotubes used will enhance the molecule's adsorption [12].

3.2 *CNMs in Catalytic Activity and Reactive Oxygen Species (ROS) Generation*

When compared to patterns seen with conventional catalyst supports, metals supported on carbon nanostructures show unique catalytic activity and selectivity. These nanoscale materials' extraordinarily high exterior surface areas greatly minimize the mass transfer constraints and a novel interaction between the deposited active phase and the support layer. The existence of an electrical contact between the outer layer of the carbon nanomaterial and the coated species may significantly influence the ultimate characteristics of the supported species. Besides their direct use as catalysts, these nanomaterials, particularly self-supported CNT, have many other beneficial uses. They may also be expected to be used in synthesizing some 1D materials (metal, oxide, or ceramic), which may have new uses in catalysis and several other fields of the developing nanotechnology field [47].

The escalating energy and environmental crisis shows how crucial effective pollutant remediation is. Due to its great effectiveness in reducing toxic contaminants and the oxidative destruction of organic pollutants, which yields H_2O and CO_2 as the end products, photocatalysis is a less-energy and highly applicable technology for ecological remediation. The photocatalytic process has generated much attention in

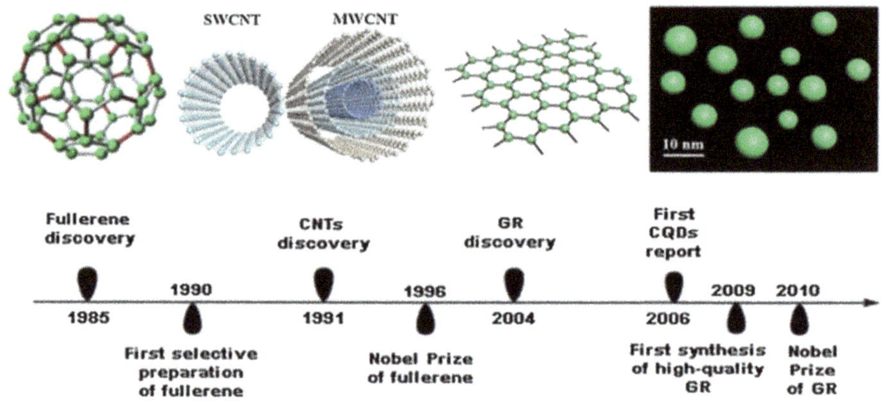

Fig. 7 Scheme of commonly used CNMs with their discovered year [78]

recent years as an effective way to remove environmental pollutants. Due to the less quantum change efficacy of the absorbed light to charge carriers and the substantial carrier recombination, most photocatalytic processes still require improvement. In order to combat the environmental pollution crisis, photocatalysis over materials modified with CNMs is a necessary procedure. CNMs, which include carbon nanofibers, graphene, fullerenes, carbon nanotubes, carbon quantum dots, and others found in various years, are currently among the most promising nanomaterials, as illustrated in Fig. 7. Since CNMs exhibit exceptional morphological, mechanical, electrical, and optical features, photocatalysis has attracted great scientific and technological attention. The structural diversity of CNMs is noteworthy because it improves photocatalytic efficiency [78].

AS Wang et al. study shows that the valence band holes (VB h +) and conduction band electrons (e^-) are created in the photocatalytic process when photocatalysts modified with CNMs are exposed to light [73, 74]. For instance, environmental contaminants like bromate can be reduced directly by electrons, or electrons can be transported to electron acceptors like CNMs and used to oxidize organic pollutants by superoxide radical species. Strongly oxidizing holes can directly participate in photocatalytic oxidation or form hydroxyl radicals when interacting with adsorbed hydroxyl ions [80]. In the photocatalytic destruction of organic pollutants, CNMs are also a good option [73, 74]. For instance, C_{60}-modified ZnAlTi-LDO demonstrated excellent photocatalytic activity when used to break down Bisphenol A (BPA) [34].

TiO_2 films doped with MWCNT were also introduced as effective photocatalytic degradation materials. Higher photocatalytic activity is made possible by the simple transmission of photoinduced electrons from TiO_2 to conductive MWCNTs, made possible by the electrical link between MWCNTs & TiO_2. MWCNT introduction into TiO_2 was restricted to less than 4 weight percent. A film's transparency would be altered by an excessive MWCNT content, which would also keep TiO_2 from coming into contact with light and decrease photocatalytic activity [45]. Many additional CNMs have also been applied to the photocatalytic method of treating organic

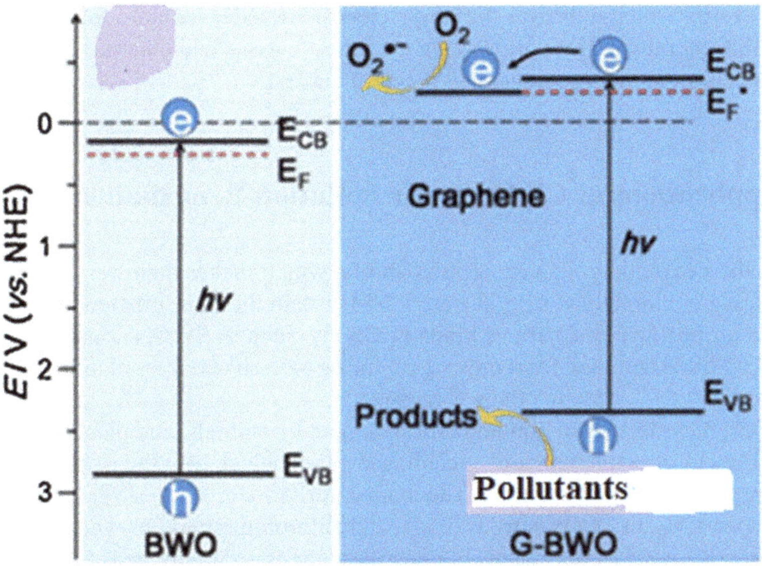

Fig. 8 Photocatalytic mechanism of organic pollutants degradation over graphene/bismuth tungstate (BWO) composite [78]

pollutants. For instance, [39] investigated a better technique for creating a core–shell 1D CNF-titanate nanotube (TiNT) composite & the CNF-TiNT demonstrated excellent photocatalytic performance for gaseous acetaldehyde oxidation. Recently, [85] also reported the development of an N-wrapping/bridging melamine-graphene-TiO_2 capsule for efficient photocatalytic elimination of gaseous formaldehyde (HCHO) [39, 85]. The general catalytic degradation of the pollutant mechanism is shown as follows in Fig. 8.

There are bacteria practically everywhere. *Escherichia coli, staphylococcus aureus, Fusarium oxysporum, and pneumococcus aeruginosa* are just a few of the bacteria that can be dangerous to humans. Through eating and inhaling, these dangerous bacteria can enter a person's body and eventually cause sickness. Due to its excellent oxidation ability, photocatalysis has been established to be an effective disinfection method [82]. Reactive oxygen species (ROS) are created after exposure to light, and they then kill bacterial cells, which is the general process of photocatalytic disinfection. CNMs-photocatalysts composite can accelerate oxygen's electron reduction and increase the production of ROS. As designed, rGO-TiO_2/rGO composite photocatalyst can increase H_2O_2 generation [53]. CNMs can also interact with photocatalysts to establish chemical bonds that increase action, like the Ti–OH and Ti–O-C created in the rGO/TiO_2 decorated MnO_x quantum dots [10]. A Carbon dots (CDs) and TiO_2 co-decorated rGO (CTR) ternary composite photocatalyst for *E. coli* inactivation was recently presented by [83]. The outcome demonstrated that after 60 min of light irradiation, the CTR slurry system achieved 1.03 log inactivation of *E. coli.* In addition to acting as an electron donor and acceptor, CDs also serve as a platform, which improves the single electron transfer required for O_2 reduction [14, 82]. Generally, CNMs for treating greenhouse gases and other air contaminants with the elimination mechanism are listed in Table 1.

4 Applications of CNMs in Air Pollution Remediation

Due to their extremely high adsorption ability, which makes them very beneficial in eliminating contaminants from the air, CNMs retain the most efficient and accurate ways of air purification. Various kinds of CNMs, such as SWCNTs and MWCNTs, can be modified and used for removing polluting gases like oxides of nitrogen, sulfur, carbon, and VOCs, among others (Fig. 9) [23].

Hence, the effect of air contamination on people, animals, and plant life is significant. The environment contains greenhouse gases, which heat the world and are the primary cause of global warming. The main culprits are greenhouse gases containing H_2O vapor, CH_4, O_3, CO_2, nitrous oxide, and chlorofluorocarbons. Diesel and petrol are primarily used as nonrenewable energy sources in urban areas, which are the main sources of air contamination. The main source of emissions from cars and other vehicles is incomplete gasoline combustion. Nitrogen oxides' interaction with the ozone layer reduces its thickness. The chemical component benzene, which is very hazardous to living things, is one of VOC's constituents. Fuel combustion and inefficient combustion both produce VOCs. According to epidemiologic studies, air pollution is a known cause of lung cancer, cardiovascular illnesses, and other conditions. Urban air pollution, the main cause of mortality, is increasing daily [33]. Therefore, Environmentalists are interested in air cleansing. They are seeking alternative technologies to deal with the problems that adversely affect both human existence and the ecosystem. Sorption has been used for storage and purification for a very long time.

Because of their distinct chemical and physical characteristics, CNTs are a revolutionary material utilized to reduce pollution emissions. Using carbon nanotubes CNTs as sorbents, environmental pollution can be reduced. Competence depends on various alternatives when using CNTs to eliminate hazardous gases. Numerous innovative filters have been created using materials like metal or ceramic trusses. The Si/ SiO_2 stainless steel or glass fiber-based CNT filters exhibit better filter productivity for various particle sizes [57]. Carbon-based materials, like active C and carbon fiber, have been widely implemented in gas (NO, VOCs) sorption practices. CNTs are a popular tool for reducing pollutant emissions today, and they also make excellent sorbents for use in environmental applications [55].

According to Xia et al. [77] study shown, the air-polluting gas SO_2 can lead to respiratory illnesses and acid rain [77]. The severity of SO_2 air contamination is increasing along with the growth of contemporary industries. It has become vital to find a technology that can detect and eliminate SO_2 from the air. CNTs are a new family of nanomaterials with many benefits, including a huge specific surface area and a powerful capacity to adsorb too many chemical components [79]. However,

Table 1 Nanoadsorptive CNMs for the treatment of greenhouse gases and other air contaminants with the elimination mechanism

Nanoadsorptives	Kinds of nanoparticles	Marked gases pollutant	Elimination mechanism	Key references
Graphene	GO/nanocomposites	H_2S, N_2, CO_2, SO_2, and NH_3	The gas adsorption is initiated by functional groups (FG) on GO surfaces; a synergetic interaction between metal centers and GO surfaces further this process	Seredych and Bandosz [69], Wu et al. [76]
Fullerene	B_{40}-Fullerene	CO_2	Tough chemisorptions with great CO_2 adsorption skill	Dong et al. [18]
	Fullerene-like boron nitride nanocage	N_2O	N_2O's adsorption & breakdown	Esrafili [19]
CNTs	MWNTs and SWNTs	The mixture of NO and NO_2	As NO travels through CNTs with O_2, it is oxidized to NO_2 and subsequently deposited on the surface of nitrate species	Zhang et al [84]
	(CNTs-APTS), Modified CNTs using 3-aminopropyltriethoxysilane (APTS)	CO_2	CNTs can absorb more CO_2 gases at lower temperatures (20–100 °C) because of the many amine groups that are present on their surface	Mohamed [52]

(continued)

Table 1 (continued)

Nanoadsorptives	Kinds of nanoparticles	Marked gases pollutant	Elimination mechanism	Key references
	SWNTs/NaClO	Isopropyl vapor	Chemical adsorption onto adsorbent surface FG and physical adsorption by van der Waals forces	Mohamed [52]
	CNTs deposited on quartz filters	VOCs	π- π interactions carried out	Amade et al. [4]
	Si and B-doped SWCNTs	CO and methanol gases	The electronic characteristics of SWCNT considerably enhance the adsorption of gas via physisorption or chemisorption	Azam et al. [7]

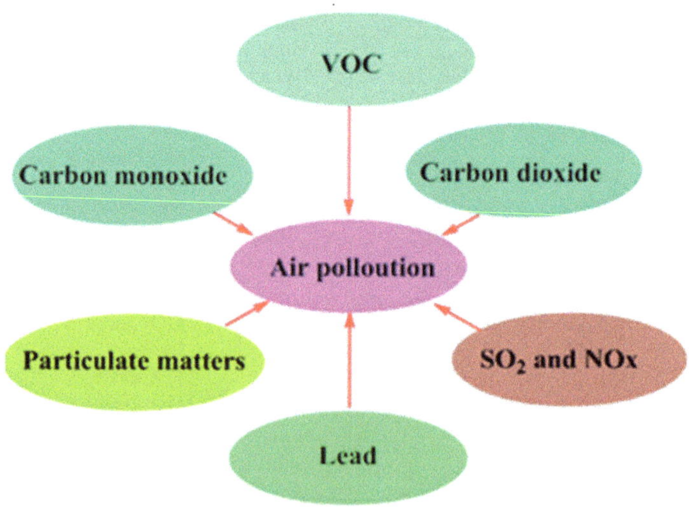

Fig. 9 Main constituents of air pollution [54]

pure CNTs function poorly when trying to adsorb SO_2 molecules. Therefore, it is essential to adjust the chemical and physical properties of CNTs in order to use them as SO_2 gas adsorbents and sensors. In order to enhance the adsorption of SO_2 gas molecules, they frequently introduce vacancy defects on their surface or dope CNTs with metal atoms [50]. Both of these methods can alter the chemical and physical characteristics of CNTs. According to Yoosefian et al. [79] study, Pt/Au doping increased the electrical conductivity of CNTs and increased the amount of charge transfer between Pt/Au and SO_2, which enhanced SO_2 detection in general. However, it is challenging to produce Pt/Au-doped CNTs in large quantities because of the high price and unavailability of Pt and Au. So, it is important to locate more appropriate metal dopants to adjust CNTs in order to make CNT-based SO_2 gas sensors accessible for a variety of applications.

Computational research has also been done by An et al. [5] on the effects of vacancy defects and Fe doping on CNTs on the adsorption of SO_2. Their calculations suggest that the adsorption of SO_2 on CNTs can be significantly improved by Fe doping based on the investigation of geometric structure, charge density, charge transfer, and adsorption energy. More particularly, the adsorption gap between Fe-CNTs and SO_2 is reduced by 44.9% compared to pristine CNTs. Fe-CNTs and SO_2 molecules both see a rise in adsorption energy and charge transfer of 1513.3% and 191.8%, respectively. The adsorption length for Fe-doped single-vacancy-defected CNTs and SO_2 is enhanced by 44.6% compared to the undoped SV-CNTs, while the charge transfer and adsorption energy are raised by 373.9% and 715.0% respectively. One may realize that doping Fe decreases the CNTs' bang gap and boosts electrical conductivity by examining the border orbit and state density. Following the SO_2 molecules adsorption, the intense coupling between Fe-SV-CNTs and Fe-CNTs reinforces their states' density at the Fermi level.

Consequently, doping with Fe will boost the compassion of CNTs to SO_2 adsorption while also increasing the adsorption of SO_2 molecules on them. According to their calculations, Fe-doped CNTs, particularly Fe-SV-CNTs, can be employed as SO_2 molecule-detecting materials [5]. CNMs possess many applications, from biosensors to remediation of environmental contaminants. Mainly, CNMs can detect hazardous substances in food, pharmaceuticals, gene and medication delivery, tissue regeneration, and gene therapy. The general application of CNMs in air pollution remediation is shown in Fig. 10.

5 Challenges in Carbon Nanomaterials Application

Even though CNMs, including CNTs, graphene, fullerene (C_{60}), and so on, are well known in a variety of fields and have significant potential for ecological remediation, there are quite several barriers preventing their widespread implementation in business, air pollution remediation, and daily life [59].

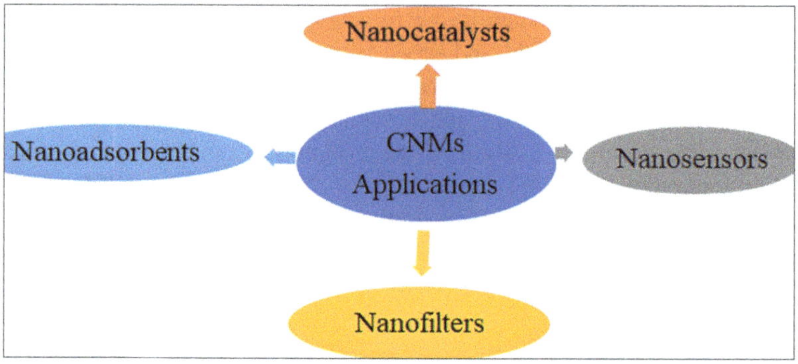

Fig. 10 Applications of CNMs in air pollution remediation (This figure illustration is not clear)

5.1 *Manufacturing Costs*

One of the challenges that prevented CNMs from being used in industrial applications is their high production cost. The manufacturing and consumption of CNMs have declined steadily from 1000.0 USD/kg to fewer than 50 USD/kg, despite recent increases in worldwide CNMs output, falling short of the anticipated yearly growth. However, it is too expensive for corporations to use CNMs in manufacturing [64]. Therefore, low-cost and selective manufacturing for high-quality CNMs is required to reduce the cost.

5.2 *Toxicity and Environmental Risks*

In addition, CNMs may pose concerns to human well-being and ecological risks to the surroundings. The shape and activation mechanisms of CNMs will impact their toxicity [72]. In general, unfunctionalized CNMs are less hazardous than raw CNMs and pose a more significant risk to the surroundings and human well-being [6]. In contrast, unfunctionalized CNMs have the potential to be carcinogenic and develop lung tumors [40]. Additionally, CNMs have been shown to harm various species, including zooplankton and earthworms. However, more knowledge and safety measures about the possible consequences on people and their surroundings [22].

5.3 *Public Acceptance*

Public acceptance has a major role in how technology advances. The development of CNMs for environmental cleanup will be challenging without widespread acceptance

[72]. However, many factors contribute to the public's reduced support. For instance, no safety guidelines exist for CNMs to specify which or what categories of CNMs may influence their toxicity. The limitations of CNMs' industrial applications are still unknown. Additionally, employees exposed to CNMs run the danger of developing hypoxemia & pulmonary illness after coming into touch with a chemical paste that encompasses unidentified NPs [13]. As a result, rules and risk assessments for CNTs should be created to ensure that the public will accept their use in environmental cleanup.

5.4 Advantages and Disadvantages

Consequently, CNMs might be considered a superb technology supporting ecological remediation. They do, nevertheless, have some shortcomings. The advantages and disadvantages of CNMs are outlined in Table 2 [61].

6 Conclusion

The significant and well-organized air pollutant remediation strategies are obvious from exacerbating the ecological and energy disaster in the "green" twenty-first century. This book chapter discusses the role of carbon nanomaterials, especially

Table 2 Advantages and Disadvantages of CNMs

Advantages	Disadvantages
– Because they are lightweight and very compact size, they are a great replacement for metallic wires	– Even after extensive research has been done on CNMs, understanding how they operate is still difficult
– Many different types of resources are required for the creation of CNMs, although only a minimal amount of those resources are needed to produce CNMs	– The nano size of CNMs makes working with them challenging
– CNMs are temperature-resistant because they can function in a wide spread of temperatures	– Production of nanotubes requires expensive operating circumstances
– Improve the mechanical conductivity of composites	– It will take some time for this new technology to substitute the outdated ones because more research is still needed to determine their possible effects on people and the environment
– Nanotube use in hybrid catalysts shortens reaction times and improves operating conditions	– Given how rapidly technology becomes outdated, placing a bet on modern technology can be dangerous

CNTs like SWCNT & MWCNT, graphene oxide, carbon nanofibers, graphene, and fullerene in air pollution remediation and their properties. These carbon nanomaterials have been employed to effectively degrade organic and inorganic air contaminants due to their excellent properties. The properties of CNMs like high surface area, good catalytic activities, heat-resistant, small size, better biocompatibility with living organisms and environment, good surface function, high reactivity, a larger pore diameter, and a large pore volume make them suitable for usage in the industry as nano adsorbents. Along with these characteristics, the CNMs are simple to manufacture in conjunction with other nanomaterials. They are simple to functionalize to give multifunctional nano-adsorbent. Based on this literature assessment, CNMs exhibit intriguing physic-chemical characteristics and have high promise for use in air pollution remediation. However, the price of manufacture, toxicity, green concerns, and public recognition have prevented it from preventing pollution and conserving the environment. It is anticipated that CNMs will have safety regulations soon and risk evaluations to determine the safety of utilizing CNMs. Consequently, this will help advance and improve their use in air pollution and environmental remediation.

References

1. Agrawal PS, Belkhode PN, Brijpuriya DS, Gouda SP, Rokhum SL (2023) Stimulation in fullerene for adsorbing pollutant gases: a review. Chem Phys Impact:100156
2. Alegre C, Modica E, Di Blasi A, Di Blasi O, Busacca C, Ferraro A, Aricò AS, Baglio V (2018) NiCo-loaded carbon nanofibers obtained by electrospinning: bifunctional behavior as air electrodes. Renew Energy 125:250–259
3. Ali N, Riead MDMH, Bilal M, Yang Y, Khan A, Ali F, Karim S, Zhou C, Wenjie Y, Sher F (2021) Adsorptive remediation of environmental pollutants using magnetic hybrid materials as platform adsorbents. Chemosphere 284:131279
4. Amade R, Hussain S, Ocaña IR, Bertran E (2014) Growth and functionalization of carbon nanotubes on quartz filter for environmental applications. J Environ Eng Ecol Sci 3(2):1–7
5. An L, Jia X, Liu Y (2019) Adsorption of SO_2 molecules on Fe-doped carbon nanotubes: the first-principles study. Adsorption 25:217–224
6. Arora B, Attri P (2020) Carbon nanotubes (CNTs): a potential nanomaterial for water purification. J Compos Sci 4(3):135
7. Azam MA, Alias FM, Tack LW, Seman RNAR, Taib MFM (2017) Electronic properties and gas adsorption behaviour of pristine, silicon-, and boron-doped (8, 0) single-walled carbon nanotube: a first principles study. J Mol Graph Modell 75:85–93
8. Azwar E, Mahari WAW, Chuah JH, Vo DVN, Ma NL, Lam WH, Lam SS (2018) Transformation of biomass into carbon nanofiber for supercapacitor application–a review. Int J Hydrogen Energy 43(45):20811–20821
9. Bayantong ARB, Shih YJ, Ong DC, Abarca RRM, Dong CD, de Luna MDG (2021) Adsorptive removal of dye in wastewater by metal ferrite-enabled graphene oxide nanocomposites. Chemosphere 274:129518
10. Cao S, Chen C, Zhang J, Zhang C, Yu W, Liang B, Tsang Y (2015) MnOx quantum dots decorated reduced graphene oxide/TiO_2 nanohybrids for enhanced activity by a UV pre-catalytic microwave method. Appl Catal B Environ 176:500–512
11. Carrales-Alvarado DH, Leyva-Ramos R, Rodríguez-Ramos I, Mendoza-Mendoza E, Moral-Rodríguez AE (2020) Adsorption capacity of different types of carbon nanotubes towards

metronidazole and dimetridazole antibiotics from aqueous solutions: effect of morphology and surface chemistry. Environ Sci Pollut Res 27:17123–17137
12. Chung JH, Hasyimah N, Hussein N(2022) Application of carbon nanotubes (CNTs) for remediation of emerging pollutants-a review. Trop Aquat Soil Pollut 2(1):13–26
13. Das R, Leo BF, Murphy F (2018). The toxic truth about carbon nanotubes in water purification: a perspective view. Nanoscale Res Lett 13:1–10
14. Deng CH, Gong JL, Zeng GM, Jiang Y, Zhang C, Liu HY, Huan SY (2016) Graphene–CdS nanocomposite inactivation performance toward Escherichia coli in the presence of humic acid under visible light irradiation. Chem Eng J 284:41–53
15. Deepti, Bachheti AJ, Bhalla P, Bachheti RK, Husen A (2022a) Growth and development of medicinal plants, and production of secondary metabolites under ozone pollution. In: Husen A (ed) Environmental pollution and medicinal plants. CRC Press, 6000 Broken Sound Parkway NW, Suite 300, Boca Raton, FL 33487–2742, pp 25–45. https://doi.org/10.1201/9781003178866-2
16. Deepti, Bachheti AJ, Chauhan K, Bachheti RK, Husen H (2022b) Impact of UV radiation on the growth and pharmaceutical properties of medicinal plants, In: Husen A (ed) Environmental pollution and medicinal plants. CRC Press, 6000 Broken Sound Parkway NW, Suite 300, Boca Raton, FL 33487–2742, pp 47–64. https://doi.org/10.1201/9781003178866-3
17. Dominski FH, Branco JHL, Buonanno G, Stabile L, da Silva MG, Andrade A (2021) Effects of air pollution on health: a mapping review of systematic reviews and meta-analyses. Environ Res 201:111487
18. Dong H, Lin B, Gilmore K, Hou T, Lee ST, Li Y (2015) B40 fullerene: an efficient material for CO_2 capture, storage and separation. Curr Appl Phys 15(9):1084–1089
19. Esrafili MD (2017) N_2O reduction over a fullerene-like boron nitride nanocage: a DFT study. Phys Lett A 381(25–26):2085-2091
20. Gabal E, Chatterjee S, Ahmed FK, Abd-Elsalam KA (2020) Carbon nanomaterial applications in air pollution remediation. In: Carbon nanomaterials for agri-food and environmental applications. Elsevier, pp 133–153
21. Geetha Bai R, Muthoosamy K, Manickam S, Hilal-Alnaqbi A (2019) Graphene-based 3D scaffolds in tissue engineering: fabrication, applications, and future scope in liver tissue engineering. Int J Nanomed:5753–5783
22. Girardello R, Tasselli S, Baranzini N, Valvassori R, de Eguileor M, Grimaldi A (2015) Effects of carbon nanotube environmental dispersion on an aquatic invertebrate, *Hirudo medicinalis*. Plos One 10(12):e0144361
23. Gulati S, Kumar S, Goyal K, Arora A, Varma RS (2022) Improving the air quality with functionalized carbon nanotubes: sensing and remediation applications in the real world. Chemosphere 299:134468
24. Gupta VK, Saleh TA (2013) Sorption of pollutants by porous Carbon, carbon nanotubes and fullerene-an overview. Environ Sci Pollut Res 20:2828–2843
25. Hu G, Zhang X, Liu X, Yu J, Ding B (2020) Strategies in precursors and post-treatments to strengthen carbon nanofibers. Adv Fiber Mater 2:46–63
26. Husen A (1997) Impact of air pollution on the growth and development of *Datura innoxia* Mill. MSc Dissertation, Jamia Hamdard (Hamdard University), New Delhi, India
27. Husen A (2021) Morpho-anatomical, physiological, biochemical and molecular responses of plants to air pollution. In: Husen A (ed) Harsh environment and plant resilience. Springer International Publishing AG, Gewerbestrasse 11, 6330 Cham, pp 203–234. https://doi.org/10.1007/978-3-030-65912-7_9
28. Husen A (2022) Environmental pollution and medicinal plants. CRC Press, 6000 Broken Sound Parkway NW, Suite 300, Boca Raton, FL 33487–2742. https://doi.org/10.1201/9781003178866
29. Husen A (2022) Plants and their interaction to environmental pollution (Damage detection, adaptation, tolerance, physiological and molecular responses). Elsevier Inc., 50 Hampshire St., 5th Floor, Cambridge, MA 02139, USA. https://doi.org/10.1016/C2020-0-03319-0
30. Husen A, Ali ST, Mahmooduzzafar IM (1999) Structural, functional and biochemical responses of *Datura innoxia* Mill. to coal-smoke pollution. Proc Acad Environ Biol 8:61–72

31. Husen A, Iqbal M (2004) Growth, biomass, pollen viability and photosynthetic efficiency of *Datura innoxia* Mill. under coal-smoke pollution. Ann For 12:182–190
32. Iravani S, Varma RS (2020) Green synthesis, biomedical and biotechnological applications of carbon and graphene quantum dots: a review. Environ Chem Lett 18:703–727
33. Jain N, Gupta E, Kanu NJ (2022). Plethora of Carbon nanotubes applications in various fields–a state-of-the-art-review. Smart Sci 10(1):1–24
34. Ju L, Wu P, Lai X, Yang S, Gong B, Chen M, Zhu N (2017) Synthesis and characterization of Fullerene modified ZnAlTi-LDO in photo-degradation of Bisphenol A under simulated visible light irradiation. Environ Pollut 228:234–244
35. Jun LY, Mubarak NM, Yee MJ, Yon LS, Bing CH, Khalid M, Abdullah EC (2018) An overview of functionalized carbon nanomaterial for organic pollutant removal. J Ind Eng Chem 67:175–186
36. Karthik V, Selvakumar P, Senthil Kumar P, Vo DVN, Gokulakrishnan M, Keerthana P, Tamil Elakkiya V, Rajeswari R (2021) Graphene-based materials for environmental applications: a review. Environ Chem Lett 19(5):3631–3644
37. Kazemzadeh H, Mozafari M (2019) Fullerene-based delivery systems. Drug Discov Today 24(3):898–905
38. Khalid MA, Mumtaz N, Izhar T, Ahmad SA, Ali N (2022) Handbook of environmental materials management. Springer
39. Kim S, Kim M, Kim YK, Hwang SH, Lim SK (2014) Core–shell-structured Carbon nanofiber-titanate nanotubes with enhanced photocatalytic activity. Appl Catal B: Environ 148:170–176
40. Kobayashi N, Izumi H, Morimoto Y (2017) Review of toxicity studies of carbon nanotubes. J Occup Health 59(5):394–407
41. Kyzas GZ, Matis KA (2015) Nanoadsorbents for pollutants removal: a review. J Mol Liquids 203:159–168
42. Lamberti M, Pedata P, Sannolo N, Porto S, De Rosa A, Caraglia M (2015) Carbon nanotubes: properties, biomedical applications, advantages and risks in patients and occupationally-exposed workers. Int J Immunopathol Pharmacol 28(1):4–13
43. Landrigan PJ (2017) Air pollution and health. Lancet Public Health 2(1):e4–e5
44. Lee C, Seo Y, Han J, Hwang J, Jeon I (2023) Perspectives on critical properties of fullerene derivatives for rechargeable battery applications. Carbon 210:118041
45. Léonard GLM, Remy S, Heinrichs B (2016) Doping TiO_2 films with carbon nanotubes to simultaneously optimize antistatic, photocatalytic, and superhydrophilic properties. J Sol-Gel Sci Technol 79:413–425
46. Li JM, Yang HY, Wu SH, Dharmage SC, Jalaludin B, Knibbs LD, Bloom MS, Guo Y, Morawska L, Heinrich J (2023) The associations of particulate matter short-term exposure and serum lipids are modified by vitamin D status: a panel study of young healthy adults. Environ Pollut 317:120686
47. Louis B, Bégin D, Ledoux MJ, Pham-Huu C (2009) Advances in the use of carbon nanomaterials in catalysis. In: Ordered porous solids. Elsevier, pp 621–649
48. Mackenzie J, Turrentine J (2016) Air pollution: everything you need to know. Natural resources defense council
49. Manisalidis I, Stavropoulou E, Stavropoulos A, Bezirtzoglou E (2020) Environmental and health impacts of air pollution: a review. Front Public Health 8:14
50. Meshginqalam B, Alaei S (2018) Transition metals adsorption and conductivity modification in carbon nanotubes: analytical modeling and DFT study. Adsorption 24(6):575–583
51. Mohajeri M, Behnam B, Sahebkar A (2019) Biomedical applications of carbon nanomaterials: drug and gene delivery potentials. J Cell Physiol 234(1):298–319
52. Mohamed EF (2017) Nanotechnology: future of environmental air pollution control. Environ Manage Sustain Develop 6(2):429
53. Moon G, Kim W, Bokare AD, Sung N, Choi W (2014) Solar production of H_2O_2 on reduced graphene oxide–TiO_2 hybrid photocatalysts consisting of earth-abundant elements only. Energy Environ Sci 7(12):4023–4028

54. Nehra S, Sharma R, Kumar D (2021) Removal of air pollutants using graphene nanocomposite. In: Environmental remediation through carbon-based nano composites, pp 275–291
55. Nunes A, Al-Jamal K, Nakajima T, Hariz M, Kostarelos K (2012) Application of carbon nanotubes in neurology: clinical perspectives and toxicological risks. Archiv Toxicol 86:1009–1020
56. Pan Y, Liu X, Zhang W, Liu Z, Zeng G, Shao B, Liang Q, He Q, Yuan X, Huang D (2020) Advances in photocatalysis based on fullerene C60 and its derivatives: properties, mechanism, synthesis, and applications. Appl Catal B Environ 265:118579
57. Park JH, Yoon KY, Na H, Kim YS, Hwang J, Kim J, Yoon YH (2011) Fabrication of a multi-walled carbon nanotube-deposited glass fiber air filter for the enhancement of nano and submicron aerosol particle filtration and additional antibacterial efficacy. Sci Total Environ 409(19):4132–4138
58. Patel KP, Kamble SS, Boraste DR, Shankarling G (2020) Green transamidation is catalyzed by graphene oxide under concentrated solar irradiation. Environ Chem Lett 18:1731–1735
59. Peng Z, Liu X, Zhang W, Zeng Z, Liu Z, Zhang C, Liu Y, Shao B, Liang Q, Tang W et al (2020) Advances in the application, toxicity and degradation of carbon nanomaterials in the environment: a review. Environ Int 134:105298
60. Pérez-Rodríguez S, Alegre C, Sebastián D, Lázaro M (2021) Emerging carbon nanostructures in electrochemical processes. In: Emerging carbon materials for catalysis. Elsevier, pp 353–388
61. Pitroda J, Jethwa B, Dave SK (2016) A critical review on carbon nanotubes. Int J Constr Res Civ Eng 2(5):36–42
62. Rahman S, Husen A (2022) Impact of sulphur dioxide deposition on medicinal plants' growth and production of active constituents. In: Husen A (ed) Environmental pollution and medicinal plants. CRC Press 6000 Broken Sound Parkway NW, Suite 300, Boca Raton, FL 33487–2742, pp 65–93. https://doi.org/10.1201/9781003178866-4
63. Rahman S, Mehta S, Husen A (2023) Plants and their unexpected response to environmental pollution: an overview. In: Husen A (ed) Plants and their interaction to environmental pollution (Damage detection, adaptation, tolerance, physiological and molecular responses). Elsevier Inc., 50 Hampshire St., 5th Floor, Cambridge, MA 02139, USA, pp 1–23. https://doi.org/10.1016/B978-0-323-99978-6.00004-2
64. Rana S, Fangueiro R (2016) Fibrous and textile materials for composite applications. Springer
65. Rao RN, Albaseer SS (2018) Nanomaterials in chromatographic sample preparations. In: Nanomaterials in chromatography. Elsevier, pp 201–231
66. Rondags A, Yuen WY, Jonkman MF, Horváth B (2017) Fullerene C60 with cytoprotective and cytotoxic potential: prospects as a novel treatment agent in Dermatology? Exp Dermatol 26(3):220–224
67. Saleem H, Zaidi SJ, Ismail AF, Goh PS (2022) Advances of nanomaterials for air pollution remediation and their impacts on the environment. Chemosphere 287:132083
68. Scoville C, Cole R, Hogg J, Farooque O, Russell A (1991) Carbon nanotubes. Notes Lect:1–11
69. Seredych M, Bandosz TJ (2012) Manganese oxide and graphite oxide/MnO_2 composites as reactive adsorbents of ammonia at ambient conditions. Microporous Mesoporous Mater 150:55–63
70. Sereni JGR (2016) Reference module in materials science and materials engineering
71. Sharma N, Agarwal AK, Eastwood P, Gupta T, Singh AP (2018) Introduction to air pollution and its control. Air Pollut Control:3–7
72. Song B, Xu P, Zeng G, Gong J, Zhang P, Feng H, Liu Y, Ren X (2018) Carbon nanotube-based environmental technologies: the adopted properties, primary mechanisms, and challenges. Rev Environ Sci Bio/Technol 17:571–590
73. Wang Y, Cao H, Chen L, Chen C, Duan X, Xie Y, Song W, Sun H, Wang S (2018) Tailored synthesis of actively reduced graphene oxides from waste graphite: structural defects and pollutant-dependent reactive radicals in aqueous organics decontamination. Appl Catal B: Environ 229:71–80
74. Wang R, Lu KQ, Zhang F, Tang ZR, Xu YJ (2018) 3D carbon quantum dots/graphene aerogel as a metal-free catalyst for enhanced photosensitization efficiency. Appl Catal B: Environ 233:11–18

75. Weinberg M, Staarmann C, Ölschläger C, Simonet J, Sengstock K (2016) Breaking inversion symmetry in a state-dependent honeycomb lattice: artificial graphene with a tunable band gap. 2D Mater 3(2):024005
76. Wu S, He Q, Tan C, Wang Y, Zhang H (2013) Graphene–based electrochemical sensors. Small 9(8):1160–1172
77. Xia L, Robock A, Tilmes S, Neely III RR (2016) Stratospheric sulfate geoengineering could enhance the terrestrial photosynthesis rate. Atmos Chem Phys 16(3):1479–1489
78. Yi H, Huang D, Qin L, Zeng G, Lai C, Cheng M, Ye S, Song B, Ren X, Guo X (2018) Selective prepared carbon nanomaterials for advanced photocatalytic application in environmental pollutant treatment and hydrogen production. Appl Catal B: Environ 239:408–424
79. Yoosefian M, Pakpour A, Etminan N (2018) Nanofilter platform based on functionalized carbon nanotubes for adsorption and elimination of Acrolein, a toxicant in cigarette smoke. Appl Surf Sci 444:598–603
80. Yu L, Ruan S, Xu X, Zou R, Hu J (2017) One-dimensional nanomaterial-assembled macroscopic membranes for water treatment. Nano Today 17:79–95
81. Zeng X, Wang G, Liu Y, Zhang X (2017) Graphene-based antimicrobial nanomaterials: rational design and applications for water disinfection and microbial control. Environ Sci Nano 4(12):2248–2266
82. Zeng X, Wang Z, Meng N, McCarthy DT, Deletic A, Pan JH, Zhang X (2017) Highly dispersed TiO_2 nanocrystals and carbon dots on reduced graphene oxide: ternary nanocomposites for accelerated photocatalytic water disinfection. Appl Catal B Environ 202:33–41
83. Zhang W, Xu C, Ma C, Wang Y, Zhang K, Li F, Liu C, Cheng HM, Du Y et al (2017) Nitrogen-superdoped 3D graphene networks for high-performance supercapacitors. Adv Mater 29(36):1701677
84. Zhang X, Yang B, Dai Z, Luo C (2012) The gas response of hydroxyl-modified SWCNTs and carboxyl-modified SWCNTs to H_2S and SO_2. Przeglad Elektrotechniczny 88(7B):311–314
85. Zhu M, Muhammad Y, Hu P, Wang B, Wu Y, Sun X, Tong Z, Zhao Z (2018) Enhanced interfacial contact of dopamine-bridged melamine-graphene/TiO_2 nano-capsules for efficient photocatalytic degradation of gaseous formaldehyde. Appl Catal B 232:182–193

Carbon Nanomaterial for Oil Spill Clean-Up

Saumya Pandey and Mayank Saxena

Abstract Oil spills are a pressing issue as they impose a risk on the marine ecosystem and have severe economic and social consequences. Conventional methods for oil spill clean-up, such as mechanical containment and chemical dispersants, have limitations regarding efficacy, ecological footprint, and economic viability. Recently, the development of technology has revolutionized new possibilities for more efficient and sustainable oil spill clean-up strategies. Carbon nanomaterials have merged as a promising candidate among advanced materials due to their unique physiochemical properties and diverse applications. Carbon nanomaterials, including carbon nanotubes, graphene, carbon nanofibers, and carbon aerogels, display exceptional superhydrophobicity, chemical stability, high surface area, and mechanical strength, making them exceedingly suited for oil spill remediation. This chapter deals with a comprehensive overview of the carbon nanomaterials application in oil spill clean-up. Also, the various types of carbon nanomaterials and their unique properties that contribute to their effectiveness in oil sorption and separation have been discussed in detail.

Keywords Superhydrophobicity · Nanocarbon aerogels · Carbon nanotubes · Oil/ water separation · Reusability

1 Introduction

Oil spill refers to the inadvertent or deliberate release of petroleum hydrocarbon in the environment, particularly the aquatic ecosystem. Winds and currents can further carry the oil spill, and distributed oil either undergoes evaporation and formation of surface slick, dispersion within the water, or sinking and deposition in sediments [54, 73]. Moreover, the combination of temperature, salinity, and wave dynamics further augments the rates of oil transportation and weathering [25]. Considering that most oils remain on the water surface, the primary victims of oil spills are sea

S. Pandey (✉) · M. Saxena
Department of Biotechnology, IILM University, Greater Noida 201306, India
e-mail: saumbhavi@gmail.com

A. Bachheti et al. (eds.), *Carbon-Based Nanomaterials*, Smart Nanomaterials Technology, https://doi.org/10.1007/978-981-97-0240-4_15

birds and marine animals [11]. However, oil reaching the shore can also threaten other organisms, including terrestrial animals [16]. Over the years, many researchers have thoroughly examined the various facets and impacts of the spill on the ecosystem and fabricated various materials and techniques for oil spill clean-up. Nevertheless, the magnitude and geography of the spill dictate the precise clean-up technology that should be implemented.

Oil spill remediation techniques can be categorized into (1) mechanical (boom, skimmers, hydrophobic meshes, sorbents) [2, 13], (2) chemical (dispersants, emulsifiers, surfactants) [63, 69], (3) biological methods (biological species such as bacteria, fungi) [61, 98], (4) thermal methods (in situ burning) [14] (Fig. 1). Sorbent materials, when compared to other approaches, offer distinct advantages for oil spill clean-up [82, 103]. They can collect and completely remove oil in situ from the water surface without causing any detrimental impact on the environment [29, 30, 38]. An ideal sorbent material should exhibit hydrophobicity and oleophilicity, high absorptivity, and be economically viable in terms of material cost [36, 55]. Nevertheless, conventional oil sorbents, such as plant-based materials, mineral products, and synthetic materials, encounter several challenges, including limited oil sorption capacity, incomplete oil/water separation, limited reusability, and ineffectiveness in harsh conditions [29, 30, 103]. These techniques also have limitations in terms of their efficiency and the generation of additional sludge during the clean-up process. Therefore, nanotechnology has garnered significant attention as a potential solution for removing oil spills [32, 58].

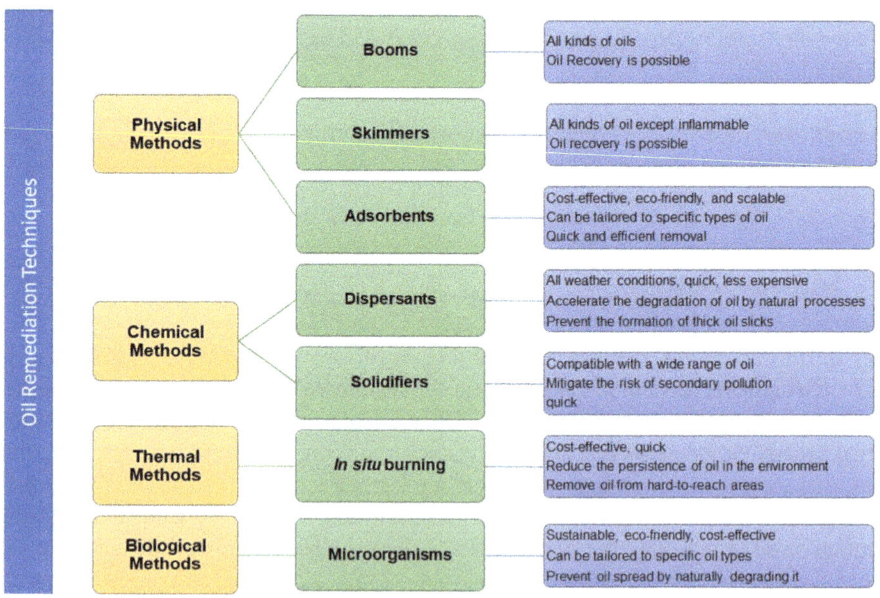

Fig. 1 Different techniques used for oil spill clean-up

Over the past decade, numerous studies have been conducted to explore and develop various techniques for creating rough surfaces with superhydrophobic properties [38, 60]. By manipulating surface morphology and chemistry, researchers have been able to achieve sorbent materials, including but not limited to organic–inorganic hybrids, crosslinked polymers, nanocomposites, and carbon-based materials with extreme superhydrophobic characteristics [57, 64]. Carbon-based materials have emerged as highly favorable candidates for developing superhydrophobic and superoleophilic surfaces.

Carbon-based nanomaterials encompass a wide range of structures, including graphene nanomaterials, carbon nanotubes, carbon nanofibers, and carbon aerogel. These carbon-based nanomaterials exhibit unique physiochemical properties such as high surface area, excellent mechanical strength, hydrophobicity, and chemical stability [29, 30, 33, 36]. These attributes make them well-suited for absorption, separation, and oil removal from contaminated sites. Moreover, their compatibility with various oil types and their ability to be tailored for specific applications make them attractive for addressing the diverse challenges posed by oil spills.

This chapter aims to provide a comprehensive overview of the recent advancements in the utilization of carbon nanomaterials for oil spill clean-up. It explores the synthesis methods, physiochemical properties, and performance of different carbon-based nanomaterials in oil sorption and separation. Furthermore, the challenges and prospects in the field of carbon-based nanomaterials for oil spill clean-up are addressed.

2 Ecosystem and Societal Impact of Oil Spill

The contamination of oil spills in the environment has an adverse effect on the population [37, 89]. The oil spills get absorbed in the soil or mixed with water and damage the ecosystem to a larger extent [22]. It is serious as it impacts the marine and coastal protected areas (MPAs) specially designated for protecting and preserving biodiversity [78].

The oil spill contamination leads to deforestation, decrement in biodiversity, and cause environmental pollution. This may disrupt the lifestyle of the people living around that ecosystem [83]. In general, pollution affects the environment and has far-reaching impacts on various aspects of human life. The consequences of pollution extend beyond environmental degradation and can significantly impact lifestyle, culture, tradition, norms, values, local economy, and overall societal development [62].

Additionally, oil spill disrupts the pattern of species leading to an imbalance in ecosystems. The toxic components in the oil can harm aquatic and terrestrial organisms, including fish, birds, mammals, and plants. The oil spill also disrupts the delicate food web dynamics and alters the ecosystem. Also, the distribution and persistence of oil sediments can cause changes in landscape and geomorphology further impacting habitats and altering natural processes [22]. Furthermore, soil contamination resulting from oil spills can impair soil quality, affecting plant growth and overall ecosystem productivity [3]. Oil spillage can indeed lead to conflict between neighboring countries in coastal regions, particularly when there is overlapping use of natural resources [5, 94]. These diverse impacts underscore the urgent need for effective prevention, response, and mitigation strategies to minimize the ecological and environmental damage caused by oil spills.

3 Superhydrophobic and Superoleophilic Carbon-Based Material

Superhydrophobic materials have recently attracted attention from researchers worldwide because of their significance in fundamental research and commercial prospects. Owing to the different oil/water interfacial phenomena, the wetting behavior of the solid surfaces has been exploited effectively to develop oil/water separation materials [45, 101]. Recently a range of carbon-based superhydrophobic materials has been fabricated by optimizing the surface chemistry and roughness of materials such as carbon nanofibers, carbon fibers, graphite nanofibers, porous carbon, core/shell/satellite composite particles, and many more [9, 74, 91]. These carbon-based materials have emerged as promising materials for oil–water separation because of their outstanding electronic and optical properties [7, 81].

The water contact angle for carbon fibers has been reported as 120° for flat surfaces. Thus, the hydrophobicity of carbon-based materials is generally enhanced by two approaches. The first approach involves developing the micro- and nano-surface roughness using different top-down methods such as annealing, sandblasting, laser ablation, lithography, and etching, and bottom-up methods such as templating, electrodeposition, casting, and replication, layer by layer assembly, rough polymer films, hydrothermal anodizing, colloidal assembly, and 3D printing [39, 72, 108]. Second, several organic, inorganic, and organic–inorganic composite chemicals are introduced to attain proper surface energy for superhydrophobic and superoleophilic wettability conditions [10, 51, 81]. Shayesteh et al. [75] have used a simple dip-coating and silicone adhesive approach to fabricate superhydrophobic/superoleophilic foamy-based porous carbon and multi-walled carbon nanotube (MWCNT). The fabricated PC/MF and MWCNT/MF are eco-friendly and reusable as well as show significant superhydrophobicity with water contact angles $159.34° \pm 1.9°$ and $156.42° \pm 1.6$, respectively. Also, the absorption capacity was found to remain unchanged in acidic (pH = 2), alkaline (pH = 12), and saline solutions (3.5% NaCl), making them potential materials for oil spill clean-up in different harsh conditions.

4 Carbon Nanomaterials for Oil Spill Clean-up

Carbon materials, with their intrinsic hydrophobic properties, offer promising solutions for clean-up oil spill. Additionally, the remarkable characteristics of these materials, including their large surface area, low density, exceptional flexibility, excellent chemical stability, eco-friendly, and significant pore volume, render them highly suitable for various oil recovery applications [65]. In recent times, an expansion in the range of carbon-based oil sorbents beyond conventional materials like activated charcoal, carbon microbes, and graphite [15, 49, 100]. This extension includes the utilization of emerging nanostructured carbons, such as carbon nanoparticles, carbon nanotubes, carbon nanofibers, and graphene-based materials [19, 90] (Fig. 2). Despite promising results exhibited by nanomaterial-based absorbents, employing nanomaterials alone for large-scale applications can be costly, intricate, and time-intensive. Conversely, a more streamlined approach can be achieved by integrating nanomaterials into a polymer matrix and developing selective sponge sorbent. This innovation allows for the amalgamation of the exceptional sorption capabilities of carbonaceous materials with the cost-effectiveness and practicality offered by polymer technology.

4.1 Carbon Nanotube-Based Sorbents

Carbon nanotubes have shown significant potential in various applications, particularly in the field of oil–water separation and gas absorption [4]. They offer several advantages that make them attractive for these applications, including their exceptional mechanical properties, high sorbent capacity, rapid sorption kinetics, and ability to engineer their surface chemistry [1, 44]. Several approaches have been established to fabricate carbon nanotubes, including chemical vapor deposition (CVD), arc discharge, pyrolysis, laser ablation, and electrochemical approaches [23, 95]. Each of these synthesis methods has its distinct benefits and challenges, and

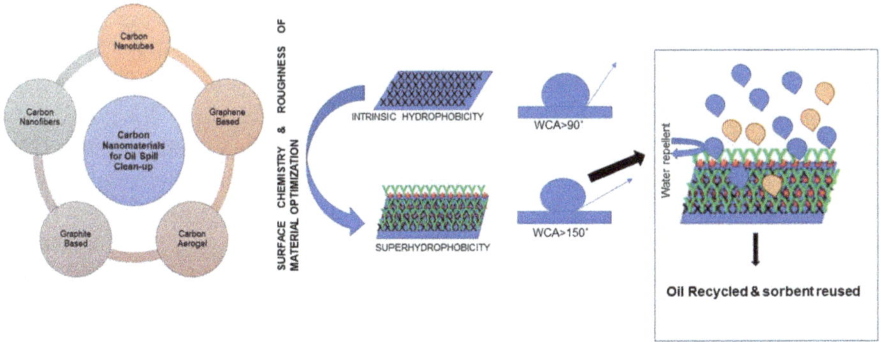

Fig. 2 The carbon nanomaterial-based sorbents used in oil spill clean-up

selecting a specific technique depends on factors such as desired nanotube properties, scalability, and cost considerations.

Carbon nanotubes demonstrate notably superior oil sorption capacity, distinguishing them from conventional sorbent materials used in oil spill clean-up. The oil sorption capacity of carbon nanotubes has been experimentally determined to reach a maximum of 93.2 g/g, which is approximately 12 times higher than that of any other materials employed in similar applications [47]. Moreover, carbon nanotubes possess remarkable superhydrophobic characteristics and exceptional affinity toward oil, enabling them to absorb oil from oil–water interfaces exclusively [19, 23]. Remarkably, the material showcases excellent tensile strength attributable to its honeycomb lattice structure comprising hexagonal carbon unit cells, ensuring its structural integrity remains intact even in turbulent water conditions. It demonstrates favorable magnetic properties that facilitate recovery [68, 77]. Notably, the absorbed oil can be effortlessly recollected by simply squeezing the sponge. Moreover, the attribute reusability makes it cost-effective and eco-friendly [1, 13]. Implementing this technique in oil spill remediation would usher in new possibilities for retrieving spilled oil from the ocean, playing a pivotal role in combating pollution.

CNTs coated sponges have also been widely used as reusable oil sorbent scaffold in water displaying high oil sorption capacity and good reusability, as reported by previous studies [76, 105]. Sponges are 3D porous polymer materials capable of absorbing both oil and water. However, when coated with CNTs, their behaviors change to absorb only oil, completely repelling water selectively. CNT coating creates micro- and nano-textured surfaces, providing superhydrophobic properties. Moreover, CNT coating significantly improves the thermal and mechanical stability of the sponges. CNT-coated sponges display an exceptional absorption capacity, capable of absorbing liquids at levels significantly exceeding their weight, typically 180 times [34]. The absorbed oil can be retrieved by employing various strategies such as mechanical squeezing, vacuum sucking, or other appropriate techniques. Zhu et al. [107] reported a significantly higher oil absorption capacity (12–13.5 times) in CNT sponges (92.30 g/g) compared to polypropylene fiber fabric (7.45 g/g) and woolen felt (6.74 g/g).

Additionally, carbon nanotubes can be functionalized to exhibit magnetic properties. This can be achieved by introducing magnetic nanoparticles onto or within the carbon nanotube structure. The magnetic properties of the modified CNTs allow for their easy manipulation and are valuable in the recovery processes, enhancing their functionality in various applications. Gui et al. [35] used the chemical vapor deposition (CVD) method to fabricate carbon nanotube sponges and magnetic CNT sponges. The resulting CNT sponges demonstrated a porous tridimensionality with interconnected network and pore sizes ranging from nanometres to a few micrometers. The CNT sponges demonstrated outstanding mechanical properties, including substantial strain deformation with nearly complete-volume restoration and structural fatigue resilience under cyclic stress conditions in oils. The carbon nanotube sponge exhibited higher absorption capacity than the magnetic nanotube due to its reduced density and enhanced hydrophobic properties.

Furthermore, the carbon nanotube sponge demonstrated remarkable reusability and can be reused over 1000 times without experiencing significant changes in structure, water contact angle, or adsorption capacities. Similarly, [28] used the hot-pressing method to synthesize superhydrophobic magnetic bulk material using PTFE, multiwall CNT, and iron oxide. The bulk material has exhibited thermal stability and superhydrophobic properties with a water contact angle (WCA) of 158° and an absorption capacity of 0.71–1.15 times its weight.

Recently several studies have been focused on leveraging the unique properties of CNTs, in developing superhydrophobic natural fibers-based sorbent materials with application in oil spill clean-up and oil/water separation, among others. Neelamegan et al. [59] successfully modified cellulose and lignin using CNTs and immobilized them with iron oxide, enabling easier separation. The water contact angles of modified materials ranged from 130 to 135°, indicating their enhanced superhydrophobicity. Moreover, the modified materials exhibited excellent absorption capacity in the range of 15–25 g/g and 10–23 g/g for CNT-g-cellulose/Fe_30_4 and CNT-g-lignin/Fe_30_4 for various oil and organic solvents. The specific sorption depends on the density of the liquid being absorbed.

4.2 Graphene-Based Sorbent

Graphene is a monolayer of sp2 hybridized carbon atoms densely packed in a hexagonal honeycomb lattice [79]. The inherent hydrophobicity, outstanding chemical, thermal, and mechanical stability, and a high specific surface area have garnered significant interest in using graphene-based material for oil spill clean-up [21, 46]. These materials have shown significant potential in effectively adsorbing and removing oil contaminants from water due to their unique properties and stability in harsh environmental conditions [20, 26]. Graphene can be developed through various techniques, including mechanical exfoliation, chemical vapor deposition [87], reduction of graphene oxide [20], carbon dioxide reduction [18], and carbon nanotube silencing [48]. These methods enable graphene production with different properties and characteristics, catering to specific applications and research requirements. For example, [71] have combined nanoporous graphene (NPG) and graphene oxide (GO) flakes with the porous structure of polyurethane (PU), resulting in the fabrication of NPG/PU and GO/PU composites, respectively. The NPG/PU sponge exhibited a higher absorption capacity than the GO/PU, thus having potential uses in oil spill clean-up. Similarly, [67] synthesized graphene composites with synthesized silane functionalized reduced graphene oxide nanoribbons (silane-f-rGONRs) and polyurethane sponges using a dip-coating approach. These composites hold great potential in addressing oil pollution in water owing to their superhydrophobicity, mechanical flexibility, and electrical conductivity.

Natural graphite rocks have been used to fabricate graphene-carbon nanotube aerogels. Introducing carbon nanotube into the interconnected network structure augments the aerogel's hydrophobicity and robustness and results in aerogels with remarkable oleophilic properties and porosity [80]. Rafiee et al. [70] have demonstrated that the combination of surface roughness and surface chemistry of graphene sheets can significantly modify the wettability of the substrate. Several novel approaches have been developed to prepare 3D graphene foam that exhibits outstanding oil adsorption capacity. Liu et al. [53] employed an ion sputtering technique to load gold nanoparticles surface of graphene, then treated the surface with 1H, 1H, 2H, and 2H—perfluorodecanethiol (PFDT). The fabricated graphene foam exhibits porosity, water repellency, and significant specific surface area. The exceptional superhydrophobic properties observed in the materials can be attributed to the synergistic impact of surface energy reduction by PFDT and the rough surface morphology of gold nanoparticles.

Graphene-based sponges with superhydrophobic and superoleophilic properties are highly efficient absorbents for various oils and organic solvents. These sponges exhibit exceptional selectivity, outstanding recyclability, and impressive absorption capacities of up to 165 times their own weight. Nguyen et al. [60] developed graphene nanosheets coated melamine sponge that exhibits considerably higher absorption capacities compared to nanowire membranes (<20 times), microporous polymers (<33 times) and comparable to those of carbon nanotubes. However, the fabrication process of the graphene-based sponge is more facile and easier to scale up.

4.3 Nanocarbon Aerogels

In recent years, carbon-based aerogels have garnered attention in diverse domains such as storage, thermal insulation, sensors, environmental clean-up, and many more owing to their distinctive physical characteristics, encompassing low apparent density, hydrophobicity, electrical conductivity, porosity, and specific surface area [17, 27].

The fabrication of carbon aerogels comprises the subsequent phases, (1) Gelation phase, involving the formation and strengthening of the gel through sol–gel aging procedures, (2) Drying involves the formation of aerogels by evaporating the solvent under ambient freeze or supercritical conditions, (3) Carbonization formation of carbon aerogel by subjecting the aerogel to a higher temperature in a continuous flow of nitrogen atmosphere [27]. Currently, a wide range of commercially available carbon aerogels is available that have been employed in real-world applications. Based on the diverse origins of precursors, carbon aerogels are typically classified into different categories, namely, polymer carbon aerogels (PCAs) [88], biomass carbon aerogels (BCAs), graphene-carbon aerogels (GCAs) [99], Carbon nanotube carbon aerogels (CNCAs) [104], and composite carbon aerogels (CCAs) [50]. Each type of carbon aerogel demonstrates distinctive characteristics and is suitable for a range of applications.

Carbon aerogels' exceptional adsorption and catalytic performance are attributed to the microscopic properties of carbon nanomaterials and the macroscopic gel structure [29, 30]. Furthermore, they possess efficient recyclability owing to their macrostructure, remarkable elasticity, and high-temperature resistance [24, 40]. Hence their deployment in environmental clean-up has rapidly increased and generated substantial scientific interest [6]. Carbon aerogels have found extensive applications in the remediation process due to their exceptional adsorption and catalytic properties, such as oil/water separation, elimination of heavy metals and volatile compounds, and other related processes [52, 106].

Aerogel typically exhibits brittleness, which poses a drawback as the mechanical properties play a crucial role in material reusability and oil recovery [56]. Insufficient mechanical strength can lead to material and reduced sorption capacity. Wu et al. [96] synthesized a carbon nanofiber aerogel through the hydrothermal carbonization technique using glucose, followed by freeze-drying and pyrolysis at 1450° and utilized it for oil/water separation. Nanofiber aerogel exhibited characteristics such as low density (10 mg/ml), substantial specific surface area (547 m^2/g), commendable mechanical strength, and high porosity (>99%). This material also demonstrated hydrophobicity (WCA of 135.8°) superoleophilicity (oil contact angle of 0°), and high adsorptive capacity of 50–140 g/g). Furthermore, the material proved recyclable with consistent performance over ten cycles, as illustrated by the successful recovery of n-hexane through distillation at 69 °C. Similarly, [35] fabricated a magnetic CNT aerogel capable of sorbing diesel with a remarkable mass sorption capacity of 56 g/g. the aerogel can be easily recycled using magnetic force and desorbed through simple heat treatment. Additionally, CNTs have been incorporated into graphene aerogels to enhance mechanical strength and specific surface area. For example, [42] have proposed a microwave-assisted technique for the fabrication of hybrid aerogels comprising multiwall carbon nanotubes (MWCNTs) and graphene. This method enables the vertical integration of MWCNTs onto the graphene wall's cell wall structure, leading to a decrease in heteroatom content and facilitating the formation of a "lotus-like" structure with significantly increased surface roughness, thereby enhancing hydrophobicity. This hybrid monolith displayed outstanding superhydrophobicity and superoleophilicity, coupled with a considerable pore volume, large pore size, and excellent compressibility (90%).

Despite the fascinating performance of polymer-, graphene-, and CNT-based aerogels in selective oil/water separation, their feasibility for industrial application is hampered by their high cost and the complexity of their preparation methods. Lately, biomass resources have been fabricated into carbon aerogels enabling their utilization in oil/water separation. Cao et al. [17] successfully fabricated lignin-based multi-structured nanofibers aerogels from co-electrospun nanofibers. Lignin-based pre-oxidized aerogels (LPA) resembling hierarchically natural wood-tracheid-like structure have been prepared by peroxidation of assembled uncrosslinked core–shell nanofibers. This fabricated LPA has demonstrated remarkable hydrophobicity and significant adsorptive capacity (reaching up to 103 g/g) for various oils and organic solvents.

4.4 Graphite Nanomaterials

Exfoliated graphite has emerged as a candidate for oil spill clean-up due to its ability to selectively adsorb oil/water mixture [43, 84, 86]. Exfoliated graphite can be obtained by subjecting natural graphite to high temperatures (around 1000–1200 °C), leading to its rapid thermal decomposition. Under high-temperature conditions, natural graphite undergoes decomposition, releasing various gaseous compounds that facilitate the exfoliation process, ultimately forming a porous structure [36]. Various methods can be employed to achieve the exfoliation of natural graphite, including microwave irradiation [93] and chemical intercalation followed by subsequent thermal treatment [12, 41]. Ghasemi et al. [31] demonstrated the successful application of expanded graphite in removing oil from seawater, highlighting its advantageous features, such as a simple synthesis process, affordability, and commendable sorption capability. Nonetheless, exfoliated graphite has certain drawbacks, including fragility, low mass, and low expanded ratio through single-step processing, so many studies have been focused on its modification [92, 102]. Various methods have been employed to modify expanded graphite, including strong acid treatment, in situ modification, surfactant modification, and surface coating modification [66, 97].

Toyoda and Inagaki [85] examined the adsorptive capabilities of commercially available exfoliated graphite, which underwent industrial processing, involving specific temperature and compression conditions. Various bulk densities of exfoliated graphite were achieved by compression, which leads to variations in pore volumes and subsequent alterations in absorption capacities. A decrease in temperature resulted in s noticeable reduction in absorption capacity, attributed to the increased viscosity of the oil at lower temperatures. The maximum absorption capacity of 80–90 g/g was measured, and a recovery ratio of 60–80% could be achieved through simple compression or suction filtration methods.

Wang et al. [92] fabricated modified expanded graphite (MEG) nanomaterials through a robust acid treatment employing H_2O_2 as the oxidant and concentrated H_2SO_4 as the intercalator. Characterization of MEG samples through field emission scanning electron microscope (FE-SEM) revealed worm-like structures featuring three distinct pore types, (1) V-shaped pores, (2) willow-leaf type pores, and (3) pores formed by undulating pore walls. The V-shaped and willow-leaf type pores within the materials facilitated internal storage for oil absorption. The modified expanded graphite demonstrated the ability to absorb a wide range of industrial oils, exhibiting absorption capacities that range from 43.25 to 84.68 g/g of MEG for an expanded volume of 100–320 mL/g. Furthermore, it was noted that the absorption capacity increased proportionally with the viscosity, as higher oil viscosity resulted in greater glutinosity.

Similarly, exfoliated graphite surrounded by N, N' dimethylformamide (DMF) membranes has been utilized to fabricate a "Tea-bag" nanocomposite system. The adsorption and absorption rate of T-B systems was evaluated using three distinct water/oil emulsions, namely fresh and used oil, which possess physical properties (viscosity and specific gravity) like heavy crude oil found in Brazil, and vacuum pump oil, which does not form an oil/water emulsion. The rate of adsorption (0.32–0.80 g/g/min) was contingent upon the surface tension of the oil, whereas the absorption rate (2.5–40 g/g) primarily relies on the surface area of the membrane/exfoliated graphite. Moreover, this system has demonstrated excellent sorptivity, recyclability, and reliable containment barrier properties [8].

5 Conclusions and Future Prospects

Carbon nanomaterials have demonstrated significant potential in the field of oil spill clean-up. They offer several advantages over traditional methods. They provide efficient and selective oil absorption, effectively removing oil from contaminated water. Various forms of carbon nanomaterials, including carbon aerogels, carbon nanotubes, porous carbon nanoparticles, graphene sponges, and carbon nanofibers, have gained significant attention and applications in oil spill clean-up. The carbon nanomaterials have demonstrated a remarkable absorption capacity ranging from 3 to 913 times their own weight. However, further research is still needed to optimize the design and synthesis of these materials, improve their reusability and recyclability, and ensure their long-term stability and effectiveness. Researchers can focus on optimizing the cost-effective carbon nanomaterial by tailoring their morphology, surface functionalization, and porosity. Also, exploring the synergistic effect of carbon nanomaterials with other technologies, such as magnetic nanomaterials or smart materials, can also lead to the development of advanced and multifunctional systems for oil spill clean-up.

References

1. Abdullah TA, Juzsakova T, Hafad SA, Rasheed RT, Al-Jammal N, Mallah MA, Salman AD, Le PC, Domokos E, Aldulaimi M (2022) Functionalized multi-walled carbon nanotubes for oil spill cleanup from water. Clean Technol Environ Policy 24:519–541
2. Abidli A, Huang Y, Cherukupally P, Bilton AM, Park CB (2020) Novel separator skimmer for oil spill cleanup and oily wastewater treatment: from conceptual system design to the first pilot-scale prototype development. Environ Technol Innov 18:100598
3. Ainsworth CH, Paris CB, Perlin N, Dornberger LN, Patterson WF III, Chancellor E, Murawski S, Hollander D, Daly K, Romero IC, Coleman F (2018) Impacts of the deepwater horizon oil spill evaluated using an end-to-end ecosystem model. PLoS ONE 13:e0190840
4. Al-Jammal N, Abdullah TA, Juzsakova T, Zsirka B, Cretescu I, Vágvölgyi V, Sebestyén V, Le Phuoc C, Rasheed RT, Domokos E (2020) Functionalized carbon nanotubes for hydrocarbon removal from water. J Environ Chem Eng 8:103570
5. Alvernia P, Soesilo TE, Herdiansyah H (2021) Social impacts of oil spills towards the people living in coastal areas. In: IOP conference series: earth environ science, vol 802. IOP Publishing, p 012001
6. Ansari MO, Kumar R, Ansari SP, Hassan MS, Alshahrie A, Barakat MA (2019) Nanocarbon aerogel composites. Woodhead Publishing, InNanocarbon and its Composites, pp 1–26

7. Asghar HMA, Hussain SN, Roberts EPL, Campen AK, Brown NW (2013) Pre-treatment of adsorbents for wastewater treatment using adsorption coupled with electrochemical regeneration. J Ind Eng Chem 19:1689–1676
8. Avila AF, Munhoz VC, de Oliveira AM, Santos MC, Lacerda GR, Gonçalves CP (2014) Nano-based systems for oil spills control and cleanup. J Hazard Mater 272:20–27
9. Baig N, Alghunaimi FI, Dossary HS, Saleh TA (2019) Superhydrophobic and superoleophilic carbon nanofiber grafted polyurethane for oil-water separation. Process Saf Environ Prot 123:327–334
10. Banerjee D, Das NS, Chattopadhyay KK (2012) Enhancement of field emission and hydrophobic properties of silicon nanowires by chemical vapor deposited carbon nanoflakes coating. Appl Surf Sci 261:223–230
11. Barron MG (2012) Ecological impacts of the deepwater horizon oil spill: implications for immunotoxicity. Toxicol Pathol 40:315–320
12. Bayat A, Aghamiri SF, Moheb A (2008) Oil sorption by synthesized exfoliated graphite (EG). Iran J Chem Eng 5(1):51–64
13. ben Hammouda S, Chen Z, An C, Lee K (2021) Recent advances in developing cellulosic sorbent materials for oil spill cleanup: a state-of-the-art review. J Clean Prod 311:127630
14. Beyke G, Fleming D (2005) In situ thermal remediation of DNAPL and LNAPL using electrical resistance heating. Remediation 15:5–22
15. Bi H, Huang X, Wu X, Cao X, Tan C, Yin Z, Lu X, Sun L, Zhang H (2014) Carbon microbelt aerogel prepared by waste paper: an efficient and recyclable sorbent for oils and organic solvents. Small 10:3544–3550
16. Brody TM, Bianca PD, Krysa J (2012) Analysis of inland crude oil spill threats, vulnerabilities, and emergency response in the midwest United States. Risk Anal: Int J 32(10):1741–1749
17. Cao M, Hu Y, Cheng W, Huan S, Bai T, Niu Z, Zhao Y, Yue G, Zhao Y, Han G (2022) Lignin-based multi-scale cellular aerogels assembled from co-electrospun nanofibers for oil/water separation and energy storage. Chem Eng J 436:135233
18. Chakrabarti A, Lu J, Skrabutenas JC, Xu T, Xiao Z, Maguire JA, Hosmane NS (2011) Conversion of carbon dioxide to few-layer graphene. J Mater Chem 21:9491–9493
19. Chen L, Du R, Zhang J, Yi T (2015) Density controlled oil uptake and beyond: from carbon nanotubes to graphene nanoribbon aerogels. J Mat Chem A 3:20547–20553
20. Chen Y, Li Y, Tai YM, N, (2013) Electromagnetic interference shielding efficiency of polyaniline composites filled with graphene decorated with metallic nanoparticles. Compos Sci Technol 80:80–86
21. Cong HP, Chen JF, Yu SH (2014) Graphene-based macroscopic assemblies and architectures: an emerging material system. Chem Soc Rev 43:7295–7325
22. de Oliveira EM, Lopes PF, de Oliveira Júnior JG, Junqueira AB, de Oliveira Santos AP, da Silva Lima JA, Malhado AC, Ladle RJ, Campos-Silva JV (2021) Immediate social and economic impacts of a major oil spill on Brazilian coastal fishing communities. Mar Pollut Bull 164:111984
23. Devi N, Kumar R, Chen YS, Singh RK (2023) Carbon-based nanomaterials: carbon nanotube, fullerene, and carbon dots. Nanomaterials. Springer, Singapore, pp 27–57
24. Dong J, Zeng J, Wang B, Cheng Z, Xu J, Gao W, Chen K (2021) Mechanically flexible carbon aerogel with wavy layers and springboard elastic supporting structure for selective oil/organic solvent recovery. ACS Appl Mater Interfaces 13:15910–15924
25. Doshi B, Sillanpää M, Kalliola S (2018) A review of bio-based materials for oil spill treatment. Water Res 135:262–277
26. Fang M, Tang Z, Lu H, Nutt S (2012) Multifunctional superhydrophobic composite films from a synergistic self-organization process. J Mater Chem 22:109–114
27. Gan G, Li X, Fan S, Wang L, Qin M, Yin Z, Chen G (2019) Carbon aerogels for environmental clean-up. Eur J Inorg Chemistry 2019(27):3126–3141
28. Ge B, Zhang Z, Zhu X, Ren G, Men X, Zhou X (2013) A magnetically superhydrophobic bulk material for oil removal. Colloids Surf 429:129–133

29. Ge J, Zhao HY, Zhu HW, Huang J, Shi LA, Yu SH (2016) Advanced sorbents for oil-spill cleanup: recent advances and future perspectives. Adv Mater 28(47):10459–10490
30. Ge J, Zhao HY, Zhu HW, Huang J, Shi LA, Yu SH (2016) Advanced sorbents for oil-spill cleanup: recent advances and future perspectives. Adv Mater 28:10459–10490
31. Ghasemi O, Mehrdadi N, Baghdadi M, Aminzadeh B, Ghaseminejad A (2020) Spilled oil absorption from Caspian sea water by graphene/chitosan nanocomposite. Energy Sources Part A 42:2856–2872
32. Ghosh S, Webster TJ (2022) Nanotechnological advances for oil spill management: removal, recovery and remediation. In: Advances in oil-water separation. Elsevier, pp 175–194
33. Godeto YG, Ayele A, Ahmed IN, Husen A, Bachheti RK (2023) Medicinal plant-based metabolites in nanoparticles synthesis and their cutting-edge applications: an overview. In: Bachheti RK, Bachheti A (eds) Secondary metabolites from medicinal plants. CRC Press, Boca Raton, USA, pp 1–34
34. Gui X, Wei J, Wang K, Cao A, Zhu H, Jia Y, Shu Q, Wu D (2010) Carbon nanotube sponges. Adv Mater 22:617–621
35. Gui X, Zeng Z, Lin Z, Gan Q, Xiang R, Zhu Y, Cao A, Tang Z (2013) Magnetic and highly recyclable macroporous carbon nanotubes for spilled oil sorption and separation. ACS Appl Mater Interfaces 5:5845–5850
36. Gupta S, Tai NH (2016) Carbon materials as oil sorbents: a review on the synthesis and performance. J Mater Chem A 4:1550–1565
37. Hatami E, Abbaspour A, Dorostkar V (2019) Phytoremediation of a petroleum-polluted soil by native plant species in Lorestan Province. Iran. Environ Sci Pollut Res 26:24323–24330
38. Hoang AT, Nižetić S, Duong XQ, Rowinski L, Nguyen XP (2021) Advanced super-hydrophobic polymer-based porous absorbents for the treatment of oil-polluted water. Chemosphere 277:130274
39. Hong YC, Uhm HS (2006) Superhydrophobicity of a material made from multiwalled carbon nanotubes. Appl Phys Lett 88:244101
40. Hou X, Zhang R, Fang D (2020) Superelastic, fatigue resistant and heat insulated carbon nanofiber aerogels for piezoresistive stress sensors. Ceram Int 46:2122–2127
41. Hristea G, Budrugeac P (2008) Characterization of exfoliated graphite for heavy oil sorption. J Therm Anal 91:817–823
42. Hu H, Zhao Z, Gogotsi Y, Qiu J (2014) Compressible carbon nanotube–graphene hybrid aerogels with superhydrophobicity and superoleophilicity for oil sorption. Environ Sci Technol Lett 1:214–220
43. Inagaki M, Konno H, Toyoda M, Moriya K, Kihara T (2000) Sorption and recovery of heavy oils by using exfoliated graphite Part II: recovery of heavy oil and recycling of exfoliated graphite. Desalination 128:213–218
44. Janqamsari Y, Ashjari M, Niazi Z (2021) Carbon nanotube promoted porous nanocomposite based on PVA and recycled PET fibers for efficient oil spill cleanup applications. Chem Pap 75:3443–3456
45. Jung YC, Bhushan B (2009) Wetting behavior of water and oil droplets in three-phase interfaces for hydrophobicity/philicity and oleophobicity/philicity. Langmuir 25:14165–14173
46. Kemp KC, Seema H, Saleh M, Le NH, Mahesh K, Chandra V, Kim KS (2013) Environmental applications using graphene composites: water remediation and gas adsorption. Nanoscale 5:3149–3171
47. Kohli K, Bhatt N (2020) Carbon nanotubes—a novel approach to oil spill cleanup. In: Advances in industrial safety: select proceedings of HSFEA. Springer Singapore, pp 73–83
48. Kosynkin DV, Higginbotham AL, Sinitskii A, Lomeda JR, Dimiev A, Price BK, Tour JM (2009) Longitudinal unzipping of carbon nanotubes to form graphene nanoribbons. Nature 458(7240):872–876
49. Lee JS, Park GS, Kim ST, Liu M, Cho J (2013) A highly efficient electrocatalyst for the oxygen reduction reaction: N-doped ketjenblack incorporated into Fe/Fe3C-functionalized melamine foam. Angew Chem Int Ed 52:1026–1030

50. Li S, Warzywoda J, Wang S, Ren G (2017) Fan Z bacterial cellulose derived carbon nanofiber aerogel with lithium polysulfide catholyte for lithium–sulfur batteries. Carbon 1(124):212–218
51. Lin Y, Ehlert GJ, Bukowsky C, Sodano HA (2011) Superhydrophobic functionalized graphene aerogels. ACS Appl Mater Interfaces 3:2200–2203
52. Liu H, Sha W, Cooper AT, Fan M (2009) Preparation and characterization of a novel silica aerogel as adsorbent for toxic organic compounds. Colloids Surf A 347:38–44
53. Liu S, Wang S, Wang H, Lv C, Miao Y, Chen L, Yang S (2021) Gold nanoparticles modified graphene foam with superhydrophobicity and superoleophilicity for oil-water separation. Sci Total Environ 758:143660
54. Liu Z, Liu J, Zhu Q, Wu W (2012) The weathering of oil after the deepwater horizon oil spill: insights from the chemical composition of the oil from the sea surface, salt marshes and sediments. Environ Res Lett 7:035302
55. Lorevice MV, Claro PI, Aleixo NA, Martins LS, Maia MT, Oliveira AP, Martinez DS, Gouveia RF (2023) Designing 3D fractal morphology of eco-friendly nanocellulose-based composite aerogels for water remediation. Chem Eng J 462:142166
56. Maleki H (2016) Recent advances in aerogels for environmental remediation applications: a review. Chem Eng J 300:98–118
57. Mehmood A, Khan FS, Mubarak NM, Mazari SA, Jatoi AS, Khalid M, Tan YH, Karri RR, Walvekar R, Abdullah EC, Nizamuddin S (2012) Carbon and polymer-based magnetic nanocomposites for oil-spill remediation—a comprehensive review. Environ Sci Pollut Res 28:54477–54496
58. Mishra S, Chauhan G, Verma S, Singh U (2022) The emergence of nanotechnology in mitigating petroleum oil spills. Mar Pollut Bull 178:113609
59. Neelamegan H, Yang DK, Lee GJ, Anandan S, Wu JJ (2020) Synthesis of magnetite nanoparticles anchored cellulose and lignin-based carbon nanotube composites for rapid oil spill cleanup. Mater Today Commun 22:100746
60. Nguyen DD, Tai NH, Lee SB, Kuo WS (2012) Superhydrophobic and superoleophilic properties of graphene-based sponges fabricated using a facile dip coating method. Energy Environ Sci 5:7908–7912
61. Okoh E, Yelebe ZR, Oruabena B, Nelson ES, Indiamaowei OP (2020) Clean-up of crude oil-contaminated soils: bioremediation option. Int J Environ Technol 17:1185–1198
62. Pegg S, Zabbey N (2013) Oil and water: the Bodo spills and the destruction of traditional livelihood structures in the Niger Delta. Community Dev J 48:391–405
63. Perfumo A, Rancich I, Banat IM (2010) Possibilities and challenges for biosurfactants use in petroleum industry. Biosurfactants 135–145
64. Pete AJ, Bharti B, Benton MG (2021) Nano-enhanced bioremediation for oil spills: a review. ACS ES&T Eng 1:928–946
65. Piperopoulos E, Calabrese L, Mastronardo E, Milone C, Proverbio E (2020) Carbon-based sponges for oil spill recovery. In: Carbon nanomaterials for agri-food and environmental applications. Elsevier, pp 155–175
66. Placke T, Siozios V, Schmitz R, Lux SF, Bieker P, Colle C, Meyer HW, Passerini S, Winter M (2012) Influence of graphite surface modifications on the ratio of basal plane to "non-basal plane" surface area and on the anode performance in lithium-ion batteries. J Power Sources 200:83–91
67. Qiang F, Hu LL, Gong LX, Zhao L, Li SN, Tang LC (2018) Facile synthesis of super-hydrophobic, electrically conductive and mechanically flexible functionalized graphene nanoribbon/polyurethane sponge for efficient oil/water separation at static and dynamic states. Chem Eng J 334:2154–2166
68. Qiao K, Tian W, Bai J, Wang L, Zhao J, Du Z, Gong X (2019) Application of magnetic adsorbents based on iron oxide nanoparticles for oil spill remediation: a review. J Taiwan Inst Chem Eng 97:227–236
69. Quintero L, Ponnapati R, Felipe MJ (2017) Cleanup of organic and inorganic wellbore deposits using microemulsion formulations: laboratory development and field applications. In: Offshore technology conference, OnePetro

70. Rafiee J, Rafiee MA, Yu ZZ, Koratkar N (2010) Superhydrophobic to superhydrophilic wetting control in graphene films. Adv Mater 22:2151–2154
71. Rahmani Z, Samadi MT, Kazemi A, Rashidi AM, Rahmani AR (2017) Nanoporous graphene and graphene oxide-coated polyurethane sponge as a highly efficient, superhydrophobic, and reusable oil spill absorbent. J Environ Chem Eng 5:5025–5032
72. Ramos SC, Vasconcelos G, Antunes EF, Lobo AO, Trava-Airoldi VJ, Corat EJ (2010) CO_2 laser treatment for stabilization of the superhydrophobicity of carbon nanotube surfaces. J Vac Sci Technol B 28:1153–1157
73. Reddy CM, Eglinton TI, Hounshell A, White HK, Xu L, Gaines RB, Frysinger GS (2002) The West Falmouth oil spill after thirty years: the persistence of petroleum hydrocarbons in marsh sediments. Environ Sci Technol 36:4754–4760
74. Sam EK, Liu J, Lv X (2021) Surface engineering materials of superhydrophobic sponges for oil/water separation: a review. Ind Eng Chem Res 60:2353–2364
75. Shayesteh H, Khosrowshahi MS, Mashhadimoslem H, Maleki F, Rabbani Y, Emrooz HB (2023) Durable superhydrophobic/superoleophilic melamine foam based on biomass-derived porous carbon and multi-walled carbon nanotube for oil/water separation. Sci Rep 13:4515
76. Siddiqa A, Shahid A, Gill R (2015) Silica decorated CNTs sponge for selective removal of toxic contaminants and oil spills from water. J Environ Chem Eng 3:892–897
77. Singh B, Kumar S, Kishore B, Narayanan TN (2020) Magnetic scaffolds in oil spill applications. Environ Sci Water Res Technol 6:436–463
78. Soares MO, Teixeira CE, Bezerra LE, Rabelo EF, Castro IB, Cavalcante RM (2022) The most extensive oil spill registered in tropical oceans (Brazil): the balance sheet of a disaster. Environ Sci Pollut Res 29:19869–19877
79. Sun S, Tang S, Chang X, Wang N, Wang D, Liu T, Lei Y, Zhu Y (2019) A bifunctional melamine sponge decorated with silver-reduced graphene oxide nanocomposite for oil-water separation and antibacterial applications. Appl Surf Sci 4731049–4731061
80. Sun Y, Xu H, Zhao Z, Zhang L, Ma L, Zhao G, Song G, Li X (2021) Investigation of carbon nanotube grafted graphene oxide hybrid aerogel for polystyrene composites with reinforced mechanical performance. Polymers 13:735
81. Tao G, Zhang L, Hua Z, Chen Y, Guo L, Zhang J, Shu Z, Gao J, Chen H, Wu W, Liu Z, Shi J (2014) Highly efficient adsorbents based on hierarchically macro/mesoporous carbon monoliths with strong hydrophobicity. Carbon 66:547–559
82. Tayeb AM, Farouq R, Mohamed OA, Tony MA (2020) Oil spill clean-up using combined sorbents: a comparative investigation and design aspects. Int J Environ Anal Chem 100:311–323
83. Tolaka W, Wardah W, Rahmawati R (2013) Sifat Fisik Tanah Pada Hutan Primer, Agroforestri dan Kebun Kakao di SUBDAS Wera Saluopa Desa Leboni Kecamatan Pamona Puselemba Kabupaten Poso. Jurnal Warta Rimba 1
84. Toyoda M, Aizawa J, Inagaki M (1998) Sorption and recovery of heavy oil by using exfoliated graphite. Desalination 115:199–201
85. Toyoda M, Inagaki M (2000) Heavy oil sorption using exfoliated graphite: new application of exfoliated graphite to protect heavy oil pollution. Carbon 38:199–210
86. Toyoda M, Moriya K, Aizawa JI, Konno H, Inagaki M (2000) Sorption and recovery of heavy oils by using exfoliated graphite Part I: maximum sorption capacity. Desalination 128(3):205–211
87. Tsai LW, Tai NH (2014) Enhancing the electrical properties of a flexible transparent graphene-based field-effect transistor using electropolished copper foil for graphene growth. ACS Appl Mater Interfaces 6:10489–10496
88. Vickery JL, Patil AJ, Mann S (2009) Fabrication of graphene–polymer nanocomposites with higher-order three-dimensional architectures. Adv Mater 21:2180–2184
89. Viesser JA, Sugai-Guerios MH, Malucelli LC, Pincerati MR, Karp SG, Maranho LT (2020) Petroleum-tolerant rhizospheric bacteria: Isolation, characterization and bioremediation potential. Sci Rep 10:2060

90. Wan W, Zhang R, Li W, Liu H, Lin Y, Li L, Zhou Y (2016) Graphene–carbon nanotube aerogel as an ultra-light, compressible and recyclable highly efficient absorbent for oil and dyes. Environ Sci Nano 3:107–113
91. Wang CF, Lin SJ (2013) Robust superhydrophobic/superoleophilic sponge for effective continuous absorption and expulsion of oil pollutants from water. ACS Appl Mater Interfaces 5:8861–8864
92. Wang L, Fu X, Chang E, Wu H, Zhang K, Lei X, Zhang R, Qi X, Yang Y (2014) Preparation and its adsorptive property of modified expanded graphite nanomaterials. J Chem
93. Wei T, Fan Z, Luo G, Zheng C, Xie D (2009) A rapid and efficient method to prepare exfoliated graphite by microwave irradiation. Carbon 47:337–339
94. Won YJ, Jang S, Jung N, Kwon Y, Moon SY, Nho H, Yoo SJ (2019) Ten years after the oil spill in Taean: the recovery of the ecosystem, the life of women, and the community. Asian Women 35:1–22
95. Wu L, Liu J, Reddy BR, Zhou J (2022) Preparation of coal-based carbon nanotubes using catalytical pyrolysis: a brief review. Fuel Process Technol 229:107171
96. Wu ZY, Li C, Liang HW, Zhang YN, Wang X, Chen JF, Yu SH (2014) Carbon nanofiber aerogels for emergent cleanup of oil spillage and chemical leakage under harsh conditions. Sci Rep 4:4079
97. Xin CU, Changhe LI, Wenfeng DI, Yun CH, Cong MA, Xuefeng XU, Bo LI, Dazhong WA, Li HN, Zhang Y, Zafar SA (2022) Minimum quantity lubrication machining of aeronautical materials using carbon group nanolubricant: from mechanisms to application. Chinese J Aeronaut 35:85–112
98. Yadav KK, Singh JK, Gupta N, Kumar VJ (2017) A review of nano bioremediation technologies for environmental cleanup: a novel biological approach. J Mater Environ Sci 8:740–757
99. Yang M, Zhao N, Cui Y, Gao W, Zhao Q, Gao C, Bai H, Xie T (2017) Biomimetic architectured graphene aerogel with exceptional strength and resilience. ACS Nano 11:6817–6824
100. Yang Y, Liu Z, Huang J, Wang C (2015) Multifunctional, robust sponges by a simple adsorption–combustion method. J Mat Chem A 3:5875–5881
101. Yu M (2013) Theoretical and experimental studies of droplets wetting on micro/nano-fibrous materials and their applications in oil-water separation (Doctoral dissertation, North Dakota State University)
102. Yue X, Zhang R, Zhang F, Wang L (2010) Decomposition of crude oil absorbed into expanded graphite/TiO_2/NiO composites. Desalination 252:163–166
103. Zamparas M, Tzivras D, Dracopoulos V, Ioannides T (2020) Application of sorbents for oil spill cleanup focusing on natural-based modified materials: a review. Molecules 25(19):4522
104. Zhan W, Yu S, Gao L, Wang F, Fu X, Sui G, Yang X (2018) Bioinspired assembly of carbon nanotube into graphene aerogel with "cabbagelike" hierarchical porous structure for highly efficient organic pollutants cleanup. ACS Appl Mater Interfaces 10:1093–1103
105. Zhang W, Wang J, Han X, Li L, Liu E, Lu C (2021) Carbon nanotubes and polydopamine modified poly (dimethylsiloxane) sponges for efficient oil–water separation. Materials 14:2431
106. Zhao H, Wang Q, Chen Y, Tian Q, Zhao G (2017) Efficient removal of dimethyl phthalate with activated iron-doped carbon aerogel through an integrated adsorption and electro-Fenton oxidation process. Carbon 124:111–122
107. Zhu K, Shang YY, Sun PZ, Li Z, Li XM, Wei JQ, Wang KL, Wu DH, Cao AY, Zhu HW (2013) Oil spill cleanup from seawater by carbon nanotube sponges. Front Mater Sci 7:170–176
108. Zhu L, Xiu Y, Xu J, Tamirisa PA, Hess DW, Wong CP (2005) Superhydrophobicity on two-tier rough surfaces fabricated by controlled growth of aligned carbon nanotube arrays coated with fluorocarbon. Langmuir 21:11208–11212

Environmental Toxicity of Engineered Carbon Nanoparticles

Mayur Mukut Murlidhar Sharma, Divya Kapoor, Atul Loyal, Rahul Kumar, Pankaj Sharma, and Azamal Husen

Abstract The employment of nanotechnological tools is increasing at a very rapid pace in modern agricultural systems. Among a gamut of nanoparticles employed, carbon nanoparticles are also being explored as potential tools for increasing plant productivity. Due to their broad spectrum biocidal activity, they are projected as future biocontrol agents. However, the multifarious approach not only reduces the chances of resistance in pathogens but also puts forward the potential off-target effects of using these nanoparticles. The practice of agriculture is highly reliant on plant–microbe symbiotic relations. Therefore, nanoparticle accumulation in agricultural systems could disrupt this relationship. The interference of nanoparticles with the genetic material of plants and microbes can lead to the generation of mutant strains and serotypes with unpredictable outcomes.

M. M. M. Sharma
Department of Agriculture and Industry, Kangwon National University, Chuncheon, Republic of Korea

D. Kapoor · P. Sharma (✉)
Department of Microbiology, CCS Haryana Agricultural University, Hisar, Haryana 125004, India
e-mail: sharmap@uic.edu

A. Loyal
Department of Genetics and Plant Breeding, CCS Haryana Agricultural University, Hisar, Haryana 125004, India

R. Kumar
Department of Zoology, CCS Haryana Agricultural University, Hisar, Haryana 125004, India

P. Sharma
Department of Ophthalmology and Visual Sciences, University of Illinois Chicago, Chicago, IL 60612, USA

A. Husen
Department of Biotechnology, Smt. S. S. Patel Nootan Science and Commerce College, Sankalchand Patel University, Visnagar, Gujarat 384315, India

Department of Biotechnology, Graphic Era (Deemed to Be University), Dehradun, Uttarakhand 248002, India

Wolaita Sodo University, Wolaita, Ethiopia

A. Bachheti et al. (eds.), *Carbon-Based Nanomaterials*, Smart Nanomaterials Technology, https://doi.org/10.1007/978-981-97-0240-4_16

Additionally, the repetitive exposure of rhizosphere microbes to carbon nanomaterials can lead to significant alterations in the rhizosphere microbiome that might significantly change plant productivity. In the current chapter, we have highlighted toxicity associated with the nanomaterials originating from carbon and potential harmful effects that might alter plant productivity. We have also tried to highlight the potential mechanism governing their off-target effects on the beneficial soil microbes.

Keywords Carbon nanomaterials · Genotoxicity · Nanoparticle toxicity · Cytotoxicity · Nanoparticle phytotoxicity

1 Introduction

The nanoparticles are classified as particles ranging from 1 to 100 nm. They are well utilized for their eminent properties, such as small size, high area-to-volume ratio, and enhanced surface potential [34, 51, 62, 66, 67, 71]. The last decade has witnessed a strong boom in the research and use of nanoparticles owing to their advantages and applications in various fields [4, 10, 28–33, 48, 50, 65, 68]. They have been widely used in medicine, agriculture, electronics, and all major modern sciences [22, 30, 31, 36, 39, 52, 60, 79]. Different kinds of metallic nanoparticles such as Ag, Au, Cu, Pt, Cd, Pt, Pd, Ru, and Rh [15, 21, 41] and non-metallic like polymers, biomolecules, micelles, dendrimers, and carbon nanoparticles, including nanotubes, graphene and fullerene [18] are being used at an ever-increasing pace. Over the last decade, more focus has been on non-metal nanoparticles, especially carbon nanomaterials (CNMs) such as carbon nanotubes, graphene sheets, and fullerene. Their unique electronic, optical, thermal, mechanical, and chemical properties have led them to be widely studied and researched nanomaterials lately. They have been used to their advantage significantly by researchers to develop disruptive technologies in pure and applied sciences [13]. For instance, their use has been prominent in developing conductive polymers, transparent electrodes, chemical sensors, high-frequency devices, opto-electronic sensors, alternative energy, and bio-inspired systems [12]. Although they are majorly used in the engineering sector, their employment in the agricultural sector is also increasing (Fig. 1).

Carbon nanomaterials are being used in agricultural systems as plant growth promoters and adjuvants for the slow release of fertilizers and pesticides, and their indirect uses for creating Nano-biosensor to detect various pathogens, biomolecules, and ions in plants are also vague [40, 46, 47, 85]. The growing use of nanomaterials in construction, electronics, medicine, optics, and agriculture has increased leaching and, therefore, exposure of nanomaterials to human beings and the environment. The nanomaterials have a high surface area to volume ratio and can be more influential due to their enhanced bioavailability compared to their conventional counterparts. Nanomaterials harm the environment in several ways. The nanomaterials get easily assimilated into water and air, for that matter, and can easily pollute these two basic resources all living beings require to survive. Secondly, they can be incorporated

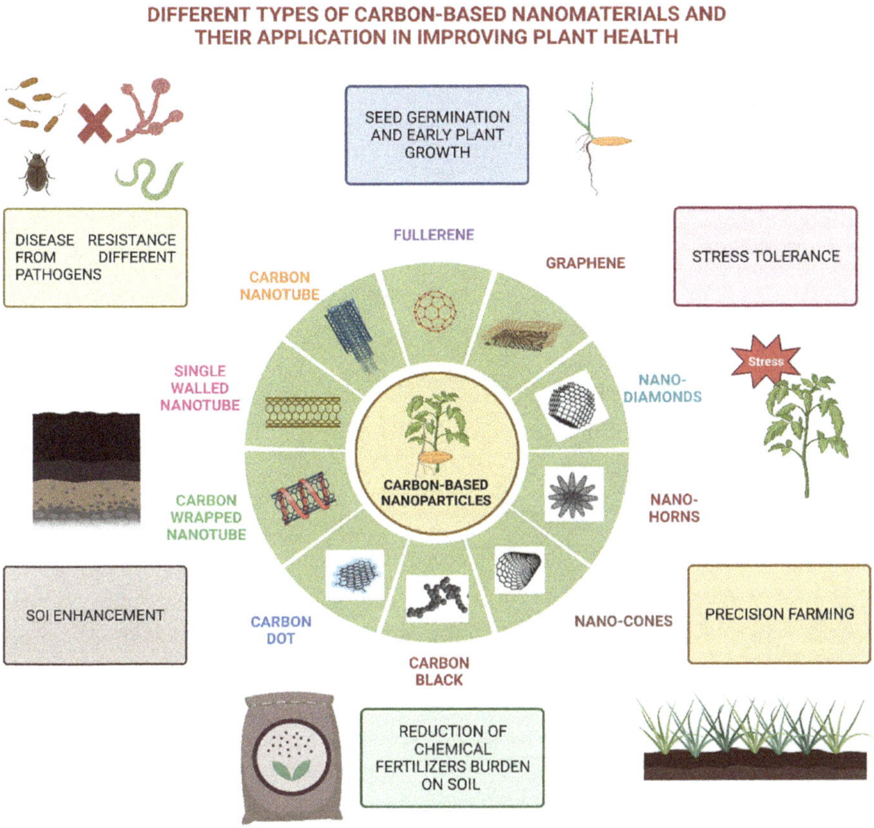

Fig. 1 A portrayal depicting the usage of different carbon nanomaterials in agricultural sectors

into biological systems such as plants, animals, fungi, and bacteria. They can have both short and long-term effects or can enable them to have adverse effects on the organism next to them in the food chain.

Additionally, the nanomaterials render the biological systems to disturb their life cycle [8]. The presence of nanomaterials in the environment can be toxic to the DNA and cell or tissues of living beings, including plants, animals, fungi, and bacteria [83]. In recent years, the mass production of carbon nanomaterials has increased to great heights making its emission to the environment more feasible, ultimately affecting all terrestrial and aquatic life, including plants, animals, and microbes. The small size of carbon nanomaterials enables them to penetrate the cell wall and cell membranes easily, where they can trigger reactions of cellular inflammation and produce direct damage to the organisms [56]. This chapter summarizes the harmfulness of carbon nanomaterials, which can interfere with the lives and adversely impact the life cycles of different organisms such as plants, bacteria, and fungi (Fig. 2).

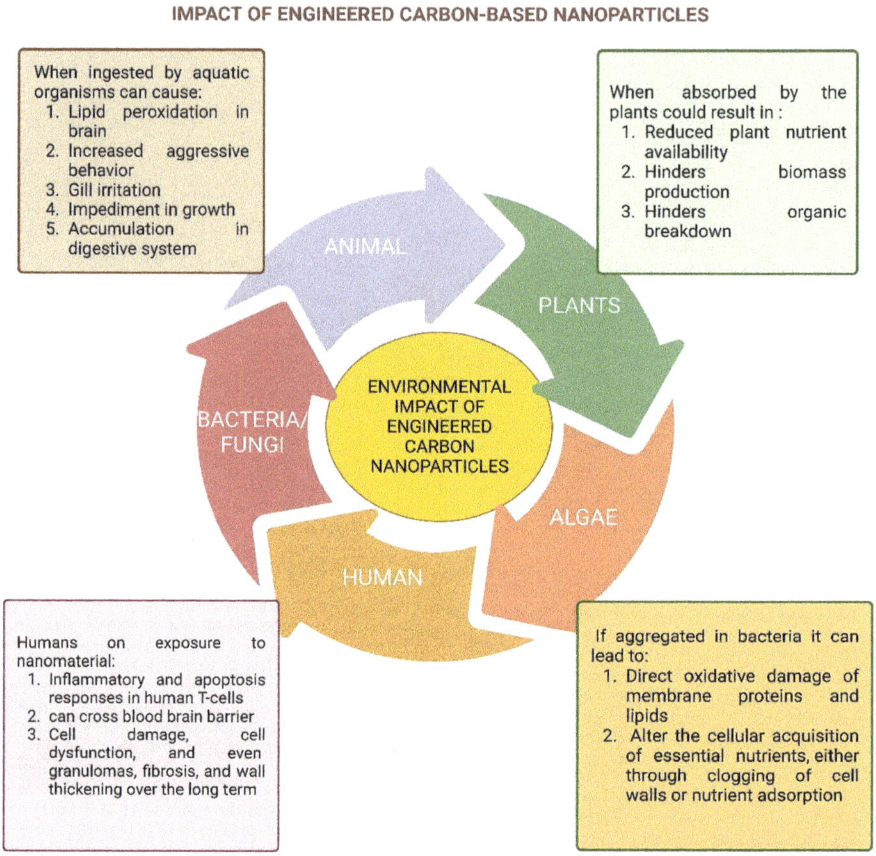

Fig. 2 Different mechanism of toxicity associated with the engineered carbon nanomaterials

2 Environment-Related Concerns of CNMs

The CNMs-oriented research has taken long ropes since last decade, and the multiple benefits they offer are quite lucrative and make their utilization more efficient. Despite their extensive current use and proposed use, the speculations of their detrimental effects are progressing. Their negative impact on the environment has been researched in different species. However, the species range is not too extensive, and the potentially detrimental effect of the CNMs on the environment and all different kinds of species sustaining in the ecosystem seems high, which might outweigh the benefits rendered by CNMs. The reactiveness of CNMs is very high, and their interference in the environment can result in unprecedented alteration in the ecological balance and the disruption of the biological chains associated with living organisms [53]. The exposure and disposal of CNMs in the environment are way too large than reported, and the exact record and chain of the source, manufacturing, and disposal, and sink consumption has not been documented in totality, which further enhances the risk and potential threats associated with the CNMs [37]. The anticipated production of CNTs and fullerene is 55 to 3,300 tons and of fullerenes from 0.6 to 1620 tons. The concern is that only a small fraction of it is utilized at the targeted sites, and a giant part is emitted to the environment by various means. Either they can be released by emission from the manufacturing unit or the field sites or accidentally disposed of in the form of waste [6]. As per the life cycle release model studies, the landfills and soil sink more than 80% of nanomaterials. Most of the CNMs are hydrophobic and are readily soluble in organic solvents. However, the presence of pore water and other contaminants in the soil also decides the fate of CNMs at the sink place. Different surfactants and natural organic matter have been reported to stabilize the CNMs in the aqueous phase, thereby increasing the stability and mobility of CNMs [7]. The soil pH, texture, clay content, porosity, and other soil-associated physical and chemical properties largely dictate the mobilization and transportation of CNMs to another component of the environment and within the soil environment. It has also been suggested that the CNMs resist mineralization and would stay in the soil for at least 2 years [3]. The emission of CNMs also threatens human health and the unpredictable range of biological deformities, which researchers have speculated for a long time. However, human exposure can be drawn in occupational settings and indirectly via different mediums, such as air and water already carrying the CNMs [54]. Of all different kinds of CNTs, the major fraction is of CNTs and fullerene composites. The composites have a higher chance of being emitted into the environment as the properties of CNMs are altered and can become more soluble after adding the composite. The author discoursed several pathways of CNT-incorporated polymer nanocomposites discharge while they are in use and for disposal in an inclusive review. It has been discussed that different natural circumstances comprising radiation, UV light, moisture, temperature, and microbes lead to the squalor of the polymer matrix holding the CNTs, causing particle release. Additionally, the events during composite such as mechanical abrasion and ignition can lead to the release of the nanomaterials in an unrestrained mode. The literature is mostly silent on the release developments of many CNMs, containing fullerenes, graphene and their fortune in different environmental grounds [57].

3 Toxicity of Carbon Nanomaterials

3.1 Cytotoxicity

The small size and the ability to penetrate the cell wall and membrane make the CNMs interfere with the cellular and molecular mechanisms wherever it gets an opportunity to be incorporated. The different type of CNMs, such as single-walled carbon nanotubes (SWCNTs) and multiwalled carbon nanotubes (MWCNTs), Graphene, and C_{60}, has been proven to be cytotoxic and genotoxic to different kind of plant animals and microbial cells [56]. MWCNTs exhibit cytotoxicity in various ways. It constrains cell proliferation and encourages oxidative damage, apoptosis, or necrosis. Moreover, the structure, diameter, and length of CNTs are also key dynamics for their toxicity. The longer the length of the CNTs, the higher the toxicity [45]. The SWCNTs have also been found toxic to human lung cells in a dose-dependent manner. Pulmonary toxicity was also associated with the mechanical blockage of large airways. The cytotoxicity is also associated with the different cytotoxicity and inflammatory markers in the lungs [35]. Graphene toxicity is also persistent in the environment.

The interaction of graphene with the membrane proteins results in serious physical damage to the cells, which may lead to cell injury or cell death [88]. The high dose of graphene nanomaterials also renders the infliction of oxidative stress, causing cell viability loss. Besides cell death, graphene nanomaterials also exhibit cytotoxicity by compromising cell-to-cell adhesion in the tissues, making the cell disintegrate from their native place, which ultimately does not survive [42]. The graphene can interact with the hydrophobic lipid and cholesterol tails of the cell membranes and can remove them, creating some pores which leads to diminishing the membrane integrity [87]. For instance, Duan et al. [16] indicated that graphene nanosheets (both pristine and oxidized) could yield holes (pores) in the membranes of A549 and Raw264.7 cells [16].

Furthermore, graphene can also damage the cells by infiltrating the cytoplasm. Once the graphene enters the cytoplasm, it can leak cytoplasmic content and disturb the osmotic potential of the cell. The production of reactive oxygen species can also lead to mitochondrial dysfunction by reducing the membrane potential. The ROSs can also interact with the fatty acids to produce lipid peroxidases such as malondialdehyde [55]. Only a few studies have been done to determine the side effects of fullerene C60, and it is known to exhibit cytotoxicity by producing ROS [17]. However, the cytotoxic effects of C60 have been used to control cancer. For instance, Didenko et al. [14] showed that the C60 exhibits an antimetastatic effect in lung carcinoma.

3.2 Genotoxicity

The carbon nanomaterials also negatively affect the environment by exhibiting genotoxic effects by interacting with the genetic material of the organisms and damaging, changing, or modifying it in harmful ways that ultimately lead to injury or death of organisms. After entering the cell membrane, the nanoparticles can diffuse inside the nucleus through the nuclear pores. If the size of nanoparticles does not fit into the nuclear pores, they can reach out to the DNA during cell division, where the nuclear membrane does not restrict them [70]. The large-sized nanomaterials may form cellular vesicles and then interfere with the segregation of chromosomes and the formation of the mitotic spindle. This may lead to errors in cell division and improper chromosome segregation. The gathered nanoparticles can also cause physical damage to the chromosomes [49]. Long-term exposure to genotoxic nanoparticles can also induce permanent mutation to the genetic makeup of that organism. However, genotoxicity largely depends upon the nanomaterials' size, shape, type, and properties. Various mechanisms have been proposed to define the nanotoxicity attributed to the CNTs. ROS production, mitochondria disruption, and cell division apparatus interference, including spindle formation. For instance, Sargent et al. [63] tested SWCNT in BEAS-2B-normal human bronchial epithelial cells. They tested the genotoxicity by Mitotic spindle analysis, Chromosome number and found that the SWCNTs were involved in mitotic spindle disruption and error in chromosome number. In the same manner, MWCNT were genotoxic to the human airway respiratory epithelial cells in a dose-dependent manner, and the genotoxicity was apparent by the aberration in chromosomes and mitotic spindle [69]. It is believed that graphene's small lateral diameter can openly intermingle with DNA and prompt genotoxicity through ROS production and oxidative DNA damage. Other means of ROS-prompted genotoxicity are mitochondrial ROS production and leucocyte inflammation. The graphene nanomaterials also exhibit genotoxicity by directly affecting the DNA by binding to it through nuclear pores or interfering with the cell division cycle. The bacteria have naked DNA so that the graphene can directly contact them, and the replicative stage is interrupted [81]. The indirect toxicity of graphenes is exhibited by the damage caused by ROS produced due to the entrance of nanomaterials inside the cells. Another means of indirect genotoxicity is epigenetic toxicity which accounts for the epigenetic modification induced by nanomaterial exposure, such as DNA methylation and histone modifications [82]. The fullerene family largely does not exhibit genotoxic effects, barring a few exceptions. The most prominent case of the genotoxicity induced by the fullerene is derived from photo-induced DNA damage by interacting with NADH [61].

4 CNMs' Toxicity to Fungi

The emission of CNMs can also target the fungi proliferating in the surroundings by exhibiting cytotoxic and genotoxic effects. However, few reports are available regarding the harm of CNMs against fungi. Sawangphruk et al. [64] examined the antifungal activities of CNMs initially. He used reduced graphene oxide (rGO) against three fungi, *Aspergillus niger* (A. niger), *Aspergillus oryzae* (A. oryzae), and *Fusarium oxysporum* (F. oxysporum). The results demonstrated that the rGO inhibited mycelium growth significantly. The nanomaterials can induce plasmolysis in the spores and inhibit the growth of mycelium and spore formation to further diminish the reproduction of fungi. Wang et al. [80] tested 6 nanomaterials single-walled carbon nanotubes (SWCNTs), multiwalled carbon nanotubes (MWCNTs), graphene oxide (GO), reduced graphene oxide (rGO), fullerene (C60) and activated carbon (AC) against two fungi; *Fusarium graminearum* (*F. graminearum*) and *Fusarium poae* (*F. poae*). The result indicated the loss of water and the plasmolysis of fungal spores.

The highest antifungal activity was carried by SWCNTs, followed by MWCNTs, GO, and rGO, C60 and AC didn't show any antifungal activity.

5 CNMs' Toxicity to Plants

Plants are indispensable parts of the environment; hence it is important to consider their well-being when it comes to the emission of different nanomaterials, which can produce detrimental effects on them. CNMs inflict toxicity on plant cells and hamper their vegetative and reproductive growth. CNMs exhibit phytotoxicity by producing ROS species and altering plants' proteins, lipids, or genetic material [77]. CNMs can be assimilated by plants in their body systems, where they can limit the seed germination, the fresh and dry shoot, and root weight, and can also compromise the quality of the produce. CNMS is also known to limit photosynthesis efficiency, the negative modulation of transpiration, and increase DNA damage and degeneration of other vital biomolecules [26]. Although plants can combat this kind of toxicity, the damage done by CNMs can still outsmart the plant's defensive strategies. Although all CNMs are not toxic to plants, for instance, the C60 improves the water-retaining capacity of the plants, which helps them to build higher biomass resulting in more growth and yield [28, 29]. The CNMs affect plants adversely by various means; one is by adhering to the root surface, where it becomes difficult for plants to absorb water and nutrient as the CNMs constantly offer hindrance [23].

MWCNTs, when examined against *Allium cepa*, led to a negative modulation in the morphology of the cells. The cells were demonstrated to lose the membrane integrity, and significant damage to the mitochondria was also apparent. Furthermore, exposure to CNMs resulted in chromosomal and nuclear aberrations [19]. Hatami [25] investigated the effects of MWCNTs on the *Cucurbita pepo*. They found that these CNMs exhibited toxicity, evident from the diminished seed germination, length of roots, and buds, biomass accretion, and vitality index of this species in a dose-dependent manner. Begum et al. [5] examined the species-dependent toxicity of water-soluble graphene (GO,'graphene oxide (GO) with sodium ions as the counter-ions) in lettuce, cabbage, red spinach, and tomato throughout a 20-day exposure period. The graphene exhibited toxicity to plants at the highest concentration (2000 mg/L), where it led to a significant reduction in plant growth (up to 78%), and biomass (up to 88%); the reduction in a number of leaves was also prominent (up to 53 and 91%, respectively). The stress molecules like reactive oxygen species were also increased, resulting in the necrosis of plants except lettuce. Hu et al. [27] found that 0.1–10 mg/L GO exposure boosted the adversative properties of As(V) in wheat seeds, leading to considerable loss of the fresh mass, shoot length, and chlorophyll content of treated plants. Furthermore, the study also indicated that the activity of peroxidase (POD) and superoxide dismutase (SOD) was amplified in a concentration-dependent mode, which indicates the stress response (Table 1).

Table 1 Review of literature highlighting the toxicological attributes of carbon nanoparticles on different plant species

Serial number	Type of carbon nanoparticle	Plant	Concentration	Effect on plants	Key references
1	Single-walled carbon nanotubes	Tomato	28, 160, 900, 5000 mg L^{-1}	Nonfunctionalized nanotubes inhibited root elongation in tomatoes	Cañas et al. [9]
2	Functional single-walled carbon nanotubes	Lettuce	28, 160, 900, 5000 mg L^{-1}	Inhibited root elongation in lettuce	Cañas et al. [9]
3	Multi-walled CNTs (MWCNTs)	Maize	5, 10, 20, 40, 60 mg L^{-1}	Depress germinative growth of maize seedlings at high concentration	Tiwari et al. [76]
4	Multi-walled carbon nanotubes (MWCNTs)	*Arabidopsis* T87	10, 60, 100, 600 mg L^{-1}	Decrease in values of cell dry weights, cell viabilities, cell chlorophyll contents, and superoxide dismutase (SOD) activities	Lin et al. [43]
5	Multiwalled carbon nanotubes (MWCNT)	Onion	5, 20, 50 mg L^{-1}	Chromosomal aberrations, DNA fragmentation and apoptosis in *Allium* root cells	Ghosh et al. [20]
6	Multiwalled carbon nanotubes (MWCNT)	Onion	5 and 10 mg L^{-1}	Significantly altered cellular morphology of root cells, Membrane integrity and mitochondrial function were also compromised. The nanotubes induced significant DNA damage, micronucleus formation and chromosome aberration	Ghosh et al. [19]
7	Multiwalled carbon nanotubes [MWCNTs	Zucchini	1000 mg/L	Bulk Cu powder reduced biomass by 69%, Cu nanoparticle exposure resulted in 90% reduction relative to control plants	Stampoulis et al. [72]
8	Multiwalled carbon nanotubes [MWCNTs	Rice	20 mg L^{-1}	Reactive oxygen species (ROS) increased, and cell viability decreased	Tan et al. [74]

(continued)

Table 1 (continued)

Serial number	Type of carbon nanoparticle	Plant	Concentration	Effect on plants	Key references
9	Carbon nanotubes CNTs	*Arabidopsis thaliana*	N/A	Photosynthesis, stomatal conductance or transpiration in Arabidopsis thaliana grown in CNTs was significantly reduced. Carbon gain was reduced by 15% in plants grown in CNTs versus control plants. A significant reduction in chlorophyll fluorescence (11% reduction) between plants grown in aged CNTs versus control media	Voleti and Wait [78]
10	Single-bilayer graphene oxide sheet	Faba bean	100, 200, 1600 mg L^{-1}	Decreases in growth parameters and the activity of H_2O_2-decomposing enzymes	Anjum et al. [1]
11	Fullerenes C_{60}	*Scenedesmus obliquus*	0.09 mg L^{-1}	Upon sub-lethal exposure, the photosynthetic products of polysaccharide, soluble protein and total lipid were decreased with exposure time, inhibition of Mg^{2+}-ATPase activity	Tao et al. [75]
12	Neutral pristine MWCNT (p-MWCNT), positively charged MWCNT-NH_2, and negatively charged MWCNT-COOH	*Soyabean*	10.0 to 50.0 mg/L	Inhibited the growth of soybean. chlorophyll accumulation in duckweed	Zhai et al. [86]
13	Single-walled carbon nanotubes (SWCNT)	*Salvia, pepper and tall fescue*	0, 20, 30, and 40 mg L^{-1})	The highest concentration of SWCNT was associated with negative effects. Perforation of the seed coat	Pourkhaloee et al. [58]

6 CNMs' Toxicity to Bacteria

CNMs have been studied to have an adverse impact on the bacteria growth in the ecosystem. For instance, the CNTs offer electrostatic interactions with the bacteria's cell membranes, resulting in significant damage to the cell membrane by the electrostatic interaction, production of oxidative species, or physical disruption [37]. For example, studies with a range of bacteria (gram-negative *Escherichia coli*, *Pseudomonas aeruginosa*, and gram-positive *Staphylococcus aureus*, *Bacillus subtilis*) assessed the toxicity of SWCNT disseminated (5 g/mL) in Tween 20 saline solution in comparison with SWCNT aggregates in saline solution. The outcomes depicted a greater antibacterial activity of dispersed SWCNT to gram-positive bacteria. The SWCNTs dispersed in Tween 20 saline solution showed antibacterial activity of 58.1 $\pm$ 5.0% for E. coli, 65.1 $\pm$ 1.8% for P. aeruginosa, 87.5 $\pm$ 6.5% for B. subtilis, and 85.6 $\pm$ 5.3% for *S. aureus*. On the other hand, SWCNT aggregates in saline solution displayed abridged antibacterial activity of 33.8 $\pm$ 4.0% for E. coli, 27.7 $\pm$ 5.9% for *P. aeruginosa*, 53.9 $\pm$ 2.8% for *B. subtilis*, and 50.3 $\pm$ 3.5% for *S. aureus* [44]. Rodrigues et al. [59] tested carboxyl functionalized SWCNTs at a concentration of 0.5 mg/L on bacteria and fungi. The treatment resulted in changes in *Pseudomonas putida* (Gram-negative) phase alteration temperatures and levels of unsaturated fatty acids. In another study, SWCNTs were found toxic against different kinds of populations, including bacteria. When examined at a concentration of 0.03–1 mg/g, it significantly decreased biomass of reduced biomass of microbial groups and fungal populations [38]. The raw form of many CNMs has antibacterial activities to some extent, but after functionalizing with other compounds, their activities increase multi-fold. For instance, the raw form of MWCNTs exhibited low antibacterial properties, after functionalization with arginine and lysine, their antibacterial activity increased largely, especially against gram-negative bacteria such as *Escherichia coli* and *Salmonella typhimurium* [84]. Graphene also exhibits antibacterial activities in a dose and time-dependent manner. SEM images of *E. coli* cells before and after incubation with GO for 2 h displayed that the bacteria are trampled, have lost their cell integrity and are enveloped by a thin layer of GO [73]. In another study by Gurunathan et al. [24], two forms of graphene, the reduced and the conventional graphene oxide, were tested to see their antibacterial activities against *Pseudomonas aeruginosa*. The exposure of the bacterial cells to these nanomaterials resulted in the production of ions and reactive oxygen species, ultimately reducing cell viability. Fullerene alone has not been reported to possess any antibacterial activity; however, when it's loaded with different other compounds, the antibacterial activities increase multi-fold. For instance, the antibacterial activities of sulfasalazine increased after it was loaded onto the fullerene compared to its conventional form [2]. The water-soluble derivatives of the fullerene produced by Cationic cyclen-functionalization exhibited antibacterial activities against gram-negative and positive bacteria at concentrations of 4.8 μg/mL and 7.4 μg/mL, respectively [11].

7 Conclusion

The increasing attention of the scientific community toward unveiling the hidden potential of nanomaterials has led to a significant boost in the research and applications of nanomaterials. The recent decade has witnessed a boost in the agricultural applications of different nanomaterials, including carbon nanomaterials. Exploring these carbon nanomaterials at the atomic scale also unveils their toxicological attributes toward different forms of life. Plants being sessile, have to depend on their microbial partners to complete their life cycle. Therefore, the toxicological attributes of nanoparticles towards any of the partners in plant–microbe symbiotic relations will end up affecting the overall productivity of plants. The explicit usage of nanoparticles results in their emission into the environment, ultimately leading to nanoparticle pollution. The direct application of nanoparticles in agricultural systems is one of many sources of nanoparticle pollution in the agricultural sector. The nanoparticles from other industrial sectors are released into the wastewater, and the farming practice of using wastewater for irrigation owing to water scarcity leads to the accumulation of nanoparticles in the agricultural soils. Their broad spectrum toxicity also kills the beneficial soil microbiota, thus significantly affecting the rhizosphere microbiome. A judiciously and need-based usage of nanoparticles can answer this concern. The toxic effects of nanoparticles are not restricted to the plant and microbial partners only. They can accumulate in the plant tissues and prove to be equally damaging to human beings on consumption of contaminated plant products. They can result in unexpected mutations in human beings with totally unpredictable outcomes. Thus, a need-based usage along with the continuous monitoring of agricultural produce and soil systems is largely required for having a continuous check on nanoparticle pollution in agricultural systems.

References

1. Anjum NA, Singh N, Singh MK, Sayeed I, Duarte AC, Pereira E, Ahmad I (2014) Single-bilayer graphene oxide sheet impacts and underlying potential mechanism assessment in germinating faba bean (Vicia faba L.). Sci Total Environ 472:834–841
2. Ansari MJ, Soltani A, Ramezanitaghartapeh M, Singla P, Aghaei M, Fadafan HK, Khales SA, Shariati M, Shirzad-Aski H, Balakheyli H, Imam SS (2022) Improved antibacterial activity of sulfasalazine loaded fullerene derivative: computational and experimental studies. J Mol Liq 348:118083
3. Avanasi R, Jackson WA, Sherwin B, Mudge JF, Anderson TA (2014) C60 fullerene soil sorption, biodegradation, and plant uptake. Environ Sci Technol 48(5):2792–2797
4. Bachheti RK, Bachheti A, Husen A (2023) Metal and metal-oxide based nanomaterials (Synthesis, agricultural, biomedical and environmental interventions). Springer Nature Singapore Pte Ltd., 152 Beach Road, #21-01/04 Gateway East, Singapore 189721, Singapore
5. Begum P, Ikhtiari R, Fugetsu B (2011) Graphene phytotoxicity in the seedling stage of cabbage, tomato, red spinach, and lettuce 49(12):3907–19

6. Bello D, Wardle BL, Yamamoto N, Guzman de Villoria R, Garcia EJ, Hart AJ, Ahn K, Ellenbecker MJ, Hallock M (2009) Exposure to nanoscale particles and fibers during machining of hybrid advanced composites containing carbon nanotubes. J Nanopart Res:231–49
7. Bennett SW, Adeleye A, Ji Z, Keller AA (2013) Stability, metal leaching, photoactivity and toxicity in freshwater systems of commercial single wall carbon nanotubes. Water Res 47(12):4074–4085
8. Bratovcic A (2019) Different applications of nanomaterials and their impact on the environment. SSRG Int J Mater Sci Eng 5(1):1–7
9. Cañas JE, Long M, Nations S, Vadan R, Dai L, Luo M, Ambikapathi R, Lee EH, Olszyk D (2008) Effects of functionalized and nonfunctionalized single-walled carbon nanotubes on root elongation of select crop species. Environ Toxicol Chem Int J 27(9):1922–31
10. Chawla S, Singh S, Husen A (2023) Smart Nanomaterials Targeting Pathological Hypoxia. Springer Nature Singapore Pte Ltd., 152 Beach Road, #21–01/04 Gateway East, Singapore 189721, Singapore. https://doi.org/10.1007/978-981-99-1718-1
11. Chen Q, Ma Z, Liu G, Wei H, Xie X (2016) Antibacterial activity of cationic cyclen-functionalized fullerene derivatives: membrane stress. Dig J Nanomater Biostruct (DJNB) 11:753–761
12. Choi W, Lahiri I, Seelaboyina R, Kang YS (2010) Synthesis of graphene and its applications: a review. Crit Rev Solid State Mater Sci 35(1):52–71
13. Choudhary N, Hwang S, Choi W (2014) Carbon nanomaterials: a review. Handb Nanomater Prop 2014:709–769
14. Didenko G, Prylutska S, Kichmarenko Y, Potebnya G, Prylutskyy Y, Slobodyanik N, Ritter U, Scharff P (2013) Evaluation of the antitumor immune response to C60 fullerene. Materialwiss Werkstofftech 44(2–3):124–128
15. Dikshit PK, Kumar J, Das AK, Sadhu S, Sharma S, Singh S, Gupta PK, Kim BS (2021) Green synthesis of metallic nanoparticles: Applications and limitations. Catalysts 11(8):902
16. Duan G, Zhang Y, Luan B, Weber JK, Zhou RW, Yang Z, Zhao L, Xu J, Luo J, Zhou R (2017) Graphene-induced pore formation on cell membranes. Sci Rep 7(1):42767
17. Franskevych D, Palyvoda K, Petukhov D, Prylutska S, Grynyuk I, Schuetze C, Drobot L, Matyshevska O, Ritter U (2017) Fullerene C60 penetration into leukemic cells and its photoinduced cytotoxic effects. Nanoscale Res Lett 12:1–9
18. Gessner I, Neundorf I (2020) Nanoparticles modified with cell-penetrating peptides: Conjugation mechanisms, physicochemical properties, and application in cancer diagnosis and therapy. Int J Mol Sci 21(7):2536
19. Ghosh M, Bhadra S, Adegoke A, Bandyopadhyay M, Mukherjee A (2015) MWCNT uptake in Allium cepa root cells induces cytotoxic and genotoxic responses and results in DNA hypermethylation. Mutat Res/Fundam Mol Mech Mutagen 1(774):49–58
20. Ghosh M, Chakraborty A, Bandyopadhyay M, Mukherjee A (2011) Multi-walled carbon nanotubes (MWCNT): induction of DNA damage in plant and mammalian cells. J Hazard Mater 197:327–336
21. Godeto YG, Ayele A, Ahmed IN, Husen A, Bachheti RK (2023) Medicinal plant-based metabolites in nanoparticles synthesis and their cutting-edge applications: an overview. In: Bachheti RK, Bachheti A (eds) Secondary metabolites from medicinal plants. CRC Press, Boca Raton, USA, pp 1–34
22. Gonfa YH, Tessema FB, Tadesse MG, Bachheti A, Bachheti RK (2023) Medicinally important plant roots and their role in nanoparticles synthesis and applications. In: Bachheti RK, Bachheti A (eds) Secondary metabolites from medicinal plants. CRC Press, Boca Raton, USA, pp 227–242
23. Gorka DE, Jeger JL, Zhang H, Ma Y, Colman BP, Bernhardt ES, Liu J (2016) Phytotoxicity of soluble graphitic nanofibers to model plant species. Environ Toxicol Chem 35(12):2941–2947
24. Gurunathan S, Han JW, Dayem AA, Eppakayala V, Kim JH (2012) Oxidative stress-mediated antibacterial activity of graphene oxide and reduced graphene oxide in Pseudomonas aeruginosa. Int J Nanomed 30:5901–5914

25. Hatami M (2017) Toxicity assessment of multi-walled carbon nanotubes on Cucurbita pepo L. under well-watered and water-stressed conditions. Ecotoxicol Environ Safety 142:274–83
26. Hossain Z, Mustafa G, Sakata K, Komatsu S (2016) Insights into the proteomic response of soybean towards Al2O3, ZnO, and Ag nanoparticles stress. J Hazard Mater 5(304):291–305
27. Hu X, Kang J, Lu K, Zhou R, Mu L, Zhou Q (2014) Graphene oxide amplifies the phytotoxicity of arsenic in wheat. Sci Rep 4(1):6122
28. Husen A, Siddiqi KS (2014) Carbon and fullerene nanomaterials in plant system. J Nanobiotechnol 12(1):1
29. Husen A, Siddiqi KS (2014) Phytosynthesis of nanoparticles: concept, controversy and application. Nanoscale Res Lett 9(229):1–24
30. Husen A (2023a) Nanomaterials and nanocomposites exposures to plants (Response, interaction, phytotoxicity and defense mechanisms). Springer Nature Singapore Pte Ltd., 152 Beach Road, #21–01/04 Gateway East, Singapore 189721, Singapore. https://doi.org/10.1007/978-981-99-2419-6
31. Husen A (2023b) Nanomaterials from agricultural and horticultural products. Springer Nature Singapore Pte Ltd., 152 Beach Road, #21–01/04 Gateway East, Singapore 189721, Singapore. https://doi.org/10.1007/978-981-99-3435-5
32. Husen A, Bachheti RK, Bachheti A (2023) Current trends in green nano-emulsions (Food, agriculture and biomedical sectors). Springer Nature Singapore Pte Ltd., 152 Beach Road, #21–01/04 Gateway East, Singapore 189721, Singapore
33. Husen A, Siddiqi KS (2023) Advances in smart nanomaterials and their applications. Elsevier Inc., 50 Hampshire St., 5th Floor, Cambridge, MA 02139, USA
34. Husen A (2020b) Carbon-based nanomaterials and their interactions with agricultural crops. In: Husen A, Jawaid M (eds) Elsevier Inc. 50 Hampshire St., 5th Floor, Cambridge, MA 02139, USA, pp 199–218. https://doi.org/10.1016/B978-0-12-817852-2.00008-1
35. Hussain MA, Kabir MA, Sood AK (2009) On the cytotoxicity of carbon nanotubes. Curr Sci 10:664–673
36. Ijaz I, Gilani E, Nazir A, Bukhari A (2020) Detail review on chemical, physical and green synthesis, classification, characterizations and applications of nanoparticles. Green Chem Lett Rev 13(3):223–245
37. Jackson P, Jacobsen NR, Baun A, Birkedal R, Kühnel D, Jensen KA, Vogel U, Wallin H (2013) Bioaccumulation and ecotoxicity of carbon nanotubes. Chem Cent J 7:1–21
38. Jin L, Son Y, DeForest JL, Kang YJ, Kim W, Chung H (2014) Single-walled carbon nanotubes alter soil microbial community composition. Sci Total Environ 466:533–538
39. Jin-Chul K, Madhusudhan A, Husen A (2021) Smart nanomaterials in biomedical applications. Springer Nature Switzerland AG, Gewerbestrasse 11, 6330 Cham, Switzerland. https://doi.org/10.1007/978-3-030-84262-8
40. Kapoor D, Yadav S, Sharma MM, Sharma P (2023) Interaction between metal nanoparticles and PGPR on the plant growth and development. In: Nanomaterials and nanocomposites exposures to plants: response, interaction, phytotoxicity and defense. Springer Nature Singapore, Singapore, pp 327–351
41. Leyu AM, Debebe SE, Bachheti A, Rawat YS, Bachheti RK (2023) Green Synthesis of gold and silver nanoparticles using invasive alien plant *Parthenium hysterophorus* and their antimicrobial and antioxidant activities. Sustainability 15(12):9456
42. Liao C, Li Y, Tjong SC (2018) Graphene nanomaterials: Synthesis, biocompatibility, and cytotoxicity. Int J Mol Sci 19(11):3564
43. Lin C, Fugetsu B, Su Y, Watari F (2009) Studies on toxicity of multi-walled carbon nanotubes on arabidopsis T87 suspension cells. J Hazard Mater 170(2–3):578–583
44. Liu Q, Chen B, Wang Q, Shi X, Xiao Z, Lin J, Fang X (2009) Carbon nanotubes as molecular transporters for walled plant cells. Nano Lett 9(3):1007–1010
45. Liu D, Wang L, Wang Z, Cuschieri A (2012) Different cellular response mechanisms contribute to the length-dependent cytotoxicity of multi-walled carbon nanotubes. Nanoscale Res Lett 7:1
46. Loyal A, Pahuja SK, Rani N, Pooja, Srivastava RK, Sharma P (2023a) Titanium oxide nanoparticles: plant response, interaction, phytotoxicity, and defence mechanisms. In: Nanomaterials

and nanocomposites exposures to plants: response, interaction, phytotoxicity and defense mechanisms. Springer Nature Singapore, Singapore, pp 263–284
47. Loyal A, Pahuja SK, Sharma P, Malik A, Srivastava RK, Mehta S (2023b) Potential environmental and human health implications of nanomaterials used in sustainable agriculture and soil improvement. In: Engineered nanomaterials for sustainable agricultural production, soil improvement and stress management. Academic Press, pp 387–412
48. Madhusudhan A, Purohit SD, Prasad R, Husen A (2023) Functional smart nanomaterials and their theranostics approaches. Springer Nature Singapore Pte Ltd., 152 Beach Road, #21–01/04 Gateway East, Singapore 189721
49. Magdolenova Z, Collins A, Kumar A, Dhawan A, Stone V, Dusinska M (2014) Mechanisms of genotoxicity: a review of in vitro and in vivo studies with engineered nanoparticles. Nanotoxicology 8(3):233–78
50. Mandal AK, Ghorai S, Husen A (2023) Functionalized smart nanomaterials for point-of-care testing. Springer Nature Singapore Pte Ltd., 152 Beach Road, #21–01/04 Gateway East, Singapore 189721, Singapore
51. Mengstu A, Esubalew S, Abate L, Husen RKBA, Bachheti A (2023) Medicinal plant-based flavonoid-mediated nanoparticles synthesis, characterization, and applications. In: Bachheti RK, Bachheti A (eds) Secondary metabolites from medicinal plants. CRC Press, Boca Raton, USA, pp 35–52
52. Mokula MR, Husen A (2023) Toxicity assessment of silver nanoparticles and silver ions on plant growth. In: Husen A (ed) Nanomaterials and nanocomposites exposures to plants. Smart Nanomaterials Technology. Springer, Singapore, pp 157–174. https://doi.org/10.1007/978-981-99-2419-6_7 ISBN: 978-981-99-2418-9
53. Mukherjee A, Majumdar S, Servin AD, Pagano L, Dhankher OP, White JC (2016) Carbon nanomaterials in agriculture: a critical review. Front Plant Sci 22(7):172
54. Nowack B, David RM, Fissan H, Morris H, Shatkin JA, Stintz M, Zepp R, Brouwer D (2013) Potential release scenarios for carbon nanotubes used in composites. Environ Int 59:1–1
55. Pelin M, Fusco L, Martín C, Sosa S, Frontiñán-Rubio J, González-Domínguez JM, Durán-Prado M, Vázquez E, Prato M, Tubaro A (2018) Graphene and graphene oxide induce ROS production in human HaCaT skin keratinocytes: the role of xanthine oxidase and NADH dehydrogenase. Nanoscale 10(25):11820–11830
56. Peng Z, Liu X, Zhang W, Zeng Z, Liu Z, Zhang C, Liu Y, Shao B, Liang Q, Tang W, Yuan X (2020) Advances in the application, toxicity and degradation of carbon nanomaterials in environment: a review. Environ Int 1(134):105298
57. Petersen EJ, Zhang L, Mattison NT, O'Carroll DM, Whelton AJ, Uddin N, Nguyen T, Huang Q, Henry TB, Holbrook RD, Chen KL (2011) Potential release pathways, environmental fate, and ecological risks of carbon nanotubes. Environ Sci Technol 45(23):9837–9856
58. Pourkhaloee A, Haghighi M, Saharkhiz MJ, Jouzi H, Doroodmand MM (2011) Carbon nanotubes can promote seed germination via seed coat penetration. Seed Technol:155–69
59. Rodrigues DF, Jaisi DP, Elimelech M (2013) Toxicity of functionalized single-walled carbon nanotubes on soil microbial communities: implications for nutrient cycling in soil. Environ Sci Technol 47(1):625–633
60. Salem SS, Husen A (2023) Zinc oxide nanoparticles: plant response, interaction, phytotoxicity, and defence. In: Husen A (ed) Nanomaterials and nanocomposites exposures to plants. Smart Nanomaterials Technology. Springer, Singapore, pp 247–261. https://doi.org/10.1007/978-981-99-2419-6_12. ISBN: 978-981-99-2418-9
61. Samadian H, Salami MS, Jaymand M, Azarnezhad A, Najafi M, Barabadi H, Ahmadi A (2020) Genotoxicity assessment of carbon-based nanomaterials; have their unique physicochemical properties made them double-edged swords? Mutat Res/Rev Mutat Res 1(783):108296
62. Sangwan S, Sharma P, Wati L, Mehta S (2023) Effect of chitosan nanoparticles on growth and physiology of crop plants. In: Engineered nanomaterials for sustainable agricultural production, soil improvement and stress management. Academic Press, pp 99–123
63. Sargent LM, Hubbs AF, Young SH, Kashon ML, Dinu CZ, Salisbury JL, Reynolds SH (2012) Single-walled carbon nanotube-induced mitotic disruption. Mutat Res/Genetic Toxicol Environ Mutagen 745(1-2):28-37

64. Sawangphruk M, Srimuk P, Chiochan P, Sangsri T, Siwayaprahm P (2012) Synthesis and antifungal activity of reduced graphene oxide nanosheets. Carbon 50(14):5156–5161
65. Sharma G, Nim S, Alle M, Husen A, Kim JC (2022) Nanoparticle-mediated delivery of flavonoids for cancer therapy: prevention and treatment. In: Kim JC, Alle M, Husen A (eds) Smart nanomaterials in biomedical applications. Nanotechnology in the Life Sciences. Springer, Cham. https://doi.org/10.1007/978-3-030-84262-8_3
66. Sharma MM, Kapoor D, Rohilla R, Sharma P (2023a) Nanomaterials and their toxicity to beneficial soil microbiota and fungi associated plants rhizosphere. In: Nanomaterials and nanocomposites exposures to plants: response, interaction, phytotoxicity and defense mechanisms. Springer Nature Singapore, Singapore, pp 353–380
67. Sharma P, Sangwan S, Mehta S (2023b) Emerging role of phosphate nanoparticles in agriculture practices. In: Engineered nanomaterials for sustainable agricultural production, soil improvement and stress management. Academic Press, pp 71–97
68. Siddiqi KS, Husen A, Sohrab SS, Osman M (2018) Recent status of nanomaterials fabrication and their potential applications in neurological disease management. Nanoscale Res Lett 13(231):1–17
69. Siegrist KJ, Reynolds SH, Kashon ML, Lowry DT, Dong C, Hubbs AF, Young SH, Salisbury JL, Porter DW, Benkovic SA, McCawley M (2014) Genotoxicity of multi-walled carbon nanotubes at occupationally relevant doses. Part Fibre Toxicol 11:1–5
70. Singh N, Manshian B, Jenkins GJ, Griffiths SM, Williams PM, Maffeis TG, Wright CJ, Doak SH (2009) NanoGenotoxicology: the DNA damaging potential of engineered nanomaterials. Biomaterials 30(23–24):3891–3914
71. Singh S, Sangwan S, Sharma P, Devi P, Moond M (2021) Nanotechnology for sustainable agriculture: an emerging perspective. J Nanosci Nanotechnol 21(6):3453–3465
72. Stampoulis D, Sinha SK, White JC (2009) Assay-dependent phytotoxicity of nanoparticles to plants. Environ Sci Technol 43(24):9473–9479
73. Szunerits S, Boukherroub R (2016) Antibacterial activity of graphene-based materials. J Mater Chem B 4(43):6892–6912
74. Tan XM, Lin C, Fugetsu B (2009) Studies on toxicity of multi-walled carbon nanotubes on suspension rice cells. Carbon 47(15):3479–3487
75. Tao X, Yu Y, Fortner JD, He Y, Chen Y, Hughes JB (2015) Effects of aqueous stable fullerene nanocrystal (nC60) on Scenedesmus obliquus: evaluation of the sub-lethal photosynthetic responses and inhibition mechanism. Chemosphere 122:162–167
76. Tiwari DK, Dasgupta-Schubert N, Villaseñor Cendejas LM, Villegas J, Carreto Montoya L, Borjas García SE (2014) Interfacing carbon nanotubes (CNT) with plants: enhancement of growth, water and ionic nutrient uptake in maize (Zea mays) and implications for nanoagriculture. Appl Nanosci 4:577–591
77. Tripathi DK, Singh S, Singh S, Pandey R, Singh VP, Sharma NC, Prasad SM, Dubey NK, Chauhan DK (2017) An overview on manufactured nanoparticles in plants: uptake, translocation, accumulation and phytotoxicity. Plant Physiol Biochem 110:2–12
78. Voleti R, Wait DA (2014) Effect of carbon nanotubes on plant growth and gas exchange using Arabidopsis thaliana
79. Wagay JA, Singh S, Raffi MM, Rahman QI, Husen A (2019) Effect of carbon-based nanomaterials on plant functioning and rhizosphere. In: Husen A, Iqbal M (eds) N*anomaterials* and plant potential. Springer International Publishing AG, Gewerbestrasse 11, 6330, Cham pp 553–575. https://doi.org/10.1007/978-3-030-05569-1_22
80. Wang X, Liu X, Chen J, Han H, Yuan Z (2014) Evaluation and mechanism of antifungal effects of carbon nanomaterials in controlling plant fungal pathogen. Carbon 68:798–806
81. Wu K, Zhou Q, Ouyang S (2021) Direct and indirect genotoxicity of graphene family nanomaterials on DNA—a review. Nanomaterials 11(11):2889
82. Xiaoli F, Qiyue C, Weihong G, Yaqing Z, Chen H, Junrong W, Longquan S (2020) Toxicology data of graphene-family nanomaterials: an update. Arch Toxicol 94:1915–1939
83. Xu Z, Long X, Jia Y, Zhao D, Pan X (2022) Occurrence, transport, and toxicity of nanomaterials in soil ecosystems: a review. Environ Chem Lett 20(6):3943–3969

84. Zardini HZ, Amiri A, Shanbedi M, Maghrebi M, Baniadam M (2012) Enhanced antibacterial activity of amino acids-functionalized multi walled carbon nanotubes by a simple method. Colloids Surf B 92:196–202
85. Zaytseva O, Neumann G (2016) Carbon nanomaterials: production, impact on plant development, agricultural and environmental applications. Chem Biol Technol Agric 3(1):1–26
86. Zhai G, Gutowski SM, Walters KS, Yan B, Schnoor JL (2015) Charge, size, and cellular selectivity for multiwall carbon nanotubes by maize and soybean. Environ Sci Technol 49(12):7380–7390
87. Zhang L, Xu B, Wang X (2016) Cholesterol extraction from cell membrane by graphene nanosheets: a computational study. J Phys Chem B 120(5):957–964
88. Zhou R, Gao H (2014) Cytotoxicity of graphene: recent advances and future perspective. Wiley Interdisc Rev Nanomed Nanobiotechnol 6(5):452–474

Role of Carbon Nanomaterials in Energy Generation, Storage, and Conversion

Noureen Amir Khan and Gul Rahman

Abstract Due to the world's rapidly increasing population and technological advancements, energy is needed. The world's energy supply is anticipated to double by 2050. Nanotechnology has opened up new possibilities in materials science and engineering, specifically in the manufacture of new materials for efficient energy conversion and storage. Carbon nanomaterials (CBNMs) possess distinctive size and surface-dependent features, such as electrical, morphological, mechanical, and optical properties, that are advantageous for improving energy conversion and storage performance compared to traditional materials. Substantial progress has been made in creating high-performance energy conversion and storage devices such as solar cells, fuel cells, batteries, and supercapacitors. This book chapter focuses on the latest developments and improvements made to the effectiveness of electrode materials used in renewable energy storage and conversion systems by utilizing graphene, carbon nanotubes (CNTs), fullerenes, and nanohybrid fillers. These materials are exceptional candidates for solar cells because of their superior capacity for photon absorption, photovoltaic characteristics, producing photocarriers, and separating charge carriers to create heterojunction. The synthetic method, pore size and distribution, and specific surface area of these materials all impact the capacitance of supercapacitor and battery materials. Additionally, these nanomaterials' high surface area and electronic conductivity enhance the rate of electrode reactions in fuel cells.

Keywords Carbon nanomaterials · CNTs · Graphene · Fullerenes · Energy conversion · Energy storage

N. A. Khan · G. Rahman (✉)
Institute of Chemical Sciences, University of Peshawar, Peshawar 25120, Pakistan
e-mail: gul_rahman47@uop.edu.pk

A. Bachheti et al. (eds.), *Carbon-Based Nanomaterials*, Smart Nanomaterials Technology, https://doi.org/10.1007/978-981-97-0240-4_17

1 Background

1.1 Energy Future: Challenges and Opportunities

The twenty-first century's growing population and rapid technological advancement have led to a surge in energy demand [1]. With global power consumption currently standing at 17 trillion watts, experts anticipate a 30 trillion-watt increase by 2050, driven by population expansion and economic growth worldwide [2]. However, fossil fuels, the primary energy source, are notorious for pollution. To mitigate the negative effects of fossil fuels, utilizing cleaner, alternative energy sources may reduce inequality and promote global peace. Nevertheless, only 10% of the world's energy resources were renewable in 2013, and generating electricity from renewable sources remains expensive [3]. Nonetheless, renewable energy costs are declining, providing hope for their widespread adoption in the near future.

Nanotechnology can aid in the development of innovative systems for generating, storing, and transferring electricity in an eco-friendly and more efficient manner while requiring less space, aiding solar energy in meeting energy demands without harming the environment or human health [4]. Research on nanostructured carbon materials, including graphene, fullerenes, and carbon nanotubes (CNTs) for energy production and storage, has been ongoing for the last decade [5]. Incorporating these materials using lightweight and flexible substrates, ambient pressure, and low-temperature manufacturing methods can significantly reduce production costs.

Nanostructured allotrope forms of carbon, with their distinctive hybridization characteristics and sensitivity to agitation during synthesis, have been extensively studied in the last two decades. This has enabled precise manipulation of the material's properties, with carbon existing in various hybridization states, each with its own properties [6]. The material's chemical, thermal, mechanical, and electrical properties are directly influenced by hybridization and structure, allowing the same material to be used for various purposes [7]. It may be possible to create electrochemical capacitors (supercapacitors) and solar cell devices that are more efficient and less expensive to manufacture than traditional silicon-based solar cells and chemical batteries using these nanostructured carbon materials.

2 Role of Carbon Nanomaterials in Energy Applications

2.1 Carbon Nanomaterials for Energy Conversion

2.1.1 Solar Cells

Solar cells have gained substantial importance among various photovoltaics due to their unique features, such as their ability to endure complex deformation, lightweight nature, integration with curved surfaces, roll-to-roll compatibility in manufacturing, and ease of transportation and storage. Moreover, solar cells are a clean energy source with no carbon footprints [8]. Carbon nanomaterials, including Graphene, Fullerenes, and CNTs, are widely used in high-performance solar cells due to their numerous benefits, including long-term stability, high abundance and conductivity, good transparency, and high mechanical flexibility. These nanomaterials significantly enhance the performance of photovoltaics by improving charge generation, collection, and transport [9]. In addition, their compatible fabrication techniques and mass-production capabilities play a crucial role in the advancement of solar cells.

Carbon Nanomaterial-Based Solar Cells

Carbon nanomaterials can improve solar cells' efficacy for sustainable technology development [10]. In this section, we will present an overview of carbon nanomaterial-based solar cells, with a particular focus on organic solar cells (OSCs), dye-sensitized solar cells (DSSCs), perovskite-based solar cells (PSCs), carbon-based silicon solar cells.

In organic solar cells, an active layer of p/n-type donor/acceptor is sandwiched between an anode and a cathode with two charge transport layers [12]. There is a hole transport layer (HTL) between the active layer and anode and an electron transport layer between the cathode and active layer [13]. The photoexcitation process generates excitons (e^- and h^+ pairs) in the active layers, which are transferred to the e^- donor/acceptor interfacial region and subsequently separated into holes and electrons (Fig. 1a). The HTL and ETL transport h^+ and e^- towards the anode and cathode, respectively, generating current and voltage [11].

In OSCs, the light conversion into electric current is enhanced by conjugated semiconducting polymers. These devices are easily made in various sizes and forms, and their design is easy to change for specific use. Additionally, they can be economically deposited onto a variety of substrates [14]. However, their practical application is limited by certain factors. In particular, their operational life is shorter than that of Si-based SCs because they are more susceptible to physical and chemical degradation and have weak interfacial contact between the electrode and active layer [15]. Nevertheless, this can be addressed by depositing an electron and hole extraction layer between the electrodes and the active layer [16]. The formation of trap states within the bandgap and the active layer's photochemical reaction cause low Photocurrent

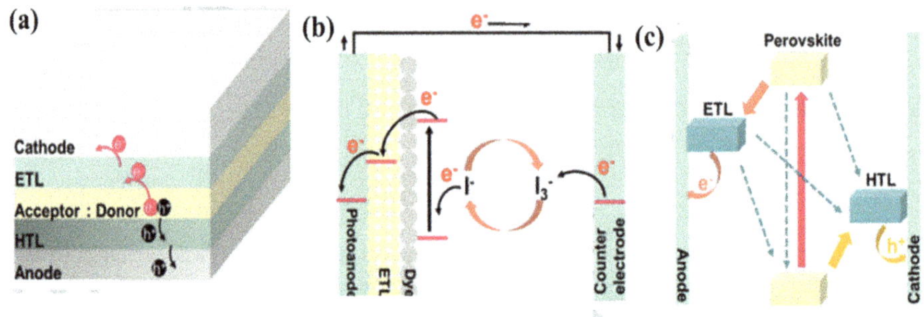

Fig. 1 Schematic illustration of structure and working mechanism of **a** OSCs, **b** DSSCs, and **c** PSCs (adopted from [11])

efficiency (PCEs) in these polymers. Poor performance may also be caused by trap-mediated recombination, decreased mobility of holes, and accumulation of charge in many trap states [17]. Another factor contributing to conjugated polymers' low efficiency is the bandgap.

A semiconducting conjugated polymer absorbs a small amount of solar energy when its bandgap is large. The incorporation of CNTs into binary heterojunctions has been proposed as a method for increasing OSC efficiency [18]. In this instance, the CNTs function as acceptors of electrons and restore the dissociation of excitons via enhancing the electric field at the interface of CNT/polymer, which further prevents the recombination of charge carriers in this heterojunction [19]. Graphene also plays a role in OSC charge transport. Park et al. [20] investigated how the hole transport properties of OSCs were affected by the different sheet sizes of orthogonal-fluorine functionalized rGO (FrGO). The sizes of the L-FrGO changed from 1.5 to 0.7 μm; for M-FrGO, it varied to 0.5 μm from 0.9 μm and S-FrGO from 0.45 to 0.15 μm. The S-FrGO-based cell had a PCE conversion of 9.27%, J_{sc} of 15.46 mA/cm^2 with Voc of 1.00 V and FF of 59.8%.

A dye-sensitized solar cell (DSSC) typically consists of a photoanode made of a semiconductor or conductive substrate, a photoactive dye, a redox couple-containing electrolyte (I_3^-/I^- or Co^{2+}/Co^{3+}), and a counter electrode made of carbon material [21]. Under illumination, excited electrons are injected by dye molecules into the photoanode's conduction band. These electrons then migrate towards the external circuit and are returned via the counter electrode to the redox electrolyte (Fig. 1b). Finally, the regeneration of dye molecules by capturing electrons from the electrolyte completes the energy conversion cycle [22]. Yu et al. [23] reported DSSCs with an excellent PCE of 10.56% designed by 3D Graphene/SWCNT carbon electrode and showed transmittance of 56.6% at 670 nm compared to bare FTO (84.71%). A bilayer graphene/MWCNTs developed by Wahyuono et al. showed a higher PCE of 4.1% than the monolayer, which could replace Pt in DSSCs [24]. The rGO sheets/Fe_3O_4 nanoparticles composite exhibited remarkable activity in the reduction of I_3 to I^-, achieving 6.65% of efficiency higher than Pt-based DSSC (6.37%) [25]. PANI/graphene composites for DSSC's Counter electrode (CE) reported by Nguyen et al.

[26] improved PV performance by lowering the CE's charge transfer resistance. In SCs fabricated with a PANI/9% graphene, CE had an electrolyte/CE interface resistance of 20.1 and an overall efficiency of 7.45%. Kilic et al. [27] used a hybrid CNT/ZnO photoelectrode in a DSSC, which amplified J_{sc} with an extended lifetime of electrons, resulting in a higher efficiency of 5.55%.

Perovskite-based solar cells (PSCs) show higher power conversions up to 22.1% compared to OSC and have gained attention in recent research [28]. In these PSCs, a photoactive perovskite is sandwiched between two active layers of charge transport (Fig. 1c), i.e., p-i-n junction as HTL/perovskite/ETL and n-i-p junction for ETL/perovskite/HTL. After photon absorption, e^-excitation occurs from the perovskite valence band to the conduction band, generating excitons. The holes and electrons are transported via HTL and ETL, respectively, followed by collection at the cathode and anode [29]. Due to their higher performance and easy manufacturing process, PSCs have received much attention in the present research. However, stability and fabrication issues restrict PSC modularization and commercialization [30]. Carbon-based materials have been utilized extensively in making PSCs to address these issues because of their superior mechanical, optical, and electrical properties. Additionally, certain carbon materials' hydrophobic properties significantly improve the stability of PSCs and protect the perovskite film from humidity [31]. HTM-free carbon-based PSCs (C-PSCs) are favorable for PV technology due to their excellent stability and affordability [32]. Paintable carbon-based perovskite solar cells (C-PSCs) continue to face challenges in achieving high efficiency due to low power conversion efficiency (PCE) and fill factor resulting from poor conductivity of the electrode and inadequate contact at the perovskite-electrode interface. To address these challenges, Wang et al. [33] incorporated SWCNTs into the perovskite and layers of carbon, forming a high-efficiency interface of perovskite/carbon and introducing a CNT bridging method for the devices. Figure 2 illustrates the carbon counter electrodes of three types used in extracting and transporting holes at the interface of $MAPbI_3$/electrode with CNT and without CNT spanning technique. The CNT bridges facilitated charge transport in the $MAPbI_3$ layer and carbon cathode and increased the electrical conductivity of the carbon electrodes.

One of the major obstacles facing the development of carbon-based PSCs is the incompatibility of the hole transport materials (HTMs) used in advanced Au-based PSCs [34]. To report this problem, Chu et al. [35] designed a composite of P_3HT/graphene using the solution-processed low-temperature technique, having good compatibility with paintable carbon electrodes. This composite demonstrated enhanced thermal tolerance and charge mobility compared to pure P_3HT, with a two-fold increase in hole mobility after annealing at 100 °C (from 8.3×10^{-3} cm^2/Vs to 1.2×10^{-2} cm^2/Vs). The composite HTL's improved charge transport and extraction led to a significant increase in efficiency compared to pristine P_3HT HTL cells. The carbon-based PSCs achieved an efficiency of 17.8%, making them the first perovskite-based cells used in stabilized testing protocol. The devices also demonstrated exceptional stability, with only a 3% decrease in efficiency after 1680 h of ambient storage (humidity: approximately 50%) and approximately 89%

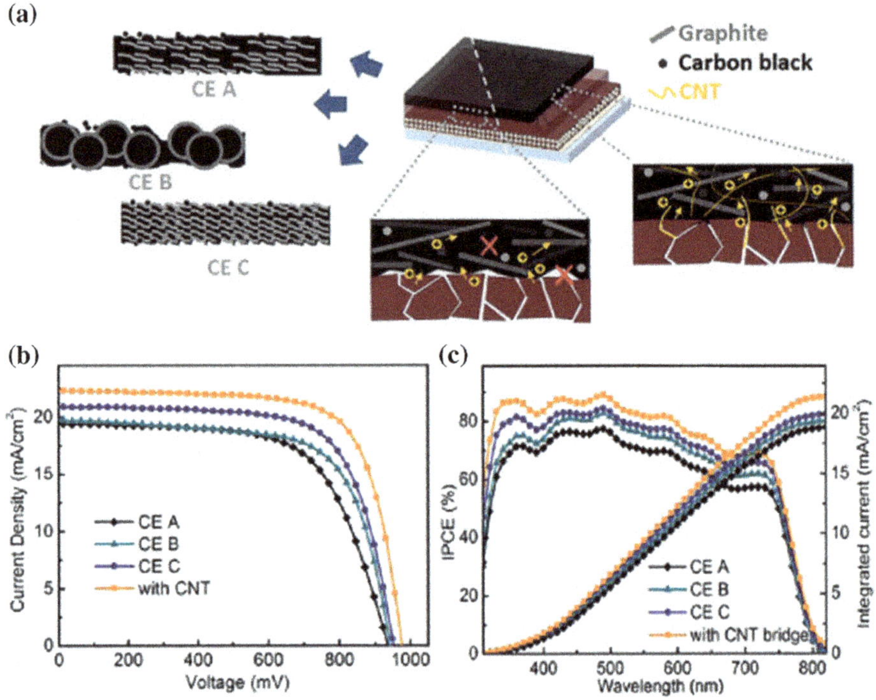

Fig. 2 **a** Schematic illustration of carbon-based CEs for extraction/transport of hole at interface of $MAPbI_3$/electrode with CNTs and without CNT bridging method, **b** representative J-V curves, and **c** corresponding IPCE spectra of C-PSCs fabricated in four types (adopted from [33])

output retention after continuous one-sun illumination for 600 h in an N_2 atmosphere at room temperature (when encapsulated) [35]. Zhou et al. [36] reported PSCs composed of WO_3-embedded carbon electrode nanoparticles. The WO_3 acted as inorganic HTM and enhanced interfacial hole extraction in perovskite/carbon composite. This approach resulted in an energy transformation effectiveness of 40%, significantly higher (10.77%) than the energy-transforming efficiencies of reference devices under normal conditions.

Early carbon nanomaterial-based solar cells typically use p/n junctions for energy conversion. Carbon-based Si solar cells have been developed with multiple junctions, achieving efficiencies of over 45% [37]. These solar cells utilize a heterojunction of Si/C formed between n-type Si and p-type carbon nanomaterials [34]. Electrons are injected into the carbon and holes into Si, creating band bending and built-in potential due to the variation in carrier diffusion concentrations. Under illumination, photons absorbed by the heterojunction generated e^-/h^+ pairs [38]. A range of carbon nanomaterials, from low-dimensional carbon nanotubes and fullerenes to graphene and polymer-based carbon nanomaterials, have been reported [39]. C_{60} is generally utilized as an electron acceptor and transporter in solar cells due to its fast photo-induced electron transfer, high mobility, and direct band gap (1.5–2.3 eV), similar to n-type semiconductors [40]. However, the power conversion efficiency (PCE) of C_{60}-based solar cells is limited by the resistance of the films. The highest photocurrents have been observed for doped n-Si/C_{60} structures, but their efficiency remains challenging due to intrinsic conductivity [41]. Hybrid cells combining p-type silicon and n-type organic C_{60} heterojunctions have shown optimal PCEs of 8.43%, and doping of C_{60} greatly influences the width of the depletion region and the electrical properties of the solar cell [42].

The electronics industry has shown great interest in graphene because of its transparency, incredible electronic mobility (2.5×10^5 cm^2/Vs), no band gap, and broad spectrum for light absorption [43]. Li et al. [44] reported a graphene-based C/Si heterojunction solar cell in which graphene was deposited on n-type Si, showing 1.65% of PCE performance. Higher PCEs can be achieved in heterojunction of graphene/Si by adjusting the graphene layer numbers and increasing the heterojunction's built-in potential by increasing the graphene's work function [45]. P-type chemical doping is a common method used to improve graphene's work function, and efficiencies of this heterojunction solar cells increased to 15.6% reported in past years [46]. According to Zhong et al. [47], the Schottky junction between graphene woven fabrics (GWFs) and n-doped Si crystals demonstrates good photoelectric conversion by maintaining a good balance amongst photoinduced carrier's generation and collection. Carbon nanotubes (CNTs) also demonstrate suitable applications in solar cells. The earliest CNT/Si heterostructure showed 1.3% PCE made by combining n-type Si and double-walled CNTs [48]. Hydrofluoric acid treatment enhanced the PCE to 7.4% by reducing SiO_2 layer thickness [49]. The CNTs and Si heterojunction make a variety of advanced materials due to their easy and simple structure production under atmospheric conditions, aiding in overcoming the limitations of pristine Si layers. Conventional solar cells derived from CNT/Si heterojunctions have limited industrial applications because of metallic frames in Si wafers. Chen et al. [50] produced a device having junctions with front and back designs, making it possible to use the complete Si wafer as a central active area. An intermixed Nafion layer served as antireflection passivation, nanotube dopant, and physically blocking layer, making it easier to maintain the standard geometry. As a result, a hybrid charge-selective CNT/Nafion passivated contact and solar cell with 1–16 cm^2 active area were constructed.

The potential of CBNMs for solar cells, in relation to photovoltaic technology's trends and advancements, makes it possible to move from the laboratory to industry-level based technologies. Advanced carbon-based solar cell applications appear to point to improved future energy harvesting, and obstacles to photovoltaic solar cells can be effectively overcome through additional research into innovative processes, novel device configurations, and emerging material design (Table 1).

2.1.2 Fuel Cells

As concerns over the depletion of fossil fuel reserves, environmental damage, and dramatic climatic changes continue to rise, researchers are focusing on finding alternative, sustainable, clean, and highly efficient energy conversion and storage systems

Table 1 Carbon nanomaterials in various types of solar cells and their performances

Type	Electrode material	V_{oc} (V)	J_{sc} (mA cm^{-2})	FF (%)	PCE (%)	Key references
Perovskite solar cell	CQDs/ TiO_2	1.136	21.36	78	18.89	Li et al. [51]
	GO	1.04	20.2	73	15.1	Li et al. [52]
	MWCNTs	0.88	15.60	75	10.30	Wei et al. [53]
	Carbon/ MWCNT	0.93	21.3	59	11.6	Cheng et al. [54]
Organic solar cells	CNT/ AgNW	0.61	7.60	48	2.21	Yang and Lee [55]
	MWCNTs	0.57	7.53	34	1.46	Ali et al. [56]
	AgNPs/ MWCNTs	0.76	15.23	60	7.00	Zhang et al. [57]
	Polymeric acid-doped CNTs	0.80	14.30	70	8.00	Hashemi et al. [58]
Dye-sensitized solar cells	TiO_2/CNTs	0.26	0.87	–	8.74	Wang et al. [59]
	rGO/TaON	0.775	11.62	38.1	3.43	Li et al. [60]
	Graphene/ nickel phosphide	0.727	12.86	61	5.72	Dou et al. [61]
	Graphene/ ZnO	0.5	17.5	45	3.89	Hsu et al. [62]
	SnS_2/rGO	0.71	14.80	67.02	7.12	Yang et al. [63]
Silicon-based solar cells	CNTs/Si	0.54	25.1	0.56	14	Xie et al. [64]
	Graphene/ HfO_2/Si	0.1	40	–	14	Kadam et al. [65]
	CNTs/ silicon	0.68	40.38	79	22.4	Yan et al. [66]
	GO/Si	0.67	40	80.6	21.6	Gao et al. [67]

[68]. Among these systems, fuel cells are gaining a lot of attention for converting and storing renewable energy because of their low emissions of greenhouse gases, superior efficiency, and environmental friendliness [69]. However, the current energy components in fuel cells use noble metals that are rare and expensive, limiting their widespread use. To address this issue, researchers are exploring the potential of carbon-based materials, which have unique properties and are receiving much attention [70]. One of the primary benefits of carbon-based materials is that they can be manufactured in various dimensional structures, ranging from 0 to 3D. Researchers are currently working to improve the stability and performance of these carbon-based electrocatalysts in fuel cells.

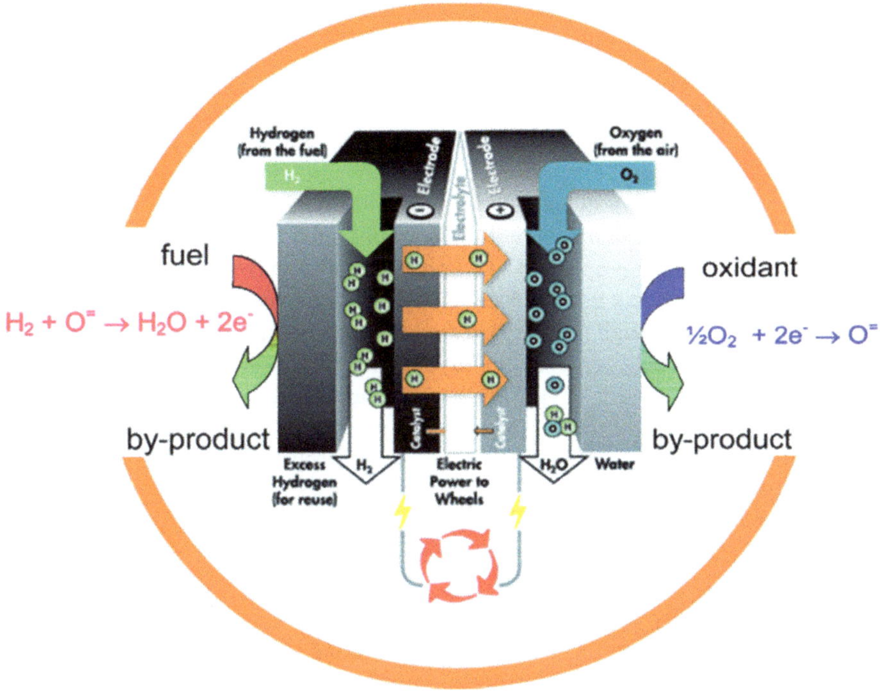

Fig. 3 Schematic illustration of components of typical fuel cell, including an anode, an electrolyte, and a cathode. The general principle remains the same, but for electrolytes with various mobile ions, the reactions can be modified (adopted from Miron et al.)

Working Principle of Fuel Cell

Fuel cells are categorized into different types based on the electrolytes used, namely polymer electrolyte membrane fuel cells (PEMFC), molten carbonate fuel cells (MCFC), phosphoric acid fuel cells (PAFC), alkaline fuel cells (AFC), and solid-oxide fuel cells (SOFC). The SOFC and MCFC require high operational temperature (>500 °C) and are appropriate for stationary applications, whereas PAFC, PEMFC, and AFC operate at moderate temperatures (300 °C) and are ideal for portable applications. When selecting electrode materials for fuel cell applications, electrochemical performance, durability, and efficiency are key factors to consider [71].

Typically, in a fuel cell, a layer of electrolyte is sandwiched between two electrodes (see Fig. 3). The electrons released via fuel oxidation at the anode are transferred through an external circuit toward the cathode to reduce the amount of oxygen. Simultaneous movement of charge carriers in electrolytes leads to the completion of the circuit. The anode part must be highly permeable, conductive, thermally and chemically stable, and compatible with other materials to improve fuel cell performance [72].

Carbon Nanomaterial-Based Fuel Cells

Carbon nanomaterials are considered the best option for electrode and support for fuel cells because they extend the three-phase boundary of the catalyst layer and can be modified for specific functions [73]. Compared to non-carbon materials like metal oxides and alumina, electronically conductive polymers, nitrides and carbides demonstrate good stability in both acidic and basic conditions. Recently, researchers have become interested in nanostructured carbons such as 0D carbon and quantum dots, 1D CNTs, and 2D graphene [74]. Previous studies have suggested that carbon nanofibers are extensively used in fuel cell applications due to their shape, size, and functionalization. The high porosity created by fiber interlocking makes it possible for species to travel between the fiber layers with little to no pressure drop, which is helpful for mass transfer [75]. The smaller diameter and rough-surfaced carbon nanofiber have shown the optimum fuel cell performance [76]. According to Xi-Zhao et al. [77], the structure of carbon nanofibers (CNF) significantly impacts how well the cell functions. For PEMFs, they were successful in depositing Pt on platelet CNFs, and the resultant structure demonstrated good performance. Furthermore, Li et al. [78] reduced the amount of costly platinum catalyst in the fuel cell by using stack-cup CNF support in the fuel cell. Because of stack-cup, CNFs maintained their electrical conductivity, leading to a reduction in platinum loading.

Several techniques are used to modify CNFs and other carbon nanomaterials, particularly important for electrocatalytic reactions. Depending on the processing methods and the amount and type of reagent used, the modified CNF has a variety of uses in fuel cell applications. Adding large amounts of oxygen functional groups can change the catalyst support surface, reducing cell performance. Nanoparticles of Pt and Pt–Ni were electrochemically deposited on CNF networks by Soundararajan et al. at cycle numbers 40, 60, and 80 [79]. FE-SEM was used to characterize the nanostructured Pt–Ni, as can be seen in Fig. 4. A raw CNF network's SEM picture displayed horizontally aligned nanofibers. It was discovered that the alloy nanoparticles deposited over the CNF network with good dispersion. Enhanced electroactivity was shown by Pt–Ni80/CNF material for ethanol electro-oxidation than PtNi40/CNF, Pt40/CNF, and PtNi60/CNF with small charge resistance of 94 Ω cm^{-2} because of reduction in size due to increased atomic ratio of Ni-Pt and increase in deposition cycles.

MWCNTs have emerged as important materials for engineering electrode surfaces in electrochemical devices due to their excellent electrical properties and unique 1D geometry with larger surface area [80]. MWCNTs are arranged in sp^2 carbon-based cylindrical shapes with lengths ranging from 40 to 50 nm and diameters from 3 to 30 nm. In addition, they are made up of hollow tubes with outside diameters of 2 nm wrapped around single graphene sheets wrapped in a manifold [81]. The MWCNTs help to increase the surface area available for the catalyst material, which increases active sites available for electrochemical reactions [82]. Composite MWCNTs have demonstrated great potential applications in fuel cells. A composite of Co_9S_8/PPANI/MWCNTs displayed a discharge capacity of 689.2 mAh g^{-1} [83]. Dogan et al. [84] introduced hexagonal-based boron nitride in conjunction with a

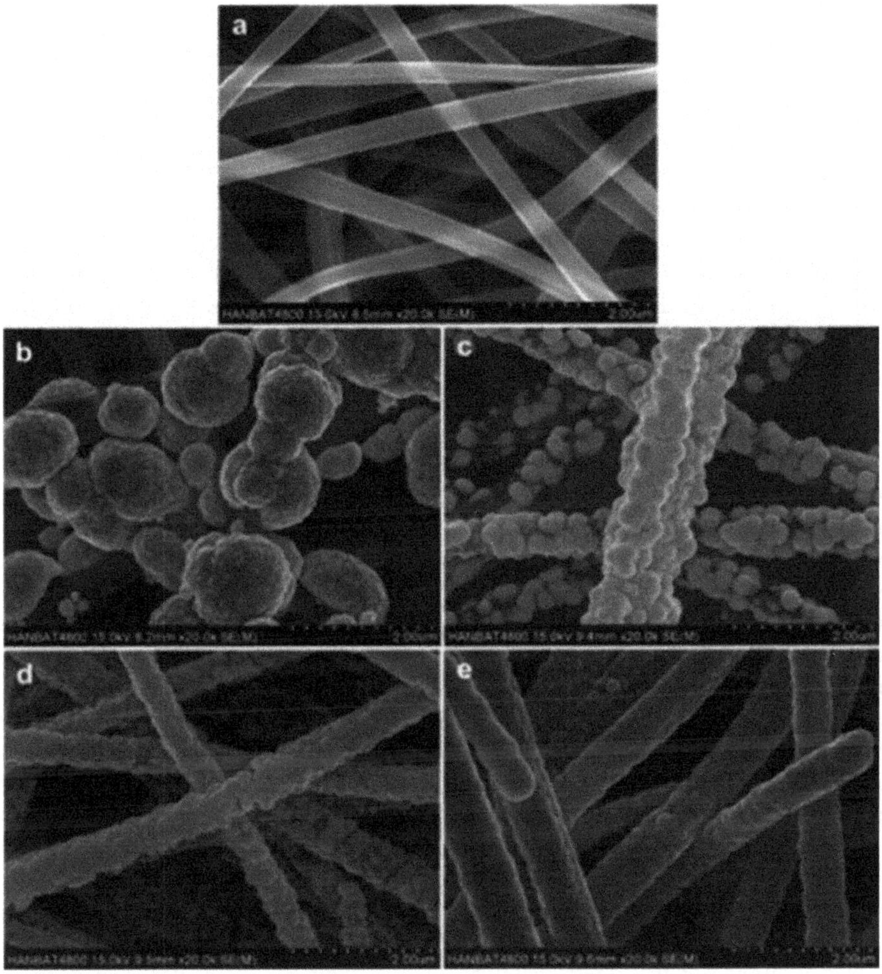

Fig. 4 FE-SEM images of **a** bare CNF, **b** Pt40/CNF, **c** Pt–Ni40/CNF, **d** Pt–Ni60/CNF, and **e** Pt–Ni80/CNF (adopted from [79])

conducting polymer-supported MWCNT (h-BN-Ph-NH-CO-MWCNT) composite through coupling oxidation and acylation. The acyl group significantly enhances the electrocatalytic activities of the electrode materials because of their extraordinary van der Waals interactions with hydrogen.

Graphene-based materials support electrocatalysts used in fuel cells due to easier electron transfer for fuel oxidation and ORR [85]. Metal-free graphene materials are low-cost, high poisoning-tolerant, and exhibit high electrocatalytic activity, making them outstanding candidates for ORR-based applications. The performance of fuel cells has been extensively studied in doping configurations, defects, electronic structural changes, and graphene functional groups [86]. Polymer membranes containing graphene-based materials have the potential to reduce fuel crossover and improve ionic conductivity [87]. These materials are used as fillers and additives due to their high proton conductivity and impermeability to methanol and H_2 [88]. Graphene-based materials can improve bipolar plate stability, fuel/air distribution, and current collection in fuel cells [89]. Combining polymer membranes with graphene also results in restricted fuel permeability, strong ionic conductivity, and high tensile strength [90]. Incorporating graphene as an electrode material can refine bipolar plate conductivity and erosion obstruction. Qiu et al. [91] reported sandwiching 3D hollow nanospheres of rGO between Pt nanoparticles using SiO_2 nanosphere templates. Due to increased mass diffusion, mechanical strength, and exposed active sites, the mass activity for MOR of GO@Pt hollow nanospheres was 1.7 and 1.3 times greater than commercial Pt/C and Pt@r-GO@Pt, respectively.

The durability and activity of ethanol electrooxidation (EOR) are significantly enhanced by modifying graphene with heteroatoms and functional groups. Yang et al. [92] reported aniline functionalized graphene that could bind Pd NPs for EOR using a diazo process. The mass action with aniline addition was several times more noteworthy than without aniline after 7200 s of testing. Researchers are currently working on modifying electrocatalysts to make them better. Doping carbon material is one type of modification that can support transition metal catalysts because of its low cost, stability, high electrical conductivity, capacity to boost ORR activity, and long life [93]. Chemical doping is a crucial component in customizing the carbon material's characteristics. It has been demonstrated that the nitrogen doping of CNTs has significantly expanded their applications in fuel cells [94]. The various nitrogen incorporations in the graphitic structure show different synergist exercises towards ORR due to their diverse electronic designs. Nitrogen in carbon material alters the electronic characteristics by making the substrate denser. Better electrochemical performances are provided by the larger density of catalytic sites on the surface of carbon in fuel cells [95]. The lone pair electron contribution by the nitrogen in this carbon structure promotes the dispersion process via interaction with carbon material pi electrons. Moreover, N-doped carbon decreases CO_{ads} on platinum by enhancing poison tolerance in the electro-catalyst. Due to their excellent electrical conductivity, nitrogen-doped carbon nanomaterials have drawn more interest as efficient electrocatalysts for ORR [96].

Carbon materials are highly preferred for fuel cells because of their high surface area, volume ratio, resistance to corrosion, and tensile strength. Moreover, carbon's high electrical conductivity facilitates the progression of electrons in fuel cells. However, in-depth understanding of the catalytic mechanism is essential for developing carbon electrocatalysts with high performance. A combination of both exploratory and hypothetical approaches would be significantly important in exploring dynamic places and considering the fundamental science behind electrocatalysis. The precise control of the content, atomic location, and catalytic active center distribution in carbon-based electrocatalysts must be achieved by developing novel synthetic methods and strategies for fuel cell applications (Table 2).

Table 2 Fuel cell application of carbon-based electrode materials and their performances

Fuel cell type	Electrode	Fuel/oxidant	Membrane	T (°C)	Maximum power density ($mWcm^{-2}$)	Key references
Proton exchange membrane fuel cell	CNT	H_2+O_2	Nafion 212	80	661	Chiang et al. [97]
	g-C_3N_4/ carbon nanofiber	H_2+O_2	AEM	50	171	Kim et al. [98]
	N-doped hierarchical carbon	H_2+O_2	FAA-3 membrane	70	228	To et al. [99]
	S-doped reduced graphene oxide	H_2+O_2	Poly(benzimidazolium)	95	2.83	Klingele et al. [100]
Direct methanol fuel cell	MC	1M methanol+O_2	Nafion 212	60	58	Bruno et al. [101]
	CNT-MEA	6M methanol+Air	Nafion	–	23.2	Deng et al. [102]
	N-doped FWCNT	1M methanol+Air	–	–	0.73	Kanninen et al. [103]
	Ni/NiO/ MWCNTs	0.7M methanol+O_2	Nafion 117	RT	41.8	Askari et al. [104]
Alkaline fuel cell	N-doped CNTs	H_2+O_2	FAA membrane	80	25.5	Higgins and Chen [105]
	N,S-doped graphene	H_2+O_2	Fumapem membrane	RT	19.8	Li et al. [106]
	CNTs/Carbon core–sheath nanostructures	H_2+O_2	Pore-filling AEM	50	280	Sa et al. [107]

2.2 Carbon Nanomaterials for Energy Storage

Carbon nanomaterials' exceptional physical and chemical properties have made them a promising candidate for energy storage applications. These materials are utilized in batteries and supercapacitors to facilitate rapid electrical energy storage and release. The large electrode–electrolyte interface, which results from their high surface area, contributes to higher capacitance and faster charging and discharging times. Additionally, their remarkable electrical conductivity improves charge transfer efficiency and enhances batteries' overall performance.

2.2.1 Supercapacitors

Supercapacitors are electrochemical energy-storing devices used to handle energy problems. They combine routine capacitors and batteries' power delivery and charge storage capabilities [108]. Their higher energy densities and specific capacitance made them a good substitute for batteries. Generally, the stored electrical energy (E) in a supercapacitor is given as [109]:

$$\mathrm{E} = \frac{QV}{2}$$

where Q is the charge stored, and V is the potential of the cell. Also, the two main efficiency parameters used for supercapacitors are the energy and power density and are given by:

$$E = \frac{CV^2}{2m}$$

$$P = \frac{E}{\Delta T}$$

Due to the double layer phenomenon, supercapacitors possess high power, but series resistance produces low power density. The leakage current and self-discharging rate determine the performance of a supercapacitor. Generally, the geometric configuration, conductivity, specific surface area, porous nature, and mass density variations affect the performance of supercapacitors [110].

Mechanism of Charge Storage in Supercapacitors

The two main types of supercapacitors are pseudo capacitors and electrochemical double-layer supercapacitors (EDLC), which store energy through different mechanisms. Pseudo capacitors use reversible redox reactions at the electrode–electrolyte interface, while EDLCs store energy through electrochemical charge accumulation and separation at the interface [111]. For high performance in pseudo capacitors, it is necessary to have fast and reversible faradic charge transfer processes at the electrode–electrolyte interface. Ions are adsorbed at the negative and positive plates, and energy is stored through a redox reaction, which does not have a fixed voltage platform but has capacitance characteristics known as pseudo capacitance [112]. Therefore, the choice of electrode material is crucial for energy storage in these supercapacitors. EDLCs, on the other hand, use a separator, typically an ion-permeable membrane, to isolate the cathode and anode electrically. An external electric field aligns negative and positive ions at the solid–liquid interface [113]. Excess charges and ions accumulate in the electrolyte during the charging process, forming an electric double layer at the interface. When discharging, the negative and positive plates transfer charge via the external circuit, which decreases the excess charge. The opposite charges at the

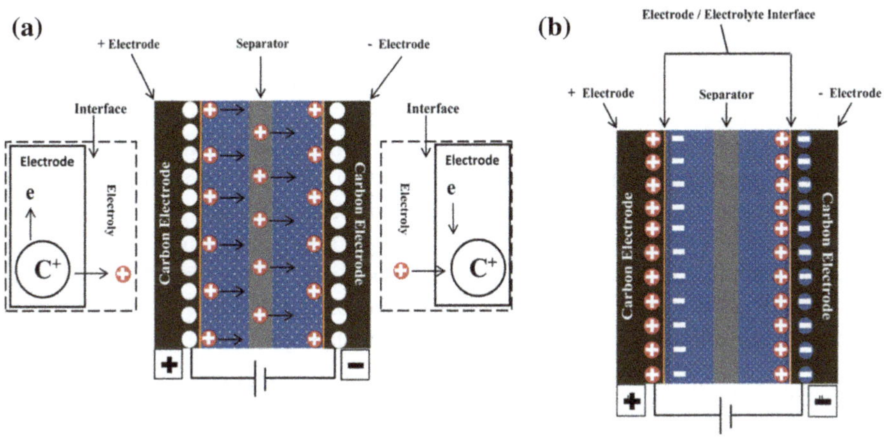

Fig. 5 Illustration of charge storage mechanisms of both **a** pseudo capacitors and **b** EDLCs (adopted from [114])

interface of solid/liquid return to the electrolyte, releasing the stored energy [114] (Fig. 5).

Carbon-Based Nanomaterials for Supercapacitors Applications

Carbon-based nanomaterials are widely studied and used as electrode materials for supercapacitors due to their high surface area and unique pore size [115]. These materials, including graphene, CNTs, fullerenes, and doped carbon materials, exhibit outstanding conductivity, chemical stability, and ultrahigh surface area [116]. Additionally, the structure of these materials can be easily regulated to create various macroscopic materials, i.e., one-dimensional fibers, two-dimensional films, and complex three-dimensional forms [117]. These carbon-based macroscopic materials are flexible and have high mechanical strength, making them suitable for use as electrodes for supercapacitors [118].

CNT films with a macroscopic nature possess superior physical flexibility and high mechanical strength, making them an attractive research subject. Among various categories, one-dimensional CNTs can easily develop van der Waals interactions with the substrate, providing good film contact, attachment, and conducting paths. This 1D morphology has been shown to have higher mechanical, electronic, and thermal characteristics [119]. Recently, a Zn_2GeO_4/CNT composite was fabricated using the solvothermal method, in which metal oxide rods were grown in the framework of CNTs, resulting in a higher specific capacitance of 164.23 F g^{-1} that decreased to 120 F g^{-1} after 200,000 cycles [120]. Although these CNT films supported on a substrate provide superior power density [121] and cyclic stability [122], they fail to deliver adequate specific capacitance and energy density. For this reason, flexible devices using composites of CNTs and pseudocapacitive materials have been constructed

[94], such as MnO_2/CNTs/textile composites [123], PANI/SWCNTs/cloth [124], polythiophene/CNTs [125], and Fe_2O_3/CNTs [126]. The helical coiled fiber structure of CNTs demonstrated remarkable elasticity and provided good alignment of CNTs with the substrate, delivering good capacitance [127]. The nanoporous structure of CNT foams provides high interconnectivity and compressibility and delivers better capacitance even under large deformation compression. They offer enhanced retention of ionic and electronic conductivity [128]. Electrostatic self-assembled Ti_3C_2/CNTs/$MnCo_2S_4$ electrodes were prepared hydrothermally. The optimized electrode showed capacitance of 823 F g^{-1} at 1 A g^{-1} current density, excellent cycling stability of 94% after 5000 charge–discharge cycles, and 63.5% of rate retention upon an increase in current density from 1 to 5 A g^{-1}. The nanoporous structure and synergistic effect provided a high power density of 350 W kg^{-1} and energy density of 49.5 Wh kg^{-1} [129]. A nitrogen-doped CNT/Co_3O_4 composite was fabricated using a hydrothermal method, in which the hollow polyhedron porous structure avoided the agglomeration of CNTs and provided a larger surface area. This composite demonstrated promising electrochemical performance with high cycling and rate performances and a specific capacitance of 894.6 F g^{-1} at 2 A g^{-1}. This supercapacitor electrode had 36.8 Wh kg^{-1} of energy density and a power density of 16326.6 W kg^{-1} [130].

Mesoporous carbon and graphene have significantly contributed to the development of supercapacitors, accelerating progress in industries such as transportation, healthcare, and wearable technology. Since Novoselov discovered graphene in 2004, this mono-crystalline film of graphene has been used in various electronics-related fields [131]. Graphene's thin layer of sp^2-bonded honeycomb-like network of carbon atoms is a good conductor of electricity and heat, with the best mix of porosity, large electrochemical surface area, and tunable electron densities [132]. The unique and desirable qualities of graphene, such as its chemical stability, high effective surface area, mechanical strength, and plane's pi-electronic charge density, make it an effective composite material for supercapacitors [133]. Even though graphene and its pores use double layer mechanism to store charge, composite graphitic materials have a variety of charge storage mechanisms, including double-layer formation and faradic charge transfer at the interface of electrode and electrolyte with materials such as conducting polymers and metal oxides [134]. Graphene is very popular in the field of supercapacitors due to its enormous ability to create composites and work with various charge storage mechanisms [135].

The electrochemical performance of carbon-based materials can be enhanced by optimizing the porous structure and improving the conductivity. For instance, a carbonized reduced graphene oxide/PVA composite structure has been used as a supercapacitor with a specific capacitance of 288 F g^{-1}, 36 Wh kg^{-1} of energy density, and power density of 3600 W kg^{-1} with 91% capacitance retention [136]. Recently, a simple and rapid microwave technique was employed to synthesize a rGO@Co_3O_4/CoO composite in just 90 s to prepare a superior supercapacitor electrode with improved performance. This electrode showed 276.1 F g^{-1} of specific capacitance and cycling stability of 10,000 with 82.7% retention [137]. In addition, a porous carbon derived from biomass and graphene composite was synthesized using a chemical self-assembly approach [138], and the prepared supercapacitor electrode material showed good rate performance (81.6%), high specific capacitance (176 F g^{-1} at 20 A g^{-1}), energy density (6.11 Wh kg^{-1}), and good cycling stability (up to 99.9% after 10,000 cycles). A hybrid 2D polypyrrole (PPy)/rGO and nickel-tungsten metal oxides electrode was designed with specific capacitance of 597 F g^{-1} and retention of 98.2% after 5000 cycles, exhibiting ideal energy density of 14.4 Wh kg^{-1} and power density of 275 W kg^{-1}, with 96.4% of its total specific capacitance preserved after 5000 charge–discharge cycles [139].

Translating 2D sheets of graphene into a 3D well-defined microscopic architecture provides high mechanical strength, larger surface area, and fast electron and mass transport by combining graphene's intrinsic properties and porous structure [140]. A 3D graphene/cellulose composite electrode was used as a supercapacitor, exhibiting high specific capacitance and mechanical flexibility, remarkable rate performances, and impressive cyclic durability, attributed to its 3D charge storing geometry, freely designed structure, and high electric conduction [141]. A flower-shaped spherical 3D $NiCo_2S_4$@Ni-Mo LDH nanocomposite in situ grown on graphene oxide was fabricated by a simple hydrothermal method [142], demonstrating excellent rate capacity and pseudo-capacitance with good reversibility, achieving highest energy density of 59.38 Wh kg^{-1} at 808.19 W power density and maintained an energy density of 25.24 Wh kg^{-1} at a high power density of 8055.32 W kg^{-1}. The capacitance of the RGO@$NiCo_2S_4$@NiMo-LDH electrode retained 80% of its initial capacitance after 10,000 cycles.

The carbon layers in fullerenes strengthen the bonds amongst molecules and enhance electronic transition. However, the low long-range conductivity of fullerene-based supercapacitors greatly hampers their performance. Therefore, it is expected that active carbons derived from fullerenes will have significant capacitances [143]. Jiang et al. [144] synthesized mesoporous carbon composites doped with different concentrations of nitrogen (N) and cobalt by controlling fullerene self-assembly with cobalt tetra methoxy phenyl porphyrin (CoTMPP) pyrolysis. CoTMPP was contained within C_{60} crystals and, after carbonization, transformed into actively bound Co–N in the carbon structures. Co–N significantly enhanced the electrochemical performance of the porous carbon composite. After 5000 cycles, the carbon composite showed no activity loss and an improved specific capacitance (416.31 F g^{-1} at 1 A g^{-1}), nearly 10 times more than pure C_{60}.

Carbon nanomaterials are significant among the materials used for SCs' electrodes. Particularly after the idea of the hybrid capacitor was put forth, asymmetric devices composed of pure carbon material with EDLCs and carbon-based composites with pseudo-capacitance properties have been used to offset the insufficient energy density of SCs partially. However, the cycle stability of carbon-based composite materials with pseudo capacitance still needs to be improved. In CBNMs, the primary strategies for enhancing supercapacitor performance and creating high-performance devices include precise control of pore structure, dimensional structure design and optimization of overall structural engineering (Table 3).

Table 3 Different carbon nanomaterial application as supercapacitor

Electrode material	Capacitance/ capacity	Energy density	Power density	Cycling stability	Key references
CNT aerogel	160.8 F g^{-1} at 0.5 mA g^{-1}	29.6 Wh kg^{-1}	27.3 kW kg^{-1}	99.7% after 3000 cycles	Li et al. [145]
Graphene fiber	36.25 F cm^{-2} at 0.1 mA cm^{-2}	0.80 μWh cm^{-2}	–	96.14% after 20,000 cycles	Ye et al. [146]
rGO hydrogel Film	206 F g^{-1} at 0.5 A g^{-1}	15.78 Wh L^{-1}	–	89% after 10,000 cycles	Feng et al. [147]
MnO_2/CQDs/GA	721 F g^{-1} at 1 A g^{-1}	38.2 Wh/ kg	1 kW kg^{-1}	92.3% at 10,000 cycles	Lv et al. [148]
CNT/graphene film	199 F g^{-1} at 0.5 A g^{-1}	110.6 Wh kg^{-1}	400 kW kg^{-1}	98.2% after 10,000 cycles	Pham et al. [149]
$NiSe_2$@CNT	980.5 F g^{-1} at 1 A g^{-1}	25.61 Wh kg^{-1}	810 W kg^{-1}	82% after 9000 cycles	Zheng et al. [150]
rGO foams	193 F g^{-1} at 1 A g^{-1}	2.86 Wh kg^{-1}	0.5 kW kg^{-1}	100% after 5000 cycles	Tao et al. [151]
Hierarchical CNT films	254.6 mF cm^{-2} at 1 mA cm^{-2}	21.9 Wh cm^{-2}	400 W cm^{-2}	91.5% after 11,000 cycles	Zhou et al. [152]
DMC/PANI	985 F g^{-1} at 0.5 A g^{-1}	32 Wh kg^{-1}	125 W kg^{-1}	82% after 5000 cycles	Pourreza et al. [153]
Graphene nanoribbons	130 F g^{-1} at 1 mV s^{-1}	55.2Wh kg^{-1}	56.8 kW kg^{-1}	–	Zheng et al. [154]
CNTs	272 mF cm^{-2} at 0.5 mA cm^{-2}	–	–	100% after 5000 cycles	Yang et al. [155]
CNTs@NiCo-LDH	176.33 mAh g^{-1} at 1 A g^{-1}	37.38 Wh kg^{-1}	800 W kg^{-1}	90.22% after 5200 cycles	Zhu et al. [156]
GO/NCs	405.6 F g^{-1} at 1 A g^{-1}	38.6 Wh kg^{-1}	180 W kg^{-1}	95.8% at 5000 cycles	Wan et al. [157]
MnO_2/AC/GR	813.0 F g^{-1} at 1 A g^{-1}	33.9 Wh kg^{-1}	319.3 W kg^{-1}	~98.4% after 1000 cycles	Tseng et al. [158]

2.2.2 Batteries

In the current market, rechargeable batteries are the dominant energy storage devices that use reversible faradic reactions to store electrical energy [159]. There are two main categories of these batteries: closed-system and open-system batteries. Closed-system batteries can transfer energy without exchanging mass with their surroundings, while open-system batteries can freely exchange energy and materials with their surroundings. Their storage mechanism involves metal desorption/dissolution inside batteries on the anode, and oxygen evolution reaction (OER) and oxygen reduction reaction on cathodes with the mass exchange of air or oxygen with the surroundings [160]. Lithium-air batteries (LABs) and Zinc air batteries (ZABs) are representative examples of metal-air batteries that offer high energy densities due to the minimal volume and mass of air electrodes. On the other hand, closed-system batteries only transfer energy without exchanging mass, such as metal-ion batteries [161] and metal-sulfur batteries [162]. Metal-ion batteries possess long cyclic life, are easy to manufacture, have higher energy efficiencies, and are more widely used in practice [163].

Charge Storage Mechanism in Batteries

Carbon nanomaterials with stable structures, high conductivity, and efficient ion diffusion pathways are desirable as electrodes for electrochemical energy storage devices [164]. Different types of carbon-based nanomaterials can be used as cathodes and anodes in batteries depending on the applied voltage and microscopic structure, and their storage mechanisms can be categorized as adsorption/desorption reactions and intercalation/deintercalation mechanisms [165].

Carbon nanomaterials such as activated carbon and graphene are frequently used electrode materials for ionic adsorption reactions due to their controllable porosity, extraordinary surface area, and relatively inert electrochemical characteristics. In physical adsorption, the electric double layer is considered the main storage mechanism at the interface of the electrode/electrolyte [166]. The Helmholtz, Gouy-Chapman, and Stern models have been proposed in carbon nanomaterials to explain the EDL process [167]. Carbon nanomaterials' porosity and surface area are critical for ion storage and achieving high EDL capacitance. Chemical adsorption also occurs in these carbon-based materials, with the adsorption reaction mainly depending on the carbon material and functional groups in heteroatom S- and N-doped carbon materials. However, ion diffusion in microporous structure of these carbon-based materials is limited by pore size and electrostatic interaction of the electrode surface and ions [167]. To address this, effective strategies include designing a suitable structure, engineering the pore size and surface area, and heteroatom doping to enhance ion adsorption have been reported [168].

For carbon nanomaterials with a layered structure, such as graphene, graphite, and CNTs, the charge storage is mainly governed by ion intercalation–deintercalation reactions [169]. Different layers in these carbon materials intercalate ions for energy storage. The larger interspacing and layered structures of soft and hard carbon can store large cations, while the redox amphiproticity of graphite makes it a good electrode for intercalated compounds. Graphite is a commonly used anode material LIBs. Li ions become lodged in graphite as a result of electrostatic attractive forces, and intercalation needs to surpass the van der Waals forces in graphene layers [170]. Li ions do not immediately intercalate into the layers; the process occurs gradually at the margins. Anions cannot always be stored in carbon compounds that can store cations via intercalation. Typically, anions with bigger sizes than cations require a more capacious interlayer space. The host materials, electrode setup, and electrolytes all significantly impact the properties of anion intercalation [167].

Carbon-Based Materials for Batteries

Carbon nanomaterials of various dimensions have great potential for enhancing their functionality and expanding their application in metal ion batteries. Recently, 0D CBNMs such as nanospheres, nanocages, quantum dots (QDs), and nanoparticles have been extensively studied and used in LIBs together with luminous materials and solar cells [171]. The characteristics of 0D CBNMs, such as their large surface area, good electrical conductivity, and mobility, play essential roles in determining the electrochemical performance of electrodes [172].

Nanoparticles coated with carbon layers of 0D are frequently used in LIBs. Zhang et al. [173] designed SnO_x quantum dots@carbon hybrid, where SnO_x was homogeneously dispersed in carbon nanocages. After surface coating of ZIF-8 with oxide and pyrolysis, an anode for LIBs was obtained that displayed a high capacity of 1824 mAh g^{-1} @ 0.2 A g^{-1}.

Recently, a hollow nanosphere of SnO_2 coated with a dual carbon shell was designed for LIBs anode [174]. This dual carbon shell enhances conductivity and retards the aggregation of SnO_2 nanospheres. This electrode material displayed good specific capacity (1483.9 mA h g^{-1} @0.2 A g^{-1}) and a reversible capacity (713.8 mAh g^{-1} after 300 cycles).

In addition to coating metal oxides and sulfides with carbon materials, some non-metallic nanomaterials with larger theoretical capacities are also being developed for better energy storage capacities. Silicon nanoparticles, which are being developed as larger-capacity materials, are likely to stand out in the post-graphite anode era. However, their poor conductivity and significant volume change severely restrict their use in energy storage materials [175]. Yolk-shell structured Si/void/C composite was designed using a self-sacrifice template [176]. The SEM and TEM imaging showed voids between the carbon layer coated on silicon nanospheres (see Fig. 6). This composite structure delivered a capacity of 854.1 mA h g^{-1} (at 0.2 A g^{-1} current density after 200 cycles) with enhanced stability. Pure 0D CBNMs have also received attention as electrode material for batteries. Using a straightforward technique of solvothermal synthesis, Zhao et al. [177] synthesized graphene hollow nanospheres (GDY-HNS), in which CuO nanospheres served as both the catalyst source and the template. At a relatively larger mass loading (2.0 mg cm^{-2}) GDY-HNSs exhibit improved electrochemical characteristics, i.e., greater reversible specific and rate capacity and cycling stability than other reported carbon materials.

One-dimensional nanomaterials, such as nanofibers, nanorods, and nanotubes, have been extensively researched. Due to their key structural properties, such as large aspect ratio, directional growth, and micro and nanoscale joint effects, they generate more efficient electrolyte contacts and can reduce diffusion lengths of metal ions. These structural benefits can efficiently buffer volume variations and

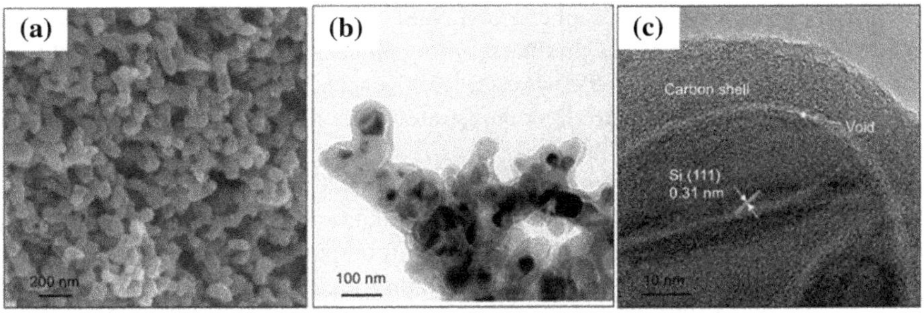

Fig. 6 **a** SEM of Si@void@C composites material, **b** corresponding TEM, and **c** TEM showing voids and shell in this composite (adopted from [176])

greatly reduce cyclic pulverization when used as electrode materials for LIBs. Moreover, the presence of carbon materials in one dimension enhances conductivity by producing confinement effects in materials. Developing and effectively using one-dimensional carbon-based nanomaterials opens up new opportunities and options for high-performance LIBs. Coating, embedding, or encapsulating the carbon layers with the main body materials are common practices for enhancing charge storage capacity in one-dimensional carbon-based nanomaterials. Similar to zero-dimensional carbon-based materials, the construction of one-dimensional carbon-based nanomaterials has frequently involved polymerization-aided synthesis using hybrid materials of organic and inorganic nature as precursors and electrospinning [178]. The fabrication of MoC/C nanowires using pyrolyzing MoO_x/p-methylaniline in an argon atmosphere at 700 °C was recently reported [179]. Nanocrystallites of MoC of about 5–6 nm were distributed in a nanowire network of carbon and showed extraordinary current densities and cyclic stability (455.4 mAh g^{-1} after 2000 cycles, with an average rate of capacity loss of 0.004% per cycle).

A low-cost technique for Li storage batteries synthesized a one-dimensional porous structure of N-doped carbon and MnO composite. This composite showed high conductivity, pseudocapacitive charge storage, and effectively accommodated volume changes during cycling because of the protection effects of the carbon layer and adequately presented internal voids in N–C@MnO. The one-dimensional unique porous features contributed towards superb rate capability of 359.5 mAh g^{-1} @ 30 A g^{-1}, high specific capacity of 971 mAh g^{-1} at 0.1 A g^{-1}, and remarkable stability of 441.5 mA h g^{-1} after 3500 cycles at 10 A g^{-1} [180].

Since the discovery of graphene, 2D carbon nanomaterials, such as carbon nanosheets and Mexene, have been successively studied. Two-dimensional carbon-based nanomaterials can greatly decrease the local current density for large specific surface areas, preventing the formation of Li dendrites. The interaction among two-dimensional structures and Li^+ or e^- could effectively minimize the transportation length and greatly enhance LIBs' electrochemical performance. Hydrothermal method designed a unique heterostructure of two-dimensional graphene and two-dimensional $SnSe_2$ [181]. The 2D/2D $SnSe_2$-on-graphene heterostructure chemically linked interaction between graphene and $SnSe_2$ nanosheets successfully avoided the accumulation of tin atoms produced during electrochemical processes and with a capacity retention rate of 59.3% after 1500 cycles at a current rate of 0.5 C, 2D/2D $SnSe_2$/graphene electrode demonstrated 490.9 mA h g^{-1} specific capacity @ 0.1 A g^{-1}.

In recent years, various non-metallic elements and their oxide composites with carbon-based materials have demonstrated significant potential for high-performance battery electrodes. For example, a graphene-Si composite with increased nitrogen content was fabricated by self-assembling melamine, citric acid, and amino carboxyl groups of Si–NH_2 [182]. Due to substantial nitrogen doping, the 2D carbon sheets could buffer volume changes effectively in silicon material, promoted electronic conductivity, and facilitated cyclic charge transfer. This anode material delivered good rate capability (572 mA h g^{-1} even at 5000 mA g^{-1}) and cyclic stability (1000 mA h g^{-1} after 300 cycles at 2000 mA g^{-1}). Graphene, the 2D CBNMs representative, is a widely used electrode material for energy storage [183]. Graphene co-doped with P/O was fabricated by thermal annealing of graphite oxide and triphenylphosphine [184]. The oxygen and P entry into the carbon skeleton caused numerous defects and vacancies in graphene sheets, which decreased the ion diffusion pathway. 2D ultra-thin sheets underwent successive transfer of electrons, and a larger surface area provided abundant sites at the electrode–electrolyte interface. This electrode material delivered a high specific capacity of 385, 235, and 160 mA h g^{-1} at current densities of 0.5, 1, and 2 A g^{-1}, respectively, and reversible capacity of 474 mA h g^{-1} at 0.05 A g^{-1} with outstanding cyclic life.

Three-dimensional CBNMs share similar characteristics with other dimensional nanomaterials, such as larger surface area and porous structure, in that a major portion of them is constructed or merged with these other dimensional materials in various ways [185]. In addition, 3D CBNMs contain three-dimensional interconnected structures, which increase the rate at which electrons and ions transfer between nanoparticles [186]. Moreover, using 3D nanostructures could significantly overcome the lack of 1D and 2D nanomaterials resulting from obvious aggregation. In particular, some 3D CBNMs are indefinitely used as free-standing electrodes, significantly increasing the process's efficiency [187].

$LiFePO_4$/C composite cathode fabricated by spherical $LiFePO_4$ nanocrystallites and 3D coralloid nitrogen-containing carbon having interpenetrating conductive framework was reported for LIBs [188]. The interpenetration of carbon branches in the carbon framework forms a 3D porous coral-like morphological structure, and crystallites of $LiFePO_4$ grew into the mesopores present at carbon support, forming a composite structure. The $LiFePO_4$/C electrode displayed high reversible capacity (144.6 mA h g^{-1} at 0.1 C and 60.4 mA h g^{-1} at 20 C) with 96.7% retention after 1000 cycles with high power (5114 W kg^{-1}). Ke et al. [189] reported carbon nanosheets decorated 3D hierarchical nanohybrids with MoS_2@SnS, which provided a reversible capacity of 713 mA h g^{-1} after 1000 cycles @ 2 A g^{-1}. Due to its abundance, clean, and non-polluting characteristics, biomass-derived carbon has recently caught researchers attention as a suitable carrier for 3D network structures designed for energy applications. Using biomass precursors, several researchers have created carbon-based materials with 3D nanostructures. Han et al. [190] investigated the use of chitosan as a carbon precursor and directing agent of 3D structure to synthesize a 3D Fe_3O_4@C core@shell composite. Calcination resulted in strong coordination of the amino and hydroxyl groups in chitosan and Fe^{3+}, resulting in an N-doped carbon composition. The composite revealed a reversible capacity of 587.2 mA h g^{-1} at 2000 mA g^{-1} and 1116.1 mA h g^{-1} at 100 mA g^{-1}, with stable cycling performance.

Carbon-based nanomaterials (CBNMs), composed of composites or pristine carbon materials, remain attractive for high-performance metal ion batteries owing to their high rate capability, excellent cycle stability, and other desirable properties. Each dimensional structure of CBNMs offers unique benefits, such as increased conductivity and specific surface area that can enhance electrode material performance in LIBs, leading to higher reaction rates. However, the increased surface area can also result in decreased tap density, increased side reactions, and safety concerns. Careful surface and structure design of CBNMs can mitigate their high reactivity and side reactions, reducing defects. Notably, one promising approach to achieve further performance breakthroughs is to design and synthesize CBNMs in combination with various dimensional structures (Table 4).

3 Conclusion

The growing demand for clean and renewable energy to foster sustainable societies has driven the development of advanced energy storage and conversion devices. Among the materials that have shown great potential for energy applications are carbon nanomaterials, including carbon nanotubes, graphene, and fullerenes. These materials possess unique properties of high surface area, tunable porosity, excellent electrical conductivity, and thermal stability, which make them suitable for various energy-related applications, such as energy storage and conversion.

Carbon nanotubes and graphene, in particular, have demonstrated exceptional energy storage performance in supercapacitors and LIBs due to their high surface area and electrical conductivity, which provide high power and energy densities. Furthermore, due to their high surface area-to-volume ratio, corrosion resistance, and tensile strength, carbon materials are preferred for fuel cells, where materials with adequately high electrical conductivity facilitate electron transfer. Graphene, in particular, has shown great promise as a catalyst in fuel cells, where its high surface area and conductivity enhance efficiency.

Additionally, carbon nanomaterials have shown promising results in solar cell applications thanks to their excellent electrical conductivity and light absorption capabilities. Graphene and fullerenes have been utilized in solar cells to enhance light absorption and improve electron transport. Overall, carbon nanomaterials have demonstrated potential in advancing energy conversion and storage technology, and ongoing research is focused on improving their performance and scalability for practical applications.

Table 4 Various carbon nanomaterials used in batteries and their performances

Battery type	Electrode material	Dimensional structure	Functions	Capacity mAh g^{-1}	Current	Cycles	Key reference
Li ion	HMCNTs	1D	Sites for Li^+ storage	1336	372 mA g^{-1}	300	Wang et al. [191]
Li ion	Porous carbon nanosheets	2D sheets	Provision of high ions accessibility	513	500 mA g^{-1}	500	Mondal et al. [192]
Li ion	MoS_2/rGO/HCS	3D Network	Electrolytes and ions reservoir and volume c alleviation	1145	100 mA g^{-1}	100	Hu et al. [193]
Na ion	HPCNS/P	3D hollow	Provide higher P loading, stability	1700	100 mA g^{-1}	300	Yao et al. [194]
Na ion	N-CNFs	1D	Na^+ storage sites	203.1	1000 mA g^{-1}	1000	Yang et al. [195]
Li ion	N-C_{60}	3D Hierarchical	Li^+ storage sites	600	2000 mA g^{-1}	800	Tan et al. [196]
K ion	S/HPCNF	1D	Electron/ion transference pathway	700	0.5 C	300	Zhao et al. [197]
K ion	NCFs	1D fibers	K^+ storage	215.2	0.2 C	100	Hao et al. [198]
Li–S	PCNFs-S	1D fibers	Polysulfide diffusion inhibition and S storage	795	0.5 C	350	Park et al. [199]
Li–S	S/GN-CNT	3D framework	Improve polysulfide adsorption, electrolyte penetration, and electron conduction	836.5	0.1 C	200	Zhang et al. [200]
Li–S	PRC/Ni/S	Hierarchical porous	ion Transportation and facilitates S loading	821	0.2 C	500	Zhong et al. [201]
Zn–Air	Meso/Micro-FeCo-N_x CN	2D	Enrichment of oxygen sites	–	–	–	Li et al. [202]
Zn–Air	P,S-CNS	3 D sponge	Fasten catalytic reaction	–	25 mA g^{-1}	500	Shinde et al. [203]

References

1. Titirici M-M, White RJ, Brun N, Budarin VL, Su DS, Del Monte F, Clark JH, MacLachlan MJ (2015) Sustainable carbon materials. Chem Soc Rev 44(1):250–290
2. Notarianni M, Liu J, Vernon K, Motta N (2016) Synthesis and applications of carbon nanomaterials for energy generation and storage. Beilstein J Nanotechnol 7(1):149–196
3. Su DS, Centi G (2013) A perspective on carbon materials for future energy application. J Energy Chem 22(2):151–173
4. Hu C, Qu J, Xiao Y, Zhao S, Chen H, Dai L (2019) Carbon nanomaterials for energy and biorelated catalysis: recent advances and looking forward. ACS Cent Sci 5(3):389–408
5. Duan Y, Weng M, Zhang W, Qian Y, Luo Z, Chen L (2021) Multi-functional carbon nanotube paper for solar water evaporation combined with electricity generation and storage. Energy Convers Manage 241:114306
6. Rondeau-Gagné S, Morin J-F (2014) Preparation of carbon nanomaterials from molecular precursors. Chem Soc Rev 43(1):85–98
7. Ni J, Li Y (2016) Carbon nanomaterials in different dimensions for electrochemical energy storage. Adv Energy Mater 6(17):1600278
8. Li Y, Meng L, Yang Y, Xu G, Hong Z, Chen Q, You J, Li G, Yang Y, Li Y (2016) High-efficiency robust perovskite solar cells on ultrathin flexible substrates. Nat Commun 7(1):10214
9. Luo Q, Ma H, Hou Q, Li Y, Ren J, Dai X, Yao Z, Zhou Y, Xiang L, Du H (2018) All-carbon-electrode-based endurable flexible perovskite solar cells. Adv Funct Mater 28(11):1706777
10. De Nicola F, Salvato M, Cirillo C, Crivellari M, Boscardin M, Scarselli M, Nanni F, Cacciotti I, De Crescenzi M, Castrucci P (2016) Record efficiency of air-stable multi-walled carbon nanotube/silicon solar cells. Carbon 101:226–234
11. Fu X, Xu L, Li J, Sun X, Peng H (2018) Flexible solar cells based on carbon nanomaterials. Carbon 139:1063–1073
12. Mabena LF, Makgopa K, Tanko-Djoubi AS, Modibane KD, Hato MJ (2021) Nanostructured carbon-based materials for fuel cell applications. In: Carbon related materials: commemoration for Nobel Laureate Professor Suzuki special symposium at IUMRS-ICAM2017. Springer, pp 357–390
13. Li Y, Xu G, Cui C, Li Y (2018) Flexible and semitransparent organic solar cells. Adv Energy Mater 8(7):1701791
14. Zhao C, Wang J, Jiao J, Huang L, Tang J (2020) Recent advances of polymer acceptors for high-performance organic solar cells. J Mater Chem C 8(1):28–43
15. Wang K, Li Y, Li Y (2020) Challenges to the stability of active layer materials in organic solar cells. Macromol Rapid Commun 41(4):1900437
16. Abdi-Jalebi M, Ibrahim Dar M, Senanayak SP, Sadhanala A, Andaji-Garmaroudi Z, Pazos-Outón LM, Richter JM, Pearson AJ, Sirringhaus H, Grätzel M (2019) Charge extraction via graded doping of hole transport layers gives highly luminescent and stable metal halide perovskite devices. Sci Adv 5(2):eaav2012
17. Lee C, Lee J, Lee S, Lee W, You H, Woo HY, Kim BJ (2020) Importance of device structure and interlayer design in storage stability of naphthalene diimide-based all-polymer solar cells. J Mater Chem A 8(7):3735–3745
18. Ramos JC, Flores JR, Turlakov G, Moggio I, Arias E, Rodríguez G (2020) Self-assembly of a poly (phenyleneethynylene) on multiwall carbon nanotubes: Correlation of structural and optoelectronic properties towards solar cells application. J Mol Struct 1222:128845
19. Subramanyam B, Mahakul PC, Sa K, Raiguru J, Mahanandia P (2020) Investigation of improvement in stability and power conversion efficiency of organic solar cells fabricated by incorporating carbon nanostructures in device architecture. JPhys Mater 3(4):045004
20. Park J-J, Heo Y-J, Yun J-M, Kim Y, Yoon SC, Lee S-H, Kim D-Y (2020) Orthogonal printable reduced graphene oxide 2D materials as hole transport layers for high-performance inverted polymer solar cells: Sheet size effect on photovoltaic properties. ACS Appl Mater Interfaces 12(38):42811–42820

21. Bella F, Lamberti A, Bianco S, Tresso E, Gerbaldi C, Pirri CF (2016) Floating, flexible polymeric dye-sensitized solar-cell architecture: the way of near-future photovoltaics. Adv Mater Technol 1(2)
22. Ye M, Wen X, Wang M, Iocozzia J, Zhang N, Lin C, Lin Z (2015) Recent advances in dye-sensitized solar cells: from photoanodes, sensitizers and electrolytes to counter electrodes. Mater Today 18(3):155–162
23. Yu F, Shi Y, Yao W, Han S, Ma J (2019) A new breakthrough for graphene/carbon nanotubes as counter electrodes of dye-sensitized solar cells with up to a 10.69% power conversion efficiency. J Power Sources 412:366–373
24. Wahyuono RA, Jia G, Plentz J, Dellith A, Dellith J, Herrmann-Westendorf F, Seyring M, Presselt M, Andrä G, Rettenmayr M (2019) Self-assembled graphene/MWCNT bilayers as platinum-free counter electrode in dye-sensitized solar cells. ChemPhysChem 20(24):3336–3345
25. Harnchana V, Chaiyachad S, Pimanpang S, Saiyasombat C, Srepusharawoot P, Amornkitbamrung V (2019) Hierarchical Fe_3O_4-reduced graphene oxide nanocomposite grown on NaCl crystals for triiodide reduction in dye-sensitized solar cells. Sci Rep 9(1):1494
26. Mehmood U, Asghar H, Babar F, Younas M (2020) Effect of graphene contents in polyaniline/graphene composites counter electrode material on the photovoltaic performance of dye-sensitized solar cells (DSSCSs). Sol Energy 196:132–136
27. Kilic B (2019) Produce of carbon nanotube/ZnO nanowires hybrid photoelectrode for efficient dye-sensitized solar cells. J Mater Sci Mater Electron 30:3482–3487
28. Meng F, Gao L, Yan Y, Cao J, Wang N, Wang T, Ma T (2019) Ultra-low-cost coal-based carbon electrodes with seamless interfacial contact for effective sandwich-structured perovskite solar cells. Carbon 145:290–296
29. Elumalai NK, Mahmud MA, Wang D, Uddin A (2016) Perovskite solar cells: progress and advancements. Energies 9(11):861
30. Wang H, Li H, Cai W, Zhang P, Cao S, Chen Z, Zang Z (2020) Challenges and strategies relating to device function layers and their integration toward high-performance inorganic perovskite solar cells. Nanoscale 12(27):14369–14404
31. Ferguson V, Silva SRP, Zhang W (2019) Carbon materials in perovskite solar cells: prospects and future challenges. Energy Environ Mater 2(2):107–118
32. Wang S, Jiang P, Shen W, Mei A, Xiong S, Jiang X, Rong Y, Tang Y, Hu Y, Han H (2019) A low-temperature carbon electrode with good perovskite compatibility and high flexibility in carbon based perovskite solar cells. Chem Commun 55(19):2765–2768
33. Wang Y, Zhao H, Mei Y, Liu H, Wang S, Li X (2018) Carbon nanotube bridging method for hole transport layer-free paintable carbon-based perovskite solar cells. ACS Appl Mater Interfaces 11(1):916–923
34. Hu X, Hou P, Liu C, Cheng H (2019) Carbon nanotube/silicon heterojunctions for photovoltaic applications. Nano Mater Sci 1(3):156–172
35. Chu Q-Q, Ding B, Peng J, Shen H, Li X, Liu Y, Li C-X, Li C-J, Yang G-J, White TP (2019) Highly stable carbon-based perovskite solar cell with a record efficiency of over 18% via hole transport engineering. J Mater Sci Technol 35(6):987–993
36. Zhou L, Zuo Y, Mallick TK, Sundaram S (2019) Enhanced efficiency of carbon-based mesoscopic perovskite solar cells through a tungsten oxide nanoparticle additive in the carbon electrode. Sci Rep 9(1):1–8
37. Green M, Dunlop E, Hohl-Ebinger J, Yoshita M, Kopidakis N, Hao X (2021) Solar cell efficiency tables (version 57). Prog Photovolt 29(1):3–15
38. Liu Y, Lv P, Zhou W, Hong J (2020) Built-in electric field hindering photogenerated carrier recombination in polar bilayer SnO/BiOX (X= Cl, Br, I) for water splitting. J Phys Chem C 124(18):9696–9702
39. Deshmukh MA, Park S-J, Hedau BS, Ha T-J (2021) Recent progress in solar cells based on carbon nanomaterials. Sol Energy 220:953–990
40. Sariciftci NS, Smilowitz L, Heeger AJ, Wudl F (1992) Photoinduced electron transfer from a conducting polymer to buckminsterfullerene. Science 258(5087):1474–1476

41. Li X, Lv Z, Zhu H (2015) Carbon/silicon heterojunction solar cells: state of the art and prospects. Adv Mater 27(42):6549–6574
42. Ohashi N, Miyadera T, Taima T, Yoshida Y (2019) Evaluation of exciton diffusion length in highly oriented fullerene films of fullerene/p-Si (100) hybrid solar cells. Jpn J Appl Phys 58(12):121004
43. Aissa B, Memon NK, Ali A, Khraisheh MK (2015) Recent progress in the growth and applications of graphene as a smart material: a review. Front Mater Sci 2:58
44. Li X, Zhu H, Wang K, Cao A, Wei J, Li C, Jia Y, Li Z, Li X, Wu D (2010) Graphene-on-silicon Schottky junction solar cells. Adv Mater 22(25):2743–2748
45. Dey A, Ghosh P, Bowen J, Braithwaite NSJ, Krishnamurthy S (2020) Engineering work function of graphene oxide from p to n type using a low power atmospheric pressure plasma jet. PCCP 22(15):7685–7698
46. Feng T, Xie D, Lin Y, Zang Y, Ren T, Song R, Zhao H, Tian H, Li X, Zhu H (2011) Graphene based Schottky junction solar cells on patterned silicon-pillar-array substrate. Appl Phys Lett 99(23):233505
47. Zhong Y, Xiao Y, Chen Q, Zhu H (2018) Heterojunction solar cells based on graphene woven fabrics and silicon. J Materiomics 4(2):135–138
48. Tune DD, Mallik N, Fornasier H, Flavel BS (2020) Breakthrough carbon nanotube–silicon heterojunction solar cells. Adv Energy Mater 10(1):1903261
49. Jia Y, Wei J, Wang K, Cao A, Shu Q, Gui X, Zhu Y, Zhuang D, Zhang G, Ma B (2008) Nanotube–silicon heterojunction solar cells. Adv Mater 20(23):4594–4598
50. Chen J, Tune DD, Ge K, Li H, Flavel BS (2020) Front and back-junction carbon nanotube-silicon solar cells with an industrial architecture. Adv Funct Mater 30(17):2000484
51. Li H, Shi W, Huang W, Yao E-P, Han J, Chen Z, Liu S, Shen Y, Wang M, Yang Y (2017) Carbon quantum dots/TiOx electron transport layer boosts efficiency of planar heterojunction perovskite solar cells to 19%. Nano Lett 17(4):2328–2335
52. Li W, Dong H, Guo X, Li N, Li J, Niu G, Wang L (2014) Graphene oxide as dual functional interface modifier for improving wettability and retarding recombination in hybrid perovskite solar cells. J Mater Chem A 2(47):20105–20111
53. Wei Z, Chen H, Yan K, Zheng X, Yang S (2015) Hysteresis-free multi-walled carbon nanotube-based perovskite solar cells with a high fill factor. J Mater Chem A 3(48):24226–24231
54. Cheng N, Liu P, Qi F, Xiao Y, Yu W, Yu Z, Liu W, Guo S-S, Zhao X-Z (2016) Multi-walled carbon nanotubes act as charge transport channel to boost the efficiency of hole transport material free perovskite solar cells. J Power Sour 332:24–29
55. Yang MK, Lee J-K (2020) CNT/AgNW multilayer electrodes on flexible organic solar cells. Electron Mater Lett 16:573–578
56. Ali A, Kazici M, Bozar S, Keskin B, Kaleli M, Shah SM, Gunes S (2018) Laminated carbon nanotubes for the facile fabrication of cost-effective polymer solar cells. ACS Appl Energy Mater 1(3):1226–1232
57. Zhang Y, He X, Babu D, Li W, Gu X, Shan C, Kyaw AKK, Choy WC (2021) Efficient semi-transparent organic solar cells with high color rendering index enabled by self-assembled and knitted AgNPs/MWCNTs transparent top electrode via solution process. Adv Opt Mater 9(8):2002108
58. Hashemi D, Ma X, Ansari R, Kim J, Kieffer J (2019) Design principles for the energy level tuning in donor/acceptor conjugated polymers. PCCP 21(2):789–799
59. Wang N, Zheng R, Chi T, Jiang T, Ding Z, Li X, Liu S, Zhang L, San H (2022) Betavoltaic-powered electrochemical cells using TiO_2 nanotube arrays incorporated with carbon nanotubes. Compos B Eng 239:109952
60. Li Y, Wang H, Feng Q, Zhou G, Wang Z-S (2013) Reduced graphene oxide–TaON composite as a high-performance counter electrode for Co (bpy) 33+/2+-mediated dye-sensitized solar cells. ACS Appl Mater Interfaces 5(16):8217–8224
61. Dou Y, Li G, Song J, Gao X (2012) Nickel phosphide-embedded graphene as counter electrode for dye-sensitized solar cells. PCCP 14(4):1339–1342

62. Hsu C-H, Lai C-C, Chen L-C, Chan P-S (2014) Enhanced performance of dye-sensitized solar cells with graphene/ZnO nanoparticles bilayer structure. J Nanomater 2014:4–4
63. Yang B, Zuo X, Chen P, Zhou L, Yang X, Zhang H, Li G, Wu M, Ma Y, Jin S (2015) Nanocomposite of tin sulfide nanoparticles with reduced graphene oxide in high-efficiency dye-sensitized solar cells. ACS Appl Mater Interfaces 7(1):137–143
64. Xie R, Sugime H, Noda S (2021) High-performance solution-based silicon heterojunction solar cells using carbon nanotube with polymeric acid doping. Carbon 175:519–524
65. Kadam KD, Rehman MA, Kim H, Rehman S, Khan MA, Patil H, Aziz J, Park S, Abdul Basit M, Khan K (2022) Enhanced and passivated Co-doping effect of organic molecule and bromine on graphene/HfO_2/Silicon Metal–Insulator–Semiconductor (MIS) Schottky junction solar cells. ACS Appl Energy Mater 5(9):10509–10517
66. Yan J, Zhang C, Li H, Yang X, Wan L, Li F, Qiu K, Guo J, Duan W, Lambertz A (2021) Stable organic passivated carbon nanotube-silicon solar cells with an efficiency of 22%. Adv Sci 8(20):2102027
67. Gao Q, Yan J, Wan L, Zhang C, Wen Z, Zhou X, Li H, Li F, Chen J, Guo J (2022) High-efficiency graphene-oxide/silicon solar cells with an organic-passivated interface. Adv Mater Interfaces 9(24):2201221
68. Li H, Zhang Y, Wan Q, Li Y, Yang N (2018) Expanded graphite and carbon nanotube supported palladium nanoparticles for electrocatalytic oxidation of liquid fuels. Carbon 131:111–119
69. Wilberforce T, Alaswad A, Palumbo A, Dassisti M, Olabi A-G (2016) Advances in stationary and portable fuel cell applications. Int J Hydrogen Energy 41(37):16509–16522
70. Bahru R, Shaari N, Mohamed MA (2020) Allotrope carbon materials in thermal interface materials and fuel cell applications: a review. Int J Energy Res 44(4):2471–2498
71. Seselj N, Engelbrekt C, Zhang J (2015) Graphene-supported platinum catalysts for fuel cells. Sci Bull 60(9):864–876
72. Dai L (2017) Carbon-based catalysts for metal-free electrocatalysis. Curr Opin Electrochem 4(1):18–25
73. Lázaro M, Calvillo L, Celorrio V, Pardo J, Perathoner S, Moliner R (2011) Study and application of carbon black Vulcan XC-72R in polymeric electrolyte fuel cells. Carbon black Prod Prop Uses 41
74. Wan K, Li Y, Wang Y, Wei G (2021) Recent advance in the fabrication of 2D and 3D metal carbides-based nanomaterials for energy and environmental applications. Nanomaterials 11(1):246
75. Zhang H, Tan Y, Luo XD, Sun CY, Chen N (2019) Polarization effects of a rayon and polyacrylonitrile based graphite felt for iron-chromium redox flow batteries. ChemElectroChem 6(12):3175–3188
76. Yuan F, Ryu H (2004) The synthesis, characterization, and performance of carbon nanotubes and carbon nanofibres with controlled size and morphology as a catalyst support material for a polymer electrolyte membrane fuel cell. Nanotechnology 15(10):S596
77. Wang X-Z, Fu R, Zheng J-S, Ma J-X (2011b) Platinum nanoparticles supported on carbon nanofibers as anode electrocatalysts for proton exchange membrane fuel cells. Acta Phys Chim Sinica 27(8):1875–1880
78. Li W, Waje M, Chen Z, Larsen P, Yan Y (2010) Platinum nanopaticles supported on stacked-cup carbon nanofibers as electrocatalysts for proton exchange membrane fuel cell. Carbon 48(4):995–1003
79. Soundararajan D, Park J, Kim K, Ko J (2012) Pt–Ni alloy nanoparticles supported on CNF as catalyst for direct ethanol fuel cells. Curr Appl Phys 12(3):854–859
80. Page A, Ding F, Irle S, Morokuma K (2015) Insights into carbon nanotube and graphene formation mechanisms from molecular simulations: a review. Rep Prog Phys 78(3):036501
81. Zhao H, Liu X, Cao Z, Zhan Y, Shi X, Yang Y, Zhou J, Xu J (2016) Adsorption behavior and mechanism of chloramphenicols, sulfonamides, and non-antibiotic pharmaceuticals on multi-walled carbon nanotubes. J Hazard Mater 310:235–245
82. Bhuvanendran N, Ravichandran S, Zhang W, Ma Q, Xu Q, Khotseng L, Su H (2020) Highly efficient methanol oxidation on durable PtxIr/MWCNT catalysts for direct methanol fuel cell applications. Int J Hydrogen Energy 45(11):6447–6460

83. Sun Y, Liu D, Liu W, Liu H, Zhao J, Chen P, Wang Q, Wang X, Zou Y (2021) Fabrication of porous polyaniline/MWCNTs coated Co_9S_8 composite for electrochemical hydrogen storage application. J Phys Chem Solids 157:110235
84. Doğan M, Selek A, Turhan O, Kızılduman BK, Bicil Z (2021) Different functional groups functionalized hexagonal boron nitride (h-BN) nanoparticles and multi-walled carbon nanotubes (MWCNT) for hydrogen storage. Fuel 303:121335
85. Yang L, Shui J, Du L, Shao Y, Liu J, Dai L, Hu Z (2019) Carbon-based metal-free ORR electrocatalysts for fuel cells: past, present, and future. Adv Mater 31(13):1804799
86. Shao Y, Jiang Z, Zhang Q, Guan J (2019) Progress in nonmetal-doped graphene electrocatalysts for the oxygen reduction reaction. Chemsuschem 12(10):2133–2146
87. Perez-Page M, Sahoo M, Holmes SM (2019) Single layer 2D crystals for electrochemical applications of ion exchange membranes and hydrogen evolution catalysts. Adv Mater Interfaces 6(7):1801838
88. Singh RS, Gautam A, Rai V (2019) Graphene-based bipolar plates for polymer electrolyte membrane fuel cells. Front Mater Sci 13:217–241
89. Farooqui UR, Ahmad AL, Hamid N (2018) Graphene oxide: A promising membrane material for fuel cells. Renew Sustain Energy Rev 82:714–733
90. Arukula R, Vinothkannan M, Kim AR, Yoo DJ (2019) Cumulative effect of bimetallic alloy, conductive polymer and graphene toward electrooxidation of methanol: An efficient anode catalyst for direct methanol fuel cells. J Alloys Compd 771:477–488
91. Qiu X, Yan X, Cen K, Sun D, Xu L, Tang Y (2018) Achieving highly electrocatalytic performance by constructing holey reduced graphene oxide hollow nanospheres sandwiched by interior and exterior platinum nanoparticles. ACS Appl Energy Mater 1(5):2341–2349
92. Yang H, Li S, Feng F, Ou S, Li F, Yang M, Qian K, Jin J, Ma J (2019) Palladium nanoparticles with surface enrichment of palladium oxide species immobilized on the aniline-functionalized graphene as an advanced electrocatalyst of ethanol oxidation. ACS Sustain Chem Eng 7(17):14621–14628
93. Samad S, Loh KS, Wong WY, Lee TK, Sunarso J, Chong ST, Daud WRW (2018) Carbon and non-carbon support materials for platinum-based catalysts in fuel cells. Int J Hydrogen Energy 43(16):7823–7854
94. Prithi J, Rajalakshmi N, Rao GR (2018) Nitrogen doped mesoporous carbon supported Pt electrocatalyst for oxygen reduction reaction in proton exchange membrane fuel cells. Int J Hydrogen Energy 43(9):4716–4725
95. Zhuang S, Nunna BB, Mandal D, Lee ES (2018) A review of nitrogen-doped graphene catalysts for proton exchange membrane fuel cells-synthesis, characterization, and improvement. Nano-Struct Nano-Objects 15:140–152
96. Promanan T, Sarakonsri T (2017) Synthesis and characterization of palladium-based nanocatalyst on n-doped graphene for direct ethanol fuel cellS. Rev Adv Mater Sci 52
97. Chiang Y-C, Hsieh M-K, Hsu H-H (2014) The effect of carbon supports on the performance of platinum/carbon nanotubes for proton exchange membrane fuel cells. Thin Solid Films 570:221–229
98. Kim O-H, Cho Y-H, Chung DY, Kim MJ, Yoo JM, Park JE, Choe H, Sung Y-E (2015) Facile and gram-scale synthesis of metal-free catalysts: toward realistic applications for fuel cells. Sci Rep 5(1):1–8
99. To JW, Ng JWD, Siahrostami S, Koh AL, Lee Y, Chen Z, Fong KD, Chen S, He J, Bae W-G (2017) High-performance oxygen reduction and evolution carbon catalysis: From mechanistic studies to device integration. Nano Res 10:1163–1177
100. Klingele M, Pham C, Vuyyuru KR, Britton B, Holdcroft S, Fischer A, Thiele S (2017) Sulfur doped reduced graphene oxide as metal-free catalyst for the oxygen reduction reaction in anion and proton exchange fuel cells. Electrochem Commun 77:71–75
101. Bruno MM, Viva FA, Petruccelli MA, Corti HR (2015) Platinum supported on mesoporous carbon as cathode catalyst for direct methanol fuel cells. J Power Sour 278:458–463
102. Deng H, Zhang Y, Zheng X, Li Y, Zhang X, Liu X (2015) A CNT (carbon nanotube) paper as cathode gas diffusion electrode for water management of passive μ-DMFC (micro-direct methanol fuel cell) with highly concentrated methanol. Energy 82:236–241

103. Kanninen P, Borghei M, Sorsa O, Pohjalainen E, Kauppinen EI, Ruiz V, Kallio T (2014) Highly efficient cathode catalyst layer based on nitrogen-doped carbon nanotubes for the alkaline direct methanol fuel cell. Appl Catal B 156:341–349
104. Askari MB, Salarizadeh P, Seifi M, Rozati SM (2019) Ni/NiO coated on multi-walled carbon nanotubes as a promising electrode for methanol electro-oxidation reaction in direct methanol fuel cell. Solid State Sci 97:106012
105. Higgins DC, Chen Z (2010) Nitrogen doped carbon nanotube thin films as efficient oxygen reduction catalyst for alkaline anion exchange membrane fuel cell. ECS Trans 28(23):63
106. Li J, Zhang Y, Zhang X, Huang J, Han J, Zhang Z, Han X, Xu P, Song B (2017) S, N dual-doped graphene-like carbon nanosheets as efficient oxygen reduction reaction electrocatalysts. ACS Appl Mater Interfaces 9(1):398–405
107. Sa YJ, Park C, Jeong HY, Park SH, Lee Z, Kim KT, Park GG, Joo SH (2014) Carbon nanotubes/heteroatom-doped carbon core–sheath nanostructures as highly active, metal-free oxygen reduction electrocatalysts for alkaline fuel cells. Angew Chem 126(16):4186–4190
108. Zhang LL, Zhao X (2009) Carbon-based materials as supercapacitor electrodes. Chem Soc Rev 38(9):2520–2531
109. Pech D, Brunet M, Durou H, Huang P, Mochalin V, Gogotsi Y, Taberna P-L, Simon P (2010) Ultrahigh-power micrometre-sized supercapacitors based on onion-like carbon. Nat Nanotechnol 5(9):651–654
110. Zhao X, Chen H, Kong F, Zhang Y, Wang S, Liu S, Lucia LA, Fatehi P, Pang H (2019) Fabrication, characteristics and applications of carbon materials with different morphologies and porous structures produced from wood liquefaction: a review. Chem Eng J 364:226–243
111. Wang G, Oswald S, Löffler M, Müllen K, Feng X (2019) Beyond activated carbon: graphite-cathode-derived li-ion pseudocapacitors with high energy and high power densities. Adv Mater 31(14):1807712
112. Jiang Y, Liu J (2019) Definitions of pseudocapacitive materials: a brief review. Energy Environ Mater 2(1):30–37
113. Sharma K, Arora A, Tripathi SK (2019) Review of supercapacitors: Materials and devices. J Energy Storage 21:801–825
114. Abbas Q, Raza R, Shabbir I, Olabi A (2019) Heteroatom doped high porosity carbon nanomaterials as electrodes for energy storage in electrochemical capacitors: A review. J Sci Adv Mater Dev 4(3):341–352
115. Wen F, Hao C, Xiang J, Wang L, Hou H, Su Z, Hu W, Liu Z (2014) Enhanced laser scribed flexible graphene-based micro-supercapacitor performance with reduction of carbon nanotubes diameter. Carbon 75:236–243
116. Xie P, Yuan W, Liu X, Peng Y, Yin Y, Li Y, Wu Z (2021) Advanced carbon nanomaterials for state-of-the-art flexible supercapacitors. Energy Stor Mater 36:56–76
117. Wu Q, Yang L, Wang X, Hu Z (2017) From carbon-based nanotubes to nanocages for advanced energy conversion and storage. Acc Chem Res 50(2):435–444
118. Shao Y, El-Kady MF, Sun J, Li Y, Zhang Q, Zhu M, Wang H, Dunn B, Kaner RB (2018) Chem Rev 118:9233–9280
119. Fleming E, Du F, Ou E, Dai L, Shi L (2019) Thermal conductivity of carbon nanotubes grown by catalyst-free chemical vapor deposition in nanopores. Carbon 145:195–200
120. Liu P, Ru Q, Zheng P, Shi Z, Liu Y, Su C, Hou X, Su S, Ling FC-C (2019) One-step synthesis of Zn_2GeO_4/CNT-O hybrid with superior cycle stability for supercapacitor electrodes. Chem Eng J 374:29–38
121. Cherusseri J, Kar KK (2016) Ultra-flexible fibrous supercapacitors with carbon nanotube/polypyrrole brush-like electrodes. J Mater Chem A 4(25):9910–9922
122. Wu P, Cheng S, Yang L, Lin Z, Gui X, Ou X, Zhou J, Yao M, Wang M, Zhu Y (2016) Synthesis and characterization of self-standing and highly flexible δ-MnO_2@ CNTs/CNTs composite films for direct use of supercapacitor electrodes. ACS Appl Mater Interfaces 8(36):23721–23728
123. Xu C, Dong L (2016) Simultaneous production of high-performance flexible textile electrodes and fiber electrodes for wearable energy storage. In: Electrochemical society meeting abstracts 230, vol 7. The Electrochemical Society, Inc., pp 1017–1017

124. Wang K, Zhao P, Zhou X, Wu H, Wei Z (2011) Flexible supercapacitors based on cloth-supported electrodes of conducting polymer nanowire array/SWCNT composites. J Mater Chem 21(41):16373–16378
125. Qureshi SS, Nimauddin S, Mazari SA, Saeed S, Mubarak N, Khan SU, Saleh TA (2021) Ultrasonic-assisted synthesis of polythiophene-carbon nanotubes composites as supercapacitors. J Mater Sci Mater Electron 32(12):16203–16214
126. Yue L, Zhang S, Zhao H, Wang M, Mi J, Feng Y, Wang D (2018) Microwave-assisted one-pot synthesis of Fe_2O_3/CNTs composite as supercapacitor electrode materials. J Alloys Compd 765:1263–1266
127. Paul R, Etacheri V, Pol VG, Hu J, Fisher TS (2016) Highly porous three-dimensional carbon nanotube foam as a freestanding anode for a lithium-ion battery. RSC Adv 6(83):79734–79744
128. Niu Z, Zhou W, Chen X, Chen J, Xie S (2015) Highly compressible and all-solid-state supercapacitors based on nanostructured composite sponge. Adv Mater 27(39):6002–6008
129. Dang A, Sun Y, Fang C, Li T, Liu X, Xia Y, Ye F, Zada A, Khan M (2022) Rational design of Ti_3C_2/carbon nanotubes/$MnCo_2S_4$ electrodes for symmetric supercapacitors with high energy storage. Appl Surf Sci 581:152432
130. Li K, Teng H, Sun Q, Li Y, Wu X, Dai X, Wang Y, Wang S, Zhang Y, Yao K (2022) Engineering active sites on nitrogen-doped carbon nanotubes/cobaltosic oxide heterostructure embedded in biotemplate for high-performance supercapacitors. J Energy Storage 53:105094
131. Olatomiwa AL, Adam T, Gopinath SC, Kolawole SY, Olayinka OH, Hashim U (2022) Graphene synthesis, fabrication, characterization based on bottom-up and top-down approaches: An overview. J Semicond 43(6):061101
132. Lemine AS, Zagho MM, Altahtamouni T, Bensalah N (2018) Graphene a promising electrode material for supercapacitors—A review. Int J Energy Res 42(14):4284–4300
133. Abeykoon A, De Silva R, Nayanajith L, Kottegoda I (2022) A review on appropriate graphene synthesis methods for diverse applications. Sri Lankan J Phys 23(2)
134. Yang W, Ni M, Ren X, Tian Y, Li N, Su Y, Zhang X (2015) Graphene in supercapacitor applications. Curr Opin Colloid Interface Sci 20(5–6):416–428
135. Ares P, Novoselov KS (2022) Recent advances in graphene and other 2D materials. Nano Mater Sci 4(1):3–9
136. Xiong C, Li B, Duan C, Dai L, Nie S, Qin C, Xu Y, Ni Y (2021) Carbonized wood cell chamber-reduced graphene oxide@ PVA flexible conductive material for supercapacitor, strain sensing and moisture-electric generation applications. Chem Eng J 418:129518
137. Kumar R, Sahoo S, Tan WK, Kawamura G, Matsuda A, Kar KK (2021) Microwave-assisted thin reduced graphene oxide-cobalt oxide nanoparticles as hybrids for electrode materials in supercapacitor. J Energy Storage 40:102724
138. Xu M, Wang A, Xiang Y, Niu J (2021) Biomass-based porous carbon/graphene self-assembled composite aerogels for high-rate performance supercapacitor. J Clean Prod 315:128110
139. Hashemi SA, Mousavi SM, Naderi HR, Bahrani S, Arjmand M, Hagfeldt A, Chiang W-H, Ramakrishna S (2021) Reinforced polypyrrole with 2D graphene flakes decorated with interconnected nickel-tungsten metal oxide complex toward superiorly stable supercapacitor. Chem Eng J 418:129396
140. Li C, Shi G (2012) Three-dimensional graphene architectures. Nanoscale 4(18):5549–5563
141. Jiang X, Gao R, Liu G, Luo H, Zhao X, Jiang L (2022) Construction of graphene-based “In-Paper” 3D interdigital microelectrodes for high performance metal-free flexible supercapacitors. Small Methods 6(5):2101454
142. Cheng C, Zou Y, Xu F, Xiang C, Sui Q, Zhang J, Sun L, Chen Z (2022) Ultrathin graphene@ $NiCo_2S_4$@ Ni-Mo layered double hydroxide with a 3D hierarchical flowers structure as a high performance positive electrode for hybrid supercapacitor. J Energy Storage 52:105049
143. Baskar AV, Ruban AM, Davidraj JM, Singh G, AaH A-M, Lee JM, Yi J, Vinu A (2021) Single-step synthesis of 2D mesoporous C_{60}/carbon hybrids for supercapacitor and Li-ion battery applications. Bull Chem Soc Jpn 94(1):133–140
144. Jiang B, Zhang G, Tang Q, Meng F, Zhou D, Zhao W, Jiang W, Ji Q (2022) Tailoring co-doping of cobalt and nitrogen in a fullerene-based carbon composite and its effect on the supercapacitive performance. Adv Mater 3(3):1539–1546

145. Li Y, Kang Z, Yan X, Cao S, Li M, Guo Y, Huan Y, Wen X, Zhang Y (2018) A three-dimensional reticulate CNT-aerogel for a high mechanical flexibility fiber supercapacitor. Nanoscale 10(19):9360–9368
146. Ye Z, Zhang T, He W, Jin H, Liu C, Yang Z, Ren J (2018) Methotrexate-loaded extracellular vesicles functionalized with therapeutic and targeted peptides for the treatment of glioblastoma multiforme. ACS Appl Mater Interfaces 10(15):12341–12350
147. Feng X, Chen W, Yan L (2015) Reduced graphene oxide hydrogel film with a continuous ion transport network for supercapacitors. Nanoscale 7(8):3712–3718
148. Lv H, Yuan Y, Xu Q, Liu H, Wang Y-G, Xia Y (2018) Carbon quantum dots anchoring MnO_2/graphene aerogel exhibits excellent performance as electrode materials for supercapacitor. J Power Sources 398:167–174
149. Pham DT, Lee TH, Luong DH, Yao F, Ghosh A, Le VT, Kim TH, Li B, Chang J, Lee YH (2015) Carbon nanotube-bridged graphene 3D building blocks for ultrafast compact supercapacitors. ACS Nano 9(2):2018–2027
150. Zheng Y, Tian Y, Sarwar S, Luo J, Zhang X (2020) Carbon nanotubes decorated $NiSe_2$ nanosheets for high-performance supercapacitors. J Power Sour 452:227793
151. Tao H-C, Zhu S-C, Yang X-L, Zhang L-L, Ni S-B (2016) Systematic investigation of reduced graphene oxide foams for high-performance supercapacitors. Electrochim Acta 190:168–177
152. Zhou S, Zeng S, Zhang S, Qiao J, Di J, Chen M, Liu N, Li Q (2017) Hierarchical carbon nanotube hybrid films for high-performance all-solid-state supercapacitors. RSC Adv 7(82):52010–52016
153. Pourreza K, Adeh NB, Mohammadi N (2020) In-situ grown of polyaniline on defective mesoporous carbon as a high performance supercapacitor electrode material. J Energy Storage 30:101429
154. Zheng C, Zhou X, Cao H, Wang G, Liu Z (2014) Edge-enriched porous graphene nanoribbons for high energy density supercapacitors. J Mater Chem A 2(20):7484–7490
155. Yang J, Guo J, Guo X, Chen L (2019) In-situ growth carbon nanotubes deriving from a new metal-organic framework for high-performance all-solid-state supercapacitors. Mater Lett 236:739–742
156. Zhu F, Liu W, Liu Y, Shi W (2020) Construction of porous interface on CNTs@ NiCo-LDH core-shell nanotube arrays for supercapacitor applications. Chem Eng J 383:123150
157. Wan L, Du C, Yang S (2017) Synthesis of graphene oxide/polybenzoxazine-based nitrogen-containing porous carbon nanocomposite for enhanced supercapacitor properties. Electrochim Acta 251:12–24
158. Tseng L-H, Hsiao C-H, Nguyen DD, Hsieh P-Y, Lee C-Y, Tai N-H (2018) Activated carbon sandwiched manganese dioxide/graphene ternary composites for supercapacitor electrodes. Electrochim Acta 266:284–292
159. Schmidt O, Hawkes A, Gambhir A, Staffell I (2017) The future cost of electrical energy storage based on experience rates. Nat Energy 2(8):1–8
160. Liu Z, Huang Y, Huang Y, Yang Q, Li X, Huang Z, Zhi C (2020) Voltage issue of aqueous rechargeable metal-ion batteries. Chem Soc Rev 49(1):180–232
161. Liang Y, Dong H, Aurbach D, Yao Y (2020) Current status and future directions of multivalent metal-ion batteries. Nat Energy 5(9):646–656
162. Gu X, Lai C (2018) Recent development of metal compound applications in lithium–sulphur batteries. J Mater Res 33(1):16–31
163. Che H, Chen S, Xie Y, Wang H, Amine K, Liao X-Z, Ma Z-F (2017) Electrolyte design strategies and research progress for room-temperature sodium-ion batteries. Energy Environ Sci 10(5):1075–1101
164. Yang S, Cheng Y, Xiao X, Pang H (2020) Development and application of carbon fiber in batteries. Chem Eng J 384:123294
165. Chen S, Kuang Q, Fan HJ (2020) Dual-Carbon batteries: materials and mechanism. Small 16(40):2002803
166. Han J, Wei W, Zhang C, Tao Y, Lv W, Ling G, Kang F, Yang Q-H (2018) Engineering graphenes from the nano-to the macroscale for electrochemical energy storage. Electrochem Energy Rev 1:139–168

167. Hou R, Liu B, Sun Y, Liu L, Meng J, Levi MD, Ji H, Yan X (2020) Recent advances in dual-carbon based electrochemical energy storage devices. Nano Energy 72:104728
168. Thompson M, Xia Q, Hu Z, Zhao XS (2021) A review on biomass-derived hard carbon materials for sodium-ion batteries. Adv Mater 2(18):5881–5905
169. Winter M, Besenhard JO, Spahr ME, Novak P (1998) Insertion electrode materials for rechargeable lithium batteries. Adv Mater 10(10):725–763
170. McCullough F, Levine C, Snelgrove R (1989) Secondary battery.
171. Zhu C-y, Ye Y-w, Guo X, Cheng F (2022) Design and synthesis of carbon-based nanomaterials for electrochemical energy storage. New Carbon Mater 37(1):59–92
172. Wu Q, Yang L, Wang X, Hu Z (2020) Carbon-based nanocages: a new platform for advanced energy storage and conversion. Adv Mater 32(27):1904177
173. Zhang Y, Yan D, Liu Z, Ye Y, Cheng F, Li H, Lu A-H (2021) A SnO x quantum dots embedded carbon nanocage network with ultrahigh Li storage capacity. ACS Nano 15(4):7021–7031
174. Cao B, Liu Z, Xu C, Huang J, Fang H, Chen Y (2019) High-rate-induced capacity evolution of mesoporous C@ SnO_2@ C hollow nanospheres for ultra-long cycle lithium-ion batteries. J Power Sour 414:233–241
175. Chae S, Choi SH, Kim N, Sung J, Cho J (2020) Integration of graphite and silicon anodes for the commercialization of high-energy lithium-ion batteries. Angew Chem Int Ed 59(1):110–135
176. Mi H, Yang X, Li Y, Zhang P, Sun L (2018) A self-sacrifice template strategy to fabricate yolk-shell structured silicon@ void@ carbon composites for high-performance lithium-ion batteries. Chem Eng J 351:103–109
177. Zhao F, Li X, He J, Wang K, Huang C (2021) Preparation of hierarchical graphdiyne hollow nanospheres as anode for lithium-ion batteries. Chem Eng J 413:127486
178. Zhai T, Yao J (2012) One-dimensional nanostructures: principles and applications. Wiley & Sons
179. Deng M, Qi J, Li X, Xiao Y, Yang L, Yu X, Wang H, Yuan B, Gao Q (2018) MoC/C nanowires as high-rate and long cyclic life anode for lithium ion batteries. Electrochim Acta 277:205–210
180. Liu D-S, Liu D-H, Hou B-H, Wang Y-Y, Guo J-Z, Ning Q-L, Wu X-L (2018) 1D porous MnO@ N-doped carbon nanotubes with improved Li-storage properties as advanced anode material for lithium-ion batteries. Electrochim Acta 264:292–300
181. Chen H, Liu R, Wu Y, Cao J, Chen J, Hou Y, Guo Y, Khatoon R, Chen L, Zhang Q (2021) Interface coupling 2D/2D $SnSe_2$/graphene heterostructure as long-cycle anode for all-climate lithium-ion battery. Chem Eng J 407:126973
182. Mu T, Zuo P, Lou S, Pan Q, Li Q, Du C, Gao Y, Cheng X, Ma Y, Yin G (2018) A two-dimensional nitrogen-rich carbon/silicon composite as high performance anode material for lithium ion batteries. Chem Eng J 341:37–46
183. Sun P, Wang K, Zhu H (2016) Recent developments in graphene-based membranes: structure, mass-transport mechanism and potential applications. Adv Mater 28(12):2287–2310
184. Ma G, Huang K, Ma J-S, Ju Z, Xing Z, Zhuang Q-c (2017) Phosphorus and oxygen dual-doped graphene as superior anode material for room-temperature potassium-ion batteries. J Mater Chem A 5(17):7854–7861
185. Fu K, Yao Y, Dai J, Hu L (2017) Progress in 3D printing of carbon materials for energy-related applications. Adv Mater 29(9):1603486
186. Zheng J, Wu Y, Sun Y, Rong J, Li H, Niu L (2021) Advanced anode materials of potassium ion batteries: from zero dimension to three dimensions. Nano-Micro Lett 13:1–37
187. Wang P, Zhu K, Ye K, Gong Z, Liu R, Cheng K, Wang G, Yan J, Cao D (2020) Three-dimensional biomass derived hard carbon with reconstructed surface as a free-standing anode for sodium-ion batteries. J Colloid Interface Sci 561:203–210
188. Cheng F, Wang S, Lu A-H, Li W-C (2013) Immobilization of nanosized LiFePO4 spheres by 3D coralloid carbon structure with large pore volume and thin walls for high power lithium-ion batteries. J Power Sour 229:249–257
189. Ke G, Chen H, He J, Wu X, Gao Y, Li Y, Mi H, Zhang Q, He C, Ren X (2021) Ultrathin MoS_2 anchored on 3D carbon skeleton containing SnS quantum dots as a high-performance anode for advanced lithium ion batteries. Chem Eng J 403:126251

190. Han C, Xu L, Li H, Shi R, Zhang T, Li J, Wong C-P, Kang F, Lin Z, Li B (2018) Biopolymer-assisted synthesis of 3D interconnected Fe_3O_4@ carbon core@ shell as anode for asymmetric lithium ion capacitors. Carbon 140:296–305
191. Wang X, Tang Y, Shi P, Fan J, Xu Q, Min Y (2018) Self-evaporating from inside to outside to construct cobalt oxide nanoparticles-embedded nitrogen-doped porous carbon nanofibers for high-performance lithium ion batteries. Chem Eng J 334:1642–1649
192. Mondal AK, Kretschmer K, Zhao Y, Liu H, Wang C, Sun B, Wang G (2017) Nitrogen-doped porous carbon nanosheets from eco-friendly eucalyptus leaves as high performance electrode materials for supercapacitors and lithium ion batteries. Chem Eur J 23(15):3683–3690
193. Hu X, Li Y, Zeng G, Jia J, Zhan H, Wen Z (2018) Three-dimensional network architecture with hybrid nanocarbon composites supporting few-layer MoS_2 for lithium and sodium storage. ACS Nano 12(2):1592–1602
194. Yao S, Cui J, Huang J, Huang JQ, Chong WG, Qin L, Mai YW, Kim JK (2018) Rational assembly of hollow microporous carbon spheres as P hosts for long-life sodium-ion batteries. Adv Energy Mater 8(7):1702267
195. Yang J, Zhou X, Wu D, Zhao X, Zhou Z (2017) S-doped N-rich carbon nanosheets with expanded interlayer distance as anode materials for sodium-ion batteries. Adv Mater 29(6):1604108
196. Tan Z, Ni K, Chen G, Zeng W, Tao Z, Ikram M, Zhang Q, Wang H, Sun L, Zhu X (2017) Incorporating pyrrolic and pyridinic nitrogen into a porous carbon made from C_{60} molecules to obtain superior energy storage. Adv Mater 29(8):1603414
197. Zhao X, Kim M, Liu Y, Ahn H-J, Kim K-W, Cho K-K, Ahn J-H (2018) Root-like porous carbon nanofibers with high sulfur loading enabling superior areal capacity of lithium sulfur batteries. Carbon 128:138–146
198. Hao R, Lan H, Kuang C, Wang H, Guo L (2018) Superior potassium storage in chitin-derived natural nitrogen-doped carbon nanofibers. Carbon 128:224–230
199. Park S-K, Park J-S, Kang YC (2018) Selenium-infiltrated metal–organic framework-derived porous carbon nanofibers comprising interconnected bimodal pores for Li–Se batteries with high capacity and rate performance. J Mater Chem A 6(3):1028–1036
200. Zhang J, Shi Y, Ding Y, Peng L, Zhang W, Yu G (2017) A Conductive molecular framework derived Li_2S/N, P-Codoped carbon cathode for advanced lithium-sulfur batteries. Adv Energy Mater 7(14):1602876
201. Zhong Y, Xia X, Deng S, Zhan J, Fang R, Xia Y, Wang X, Zhang Q, Tu J (2018) Popcorn inspired porous macrocellular carbon: rapid puffing fabrication from rice and its applications in lithium–sulfur batteries. Adv Energy Mater 8(1):1701110
202. Li W, Fang R, Xia Y, Zhang W, Wang X, Xia X, Tu J (2019) Multiscale porous carbon nanomaterials for applications in advanced rechargeable batteries. Batter Supercaps 2(1):9–36
203. Shinde SS, Lee C-H, Sami A, Kim D-H, Lee S-U, Lee J-H (2017) Scalable 3-D carbon nitride sponge as an efficient metal-free bifunctional oxygen electrocatalyst for rechargeable Zn–air batteries. ACS Nano 11(1):347–357
204. Miron C, Mele P, Kaneko S, Endo T. Carbon-related materials

GPSR Compliance
The European Union's (EU) General Product Safety Regulation (GPSR) is a set of rules that requires consumer products to be safe and our obligations to ensure this.

If you have any concerns about our products, you can contact us on

ProductSafety@springernature.com

In case Publisher is established outside the EU, the EU authorized representative is:

Springer Nature Customer Service Center GmbH
Europaplatz 3
69115 Heidelberg, Germany

www.ingramcontent.com/pod-product-compliance
Ingram Content Group UK Ltd.
Pitfield, Milton Keynes, MK11 3LW, UK
UKHW022253091026
15624UKWH00015B/181
* 9 7 8 9 8 1 9 7 0 2 3 9 8 *